THE CHEMICAL SENSES AND NUTRITION

THE NUTRITION FOUNDATION

A Monograph Series

HORACE L. SIPPLE AND KRISTEN W. MCNUTT, EDS.: *Sugars in Nutrition,* 1974

ROBERT E. OLSON, ED.: *Protein-Calorie Malnutrition,* 1975

ANANDA S. PRASAD, ED.: *Trace Elements in Human Health and Disease, Volume I, Zinc and Copper, 1976; Volume II, Essential and Toxic Elements,* 1976

MORLEY R. KARE AND OWEN MALLER, EDS.: *The Chemical Senses and Nutrition,* 1977

THE CHEMICAL SENSES AND NUTRITION

Edited by

MORLEY R. KARE
OWEN MALLER

Monell Chemical Senses Center
University of Pennsylvania
Philadelphia, Pennsylvania

ACADEMIC PRESS New York San Francisco London 1977

A Subsidiary of Harcourt Brace Jovanovich, Publishers

ACADEMIC PRESS, INC.
111 Fifth Avenue, New York, New York 10003

United Kingdom Edition published by
ACADEMIC PRESS, INC. (LONDON) LTD.
24/28 Oval Road, London NW1

Library of Congress Cataloging in Publication Data

Main entry under title:

The Chemical senses and nutrition.

(Nutrition Foundation series)
Includes bibliographies and index.
1. Chemical senses–Congresses. 2. Nutrition –Congresses. I. Kare, Morley Ricard, Date II. Maller, Owen. III. Series: Nutrition Foundation, New York. Nutrition Foundation monograph series. [DNLM: 1. Nutrition–Congresses. 2. Taste–Congresses. 3. Smell–Congresses. W3 IN1946N 1976c /QU145 I6025 1976c]
QP455.C474 591.1'8 77-74054
ISBN 0–12–397850–5

PRINTED IN THE UNITED STATES OF AMERICA

Contents

Chapter 3
Nutritional States, Food Odors, and Olfactory Function

JEANNE PAGER

SENSORY QUALITY AND INGESTION

Chapter 4
Intensity and Hedonic Functions for Chemosensory Stimuli

HOWARD R. MOSKOWITZ

Chapter 5
Influence of Internal Factors on the Perceived Intensity and Pleasantness of Gustatory and Olfactory Stimuli

GEORGE D. MOWER, ROBERT G. MAIR, AND TRYGG ENGEN

Chapter 6
Sensory–Neuroendocrine Reflexes and Their Anticipatory and Optimizing Role on Metabolism

STYLIANOS NICOLAIDIS

Chapter 11
The Control of Food Intake: When and How Are Amino Acids Involved?

Q. R. ROGERS AND P. M. B. LEUNG

PERCEPTION AND EXPERIENCE

Chapter 12
The Ontogeny, Evolution, and Stimulus Control of Feeding in Humans and Reptiles

GORDON M. BURGHARDT

Chapter 13
Conditioned Responses to Food Odor and Taste in Rats and Wild Predators

JOHN GARCIA AND LINDA P. BRETT

Chapter 14
The Development of Flavor Preferences in Humans: A Review

GARY K. BEAUCHAMP AND OWEN MALLER

FLAVOR RECOGNITION

Chapter 15
Food Preference Ratings of Congenitally Anosmic Humans

RICHARD L. DOTY

Chapter 16
Sweet Taste Sensitivity in Japanese Macaques

MASAYASU SATO, YASUTAKE HIJI, HIROSUMI ITO, AND TOSHIAKI IMOTO

Chapter 17
A Framework for the Mechanisms of Action of Special Taste Substances: The Example of Monosodium Glutamate

ROBERT H. CAGAN

DYSFUNCTIONS

Chapter 18
Disordered Oral Sensation and Appetite

VERNON J. BRIGHTMAN

Chapter 19
Changes in Taste Sensation in Cancer Patients: Correlation with Caloric Intake

WILLIAM D. DEWYS

Chapter 20
Palatability and Oral Factors and Their Role in Obesity

HENRY A. JORDAN AND THERESA A. SPIEGEL

OVERVIEW

Chapter 21
The Role of the Chemical Senses in Nutrition

SAMUEL LEPKOVSKY

Chapter 22
Socioeconomic Factors in Fat and Sugar Consumption

S. M. CANTOR AND M. B. CANTOR

Chapter 23
The Nimrod Connection: Myth and Science in the Hunting Model

TIMOTHY PERPER AND CARMEL SCHRIRE

SUMMATION

Chapter 24
Concluding Comments

C. PFAFFMANN, V. G. DETHIER, AND D. M. HEGSTED

List of Contributors

Numbers in parentheses indicate the pages on which the authors' contributions begin.

GARY K. BEAUCHAMP (291), Monell Chemical Senses Center, and Department of Otorhinolaryngology and Human Communication, University of Pennsylvania, Philadelphia, Pennsylvania

LINDA P. BRETT (277), Departments of Psychology and Psychiatry, University of California, Los Angeles, California

VERNON J. BRIGHTMAN (363), University of Pennsylvania School of Dental Medicine, Philadelphia, Pennsylvania

GORDON M. BURGHARDT (253), Department of Psychology, University of Tennessee, Knoxville, Tennessee

ROBERT H. CAGAN (343), Veterans Administration Hospital, Philadelphia, Pennsylvania, and Monell Chemical Senses Center, University of Pennsylvania, Philadelphia, Pennsylvania

M. B. CANTOR (429), Department of Psychology, Columbia University, New York, New York

S. M. CANTOR (429), Sidney M. Cantor Associates, Haverford, Pennsylvania

V. G. DETHIER (463), Department of Zoology, University of Massachusetts, Amherst, Massachusetts

WILLIAM D. DEWYS (381), Section of Medical Oncology, Department of Medicine, Northwestern University Medical School, Chicago, Illinois

RICHARD L. DOTY (315), Monell Chemical Senses Center, and Department of Otorhinolaryngology and Human Communication, University of Pennsylvania, Philadelphia, Pennsylvania

S. DUA-SHARMA (167), Department of Physiology, St. John's Medical College, Bangalore, India

TRYGG ENGEN (103), Walter S. Hunter Laboratory of Psychology, Brown University, Providence, Rhode Island

HANS FISHER (189), Department of Nutrition, Cook College, Rutgers University, New Brunswick, New Jersey

JOHN GARCIA (277), Departments of Psychology and Psychiatry, University of California, Los Angeles, California

STANLEY N. GERSHOFF (201), Department of Nutrition, Harvard School of Public Health, Boston, Massachusetts

V. GOPAL[a] (167), Department of Physiology, St. John's Medical College, Bangalore, India

D. M. HEGSTED (471), Department of Nutrition, School of Public Health, Harvard University, Boston, Massachusetts

YASUTAKE HIJI (327), Department of Physiology, Miyazaki Medical College, Miyazaki, Japan

TOSHIAKI IMOTO (327), Department of Physiology, Kumamoto University Medical School, Kumamoto, Japan

HIROSUMI ITO (327), Department of Physiology, Kumamoto University Medical School, Kumamoto, Japan

H. L. JACOBS (167), Behavioral Sciences Division, Food Sciences Laboratory, United States Army Natick Research and Development Command, Natick, Massachusetts

HENRY A. JORDAN[b] (393), Department of Psychiatry, School of Medicine, University of Pennsylvania, Philadelphia, Pennsylvania

MORLEY R. KARE (145), Monell Chemical Senses Center, University of Pennsylvania, Philadelphia, Pennsylvania

SAMUEL LEPKOVSKY (413), Department of Poultry Husbandry, University of California, Berkeley, California

P. M. B. LEUNG (213), Department of Physiological Sciences, School of Veterinary Medicine, University of California, Davis, California

ROBERT G. MAIR (103), Walter S. Hunter Laboratory of Psychology, Brown University, Providence, Rhode Island

OWEN MALLER (291), Veterans Administration Hospital, Philadelphia, Pennsylvania, and Monell Chemical Senses Center, University of Pennsylvania, Philadelphia, Pennsylvania

HOWARD R. MOSKOWITZ (71), MPI Sensory Testing, Inc., New York, New York

GEORGE D. MOWER (103), Walter S. Hunter Laboratory of Psychology, Brown University, Providence, Rhode Island

ROGER A. MUGFORD (25), Animal Studies Center, Pedigree Petfoods, Melton Mowbray, Leicestershire, England

MICHAEL NAIM (145), Monell Chemical Senses Center, University of Pennsylvania, Philadelphia, Pennsylvania

STYLIANOS NICOLAIDIS (123), Laboratoire de Neurobiologie de Regulations, College de France, Paris, France

[a] Present address: Division of Neurophysiology and Behavior, Department of Zoology, Madras University PG Centre, Coimbatore, India

[b] Institute for Behavioral Education, King of Prussia, Pennsylvania

JEANNE PAGER (51), Laboratoire d'Electrophysiologie, Université Claude Bernard, Villeurbanne, France

TIMOTHY PERPER (447), Biology Department, Livingston College, Rutgers University, New Brunswick, New Jersey

C. PFAFFMANN (467), The Rockefeller University, New York, New York

Q. R. ROGERS (213), Department of Physiological Sciences, School of Veterinary Medicine, University of California, Davis, California

MASAYASU SATO[c] (327), Department of Physiology, Kumamoto University Medical School, Kumamoto, Japan

CARMEL SCHRIRE (447), Department of Human Ecology and Social Sciences, Cook College, Rutgers University, New Brunswick, New Jersey

DAVID M. SENSEMAN (3), Monell Chemical Senses Center, University of Pennsylvania, Philadelphia, Pennsylvania

K. N. SHARMA[d] (167), Department of Physiology, St. John's Medical College, Bangalore, India

THERESA A. SPIEGEL (393), Department of Psychology, Beaver College, Glenside, Pennsylvania

[c] Present address: Tokyo Metropolitan Institute for Neurosciences, Fuchu City, Tokyo, Japan

[d] Present address: Department of Physiology, University College of Medical Sciences, New Delhi, India

List of Participants

THE SECOND INTERNATIONAL CONFERENCE
ON THE CHEMICAL SENSES AND NUTRITION
Monell Chemical Senses Center
June 14–16, 1976

DR. GARY K. BEAUCHAMP
DR. JAMES F. BOSMA
DR. JOSEPH G. BRAND
DR. VERNON J. BRIGHTMAN
DR. GORDON M. BURGHARDT
DR. ROBERT H. CAGAN
DR. SIDNEY M. CANTOR
DR. MABEL M. CHAN
DR. WILLIAM J. DARBY
DR. VINCENT G. DETHIER
DR. WILLIAM D. DEWYS
DR. RICHARD L. DOTY
DR. TRYGG ENGEN
DR. HANS FISHER
DR. DAN GALILI
DR. JOHN GARCIA
DR. STANLEY N. GERSHOFF
DR. D. MARK HEGSTED
DR. HENRY A. JORDAN
DR. MORLEY R. KARE
DR. SAMUEL LEPKOVSKY
DR. OWEN MALLER
DR. SANDFORD A. MILLER
DR. HOWARD R. MOSKOWITZ
DR. HENRY MOSS
DR. ROGER A. MUGFORD
DR. MICHAEL NAIM
DR. STYLIANOS NICOLAIDIS
DR. JEANNE PAGER
DR. ROSE MARIE PANGBORN
DR. CARL PFAFFMANN
DR. MICHAEL QUINN
DR. QUINTON R. ROGERS
DR. MASAYASU SATO
DR. CARMEL SCHRIRE
DR. DAVID M. SENSEMAN
DR. K. N. SHARMA

Foreword

It was a decade ago that the cooperation between the Monell Chemical Senses Center and The Nutrition Foundation resulted in the development of a monograph on the chemical senses, which has become a standard reference work for those in this field. Since the appearance of that work, an increased awareness has developed among the medical profession of the importance of understanding these specialized senses and the psychological and behavioral phenomena to which they are related or which are influenced or governed by these senses. This past decade has seen much progress in understanding the basic physiology, molecular biology, and biochemistry of the chemical senses. Accordingly, The Nutrition Foundation and The Monell Chemical Senses Center have again joined in sponsoring a current examination of the field and of the decade of progress against the complex background of considerations of the functions of and reactions to the chemical senses.

In order to accomplish this expeditiously and to produce an authoritative, current review, this effort was planned in two stages. First, there was a workshop of scientists engaged actively in advancing knowledge in this field for the purpose of providing an opportunity to freely exchange information and experiences and to critically examine the views of individual workers. Second, following these exchanges, the prepared manuscripts were modified and edited for prompt publication. Therefore, this volume reflects the beneficial exchange that occurred in the workshop and provides a particularly well-considered update of the subject of the chemical senses in relation to nutrition.

The conference and the publication of this monograph were supported jointly by The Nutrition Foundation and The Monell Chemical Senses Center, and assisted by a grant from the National Institutes of Health (National Institute for Dental Research 1 R13 DE 04533).

The assistance of the staff of The Monell Center for the arrangements of the conference and the preparation of this book is gratefully acknowledged. In the desire for rapid publication, there has been limited opportunity, particularly for the foreign contributors, for proof correction. The discussions have been condensed. The editors assume responsibility for any changes in these sections.

William J. Darby
President
The Nutrition Foundation, Inc.
New York and
Washington, D.C.

Preface

Ten years after the appearance of the first book on the chemical senses and nutrition, additional research data justified a second volume. These papers constitute an overview by leading scientists in the field of the growing understanding of the role of the chemical senses in nutrition.

It is extremely important to be aware of the involvement of the phenomena of taste and smell in many aspects of human behavior. It is, however, critically important to the entire field of nutrition to be able to identify and evaluate olfactory and gustatory interactions. Taste and smell are the two senses that basically control a vast number of human reactions. One or another of these senses may reach a high level of acuity among animal species. A vast array of materials and environments is identifiable by a combination of taste and smell factors.

The world is currently facing a burgeoning population, a problem with which we have thus far not been able to cope successfully; population stabilization has been effective in only a very few nations. Thus, while the world continues to advance in its technology, its educational systems, and other social objectives, it has not yet been possible to establish "universal family planning." In former times, when the world's population was relatively small, the increase in numbers was hardly perceptible. One result of the rapid growth in modern times has been the obvious pressures of people upon environment. Through neglect, mismanagement, and other forces, a vast array of disruptive factors has been brought to bear on our natural environment. These include floods, droughts, earthquakes, forest fires, erosion, and the full complement of those phenomena that have taken great toll of the world's resources. Moreover, under continuing pressure, man's inroads into forests, fuel, minerals, and other reserves steadily increase the growing threat to our environment.

Obviously, if we are unable to provide the essential foods for all the world's population, there will be a gradual erosion of the social scene.

Starving people become restless and often violent when the needs of their families continue to be unmet. This can lead to the breakdown of law and order, resulting in chaos. Thus there is a crying need to stabilize population growth (a long-term undertaking) and to increase the efficiency of food production for the increasing numbers of people who will need to be fed in the near future. Research in progress on the chemical senses may prove to be extraordinarily valuable in attaining these goals.

In simplest terms, the fundamental purpose of understanding the chemical senses is to assist in solving the problems of population and nutrition. The science of agriculture has made enormous strides during the past sixty years. The yields of cereal and other crops have steadily increased and until recently have set new records on an almost continuous basis. Nevertheless, in spite of the intensive efforts that have been made to increase production of the basic cereal grains, we still fall short of satisfying world needs for these foods. Twenty-five percent of all foodstuffs produced are unavailable due to losses in the field, in storage, and in transportation, and the lack of adequate distribution facilities. Presumably it will be necessary to increase and intensify the production of a variety of plant and animal materials suitable for human consumption.

In the minds of many people, the production of foodstuffs is a quantitative phenomenon, and under some circumstances this is largely the case. That is to say, many crops and animals best adapted to the local ecology prove also to be well-adapted to maximum production and to have highly desirable nutritional content. Taste and smell in concert are often responsible for attracting the individual to foodstuffs which are palatable and nutritious. It is largely through the organs of taste and smell that discrimination occurs with respect to a variety of foodstuffs.

Taste can influence activity along the digestive tract, the level of circulating metabolic hormones, and both the quantity and the character of secretions of the exocrine pancreas. Recently, it was demonstrated that taste could modify the physiological factors that can determine growth efficiency. Taste could be used to extend the world food supply for both animals and man by rendering the unpalatable less so. By commanding taste, the range of cultivatable plants could be extended. There are many plant sources that contain high quality protein but unfortunately are unpalatable; e.g., the proteins of soy and guar gum. The possibility of improving these and other flavors represents a simple way of increasing the world's supply of acceptable proteins.

The chemical senses play a crucial role in reproduction in many mammals. Animal research has demonstrated that chemical signals (primarily olfactory) serve to bring members of opposite sexes together for mating, control aggressive behavior, and even regulate sexual maturation rates. In programs to maximize breeding efficiency of beneficial animals, a knowledge of how the chemical senses work is thus imperative. It is possible that many of the reproductive and endocrine effects of chemical signals in animals may also exist in humans.

Some reports suggest that odor may serve to synchronize the time of ovulation in human females living together. Consequently, odors may affect the hypothalamic-pituitary-gonadal axis. In addition, if females excrete odorous compounds characteristic of ovulation, the compounds can be easily detected via instrumental and chemical methods to serve as diagnostic indicators. The practical effects could be profound.

The population explosion is commonly thought of primarily in terms of pressures on man. As the shortage of food becomes aggravated, man's direct competition with animals, both wild and domestic, will become more acute. A knowledge of the sense of taste and smell of each species would make the available resources more elastic. For example, it is conceivable that plants and plant parts discarded as unpalatable by one species could be used for another, or after modest chemical alteration, could be made palatable. This would include most exotic plants considered as potentially cultivatable candidates, but eliminated at some point because of their taste or odor to man.

Thus, although the phenomena of taste and smell in humans present a series of challenges to the investigator, they also contain the clues to a wide spectrum of behavioral patterns. This volume is most timely in that it explores and employs new knowledge in the relatively undeveloped field of the chemical senses to improve our perspective on nutrition problems.

J. G. Harrar
Chairman, National Advisory Board
The Monell Chemical Senses Center
and
Chairman, Board of Trustees
The Nutrition Foundation

PALATABILITY AND INGESTION

Chapter 1

Gastropod Mollusks as Model Systems for the Study of Integrative Mechanisms Controlling Feeding Behavior

DAVID M. SENSEMAN

The link between palatability and nutrition is a direct and obvious one. An animal eats what is available and palatable in its environment; an animal's nutritional state is directly dependent upon the quantity and nutritional quality of the foods it eats. What is palatable on this planet depends in large measure on what species, and to a certain extent upon which individual of that species, is making the assessment. Although palatability, like beauty, is in the eye of the beholder, there is often a consensus among quite diverse species as to what is palatable. Witness man's escalating war with the insects over the rights to the world's dwindling food supplies (Dethier, 1976). How palatabilities are established during the course of evolution, and how they are modified by dietary experience, remain an active and productive area of scientific inquiry as documented by several chapters in this volume.

Palatability may be defined as the efficacy of external food-related stimuli to initiate and sustain an ingestional response (e.g., LeMagnen, 1971). This relationship between the physical and chemical characteristics of a potential food item and its acceptance (or rejection) is not a simple one, but rather depends upon the final signal produced by the simultaneous integration within the central nervous system (CNS) of multiple exteroceptive and interoceptive inputs. Ten years ago at the previous conference on the chemical senses and nutrition, de Ruiter (1967) proposed a model (Fig. 1) to serve as a conceptual framework in which to study the role of the chemical senses in nutritional homeostasis. The model consists of five black boxes, each representing a discrete integrative function that will be described below.

Ultimately, cellular neurophysiologists would like to replace each black box with its neuroanatomical substrates in the human CNS, along with a detailed description of the neuronal mechanisms underlying each integrative function. Ideally, each neuron participating in the feeding circuit should be identified and the nature of its synaptic interactions with other neurons in the circuit determined. Such a complete mechanistic account of the feeding response is still in the distant future for any mammalian species. It would be of considerable advantage to have some preconceived notions concerning the neuronal mechanisms that might subserve the various integrative functions shown in Fig. 1.

Within recent years, cellular neurophysiologists have begun the study of restricted neuronal networks in several invertebrate species. Much of this research has been directed toward a detailed understanding of the ways in which the simpler nervous systems of these animals solve a variety of integrative tasks. Of special interest in the present context is the attention being focused on the feeding behavior of gastropod mollusks from the subclasses Opisthobranchiata and Pulmonata. As will be described below, the nervous systems of these animals provide particularly favorable mate-

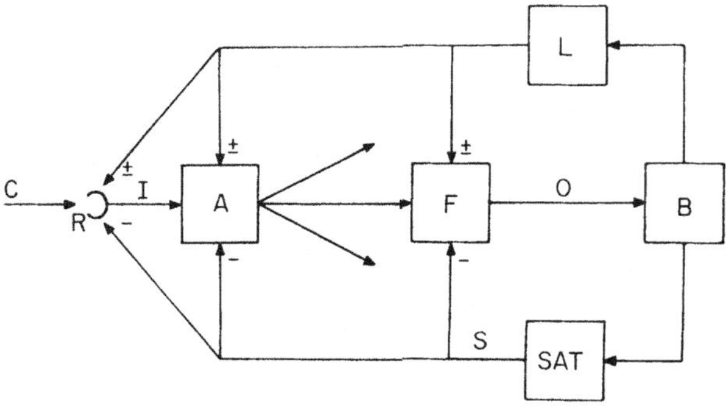

Fig. 1. Simplified diagram of the relationship between chemosensory stimulation and feeding behavior. "Black boxes" represent neuronal circuits, each performing a discrete integrative function as follows: (A) afferent analyzer; (B) generation of consummatory behavior; (F) decision making to feed; (SAT) monitoring of satiety; (L) food-related learning. See text for further details. From de Ruiter, 1967. (Reprinted by permission of The Johns Hopkins University Press.)

rial for studies on neural integration. A summary of the neuronal mechanisms generating consummatory behavior in the pulmonate snail *Helisoma* will be presented to illustrate the type of analysis that it is possible to achieve in gastropods with current electrophysiological techniques. It will also be shown that most of the phenomena considered in de Ruiter's model (i.e., food recognition, satiation, learning) can be observed and therefore studied in one or more of the gastropod species presently under electrophysiological investigation. Finally, the significance of molluscan neurobiology for future studies on the role of the chemical senses in nutritional homeostasis in man and other vertebrate species will be discussed.

EXPERIMENTAL ADVANTAGES OF GASTROPODS

The principal advantage of gastropods is the facility with which intracellular recordings can be made from neurons within the central nervous system. Several factors, which have been more fully discussed elsewhere (Kandel and Kupfermann, 1971), contribute to this characteristic. First, the neuron cell bodies (or somata) of gastropods can reach extraordinary size, up to 1 mm in diameter in some opisthobranch species. Second, unlike vertebrates, somata lie at or near the surface of the brain (or ganglion*). Third, the individual neurons are brightly pigmented in many gastropod species. The large size, superficial location, and cellular pigmentation allow individual neurons to be readily observed and identified in the unstained living ganglion. A semidiagrammatic view of the left buccal ganglion from the pulmonate slug *Ariolimax* is shown in Fig. 2 as it might appear under a low power dissection microscope.

It is a relatively simple matter to penetrate two or more neurons with intracellular microelectrodes under visual control. Since intracellular rather than extracellular microelectrodes are used, it is possible to control the activity of the penetrated cell by passing hyperpolarizing or depolarizing currents directly into the cell from the microelectrode. Cell potential can be monitored during current injection, either by employing a double-barrel microelectrode or a single-barrel microelectrode in conjunction with an amplifier equipped with a bridge circuit.

The essence of integrative research is the elucidation of the synaptic interactions that occur between two neurons. To determine whether a

* The gastropod CNS is comprised of discrete aggregates of neurons called ganglia. In general, gastropods possess five bilaterally paired ganglia (i.e., the buccals, cerebrals, parietals, pleurals, and pedals) and an unpaired abdominal ganglion. In some species, fusion between various ganglia may occur.

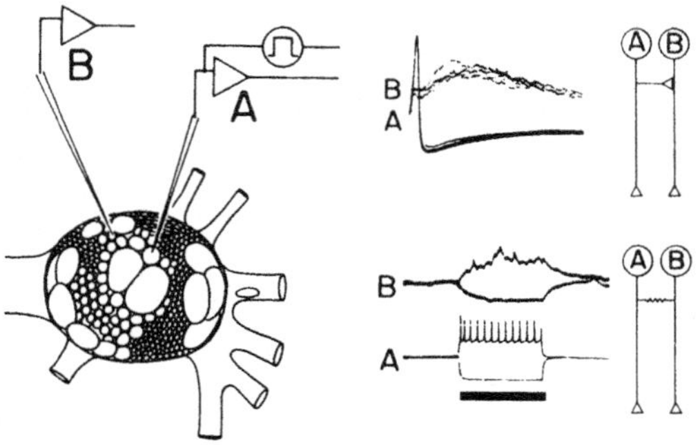

Fig. 2. Semidiagrammatic representation of the right buccal ganglion of a terrestrial slug *(Ariolimax californicus)*. Ovoid shapes denote individual neuron cell bodies, the largest being approximately 250 μm in diameter. Electrophysiological records are intracellular recordings to illustrate the evidence used to assess synaptic connections that are chemical (upper traces) or electrical (lower traces). In both sets of traces, cell "A" is shown at low gain, cell "B" at high gain. Above, several oscilloscope sweeps, triggered by the spike in "A," have been superimposed to show that the depolarizing psp in "B" occurs in an all-or-none fashion at a short constant latency. These data suggest that "A" makes a chemical synapse on to "B" schematically represented by an open triangle. Below, two oscilloscope sweeps have been superimposed. During the first sweep, depolarizing current was injected into "A" evoking a train or 12 action potentials in "A" and a series of depolarizing psp's in "B" riding on a longer depolarizing wave. During the second sweep, hyperpolarizing current was injected into "A" producing a hyperpolarization in both "A" and "B." Duration of current injection denoted by solid bar. These data suggest that "A" and "B" are electrotonically coupled (schematically represented by a resistor symbol).

particular neuron (e.g., cell "A") makes a synaptic connection with some other neuron (e.g., cell "B"), cell "A" is made to fire a series of action potentials. Presence of a synaptic connection between "A" and "B" will be revealed as a series of postsynaptic potentials (psp's) in "B." If a psp in B occurs after each spike in A at a short constant latency, and this one for one following of the psp persists under various pharmacological conditions, it is likely that cell "A" makes a direct monosynaptic connection with cell "B" (Fig. 2, upper traces).

Synaptic connections between neurons can be chemically and/or electrically mediated. In both instances, an action potential in the presynaptic neuron can evoke a unitary psp in the postsynaptic cell. Chemical transmission can be differentiated from electrical coupling on two grounds. First, chemical (but not electrical) transmission fails if the ganglion is bathed in a saline solution containing a high concentration of magnesium. Second, with an electrical (but not chemical) synapse, hyperpolarizing currents injected into one neuron will cause an electrically coupled cell to become hyperpolarized (Fig. 2, lower traces).

Variations in relative size, coloration, ganglionic position, and electrophysiological properties can serve as cues to identify the same neuron in each member of a particular species. Using the techniques described above it has been found that the synaptic connections made by an identified neuron are virtually identical among conspecifics. This fact enables the cellular neurophysiologist to construct a detailed "wiring diagram" of a particular region of the nervous system based upon the results obtained in a number of animals.

Complementing these electrophysiological techniques are histological

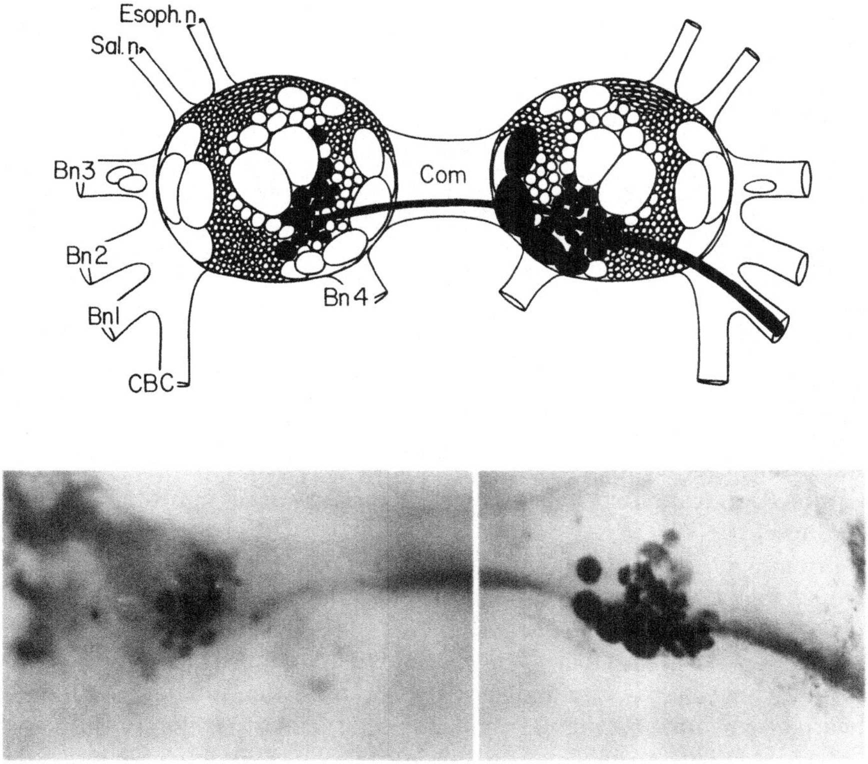

Fig. 3. Example of cobalt "back-filling" technique. Photomicrograph shows buccal ganglia from the slug *Ariolimax californicus* in which cobalt was diffused into nerve root designated rBn1. Somata of neurons having axons in rBn1 are visible in both the ipsilateral and contralateral ganglia. Semidiagrammatic representation of buccal ganglia shows the approximate location of stained neurons relative to major nerve roots: cerebrobuccal connective (CBC), buccal commissure (Com), esophogeal nerve (Esoph. n), salivary nerve (Sal. n), and nerves innervating the intrinsic musculature of the buccal mass (Bn1, Bn2, Bn3, Bn4).

methods which have been developed to study the morphology of individual neurons (cf. Kater and Nicholson, 1973). Dye molecules can either be directly injected into single neurons through intracellular microelectrodes (e.g., see Fig. 10), or they can be allowed to diffuse up cut axonal processes. The latter "back-filling" technique, as it has been called, is extremely useful in localizing neurons that innervate a particular body region. The cut distal end of the nerve of interest is placed in a pool of dye (e.g., cobalt ions or procion yellow). After a period of time, the ganglion is histologically prepared and viewed under a microscope. The cell bodies of neurons sending axonal processes into the stained nerve can be readily discerned (Fig. 3).

Neurons localized by this procedure that innervate structures involved in feeding (e.g., mouth musculature, salivary glands, etc.) are prime targets for initial electrophysiological investigations on the "motor side" of the feeding circuit.

BLACK BOX "B"

Black box "B" represents the neuronal network coordinating the output (O) of de Ruiter's circuit, i.e., consummatory behavior. This neuronal network must perform two basic functions. First, it must generate the spatial and temporal pattern of muscular and glandular activity required for mastication and deglutition. The sum of this activity will be called the *feeding motorprogram*. Second, the neuronal network must be able to integrate proprioceptive inputs to modify the feeding motorprogram in response to variations in the physical properties of the ingesta (e.g., hardness).

From a mechanistic standpoint, the ingestional process in many gastropod species is somewhat similar to that found in man and other vertebrate species. In man, mastication involves the occlusion of a mobile lower mandible with a stationary upper mandible. In most herbivorous gastropods, ingestion is accomplished by the occlusion of a mobile, tooth-bearing structure called the radula, with a stationary horny structure called the medial tooth (Fig. 4). During each bite, the radula is forceably *protracted* into the food and *retracted* against the medial tooth back into the buccal cavity along with a small quantity of food. As in mammals, saliva is secreted during mastication to facilitate bolus formation and deglutition.

Detailed investigations of the synaptic interactions underlying the consummatory feeding response are presently under way in several gastropod species (e.g., *Aplysia*, Kupfermann and Cohen, 1971; *Ariolimax*, Sense-

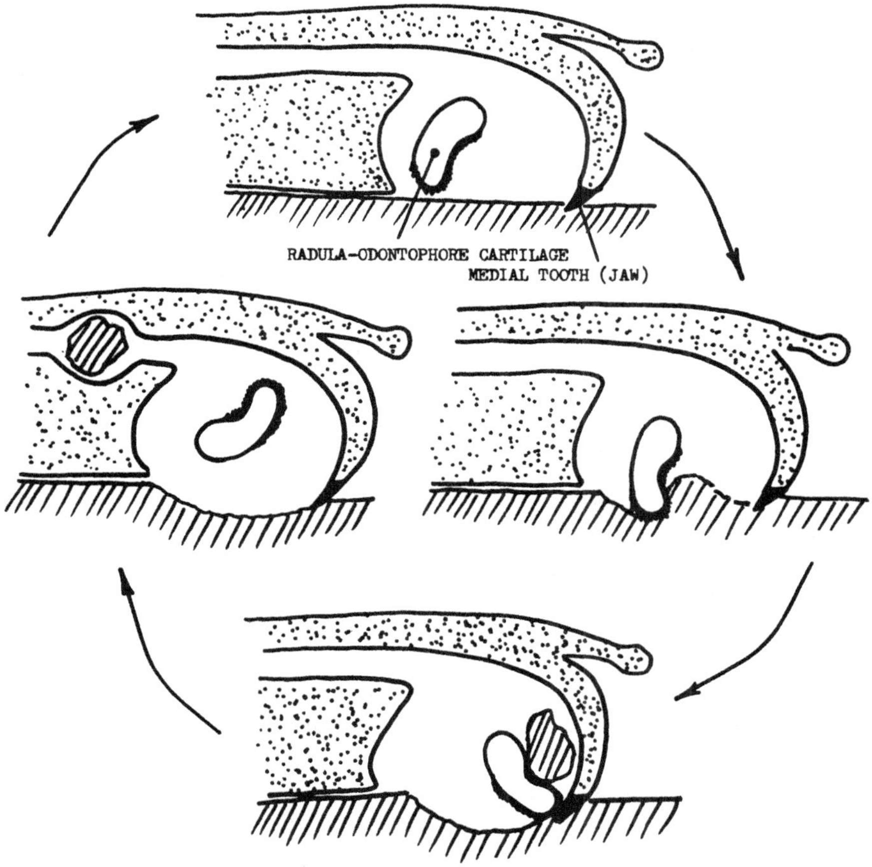

Fig. 4. Diagram illustrating the mechanics of mastication for herbivorous pulmonates showing a midsagittal section of the head. The tooth-bearing radula is borne on a hard internal structure called the odontophore cartilage. During a bite, the radula–odontophore complex is driven into the food substrate (protraction phase), moved anteriorally until it occludes the upper mandible (jaw), and finally returned into the buccal cavity (retraction phase) whence deglutition occurs and the cycle is repeated. Modified after Kater, 1974.

man, 1976; *Helisoma,* Kater, 1974; *Limax,* Gelperin and Forsythe, 1976; *Planorbis,* Berry, 1972; *Pleurobranchaea,* Siegler *et al.,* 1974; *Navanax,* Woollacott, 1974; Spray and Bennett, 1975; and *Tritonia,* A. O. D. Willows, personal communication).

The most extensive analysis to date has been made in the pulmonate snail *Helisoma trivolvis* (Kater, 1974). Neurons responsible for the control of the intrinsic musculature of the buccal mass, the primary feeding organ, are clustered in the bilaterally paired buccal ganglia. In *Helisoma,* each ganglion contains approximately 200 neurons of which seven have been identified as retractor motoneurons, six as protractor motoneurons, one as a salivary effector neuron, and eight as premotor interneurons. This last population of interneurons has been called "cyberchrons,"* since they are believed to be the principal elements timing and driving the feeding response. Figure 5 shows a summary of the neural connections underlying the generation of the consummatory response in *Helisoma*. Reciprocal activation of protractor (P) and retractor motoneurons (R) is accomplished in the following manner. At the beginning of the bite, the cyberchron neurons simultaneously fire a barrage of action potentials. Extensive electrotonic coupling between individual cyberchrons accounts for this simultaneous activity. One member of the cyberchron circuit (i.e., cell 52) makes inhibitory monosynaptic connections with protractor motoneurons (e.g., cells 19 and 18), resulting in suppression of all protractor motoneuron activity. Another cyberchron neuron (i.e., cell 41) makes excitatory monosynaptic connections with retractor motoneurons (e.g., cells 28 and 38), resulting in the activation of the retracter motoneuron population. During this phase of the bite cycle, the radula is retracted against the medial tooth and back into the buccal cavity. Termination of the cyberchron burst leads to the cessation of retractor activity due to the loss of the excitatory synaptic input. Protractor motoneurons now fire a burst of action potentials as a consequence of postinhibitory rebound (i.e., "anode break" excitation). This protractor activity causes the radula to be protracted out of the buccal cavity and into the food substrate, at which time the cycle is repeated.

Kater and Rowell (1973) have further investigated the role played by proprioceptive feedback on this centrally generated feeding motorprogram. Muscle receptors (MR in Fig. 5) provide information about the resistance being encountered by the radula. Increasing radular load increases the output of the muscle receptors during the retraction phase of the bite cycle. These receptors make excitatory monosynaptic connections with retractor mononeurons providing these motoneurons with an additional excitatory drive. Muscle receptors also make an inhibitory synapse with protractor motoneurons. These reciprocal synaptic inputs function to extend the duration of the retraction at the expense of the protraction phase. This insures that both the strength and the duration of retraction will be commensurate with the encountered load.

To summarize, the snail *Helisoma* solves the task of generating a reciprocal motorprogram by simultaneously exciting and inhibiting antagonistic motoneurons and then allowing the inhibited motoneurons to fire out of

* From Greek, *cyber*, meaning to drive, and *chron*, meaning time.

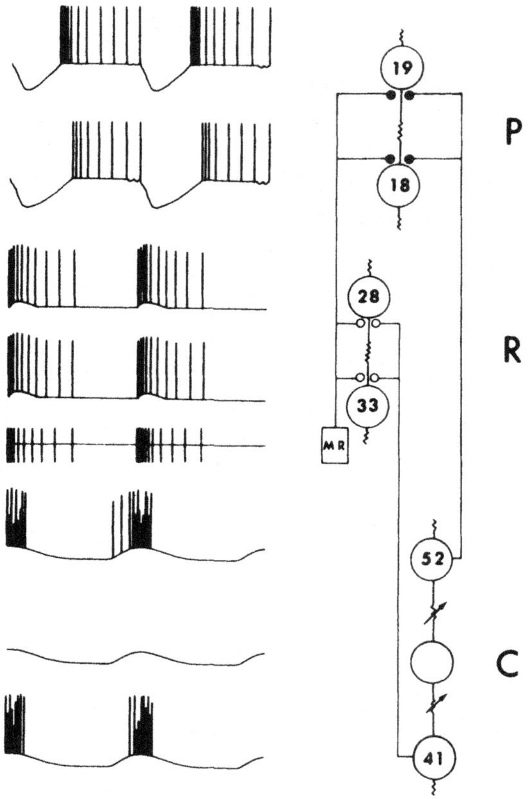

Fig. 5. Neuronal circuit proposed by Kater to explain the generation of the consummatory response in *Helisoma trivolvis*. Individual neurons are represented as circles and buccal mass mechanoreceptors as a box. Schematized records opposite each neuronal element depict characteristic activity observed in these elements during the generation of feeding activity with a cycle period of about 3 sec. Protractor motoneurons (P) fire in antiphase to retractor motoneurons (R) as a result of synaptic input from neurons of the cyberchron network (C). Resistor symbols denote electrical coupling between neurons with the presence of additional elements implied by resistors leaving each group (the protractor group actually contains at least seven neurons per ganglion; the retractor group, at least 11 neurons per ganglion; the cyberchron, at least eight neurons per ganglion). Open circles are excitatory chemical synapses, closed circles are inhibitory chemical synapses, and the variable resistor symbol between cyberchron neurons refers to the probable ability of higher-order inputs to decrease the coupling between individual neurons of the network and thus inhibit generation of feeding output. From Kater, 1974. (Reprinted by permission of the American Society of Zoologists.)

postinhibitory rebound. The basic motorprogram can be modified in response to loading of the radula by having the proprioceptive input provide an additional excitatory driving force to one population of motoneurons (i.e., retractors) and simultaneously inhibiting the antagonistic motoneuron population (i.e., protractors). The neuronal wiring diagram shown in Fig. 5 schematically expresses these integrative mechanisms and replaces black box "B" in the feeding circuit for *Helisoma*.

As yet, the neuronal circuitry is only understood in a general sense and further work, particularly on the properties of the cyberchron network, is required. Even in the present preliminary state, the results presented for *Helisoma* and the similar results that have been obtained in other gastropod species (see references cited above) represent a substantially greater insight into the integrative processes underlying feeding behavior than it has been possible to obtain in any other group of animals. These data indicate the type of analysis that should be possible for other integrative functions represented in de Ruiter's model.

BLACK BOX "A"

Black box "A" represents a neural network which de Ruiter called an "afferent analyzer." This neuronal network performs the integrative function of determining the qualitative nature of a chemosensory input (I) arising from receptors (R) in response to a chemical stimulus (C). In contrast to the wide attention that has been directed toward the behavioral output of the feeding circuit, relatively little effort has been focused on the chemosensory input, except in the marine sea slug *Aplysia californica*.

Aplysia is an oligophagus herbivore that feeds on various species of marine algae (Kupfermann and Carew, 1974). Behavioral studies have shown that amino acids which occur in algae can stimulate the mouth-opening reflex of the feeding response (Jahan-Parwar, 1972). The most potent of the 20 L-amino acids assayed were glutamic and aspartic acids, which gave a threshold response between 10^{-7} and 10^{-6} M, and 10^{-5} and 10^{-6} M, respectively. Electrophysiological studies found chemosensory afferents sensitive to these amino acids in the oral veil nerves that innervate the anterior region of the animal (Fig. 6). Thresholds measured electrophysiologically were approximately one order of magnitude higher than thresholds measured behaviorally.

To study the central projections of these chemosensory afferents, Jahan-Parwar has developed a preparation that consists of the isolated nervous system attached to an excised region of oral veil containing the chemosensory receptors. The receptor tissue is isolated in a separate

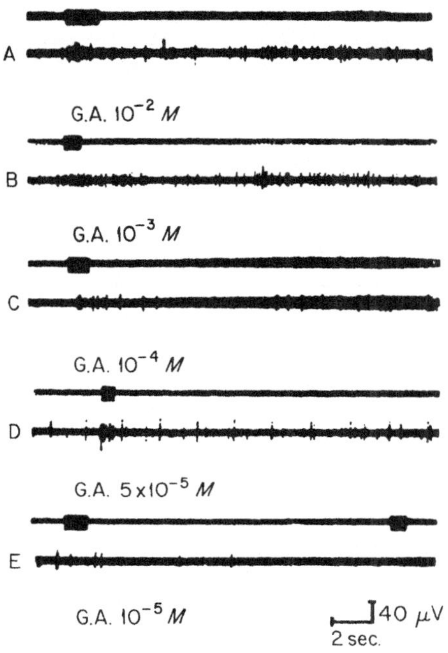

Fig. 6. Extracellular unit responses from the anterior tentacular nerve of *Aplysia californica* following stimulation of the oral veil with various concentrations of glutamic acid (G. A.). Stimulus presentation is marked on the upper beam of each record. From Jahan-Parwar, 1972. (Reprinted by permission from the American Society of Zoologists.)

watertight compartment with nerves passing through vaseline seals to the CNS. Excitatory synaptic responses can be recorded in certain neurons in cerebral ganglia when seaweed extract or amino acids are applied to the receptors (Jahan-Parwar, 1972). Further work along these lines has concentrated on identifying the synaptic connections made by these neurons with other neurons located in the cerebral ganglia (Jahan-Parwar and Fredman, 1976).

The use of a minimal preparation as described above offers much greater stimulus control than is possible in more intact preparations. However, it is difficult to know if in fact the isolated nervous system is processing the chemical stimulus as a feeding stimulant. Prior and Gelperin (1977) have shown that application of food extract to the excised lips of the terrestrial slug *Limax maximus* evokes patterned neural activity in the isolated central nervous system. This neural activity is virtually identical to that recorded in intact feeding slugs with surgically implanted extracellular electrodes (Gelperin and Forsythe, 1976). Similarly, "stimulus-bound feeding" in an isolated CNS/excised lip preparation has also been

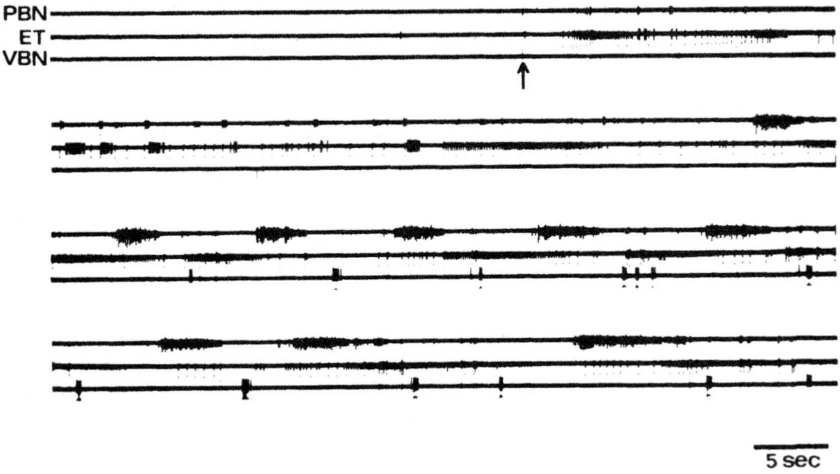

Fig. 7. Continuous extracellular recordings demonstrating "stimulus-bound feeding" in the isolated central nervous system/excised lip preparation of the pulmonate snail *Helisoma trivolvis*. Application of an aqueous extract from a palatable food plant (*Taraxacum officinale*) to the excised lips (at arrow) evoked the feeding motorprogram as monitored by reciprocally patterned activity in the posteriobuccal nerve (PBN), the esophageal trunk (ET), and the ventrobuccal nerve (VBN). (Original unpublished figure by this author.)

observed in the aquatic snail *Helisoma trivolvis* (Fig. 7). In both *Limax* and *Helisoma*, the initiation of the feeding motorprogram following the application of food extracts to the oral chemoreceptors indicates that the isolated CNS can recognize and process feeding stimulants in a behaviorally appropriate manner. Minimal preparations that display "stimulus-bound feeding" should prove to be extremely useful in future intracellular studies designed to elucidate the integrative processes underlying food recognition.

BLACK BOX "SAT"

Black box "SAT" in de Ruiter's model represents the neural circuit that provides information as to the animal's level of satiety. To accomplish this task, the circuit must be able to monitor both short-term cues related to meal size, and long-term cues related to the nutritional adequacy of the diet.

Behavioral evidence shows that at least three gastropod species control their meal size in the presence of a superabundance of food (*Aplysia*, Susswein and Kupfermann, 1975; *Pleurobranchaea californica*, Davis *et al.*, 1974; *Ariolimax*, Senseman, 1976). In the marine slug *Aplysia*, which

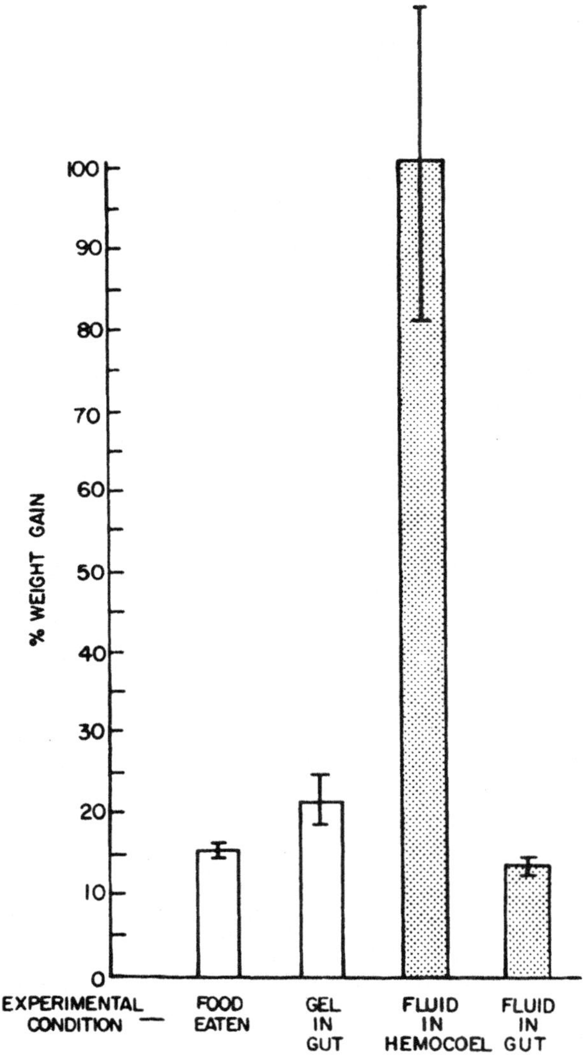

Fig. 8. One piece of evidence supporting the hypothesis that bulk stimuli from the anterior gut induces satiety in *Aplysia californica*. Percent weight gain of animals which were satiated with seaweed or gel, and of animals injected with fluid in the hemocoel or in the gut, are shown. Clear bars indicate that the procedures effectively ended feeding behavior. Stippled bars indicate that the procedure did not end feeding behavior. Standard errors are shown. From Susswein and Kupfermann, 1975. (Reprinted by permission from the Journal of Comparative Physiology.)

has been the most carefully studied species in this regard, short-term control of meal size appears to result from sensory cues evoked by the bulk distension of the anterior gut (Susswein and Kupfermann, 1975). Loading this region of the animal with nonnutritive bulk (e.g., polyacrylamide gel) in amounts that are comparable to normal meal sizes induces satiation, whereas much larger volumes of saline injected into the intestine or body cavity do not (Fig. 8).

Evidence for long-term regulation of caloric intake is lacking in gastropods. Susswein (1975) fed *Aplysia* a calorically diluted diet with no apparent compensatory increase in food intake. It is possible that gastropods do not regulate caloric intake and therefore represent poor models for the study of this phenomenon.

BLACK BOX "L"

Many vertebrate species can learn to associate a particular food item with postingestional malaise (e.g., Garcia, Chapter 13) or with an amelioration of symptoms arising from dietary insufficiencies (cf. Rozin and Kalat, 1971). Black box "L" in de Ruiter's model represents the neural circuit that establishes a "memory" that can modify the palatability of a previously ingested food.

At least two species of gastropod mollusks that are presently under electrophysiological investigation display the capacity for food-related learning. The opisthobranch carnivore *Pleurobranchaea californica* can be classically conditioned to strike a clean glass rod after a period of training in which the animal is presented with a glass rod coated with a feeding stimulant, i.e., squid extract (Mpitsos and Davis, 1973).

Perhaps of greater interest are the reports of taste aversion learning in *Pleurobranchaea* (Mpitsos and Collins, 1975) and in the terrestrial slug *Limax maximus* (Gelperin, 1975). For example, *Limax* rapidly learns to avoid a palatable food (mushroom) when ingestion is paired with a noxious exposure to CO_2 gas. Forty percent of the animals that have been trained to date develop the aversion after one or two exposures to the gas (Gelperin and Forsythe, 1976). Animals learning the aversion continue to eat food (potato) that have not been associated with the gas treatment (Fig. 9).

It has yet to be determined whether gastropods learn to prefer specific foods that ameliorate the effects of vitamin or mineral deficiencies. Anecdotal observations suggest that this might be a profitable area for further research. Ridgway and Walker (cited in Runham and Hunter, 1970) reported that terrestrial slugs *(Arion ater)* fed a synthetic diet deficient in B vitamins displayed a "gradual falling off in the growth rate, *frequent can-*

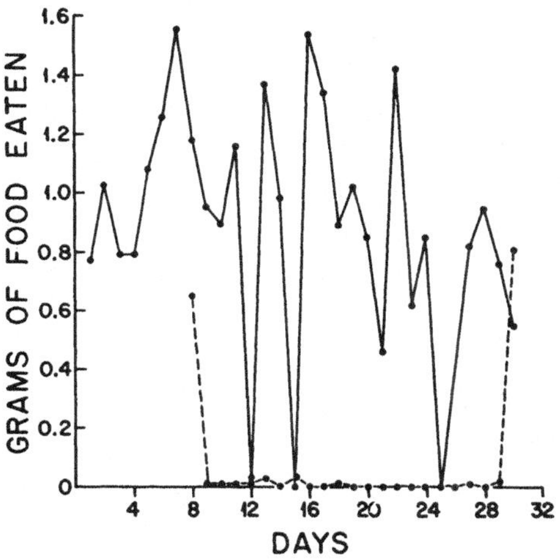

Fig. 9. Example of one trial food aversion learning in the terrestrial slug *Limax maximus*. Points connected by solid lines show daily consumption of potato. Points connected by dashed lines show daily intake of mushroom. Mushroom first presented on day 8, 4 hr after potato meal. Punishment of mushroom consumption by CO_2 exposure suppressed further mushroom consumption until day 30. From Gelperin, 1975. (Copyright 1975 by the American Association for the Advancement of Science.)

nibalism (italics added by this author) and a marked reduction in the length of life." Similarly, terrestrial snails grown in low calcium environments have been observed to aggregate on pieces of bivalve shells presumably to obtain calcium carbonate (Schmidt, 1955). In both cases, the shift in food selection provoked by dietary insufficiencies may represent examples of learned food preferences or "specific hungers."

BLACK BOX "F"

At the center of his model de Ruiter placed black box "F," in which the final decision was made to initiate or terminate feeding, depending upon the inputs received from black boxes "A," "SAT," and "L." The neuronal circuit represented by black box "L" must therefore act as a summation point for synaptic inputs generated by the various neuronal circuits described above. It must also exert an excitatory and/or inhibitory influence over the neuronal circuit generating the feeding motorprogram (black box "B").

The cerebral ganglia of many (if not all) opisthobranch and pulmonate mollusks contain a prominent pair of neurons that have been called the

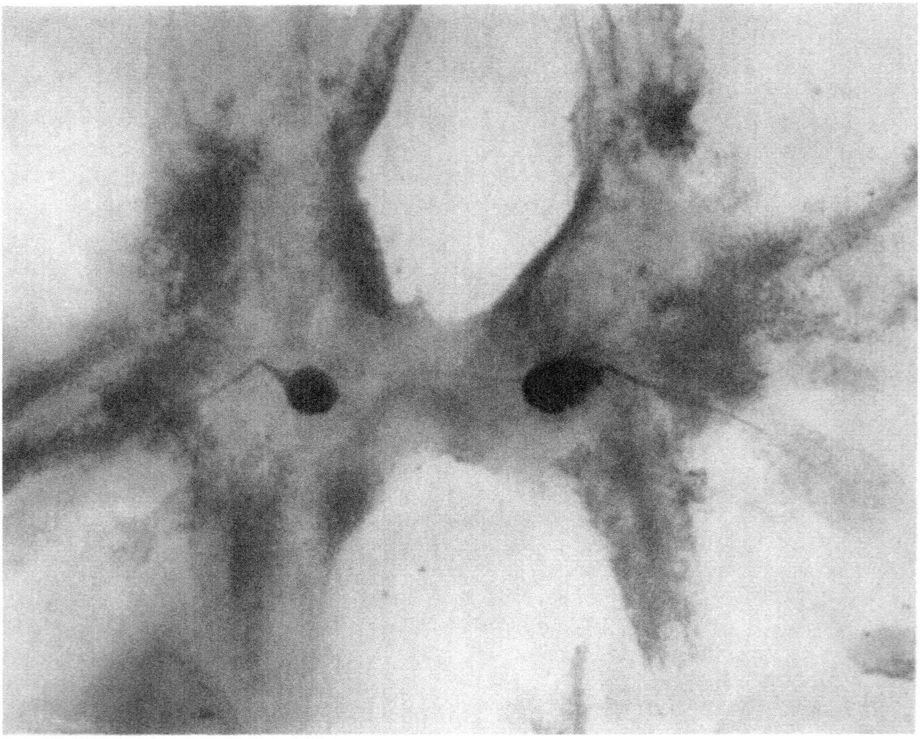

Fig. 10. Cerebral ganglia of the terrestrial slug *Ariolimax californicus* in which the left and right metacerebral giant cells (MCGC's) were stained by intrasomatic injection of cobalt ions. Axonal processes can be observed centrally in the cerebral commissure and laterally as they pass to the cerebrobuccal connectives and external lip nerves. Diameter of cell body is approximately 300 μm.

metacerebral giant cells (MCGC's or MCG's, shown in Fig. 10). The marked similarities between the morphological, biochemical, and electrophysiological characteristics of the MCGC's in the various species in which they have been studied suggest that they are homologous (Senseman and Gelperin, 1974; Weiss and Kupfermann, 1974; Gillette and Davis, 1975).

Three properties of the MCGC's make them potential elements in the neuronal circuitry underlying black box "F." First, the MCGC's receive synaptic inputs from a variety of sources, including nerves innervating chemosensory receptor areas (e.g., Kandel and Tauc, 1966; Gillette and Davis, 1975). Second, they are active during feeding in intact animals with chronically implanted electrodes (Gelperin and Forsythe, 1976) and in isolated preparations generating the feeding motorprogram (Gillette and Davis, 1975; Prior and Gelperin, 1977). Third, on some occasions, intracellular stimulation of a single MCGC can evoke several cycles of bursting activity in buccal motoneurons (Fig. 11).

While these characteristics implicate a role for the MCGC in feeding behavior, the exact nature of this role is as yet not well defined. The low level of MCGC activity observed during feeding in intact animals, and the lability at which the MCGC drives buccal motoneurons, have suggested to Prior and Gelperin (1977) that the MCGC acts to modulate rather than drive feeding behavior in the slug *Limax*. Alternatively, Gillette and Davis (1975) believe that the MCGC's in *Pleurobranchaea* are elements of a neuronal population that initiates and terminates feeding. The lability of the MCGC's efficacy to drive feeding is easily explained by assuming that feeding occurs only when there is a "consensus" among the population. Further characterization of the postsynaptic effects of the MCGC's and

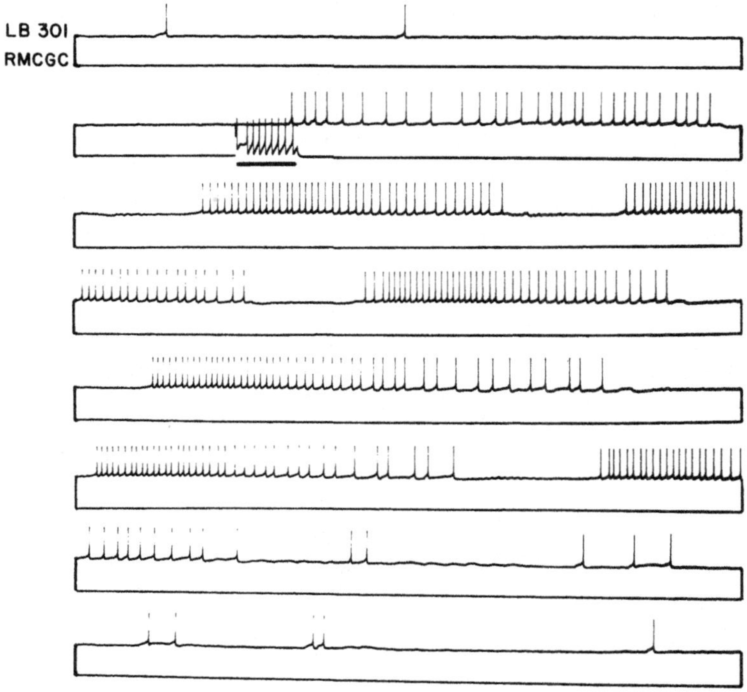

Fig. 11. Continuous intracellular recordings from the right metacerebral giant cell (RMCGC) and an identified buccal motoneuron (LB 301) in the terrestrial slug *Ariolimax californicus*. Brief intrasomatic stimulation of the RMCGC (denoted by solid bar) evoked several cycles of bursting in LB 301. Calibration for both cells: 150 mV, 5 sec.

other unidentified cerebral and buccal neurons will be required before any strong statements can be made in this regard.

THE SIGNIFICANCE OF GASTROPOD NEUROBIOLOGY

I have attempted to make two basic points. First, gastropod mollusks are particularly favorable systems in which to study neuronal integration. Second, the feeding behavior of gastropods is similar in a number of aspects with that of mammalian species. There is one crucial question left to be answered. Even if behavioral similarities exist between mammals and mollusks, must the underlying neural mechanisms be similar as well?

The phylogenetic distance between mammals and mollusks is exceedingly great. Yet at the level of the single cell, there does not appear to be any fundamental differences between the functional capacity of vertebrate and invertebrate neurons (cf. Bullock and Horridge, 1965). It seems reasonable to suspect that the functional capacities of single nerve cells would impose constraints on the functional capacities of neuronal networks. Such constraints would in turn be expected to limit the degrees of freedom available to natural selection in the design of new types of neuronal circuits to solve specific integrative problems. The greater this restriction has been, the more reasonable the proposition that similar neuronal mechanisms will be found performing analogous functions in genetically unrelated species.

Experimental support for this proposition is presently limited to a precious few examples in which cellular neurophysiological analyses of a behavioral act are sufficiently coadvanced in a vertebrate and an invertebrate preparation. One of the best studied and most striking examples of similar neural processes subserving analogous integrative functions has been found in the control of locomotory behavior in the cat and the cockroach (Pearson, 1977). Within the context of feeding behavior, recent studies (Luschei and Goodwin, 1974) suggest that the masticatory rhythm of the monkey is produced by a central pattern-generating mechanism as in gastropod mollusks, rather than reflexly driven as originally believed (Sherrington, 1917). For more complex integrative tasks such as olfactory and gustatory coding, maintenance of body weight, and conditioned food preferences, we are still too ignorant even to speculate on what similarities or differences might exist in the brains of "higher" and "lower" animals. Because of our ignorance we must accept or reject the utility of simpler animals as models systems based only on our personal biases and scientific intuitions. For those of us who embrace the simple system approach, the snail has recently assumed a larger role in the areas of nutrition and the chemical senses than simply as an appetizer to a fine French meal.

REFERENCES

Berry, M. S. (1972). A system of electrically coupled small cells in the buccal ganglia of the pond snail *Planorbis corneus*. *J. Exp. Biol.* **56,** 621–637.

Bullock, T. H., and Horridge, G. A., eds. (1965). "Structure and Function in the Nervous Systems of Invertebrates," Vol. 2, Chapter 23. Freeman, San Francisco, California.

Davis, W. J., Mpitsos, G. J., and Pinneo, J. M. (1974). The behavioral hierarchy of the mollusk *Pleurobranchaea*. 1. The dominant position of the feeding behavior. *J. Comp. Physiol.* **90,** 207–224.

de Ruiter, L. (1967). 6: Critic's comments. *In* "The Chemical Senses and Nutrition" (M. Kare and O. Maller, eds.), p. 83. Johns Hopkins Univ. Press, Baltimore, Maryland.

Dethier, V. G. (1976). "Man's Plague? Insects and Agriculture." Darwin Press, Princeton, New Jersey.

Gelperin, A. (1975). Rapid food-aversion learning by a terrestrial mollusk. *Science* **189,** 567–570.

Gelperin, A., and Forsythe, D. (1976). Neuroethological studies of learning in mollusks. *In* "Simpler Networks: An Approach to Patterned Behavior and Its Foundations" (J. C. Fentress, ed.), Chapter 16, Sinauer Assoc., Sunderland, Massachusetts.

Gillette, R., and Davis, W. J. (1975). Control of feeding by the metacerebral giant neuron of *Pleurobranchaea. Neurosci. Abstr., Soc. Neurosci. 5th Annu. Meet.* p. 571.

Jahan-Parwar, B. (1972). Behavioral and electrophysiological studies on chemoreception in *Aplysia. Am. Zool.* **12,** 525–537.

Jahan-Parwar, B., and Fredman, S. M. (1976). Cerebral ganglion of *Aplysia:* Cellular organization and origin of nerves. *Comp. Biochem. Physiol. A* **54,** 347–358.

Kandel, E. R., and Kupfermann, I. (1970). The functional organization of invertebrate ganglia. *Annu. Rev. Physiol.* **32,** 193–258.

Kandel, E. R., and Tauc, L. (1966). Input organization of two symmetrical giant cells in the snail brain. *J. Physiol (London)* **183,** 269–286.

Kater, S. B. (1974). Feeding in *Helisoma trivolvis:* The morphological and physiological bases of a fixed action pattern. *Am. Zool.* **14,** 1017–1036.

Kater, S. B., and Nicholson, C. (1973). "Intracellular Staining in Neurobiology." Springer-Verlag, Berlin and New York.

Kater, S. B., and Rowell, C. H. F. (1973). Integration of sensory and centrally programmed components in the generation of cyclic feeding activity of *Helisoma trivolvis. J. Neurophysiol.* **36,** 142–155.

Kupfermann, I., and Carew, T. J. (1974). Behavioral patterns of *Aplysia californica* in its natural environment. *Behav. Biol.* **12,** 317–337.

Kupfermann, I., and Cohen, J. (1971). The control of feeding by identified neurons in the buccal ganglion of *Aplysia. Am. Zool.* **11,** 667.

LeMagnen, J. (1971). Advances in studies on the physiological control and regulation of food intake. *Prog. Physiol. Psychol.* **4,** 203–261.

Luschei, E. S., and Goodwin, G. M. (1974). Patterns of mandibular movements and jaw muscle activity during mastication in the monkey. *J. Neurophysiol.* **37,** 954–966.

Mpitsos, G. J., and Collins, S. (1975). Learning: Rapid aversive conditioning in the gastropod mollusk *Pleurobranchaea*. *Science* **188,** 954–957.

Mpitsos, G. J., and Davis, W. J. (1973). Learning: Classical and avoidance conditioning in the mollusk *Pleurobranchaea*. *Science* **180,** 317–320.

Prior, D. J., and Gelperin, A. (1977). Autoactive molluscan neuron: Reflex function and synaptic modulation during feeding in the terrestrial slug, *Limax maximus*. *J. Comp. Physiol.* **114,** 217–232.

Pearson, K. G. (1977). The control of walking. *Sci. Am.* **235,** 72–87.

Rozin, P., and Kalat, J. W. (1971). Specific hungers and poison avoidance as adaptive specializations of learning. *Psychol. Rev.* **78,** 459–486.

Runham, N. W., and Hunter, P. J. (1970). *In* "Terrestrial Slugs," p. 38. Hutchinson, LTD, London.

Schmidt, H. A. (1955). Zur Abhangigkeit der Entwicklung von Gehäuseschnecken vom Kalkgehalt des Bodens. Dargestelt bei Oxychilus draparnaldi. *Arch. Molluskenkd.* **84,** 167–177.

Senseman, D., (1976). Neurobehavioral studies on feeding in the terrestrial slug *Ariolimax californicus* Cooper. Ph.D. Dissertation, Princeton University, Princeton, New Jersey.

Senseman, D., and Gelperin, A. (1974). Comparative aspects of the morphology and physiology of a single identifiable neuron in *Helix aspersa, Limax maximus* and *Ariolimax californica*. *Malacol. Rev.* **7,** 51–52.

Sherrington, C. S. (1917). Reflexes elicitable in the cat from pinna, vibrisse and jaws. *J. Physiol. (London)* **51,** 404–431.

Siegler, M. V. S., Mpitsos, G., and Davis, W. J. (1974). Motor organization and generation of rhythmic feeding outputs in the buccal ganglion of *Pleurobranchaea*. *J. Neurophysiol.* **37,** 1173–1196.

Spray, D. C., and Bennett, M. V. L. (1975). Pharyngeal sensory neurons and feeding behavior in the opisthobranch mollusk *Navanax*. *Soc. Neurosci. 5th Annu. Meet.* p. 570.

Susswein, A. J. (1975). Studies on satiation of feeding in the marine mollusc *Aplysia californica*. Ph.D. Dissertation, New York University, New York.

Susswein, A. J., and Kupfermann, I. (1975). Localization of bulk stimuli underlying satiation in *Aplysia*. *J. Comp. Physiol.* **101,** 309–328.

Weiss, K. R., and Kupfermann, I. (1974). Homology of giant cerebral cells in *Aplysia* to the metacerebral cells of pulmonate molluscs. *Soc. Neurosci., 4th Annu. Meet.* p. 471.

Woollacott, M. (1974). Patterned neural activity associated with prey capture in *Navanax*. *J. Comp. Physiol.* **94,** 69–84.

DISCUSSION

Nicolaides: Do gastropods have any nerves analogous to the recurrent nerve found in flies? Have you tried to cut it and see if you could make hyperphagic slugs as has been done in the blowfly *Phormia?*

Senseman: There is a nerve in gastropods, the esophageal or stomatogastric nerve, that innervates the esphagus, crop, and stomach. It seems most likely that this nerve would carry any afferent input from stretch-sensitive neurons responding to gut distention to the central nervous system. However, transection of this nerve in the marine gastropod *Aplysia* decreased rather than increased meal size (Susswein, 1975). Susswein explained this paradox by noting that the esophageal nerve in *Aplysia* is a mixed nerve carrying both sensory and motor impulses. By eliminating motor input to the esophogus, ingested food was not efficiently moved from the buccal cavity into the stomach. Presumably an accumulation of food in the buccal cavity "prematurely" terminated feeding, thus explaining the decreased meal size.

Garcia: Do gastropods use olfactory cues to locate food?

Senseman: Yes. It has been repeatedly shown that marine, aquatic, and terrestrial gastropods can use chemical cues to find food items at some distance. In terrestrial snails and slugs, the olfactory sense appears to be especially well developed based on the complexity of the presumptive sensory structures and the central integrative areas (see Bullock and Horridge, 1965; Gelperin, 1975).

Cagan: When a slug has learned a specific food aversion, is the food recognized by taste or olfactory cues?

Senseman: In the terrestrial slug *Limax,* Gelperin and Forsythe (1976) reported that trained animals usually did not approach the aversive food and therefore must utilize olfactory cues to make this discrimination. As far as I know, no one has determined whether slugs made anosmic by tentacle removal can make the discrimination based on taste cues alone. That would be an interesting experiment.

Chapter 2

External Influences on the Feeding of Carnivores

ROGER A. MUGFORD

INTRODUCTION

As subjects for experimental studies of feeding behavior, most carnivores present the obvious problems of being expensive to feed and house; they breed rather slowly and can be dangerous to handle. It is fortunate, therefore, that at least two domesticated representatives of the order Carnivora are readily available for study—the dog *Canis familiaris* and the cat *Felis catus*. In this chapter I present data on the feeding behavior of these species, and discuss the potential implications of these data for the behavioral ecology of other carnivores. Since large individual differences in the feeding behaviors of cats and dogs are commonly observed, and since the extrapolation of findings from highly domesticated forms to wild relatives can be quite risky, verification of some of my hypotheses in wild carnivores is ultimately required.

The emphasis of this chapter is on environmental (e.g., social context of a meal), experiential (e.g., prior diet), and sensory factors (e.g., role of food odors) which affect ingestive behavior. Since the significance of these can only be properly understood in the context of the homeostatic systems for food intake which act via internal metabolic signals for hunger and satiety, a brief review of these influences follows in the next section.

CHARACTERISTICS OF INGESTIVE BEHAVIOR

Experimentally supported models of ingestive behavior generally emphasize energy relationships between the animal and its food, as when caloric intake is adjusted by modification to consumption rate, meal size, duration, and frequency of feeding (e.g., Hatton, 1975; Toates and Booth, 1974). Most of these consumption data have been related to theoretical mechanisms which might regulate food intake via gastric (Sudakov and Rogacheva, 1963), glucostatic (Mayer and Thomas, 1967; Nagamachi, 1972), lipostatic, and oral signals. Many of the data have been obtained by exclusive reference to the rat and man, but a small body of comparative evidence suggests that different mechanisms have evolved in response to the particular circumstances facing free-living carnivores.

The neuroanatomical structures that mediate eating have been explored in some detail in the rat and cat, and a similar involvement of the lower brainstem and hypothalamus has been confirmed for both species by lesioning and direct electrical stimulation techniques (Bernston and Hughes, 1974; Skultety, 1969). However, the same brain areas do not necessarily respond to identical blood-borne chemical factors characteristic of the satiated versus hungry animal. Thus, whereas injection of 2-deoxy-D-glucose initiates eating in both rats and monkeys by inhibiting intracellular glucose utilization (Smith and Epstein, 1969), it depresses food intake in the cat (Jalowiec *et al.*, 1973). Whereas rats are able to maintain a relatively constant caloric intake from diets diluted to give a wide range of caloric densities (Booth, 1972; Jacobs and Sharma, 1969), the dog and cat are probably more influenced by the bulking qualities of food within the stomach (for data on the dog and cat, respectively, see Janowitz and Grossman, 1949; Kanarek, 1975). Such interspecies differences in ingestive behavior are to be expected among animals adapted to such different nutritional media as are available to an omnivore and a predaceous carnivore, and whose availability makes such different demands upon their ability to procure food.

The position facing most predators can be summarized as alternating periods of famine and feast, under conditions that favor individuals capable of rapidly consuming and digesting their kills. For instance, the wolf has developed a remarkable gorging ability; in a single meal, 7-day fasted wolves were reported by Young (1944) to consume up to 17% of their own body weight in meat.

Among domestic dogs, some individuals will also take relatively enormous meals. We have measured free access meal sizes of dogs normally fed only to their theoretical caloric requirements once per day. A male Labrador was remarkable for having once eaten 10% of its own body

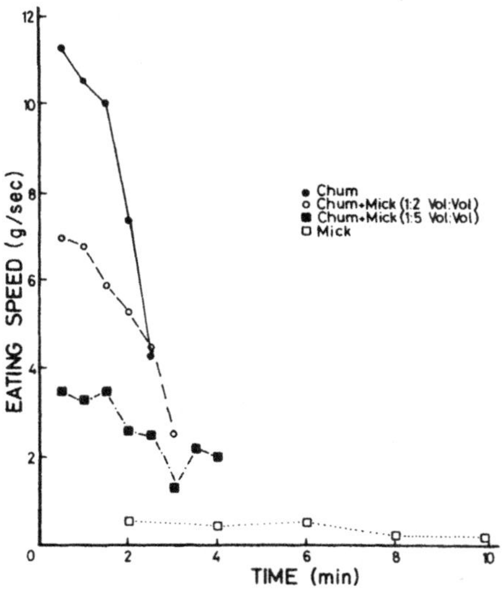

Fig. 1. Mean consumption rate of dog foods during free access meals by Beagles.

weight of a complete canned food, which it subsequently regurgitated! Eating speeds and meal sizes among a group of 16 Beagles were characteristic of both the diet and of the individual dog. Figure 1 shows consumption rate of Pedigree Chum* (which is a highly palatable canned food providing about 82 kcal/100 g), Mick (a dry biscuit dog food providing about 360 kcal/100 g), and various admixtures of the two. The more palatable and calorically dilute Pedigree Chum was consumed significantly more rapidly ($p < 0.001$) than the 1 : 2 mixture of Pedigree Chum and Mick, which itself occupies an intermediate position between Pedigree Chum and Mick when presented alone. Interestingly, this meal dynamic of Pedigree Chum is very like that reported by Le Magnen and Devos (Le Magnen, 1971, Fig. 7) for consumption of a dry diet by the rat.

When the same group of dogs was monitored for meal size and eating speed with other types of dog food (Fig. 2), it was found that both data were largely determined by the diet's moisture content (and thus, indirectly, its caloric density). Meal size and eating speed were found to be highly correlated ($r = 0.85$).

Domestic cats are also influenced by the moisture content of the diet, which determines meal size and eating speed in the manner described for dogs. The data in Fig. 3 are means from a group of nine cats accustomed

* Diets named in the text were manufactured by Pedigree Petfoods, England, and Kal Kan Foods, Inc., California.

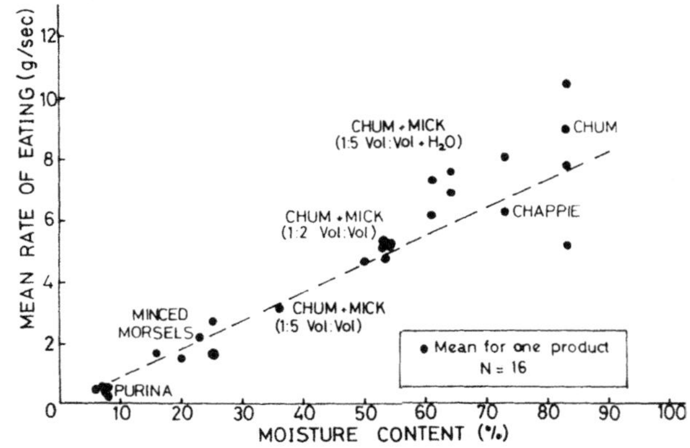

Fig. 2. Average consumption rates for dog foods of varying moisture contents.

to once-daily feeding. Whiskas is a canned cat food comprising meat chunks similar to Pedigree Chum, while Munchies is a dry food which (like Mick) is consumed at a slow and constant rate.

Whether the wolf or other predaceous carnivore can sustain a level of food intake much in excess of its metabolic requirements has probably not been investigated under rigorous experimental conditions. Our studies

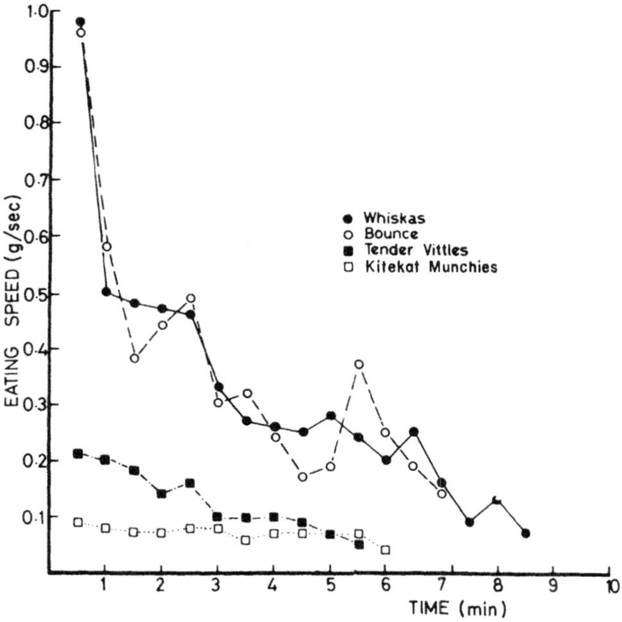

Fig. 3. Mean consumption rate of cat foods during free access meals.

with the domestic dog would lead one to anticipate that related wild canids might also tend to become obese when food is plentiful, though cats form a better balance between intake and expenditure.

We have observed enormous variation in the efficiency with which dogs regulate their food intake, both between breeds and individuals within the same breed. A group of 16 Terriers (eight Cairn and eight West Highland White, equal numbers of males and females) were fed twice daily free access meals of commercial dog foods. The diets, which were subject to large variations in both caloric density and palatability, were varied for each meal. The dogs were, therefore, denied the opportunity of adapting to the calorie repleting effect of individual diets, a situation that provokes overeating in both the rat (Le Magnen, 1956) and man (Siegel, 1957). The Terriers' body weights all remained remarkably constant over a 2-year period, individuals varying by $< \pm 10\%$ and the group mean by $< \pm 4\%$. None of the dogs would be regarded as obese, being well within the range specified by the breed standards (Jones and Hamilton, 1971).

However, clinically defined obesity is commonly encountered among the pet dog population of the United Kingdom (Anderson, 1974) and is often attributed to the feeding conditions (indulgent of both quantity and variety) and limited exercise which failed to provoke weight gains among our Terriers (Joshua, 1970). A separate study of food intake by Beagles during daily free access meals was undertaken and confirms the existence of very large individual differences in their predisposition to obesity. Prior to the experiment, the 16 Beagles had been maintained in good condition for 3–5 years on a fixed intake of Chappie (a complete canned food), equivalent to a daily intake of approximately 800 kcal. When allowed free access to the same diet, all of the dogs immediately increased their initial caloric intakes to between two and five times their former levels.

It was expected that this excessive intake would be only transitory (positive overshoot: Hatton, 1975) until set point for body weight was achieved at a slightly higher plane than that obtained by restricted feeding. Indeed, 13 of the 16 dogs did eventually achieve relatively stable food intakes and body weights, but over different periods (1–3 months) and at body weights from 10 to 60% of starting levels. The other three dogs increased in weight so rapidly that consideration of their welfare led us to place them on caloric restriction after 4 months.

Records from two dogs sharing a common history of housing and feeding are shown in Fig. 4. Whereas no. 307 self-regulated at an appropriate 10 kg, no. 334 became clinically obese after a nearly 40% rise in body weight.

Individual differences between the caloric intakes and weight gains of mongrel cats were noted by Skultety (1969). However, obesity is uncom-

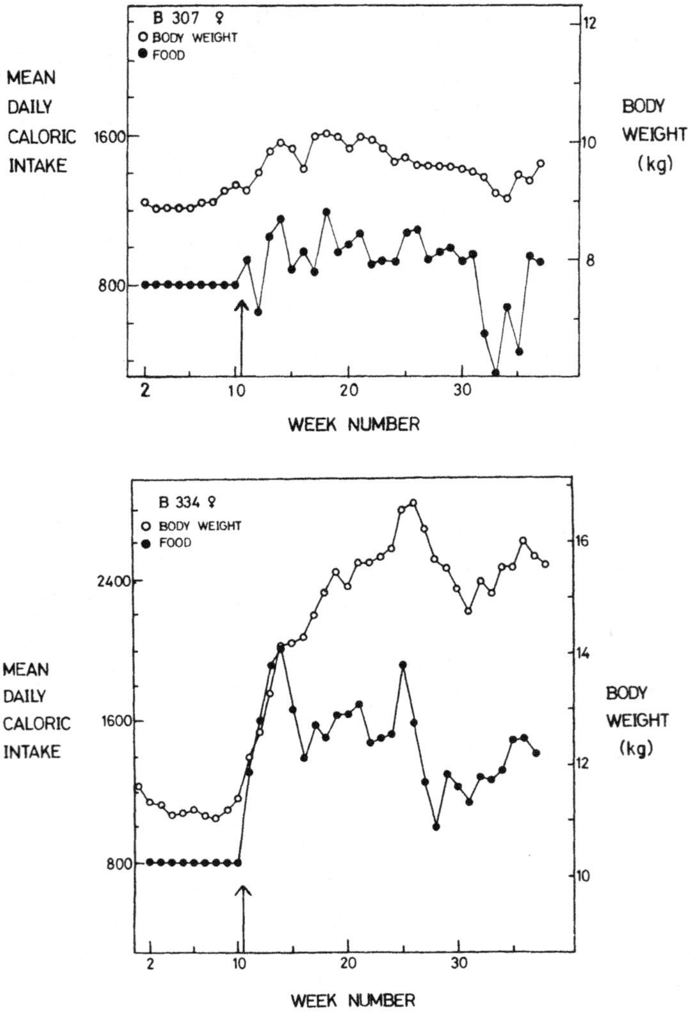

Fig. 4. Weekly means of daily caloric intake and body weight in two Beagles.

mon among domestic cats (Anderson, 1974) and rarely encountered among our (indulgently fed) colony of 400 cats.

The physiological basis for the previously discussed species, breed, and individual differences in regulation of body weight via controls over food intake has not been investigated much, but promises to be a rewarding research area for the future.

SOCIAL FACTORS AFFECTING INGESTIVE BEHAVIOR

With the exception of the lion, the wild felids are mostly solitary hunters relying upon stealth and speed to capture their prey (Ewer, 1973). In contrast, canids such as the wolf and Cape Hunting Dog (Mech, 1975) have adopted highly organized social groupings which enable them to prey upon species both larger and faster than could be killed by a single individual. Social facilitation of feeding *by puppies* (i.e., a greater food consumption when offered meals in groups rather than singly) has been reported by numerous workers (Compton and Scott, 1971; James, 1953, 1961; Ross and Ross, 1949a,b). One might speculate that this phenomenon could be an adaptation by a cooperative pack-living predator to ensure that maximum amounts of a large kill are consumed before they are lost to competing predators sharing the same ecosystem (Kruuk, 1972; Schaller, 1972). However, this hypothesis does not account for Platt and James' (1966) demonstration of a marked social facilitation of feeding among young opossums *(Didelphis virginiana),* marsupials whose adults are highly antagonistic to one another (Crandall, 1964). Social facilitation has not (to the author's knowledge) been reported as occurring among felids or domestic kittens, but its absence should not necessarily be construed as being consistent with their generally solitary hunting habits.

It must be emphasized that the foregoing discussion of social facilitation of feeding is entirely based upon observations of young animals *not* having continuous access to food (i.e., they were meal fed). Puppies tend to center their dominance relationships about the food pan when feeding is limited to discrete meals (James, 1949) to a greater extent than when they are given a continuous *(ad libitum)* food supply. Small groups of both adult dogs and cats rarely fight for possession of food that is always available, and their intakes are no higher than when they are housed individually (author's unpublished observations). The generality of the published data referring to litters, therefore, seemed to be open to question, and so we decided to look for social facilitation among adult animals accustomed to an *ad libitum* feeding regime.

The subjects were continuously monitored by a low light/infrared sensitive camera linked to a time-lapse video tape recorder (Sanyo 1360). This enabled rapid television review of infrequently occurring behaviors on replay at ×60 real-time. The dogs consisted of a litter of fully grown Miniature Poodles having continuous access to a complete, semimoist dog food (Bounce Minced Dinner), whose consumption could be calculated to the nearest gram by reference to a digitized-head scale situated within the field of the camera. A similar arrangement existed for a group of six adult cats, which were given an *ad libitum* supply of a complete dry cat food (Munchies).

Feeding visits by individual dogs or cats were recorded together with the meal size, and these could be related to the behavior of the rest of the group. The most striking feature of the results presented as cumulative intakes in Fig. 5 was the distinctly diurnal pattern of food intake by the dogs, whereas cats distributed their meals evenly throughout the 24 hr. Both species tended to take a large number of small meals (10.4 and 13.3/day, means per dog and cat, respectively), in a manner more reminiscent of the laboratory rat (e.g., Kissileff, 1970) than a predaceous carnivore. The frequency and spacing of meals in both the dogs and cats of this experiment were not much affected by the diet: meal sizes of palatable canned, semimoist, and dry foods being more or less in proportion to their caloric densities. Individual differences in meal size and frequency were apparent in both species, the most and least frequent eaters differing by a factor of more than two. Nevertheless, the suggestion by Scott (1968, p. 21) that cats are "intermittent" feeders, or stated more categorically by the same author (1971) that they are ". . . naturally occasional feeders" and "adult cats feed only once in every 24 hours or sometimes less frequently . . ." is obviously fallacious. The staple diet of all species of *Felis* is small rodents (Coman and Brunner, 1972; Ewer, 1973, p. 214), which would necessitate frequent kills providing several small meals (contrasting with feeding habits of the greater cats such as the lion or tiger, capable of taking prey larger than themselves). Interestingly, a single mouse would provide about the same number of calories as the typical meal of our laboratory cats (30 kcal).

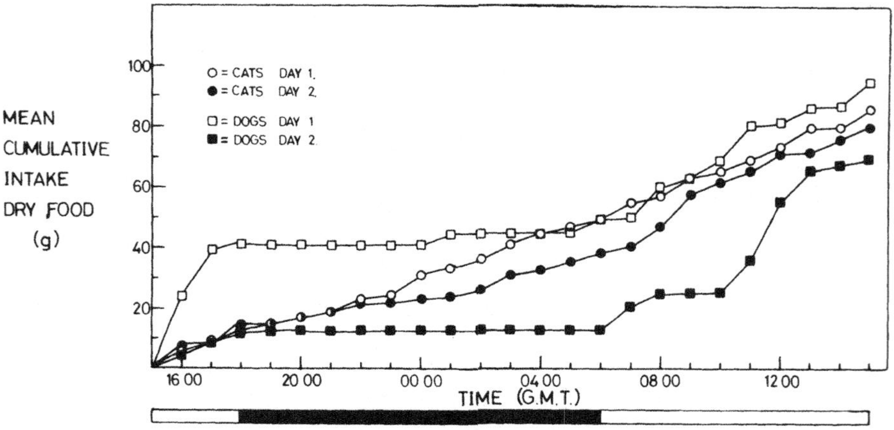

Fig. 5. 24-hr cumulative food intakes by dogs and cats. G.M.T., Greenwich mean time.

Meals were classified as being either "social" or "nonsocial" on the basis that feeding of an individual animal was or was not associated with time by the feeding of another individual. The criterion of association was that meals of two or more animals should either overlap in time or occur within 1 min of the other terminating. Of 82 dog meals and 60 cat meals observed, only 12% and 20%, respectively, were classified as social, insofar as they were associated with another animal's feeding. Less than 1% of total available time was accounted for by feeding, so it would be difficult for any single dominant animal to retain exclusive possession of the food pan (and thus give a spurious impression that feeding was an activity having little consequence for other group members). Indeed, cats were selected for these experiments on the basis of their known compatability with the rest of the group, and little fighting was observed among either the dogs or cats that might suggest food pan possession by dominant animals of the type noted by James (1949) with puppies.

EFFECTS OF PRIOR EXPERIENCE UPON FOOD SELECTION AND INTAKE

In order to satisfy their nutritional requirements, all animals must face the challenge of discriminating useful (i.e., nutritionally reinforcing) food sources from a usually more abundant array of either micronutrient-imbalanced (Harris *et al.*, 1933; Simson and Booth, 1973) or even downright toxic substances (Garcia *et al.*, 1974; Rozin and Kalat, 1971) possibly sharing many of the same sensory qualities. The learned strategies needed to overcome this challenge must doubtlessly be maintained throughout life, but feeding experiences during early development would be expected to exert particularly potent influences upon selection habits of the adult. However, the published data relevant to this issue indicate that two (apparently contradictory) feeding tendencies are established by prior dietary experiences. On the one hand, precocial birds and possibly other vertebrates are susceptible to a process that has been variously called "food imprinting" (Hess, 1964) and "fixation of food habits" (Kuo, 1967, pp. 67–72), while on the other hand, there is a tendency for laboratory raised rats (at least) to prefer foods having a novel flavor or physical characteristic (Bronson, 1966; Morrison, 1974; Welker and King, 1962; Young, 1940).* Which process is the more important for molding the food

* Such elevated preferences for novel diets and flavors should, however, be considered in the context of a marked neophobic response of the wild rat toward novel food and fluid (Barnett, 1963), though the albino rat only exhibits a comparatively weak and transitory neophobia (Carroll *et al.*, 1975).

and prey selection habits of wild carnivores must, in the absence of quantitative data, be a matter for conjecture unlikely to justify bold cross-species generalization.

Working with dogs, cats, and Mynah birds, Kuo (1967) reported that surprisingly "fixed" food preferences could be induced in all three species by raising them from birth on distinctive diets. Great difficulty must have been encountered in feeding day-old puppies and kittens on these diets without any supplementation by mothers' milk, but Kuo does not mention any mortalities among his subjects. We have conducted a number of studies with puppies and kittens along similar lines to those of Kuo, except that they were introduced to fixed diets at weaning rather than from birth. The results are consistently at variance with Kuo's, in that the adult animals never developed an enhanced preference for their rearing diet, but rather they preferred to eat a novel alternative.

In the first experiment, two groups of five 6-week-old Basenji puppies were assigned to one of two different experimental dry dog foods (P and R/CC) which they received as their sole diet for the ensuing 16 weeks. All grew satisfactorily during this period. At the age of 22 weeks they were allowed a choice of the two diets in their usual individual feeding boxes in morning and evening meals over ten consecutive days. The results (Fig. 6) indicate a marked preference for the novel dry diet on the first day

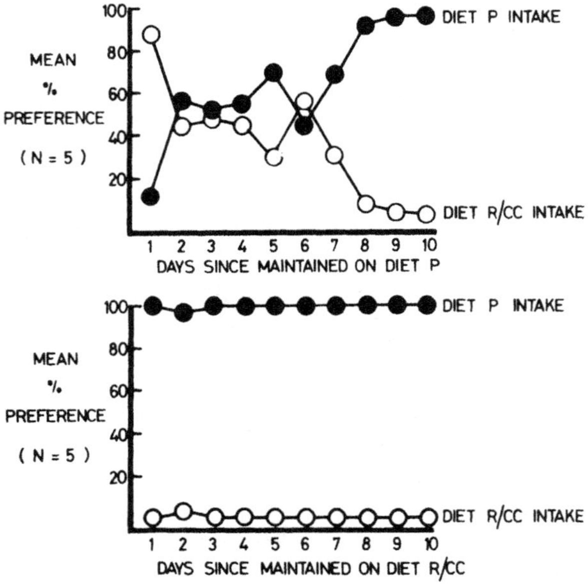

Fig. 6. Preference of Basenji puppies for novel dry foods.

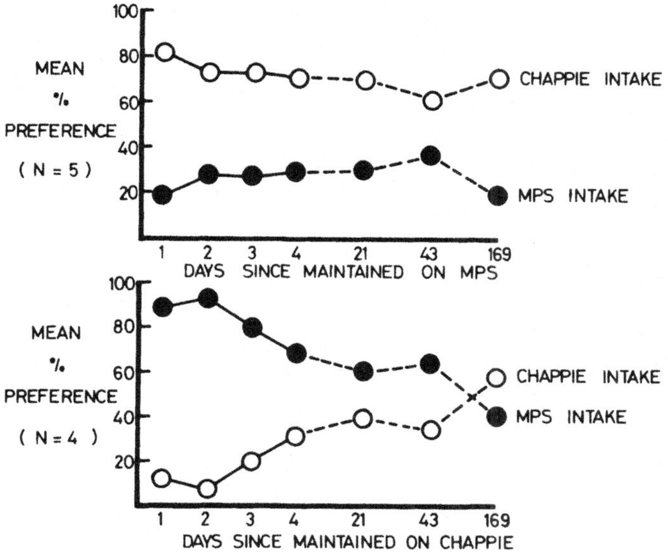

Fig. 7. Preference of Terrier puppies for novel canned foods.

($p < 0.01$, between-groups comparison), which for the group previously maintained on diet P declined on subsequent tests. In contrast to these diet P dogs, those reared on R/CC exclusively preferred diet P on all tests. This reversal of an initial preference (induced only by long established exclusive feeding) is consistent with Young's (1940) now classic observation of a similar phenomenon in rats, since we had previously established that diet P was more palatable than R/CC among dogs naive to both diets.

A second experiment with puppies was conducted with a different breed (nine West Highland White Terriers) and more palatable diets (Chappie versus Kal Kan M.P.S.) than in the experiment reported above. Prefeeding of the diets was again maintained for 16 weeks, followed by 12 consecutive preference tests conducted over 4 days, and further tests after 3, 6, and 24 weeks. Both the Chappie- and M.P.S.-reared dogs showed a consistent preference for the novel diet. The data in Fig. 7 were broken down by analysis of variance, and showed significant ($p < 0.001$) differences between the two diet groups.

Experiments with kittens have yielded comparable results. Twenty-four weanling kittens were assigned to three groups, to be fed exclusively either C26 chicken variety, C26 liver variety, or Whiskas (all palatable canned cat foods). The two groups of C26-reared kittens had precisely reversed initial preferences after the 16-week maintenance period. However, unlike the puppies in the second experiment, the kittens' initial preference for the novel variety quickly decreased across time so that the groups did not significantly differ from one another after the second and subsequent days' preference tests (Fig. 8a). This outcome was consistent with our prior expectations that the two varieties were equally palatable.

The kittens prefed Whiskas were allowed the choice of Whiskas and Kitekat (the novel food) over 40 days as shown in Fig. 8b. Again, the latter was initially preferred ($p < 0.05$), but after 10 days the more palatable Whiskas assumed the usual preference advantage ($p < 0.01$) that is observed among naive cats (author's unpublished observations).

To summarize, we have found that massed feeding of a single nutritionally balanced diet to young dogs and cats induces a transient depression of its relative palatability. This outcome is not unique to young animals, since novel foods are also favored by adult cats (Hegsted *et al.*, 1956) and dogs (author's unpublished observations). The concept of "oral satiety"

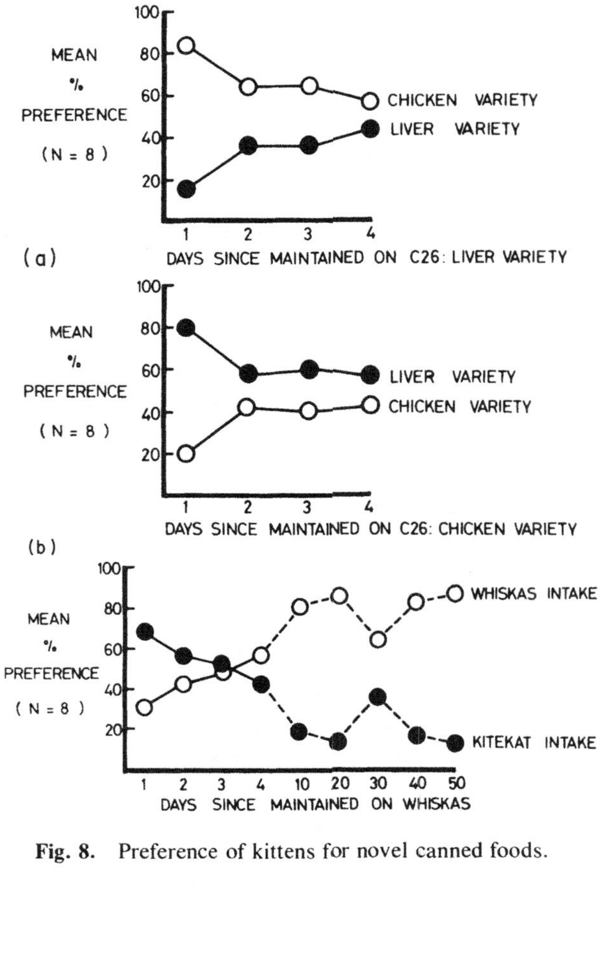

Fig. 8. Preference of kittens for novel canned foods.

TABLE I

Influence of Meal Frequency and Between-Meal Variety upon Caloric Intake of Cats

Diet	Caloric intake (kcal/day)
Fed once per day	
Whiskas	227
Sam	181
Kitekat	224
Average	211
Fed three times per day, same diet	
Whiskas–Whiskas–Whiskas	401
Sam–Sam–Sam	287
Kitekat–Kitekat–Kitekat	236
Average	308
Fed three times per day, different diets	
Whiskas–Sam–Kitekat	435
Sam–Kitekat–Whiskas	508
Kitekat–Whiskas–Sam	413
Average	452

(Le Magnen, 1971) is probably relevant in this context. Thus, both dogs and cats can be induced to increase the duration and size of their meals by successive presentations of three distinctively different diets, compared to any regularly replenished single diet (author's unpublished observations). Similar effects due to regular variation in flavor of rat diets were reported by Le Magnen (1956).

A demonstration of how presentation of contrasting diet types can enlarge total daily caloric intakes of cats is illustrated by the data in Table I. Twenty-seven cats were assigned in equal numbers to one of three groups on the basis of best match for age, sex, body weight, and prior food intakes.

Each group of nine cats was itself subdivided into three sets of three, which were then fed one of the diets or diet combinations indicated in Table I. Over three consecutive days, each set of three cats experienced all of the three diets or diet combinations. Data from this modified factorial design was treated to an analysis of variance. Highly significant diet and group effects ($p < 0.001$) can be attributed to enlarged caloric intakes of cats fed three times per day (46% greater than single-meal feeding), which was still further increased among those cats offered a succession of *different* diets over three meals on the same day.

The doubling in caloric intake (from 211 to 452 kcal/day) associated with providing both increased feeding opportunities and between-meal variety confirms what has already been stated about the plasticity of the regulatory mechanisms over food intake. Clearly, stimulus change or sensory contrast is an important factor affecting a diet's acceptance, which can doubtlessly be attributed to change along more than one dimension and within more than any one sensory modality. The particular role of olfaction as a determinant of a diet's palatability and its contrast with diets having a different odor are explored in the next section.

OLFACTORY INFLUENCES OVER INGESTIVE BEHAVIOR

It is our everyday human experience that the "normal" appreciation of food can be greatly affected by our sense of smell (Henkin *et al.*, 1971). Perhaps experimenters have been deterred from "discovering" the obvious, or is it that they are still slaves to the misconception that "taste" is synonymous with "flavor" and odor provides just a little something extra? There are certainly methodology difficulties associated with studying the relationship between olfaction and feeding, but it is hoped that the present chapter will show that many of them can be overcome by the simple expediency of dissociating the food itself from odors introduced through or over the food via a clean airstream.

Previous studies exploring the relationship between food palatability and odors can be criticized on the grounds that the supposed "odor" was actually incorporated into the test water or diet (e.g., Simson and Booth, 1973; Soane and Clarke, 1973; Le Magnen and Tallon, 1968). This raises the possibility that the introduced "odor" might be perceived via a nonolfactory route, as by direct absorption through the roof of the oral cavity (Kare *et al.*, 1969) or by alteration to the taste of the ingested food or water (Doty, 1975, p. 397).

Olfactory deficits induced by lesions or surgical removal of the olfactory bulb (Kovach and Kling, 1967; Larue and Le Magnen, 1972) also make for problems of interpretation since destructive intervention of the nervous system may produce unwanted secondary effects additional to anosmia (Whitten, 1956). The same can be said of techniques intended to effect a more specific interference with the sense of smell, as by irrigation of the olfactory mucosa with zinc sulphate (Alberts and Galef, 1971) or procaine hydrochloride (Doty and Anisko, 1973).

If surgical or chemical destruction of components of the olfactory system does not provide the methodology with which to make unambiguous demonstrations of the role of odors in feeding behavior, then such severe

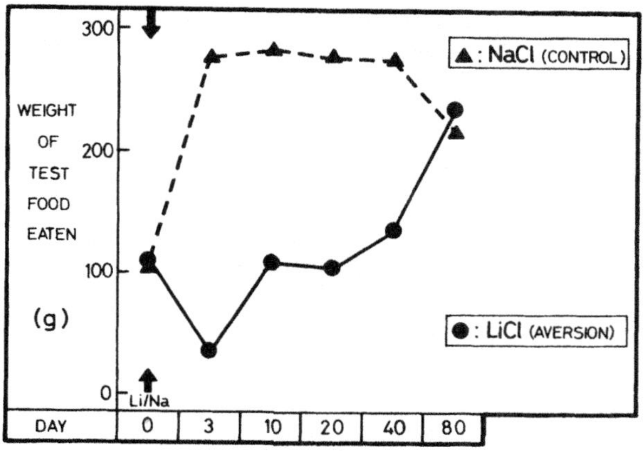

Fig. 9. Mean intakes of canned meat by cats, demonstrating dietary aversion learning.

impositions upon the animals' welfare are difficult to defend upon purely scientific grounds (Ryder, 1975). We have, therefore, adopted an alternative approach in which food-derived odors or the volatile components of a flavor are localized in their presentation to the animal by being suffused over or through a pelleted dry cat food. A three-part series of studies will be described in this section which demonstrates a crucial role of olfaction in establishing the palatability of food for the domestic cat. Thus, odors can (a) mediate the recognition of a diet to which a learned aversion has been formed, (b) direct selection of one food over an alternative that differs only by odor quality, and (c) both initiate and sustain feeding.

Demonstration of Dietary Aversion Learning*

Most readers will be familiar with the phenomenon of "bait shyness" in rats and other vertebrates, induced as a result of temporal association between sickness and a characteristically flavored diet or drink (see reviews by Garcia *et al.*, 1974; Rozin and Kalat, 1971). The experiment to be described demonstrates this phenomenon in cats.

Sixteen cats were assigned to two groups of eight on the basis of matched sexes and body weight. On the first day of the experiment (Day 0, Fig. 9) they were all individually fed 100 g of an experimental meat-based canned diet (M, based mainly upon cow lungs). The cats' normal maintenance diet was a complete dry cat food, so that M was quickly eaten because it was both palatable and novel. Immediately after they had fin-

* These studies of dietary and odor aversion learning were conducted while the author was a visiting scientist at the Monell Chemical Senses Center, Philadelphia.

ished eating, one group (aversion) was given lithium chloride (1% body weight of a 0.15 N solution, as in Nachman and Ashe's 1973 study with the rat). The other (control) group was treated with an equimolar solution of sodium chloride.

After two days' maintenance on the dry food, all cats were again offered M to appetite, as also on the tenth, twentieth, and fortieth days of the study. Figure 9 shows mean intakes of M during their six meals.

Analysis of variance (between groups, days 3–80) revealed a significant ($p < 0.001$) depression of intake by the aversion group of M, but not of their usual maintenance diet. Cats in the control group ate meals of M that were three times larger than aversion cats' meals of the same diet, whereas both groups continued to eat the maintenance dry cat food as before. Comparisons of intakes on each of days 3, 10, 20, and 40 confirmed that the dietary aversion was sustained for a period in excess of 1 month after only a single exposure to the U.C.S. (LiCl). By day 80, the two groups ate equally large meals of M. However, this interval of between 40 and 80 days probably does not encompass the time limit for retention of a dietary aversion, since the learned response would have suffered interference (or M-acquired properties of "learned safety") in five meals when M was offered (see Rozin and Kalat, 1971, p. 477). Nevertheless, the generality of the phenomenon of dietary aversion learning has been extended to include the cat on the basis of the results from this experiment. One could speculate that this ability might have a parallel in nature, serving to protect wild felids from repeated ingestions of prey or organs of prey that might induce gastric distress. The widely reported avoidance of insectivorous rodents by predaceous carnivores (e.g., Macdonald, 1977; Pearson, 1966) might be an example of aversion learning occurring under natural conditions. Garcia *et al.* (1974) and Chapter 13 of this volume have reported that aversions to mutton can be induced in coyotes by lithium chloride with a consequent reduction in their predation of sheep.

Influence of Odors upon Food Selection and Food Avoidance

There is some controversy as to whether rats can use olfactory stimuli to sustain a learned dietary aversion. Garcia *et al.* (1974) failed to find any diminution of the ability of "anosmic" rats to form an aversion to apple juice, and Barnett, Cowan, and Radford (1975) came to the same conclusion after finding that wild *Rattus rattus* made anosmic by the zinc sulphate method (Alberts and Galef, 1971) still avoided millet treated with poisonous zinc phosphide. However, in the Garcia *et al.* study one could argue that the most salient cue available to form the basis of an aversion to apple juice would be an acid taste. Where an easily discriminable or unusual odor is paired with a poisoning episode, quite effective aversions to the odor can be learned (Lorden *et al.*, 1970; Supak *et al.*, 1971). Furthermore, pairing an odor with a flavored water supply can then provide the basis for a learned odor-aversion being generalized to the flavored liquid (Domjan, 1973; Taukalis, 1974), which certainly hints strongly that odors have a more important role in "setting the hedonic tone of food" than is suggested by Garcia *et al.* (1974, p. 828).

The influence of odors upon food selection has been studied in our laboratory by arranging a simple two-choice preference situation (Hegsted *et al.*, 1956) of two food bowls, each fitted with a perforated base through which an airstream (3 liters/min) could be passed. If one of the airstreams is bubbled through homogenized meat, the aroma can be transferred by the airstream to the food. Ten cats were tested for preferences of their usual dry diet, either odorized by an airstream that had been bubbled through M, or just suffused with the clean carrier airstream. Reference to the data for days 1 + 2 in Fig. 10 shows that their average preferences were 87% in favor of the M odor. Two groups of five cats were then matched on the basis of their equal preference for M odor. Both were *fed* M on day 3; one was treated with LiCl (aversion) and the other with NaCl (control), as in the previous experiment. A succession of preference tests between air and M odor followed on days 5–10, and the results are presented as 2-day means from alternated bowl positions (because of the need to control for idiosyncratic bowl biases).

On day 5 there was a *total* avoidance by all cats in the aversion group of the M-odorized food, but an 80% preference by controls. This avoidance of the M odor was not quite so marked on the sixth and succeeding days, though intakes of the odorized food were never more than half that of the

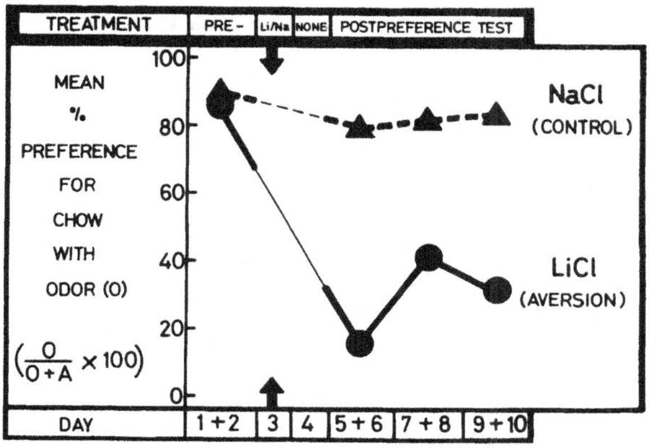

Fig. 10. Dry food/odor preferences of cats, demonstrating odor toxicosis learning.

control group (overall days, $p < 0.001$). The total food intakes by the two groups was rather similar (aversion: 16.8 g/meal, controls: 17.4 g/meal), so that the main effect of the M odor was to *direct* selection.

The Initiation of Feeding by Odors

If one recalls that M was itself a novel diet for the cats used in these studies, one could argue that its odor provided the basis for an enhanced "palatability response" toward their familiar maintenance food. However, a more parsimonious explanation would be that the M odor had reinforcing properties not connected with its presentation around the dry food (cf. Long and Tapp, 1967, 1968) and may not contribute much to its perceived palatability. Thus, the odor may just have attracted investigation (and so, physical movement) by the cats toward its source around the food, and thereby biased intake in favor of the nearest bowl. Many other classes of olfactory stimuli will initiate investigation (Doty, 1972, 1975), but in the context of social and sexual selection rather than feeding behavior. Interestingly, a study by Shumake, Thompson, and Bullard (1973) demonstrated that rats will greatly increase their entries and occupation times of the one arm of a four-arm maze suffused with a food odor. Although these authors did not present the odor in the same context as food, it was probably affected by the rats' motivation to find food since hungry rats were significantly more responsive than satiated rats.

After demonstrating an influence of odors upon food selection, the relevant question seemed to be whether food-derived odors can initiate and sustain consumption of a diet that would not otherwise be eaten. This was investigated in two experimental situations. The first utilized meal-fed cats, which were introduced one at a time into an observation cage to be fed their regular dry diet. The second situation permitted long-term (13-hr) observation of groups of cats, always provided with an *ad libitum* supply of food and whose behavior could be recorded by time-lapse video tape (see section on social factors affecting ingestive behavior). Both methods have proved suitable for evaluating the ingestion stimulatory qualities of odors derived from highly palatable foods.

The results of two experiments will be presented here, to illustrate this action of odors. Working with 20 cats usually fed individual meals, we have consistently found that both the duration and size of the meals can be increased by blowing a succession of "attractive" food odors through their maintenance diet after satiety or voluntary termination of feeding. Figure 11 illustrates one aspect of the results from an experiment in which a continuous airstream (3 liters/min) was substituted by the odor of M, after the cats had stopped eating during both of the preceding phases of the

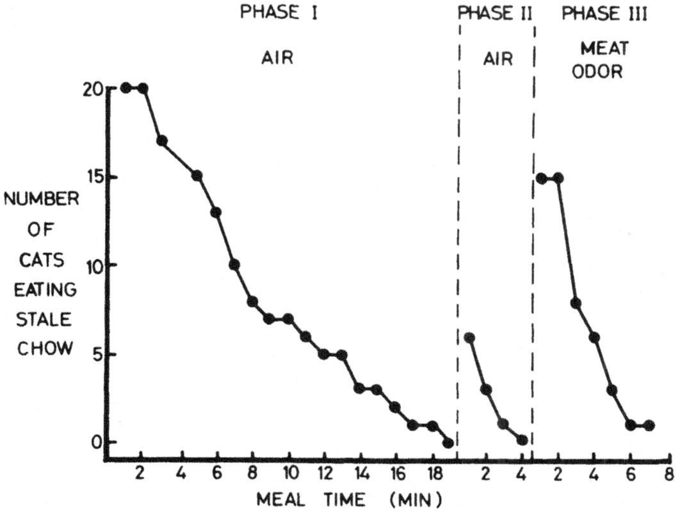

Fig. 11. Duration and incidence of feeding during single free access meals of cats, showing modification by the odor of a palatable canned meat diet.

meal. A "phase" consisted of the time during which the cat ate its food, plus an additional time-out of 60 sec after feeding ceased.

M-odor provoked a rapid revival of interest in the food, to the extent that eating recommenced in 15 of the 20 cats. If total food consumption of each cat is rated as 100, then the proportions eaten during the three consecutive phases AIR–AIR–M ODOR were on the average 45–17–38. This profile of food consumption has been significantly altered ($p < 0.01$) from the pattern obtained in control experiments run as a consecutive sequence AIR–AIR–AIR (65–22–13).

An essentially similar odor-mediated phenomenon was demonstrated to initiate and sustain bouts of feeding among nine cats having 24-hr access to dry cat food. In this experiment alternating half-hour sessions of AIR and ODOR derived from cooked rabbit meat (a highly preferred food item for the cat) were presented. Both the number of meals and food consumption were significantly increased by the rabbit odor (Fig. 12). This substantially altered pattern of food intake emphasizes that odors do indeed contribute to the hedonic tone of food (i.e., its palatability) in a way that is closer to the conceptualization suggested by Le Magnen (1971) than that of Garcia *et al.* (1974).

We have only recently begun to extend the methodology described here for cats to dogs. Preliminary results indicate that odors are equally influential over patterns of food selection and intake in this species. The profi-

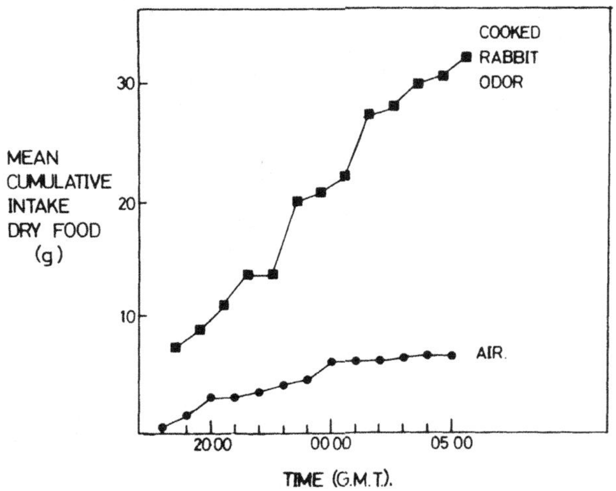

Fig. 12. Mean cumulative intakes of dry cat food during alternating half-hour presentations of air and odor over the food. G.M.T., Greenwich mean time.

ciency of the dog in olfactory detection tasks has been extensively studied in both an applied and scientific context (Moulton and Marshall, 1977). From naturalistic studies, it is evident that wolves also possess a highly developed olfactory sense, employed as both a means for receiving odor signals from conspecifics and detecting and trailing prey (Mech, 1970). Other authors mention the use made by canids of olfactory cues for detecting prey (e.g., Macdonald, in press, for the red fox *Vulpes vulpes*), but few reports on other families of carnivores (including the Felidae) are to be found in the literature. It is obvious that many more comparative studies are needed before the significance of the results reported here can be extended to the circumstances facing pet or feral dogs and cats, let alone the truly wild carnivores. However, sufficient evidence exists to surmise that olfaction is intimately involved in the processes of both prey location and close evaluation of food by many more than just the two species described here.

While the role of olfaction in feeding behavior has been emphasized in this paper, this is not intended to suggest that gustatory influences are any less significant. Indeed, an enormous literature attests to the ease with which the palatability of both foodstuffs and liquids can be modified by alteration along one of the dimensions of taste (e.g., Bartoshuk *et al.*, 1975; Carpenter, 1956; Kitchell and Baker, 1972), making further discussion of this point in relation to dogs and cats unnecessary. Little, if any, sound experimental data exist on taste or odor preferences of carnivores that exploit contrasting food sources, and which could form the basis for a comparative ecological model of sensory control over feeding behavior. It seems unlikely that this situation will be remedied in the foreseeable future, since it is mainly the so-called "pest" species that presently attract much attention, and regrettably that is often with the objective of developing more effective media with which they may be baited and destroyed.

REFERENCES

Alberts, J. R., and Galef, B. G. (1971). Acute anosmia in the rat. A behavioral test of a peripherally-induced olfactory deficit. *Physiol. Behav.* **6,** 619–621.

Anderson, R. S. (1974). Obesity in the dog and cat. *Vet. Annu.* pp. 182–186.

Barnett, S. A. (1963). "A Study in Behaviour," pp. 28–33, Methuen, London.

Barnett, S. A., Cowan, P. E., and Radford, G. G. (1975). Peripheral anosmia and the discrimination of poisoned food by *Rattus rattus* L. *Behav. Biol.* **13,** 183–190.

Bartoshuk, L. M., Jacobs, H. L., Nichols, T. L., Hoff, L. A., and Ryckman, J. J. (1975). Taste rejection of nonnutritive sweeteners in cats. *J. Comp. Physiol. Psychol.* **89,** 971–975.

Bernston, A. A., and Hughes, H. C. (1974). Medullary mechanisms for eating and grooming behavior in the cat. *Exp. Neurol.* **44,** 255–265.

Booth, D. A. (1972). Caloric compensation in rats with continuous or intermittent access to food. *Physiol. Behav.* **8,** 891–899.

Bronson, A. (1966). Evidence of the lack of influence of early diet on adult food preference in rats. *J. Comp. Physiol. Psychol.* **62,** 162–164.

Carpenter, J. A. (1956). Species differences in taste preferences. *J. Comp. Physiol. Psychol.* **49,** 139–144.

Carroll, M. E., Dinc, H. I., Levy, C. J., and Smith, J. C. (1975). Demonstrations of neophobia and enhanced neophobia in the albino rat. *J. Comp. Physiol. Psychol.* **89,** 457–467.

Coman, B. J., and Brunner, H. (1972). Food habits of the feral house cat in Victoria. *J. Wildl. Manag.* **36,** 848–853.

Compton, J. M., and Scott, J. P. (1971). Allelomimetic behavior system: Distress vocalisation and social facilitation of feeding in telomian dogs. *J. Psychol.* **78,** 165–179.

Crandall, L. S. (1964). *"Management of Wild Animals in Captivity."* Univ. of Chicago Press, Chicago, Illinois.

Domjan, M. (1973). Role of ingestion in odor-toxicosis learning in the rat. *J. Comp. Physiol. Psychol.* **84,** 507–521.

Doty, R. L. (1972). Odor preferences of female *Peromyscus maniculatus bairdi* for male mouse odors of *P.m. bairdi* and *P.leucopus noveboracensis* as a function of estrous state. *J. Comp. Physiol. Psychol.* **81,** 191–197.

Doty, R. L. (1975). Determination of odor preferences in rodents: Methodological review. *In* "Methods in Olfactory Research" (D. G. Moulton, A. Turk, and J. W. Johnston, Jr., eds.), pp. 395–406. Academic Press, New York.

Doty, R. L., and Anisko, J. J. (1973). Procaine hydrochloride olfactory block eliminates mounting in the male golden hamster. *Physiol. Behav.* **10,** 395–397.

Ewer, R. (1973). "The Carnivores." Cornell Univ. Press, Ithaca, New York.

Garcia, J., Hankins, W. G., and Rusiniak, K. W. (1974). Behavioral regulation of the milieu interne in man and rat. *Science* **185,** 824–831.

Harris, L. J., Clay, J., Hargreaves, F. J., and Ward, A. (1933). Appetite and choice of diet.

The ability of the vitamin B deficient rat to discriminate between diets containing and lacking the vitamin. *Proc. R. Soc. London, Ser. B* **113,** 161–190.

Hatton, G. I. (1975). Ingestive mechanisms and behaviours. *In* "The Behaviour of Domestic Animals" (E. S. E. Hafez, ed.), pp. 73–107. Baillière, London.

Hegsted, D. M., Gershoff, S. N., and Lentini, E. (1956). The development of palatability tests for cats. *Am. J. Vet. Res.* **17,** 733–737.

Henkin, R. I., Schechter, P. J., Hoye, R., and Mattern, C. F. T. (1971). Idiopathic hypogeusia with dysgeusia, hyposmia, and dysomia. *J. Am. Med. Assoc.* **217,** 434–440.

Hess, E. H. (1964). Imprinting in birds. *Science* **146,** 1128–1139.

Jacobs, H. L., and Sharma, K. N. (1969). Taste versus calories: Sensory and metabolic signals in the control of food intake. *Ann. N.Y. Acad. Sci.* **157,** 1084–1125.

Jalowiec, J. E., Panksepp, J., Shabshelowitz, H., Zolovick, A. J., Stern, W., and Morgane, P. J. (1973). Suppression of feeding in cats following 2-deoxy-D-glucose. *Physiol. Behav.* **10,** 805–807.

James, W. T. (1949). Dominant and submissive behavior in puppies as indicated by food intake. *J. Genet. Psychol.* **75,** 33–43.

James, W. T. (1953). Social facilitation of eating behavior in puppies after satiation. *J. Comp. Physiol. Psychol.* **46,** 427–428.

James, W. T. (1961). Relationship between dominance and food intake in individual and social eating in puppies. *Psychol. Rep.* **6,** 478.

Janowitz, H. D., and Grossman, M. I. (1949). Some factors affecting the food intake of normal dogs and dogs with esophagostomy and gastric fistula. *Am. J. Physiol.* **159,** 143–148.

Jones, A. F., and Hamilton, F. (1971). "The World Encyclopedia of Dogs." World Publ. Co., New York.

Joshua, J. O. (1970). The obese dog and some clinical repercussions. *J. Small Anim. Pract.* **11,** 601–606.

Kanarek, R. B. (1975). Availability and caloric density of the diet as determinants of meal patterns in cats. *Physiol. Behav.* **15,** 611–618.

Kare, M. R., Schechter, P. J., Grossman, S. P., and Roth, L. J. (1969). Direct pathway to the brain. *Science* **163,** 952–953.

Kissileff, H. (1970). Free feeding in normal and "recovered lateral" rats monitored by a pellet-detecting eatometer. *Physiol. Behav.* **5,** 163–173.

Kitchell, R. L., and Baker, G. G. (1972). Taste preference studies in domestic animals. *In* "Nutrition Conference for Feed Manufactures: 6" (H. Swan and D. Lewis, eds.), pp. 158–202. Livingstone, Edinburgh.

Kovach, J. K., and Kling, A. (1967). Mechanisms of neonate suckling behaviour in the kitten. *Anim. Behav.* **15,** 91–101.

Kruuk, H. (1972). "The Spotted Hyena: A Study of Predation and Social Behavior." Univ. of Chicago Press, Chicago, Illinois.

Kuo, Z. Y. (1967). "The Dynamics of Behavior Development, An Epigenetic View." Random House, New York.

Larue, C. G., and LeMagnen, J. (1972). The olfactory control of meal pattern in rats. *Physiol. Behav.* **9,** 817–821.

Le Magnen, J. (1956). Hyperphagie provoquée chez le rat blanc par altération du méchanisme de satiété périphérique. *C. R. Seances Soc. Biol. Ses. Fil.* **150,** 32.

Le Magnen, J. (1971). Advances in studies on the physiological control and regulation of food intake. *Prog. Physiol. Psychol.* **4,** 203–261.

Le Magnen, J., and Tallon, S. (1968). Préférence alimentaire du jeune rat induit par l'allaitement maternal. *C. R. Seances Soc. Biol. Ses Fil.* **162,** 387–390.

Long, C. J., and Tapp, J. T. (1967). Reinforcing properties of odors for the albino rat. *Psychon. Sci.* **7,** 17–18.

Long, C. J., and Tapp, J. T. (1968). An apparatus for the assessment of the reinforcing properties of odors in small mammals. *J. Exp. Anal. Behav.* **11,** 49–51.

Lorden, J. F., Kenfield, M., and Braun, J. J. (1970). Response suppression to odors paired with toxicosis. *Learn. Motiv.* **1,** 391–400.

Macdonald, D. W. (1977). On food preference in the red fox. *Mamm. Rev.* **7,** 7–24.

Mayer, J., and Thomas, D. (1967). Regulation of food intake and obesity. *Science* **156,** 328–337.

Mech, L. D. (1970). "The Wolf" Nat. Hist. Press, Garden City, New York.

Mech, L. D. (1975). Hunting behavior in two similar species of social canids. *In* "The Wild Canids. Their Systematics, Behavioral Ecology and Evolution" (M. W. Fox, ed.), pp. 363–368. Van Nostrand-Reinhold, Princeton, New Jersey.

Morrison, G. R. (1974). Alterations in palatability of nutrients for the rat as a result of prior tasting. *J. Comp. Physiol. Psychol.* **86,** 56–61.

Moulton, D. G., and Marshall, D. A. (1977). The performance of dogs in detecting α-ionone in the vapor phase. *J. Comp. Physiol.* **110,** 287–306.

Nachman, M., and Ashe, J. H. (1973). Learned taste aversions in rats as a function of dosage, concentration, and route of administration of LiCl. *Physiol. Behav.* **10,** 73–78.

Nagamachi, Y. (1972). Effect of satiety center damage on food intake, blood glucose and gastric secretion in dogs. *Am. J. Dig. Dis.* **17,** 139–148.

Pearson, O. P. (1966). The prey of carnivores during one cycle of mouse abundance. *J. Anim. Ecol.* **35,** 217–233.

Platt, J. J., and James, W. T. (1966). Social facilitation of eating behavior in young opossums. I. Group vs. solitary feeding. *Psychon. Sci.* **6,** 421–422.

Ross, S., and Ross, J. G. (1949a). Social facilitation of feeding. *J. Genet. Psychol.* **74,** 97–108.

Ross, S., and Ross, J. G. (1949b). Social facilitation of feeding behavior in dogs. II. Feeding after satiation. *J. Genet. Psychol.* **74,** 293–304.

Rozin, P., and Kalat, J. W. (1971). Specific hungers and poison avoidance as adaptive specializations of learning. *Psychol. Rev.* **78,** 459–486.

Ryder, R. D. (1975). "Victims of Science: Use of Animals in Research." Davis Poynter, London.

Schaller, G. B. (1972). "The Serengeti Lion: A Study of Predator-Prey Relations." Univ. of Chicago Press, Chicago, Illinois.

Scott, P. P. (1968). The special features of nutrition of cats with observations on wild Felidae nutrition in the London Zoo. *Symp. Zool. Soc. London* **21,** 21–36.

Scott, P. P. (1971). Dietary requirements of the cat in relation to practical feeding problems. *Pap. Small Anim. Nutr. Workshop, Univ. Ill. Coll. Vet. Med., 1971.*

Shumake, S. A., Thompson, R. D., and Bullard, R. W. (1973). An automated odor test for rats. *Behav. Res. Methods & Instrum.* **5,** 279–282.

Siegel, P. S. (1957). The repetitive element in the diet. *Am. J. Clin. Nutr.* **5,** 162–164.

Simson, P. C., and Booth, D. A. (1973). Olfactory conditioning by association with histidine-free or balanced amino acid loads in rats. *Q. J. Exp. Psychol.* **25,** 354–359.

Skultety, F. M. (1969). Alterations of caloric intake in cats following lesions of the hypothalamus and midbrain. *Ann. N.Y. Acad. Sci.* **157,** 861–874.

Smith, G. P., and Epstein, A. N. (1969). Increased feeding in response to decreased glucose utilization in the rat and monkey. *Am. J. Physiol.* **217,** 1083–1087.

Soane, I. D., and Clarke, B. (1973). Evidence for apostatic selection by predators using olfactory cues. *Nature (London)* **241,** 62–63.

Sudakov, K. V., and Rogacheva, S. K. (1963). The afferent and efferent activity of the gastric fibers of the vagus nerve during fasting and after taking food. *Fed. Proc., Fed. Am. Soc. Exp. Biol.* **22,** T306–T310.

Supak, T., Macrides, F., and Chorover, S. (1971). The bait shyness effect extended to olfactory discrimination. *Commun. Behav. Biol.* **5,** 321–324.

Taukalis, H. K. (1974). Odor aversions produced by long CS–US delays. *Behav. Biol.* **10,** 505–510.

Toates, F. M., and Booth, D. A. (1974). Control of food intake by energy supply. *Nature (London)* **251,** 710–711.

Welker, W. I., and King, W. A. (1962). Effects of stimulus novelty on gnawing and eating by rats. *J. Comp. Physiol. Psychol.* **55,** 838–842.

Whitten, W. K. (1956). The effect of removal of the olfactory bulbs on the gonads of mice. *J. Endocrinol.* **14,** 160–163.

Young, P. T. (1940). Reversal of food preferences of the white rat through controlled prefeeding. *J. Genet. Psychol.* **38,** 129–164.

Young, S. P. (1944). "The Wolves of North America," Part I. Am. Wildl. Inst., Washington, D.C.

DISCUSSION

Doty: Did I understand you correctly: once an animal has had an early experience of a food, and was then switched to another which was novel, it preferred the novel alternative for a long period of time? In other words, an imprinting occurs.

Mugford: If one must use the term imprinting, it could be described as a form of negative imprinting whose effect upon food selection is usually only transient. However, in the experiment with West Highland White puppies, the effect seemed to be more persistent—up to 6 months.

Doty: But these effects were due to early feeding experiences.

Mugford: Not necessarily, since the puppies would all have had some contacts with a diversity of solid foods up to the age of 8 weeks when they were weaned. Thus, they would share the same diet as their dam prior to joining the experiment. We have worked with adults and obtained exactly the same potentiation of interest in the new food. Sometimes the effect is transient, at other times more long lasting, depending upon the relationship between palatability of the familiar and novel foods. The effects are particularly transient if the familiar food is more palatable than the unfamiliar food.

Rogers: I would like to describe some observations we have made of kittens and purified diets, which they normally will not accept very readily. If you take an adult cat and give it a purified diet containing specified amino acids, it may take from 2 to 4 weeks before it will eat properly. If the same diet is given to young kittens, they will accept it quite readily, eat, and grow. Now, after we have had kittens on this diet for a couple of months we have changed them to a regular kibble for 6 months. On the first re-presentation of the amino acid diet to the now adult cat, it is readily accepted. We thought that this was interesting because it demonstrates how acceptance of a generally unpalatable diet can be improved by prior feeding during early life.

Mugford: This is indeed a most interesting observation.

Moskowitz: I have two questions. First, in your data on eating speeds of dogs you showed a combination of two foods, Chum and Mick, which seemed to show an intermediacy rather as Senseman had suggested before with the two kinds of food for his slugs. Is there an algebra whereby you could predict the rate of eating as a function of the initial rate of the two components, or is it just plain intermediate?

Mugford: The initial rates are highly correlated with the overall rate.

Moskowitz: Well, do you think it is possible to predict the rate of the mixtures in varying combinations if you knew the initial rates of consumption of the two or three components?

Mugford: I would think that it would be possible to do so, because the responses are extremely stable for particular individuals and are characteristic of each diet studied.

Moskowitz: I know that in humans you cannot predict the hedonic response to a mixture of odors; only the hedonic responses of each component. My second question concerns this very interesting novelty effect. If you watch human children, there is a period at which they will accept almost anything, then they grow into a period of recalcitrance when they will not accept anything that they do not know, and then as they get still older they diverge into adventurous eaters and normal eaters; along two completely different paths. Some people like anything new while others will never eat anything new. Have you any observations in animals analogous to this situation in humans?

Mugford: Our data do not support the notions of either dietary imprinting or neophobia for the dog or cat. However, we have really only worked with young animals of less than 4 years of age. It is conceivable, though improbable, that animals in their dotage may change their selection strategies in the way you have suggested occurs among humans.

Pangborn: Do cats and dogs eat differently in a group compared to a solitary situation?

Mugford: No, with qualifications. The surprising outcome of our observations with mature animals was that the timing, duration, and frequency of meals was similar when several individuals were compared to the same animals living as a compatible group. Much depended upon the social relationships between members of the group, and obviously one could select a pair of dogs in which one was always dominant and aggressive toward the other. In such a situation, food intake by the subordinate would decline.

Garcia: We have done a number of taste and odor studies, and the problem is one of telling whether the animal is tasting the odor. This really is a problem, especially since animals can make such incredibly subtle taste discriminations. Does that bother you as much as it bothers me?

Mugford: Yes, it does bother me, especially since in many studies that I have cited, as Soane and Clarke (1973) the odorant was incorporated into the diet at fairly high concentrations, making it quite likely that it could have been detected via a nonolfactory route. In our experiments, absorption of an airborne odor onto dry food, having a moisture content of only 8%, would probably be minimal. It would seem to me to be more parsimonious to hypothesize that there may be an olfactory mediation at work, rather than absorption prior to gustatory reception.

Garcia: Well, the problem is that when we used anosmic animals they showed no loss in ability to discriminate the poisoned food—absolutely none. In fact, most of the time they were a little better than the ones having senses of both taste and smell.

Mugford: Surely, one would expect the animals to be attending to the most salient cue emanating from the food. In the case of the experiment where you were using apple juice (see main text, Garcia *et al.*, 1974) the overwhelming sensation was probably acid, i.e., taste. Doubtlessly, animals make the best of a situation in which they are challenged by the need to avoid poison, and accomplish protective strategies of food discrimination by employing more than any single sensory modality.

Sato: In the experiment with Basenjis, what was the difference between diets offered to the two groups?

Mugford: They contained different ingredients and were manufactured by different processes, so they would be expected to differ along several sensory dimensions.

Sato: Is there a possibility that nutritional factors were involved in the switch from one diet to the other?

Mugford: I think that this might have been the case, except that in all of the experiments I have just reported, the animals grew well and at rates appropriate to the particular breed. In sustaining adequate growth, the diets were presumably nutritionally balanced and not lacking in an essential micronutrient.

Sato: Have you done any chemical characterization of the rabbit odor which might explain its attractiveness?

Mugford: Meat aromas are all of immense chemical complexity, so we have not fractionated or identified compounds in the odor of rabbit meat that might change cats' feeding behavior.

Chapter 3

Nutritional States, Food Odors, and Olfactory Function

JEANNE PAGER

Endowed with an amazing discriminating power, the olfactory system of vertebrates seems to be able to identify an infinite range of volatile substances. Many tentative classifications of odors have been proposed, according to the biological responses which they elicit. Although the coding keys of olfactory messages are still unknown, the general agreement is that several physicochemical properties of the molecules are recognized by the receptor sites with which they interact at the surface of the nasal sensory epithelium. In the olfactory receptor cells of the frog, a factor analysis of the unit responses to 20 chemically pure stimuli has revealed that five molecular features may account for 60% of the variability of the results (Holley *et al.*, 1974). It seems, for example, that the presence of a benzene ring may to some extent characterize a group of olfactorily active substances. Thus, the properties of the olfactory system as a coder of chemical information are on the way to being better understood.

The olfactory system as a whole nevertheless differs from a mere chemical analyzer in that it yields information useful to the animal and to survival of the species. Independent of any specific behavioral arousal, a pleasant–unpleasant axis emerges from olfactory psychophysics data (Schiffman, 1974). The multidimensional scaling analysis of human olfactory responses does not lead to a satisfactory model linking physicochemical features of the stimulating molecules and odor quality assessments. As Schultz and Tapp (1973) point out, fruitful studies of olfactory guiding of behavior require using biologically meaningful odors as stimuli. Odors become effective parameters in the management of adapted behavior through their affective character. In the particular case of nutrition, the affective tone of odors corresponds to the olfactory dimension of palatability, a parameter modulated by the nutritional state of the subjects. According to this phenomenon, the so-called alliesthesia of Cabanac (1971), the same gustatory or olfactory nutritional stimulus evokes a pleasant feeling in hungry human subjects, and an unpleasant one in satiated subjects. Pleasure thus appears to be felt when the biological stimulus corresponds to the specific behavioral arousal of the subject. Odors seem to be analyzed not only in terms of chemical properties, but also in terms of incentive properties. At the high level of neural integration reached by human sensations, the hedonic axis would probably dominate the others.

Theoretically, food odors can be distinguished from other odors along the ascending olfactory pathway by at least two means. One means is the coding process, in which discrimination is the fundamental function. Another is that some discrimination between the two groups of stimuli could be expected at any stage where nutritional olfactory information is integrated into the neural processes underlying food intake behavior. A survey of the general organization of olfactory pathway in mammals may help to some extent to conceive an experimental approach to the problem.

Figure 1 is a scheme of the main anatomical data obtained and discussed recently (MacLeod, 1971; Price and Powell, 1970; Pinching, 1971; Dennis and Kerr, 1975). In the olfactory pathway, only the receptor neurons are located in the receptor organ, namely, the olfactory mucosa. Their activity depends on the availability of odoriferous molecules, whose concentration may be modulated by the rhythmical, alternative broadening of nasal passages (Bojsen-Møller and Fahrenkrug, 1971), and obviously by the respiratory activity. Moreover, humoral and nervous sympathetic factors (Tucker and Beidler, 1956) may influence the mucus coat properties of the receptor neurons, and their responses. The axons of the receptor cells enter the olfactory bulb, and their terminations reach the glomerular layer where they establish synapses on the second-order neuron dendrites.

At the glomerular level, a notable concentration and redistribution of information occurs (Allison, 1953; Leveteau and MacLeod, 1969). Mitral cells, the second-order neurons regarded as the processors of qualitative information, actually constituting a primitive cortical layer, are subjected to multiple neural controls of their activity. The two distinct stages where these modulations occur implicate distinct interneurons: the periglomerular cell at the input stage, and the granular cell at the output stage. Both kinds of interneurons establish reciprocal inhibitory synapses with mitral cell dendrites, and receive collaterals from the mitral cell axons. They also

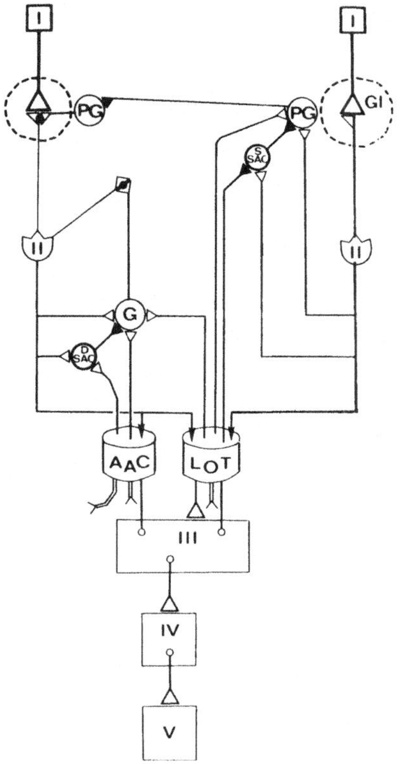

Fig. 1. General organization of the olfactory pathways, from the receptor cell (I) to the thalamic and hypothalamic (IV) and neocortical projections (V). Only mitral cells are represented as second-order neurons (II). The third-order areas include projections to all the olfactory cortex, the olfactory tubercle, the amygdala, the anterior olfactory nucleus, and the bed nucleus of the lateral olfactory tract. White and black triangles correspond to excitatory and inhibitory synapses, respectively. Black dots signal reciprocal synapses. The direct lateral inhibition between mitral cells is not shown in the figure. Gl, glomerulus; PG, periglomerular cell; G, granular cell; SSAC and DSAC, superficial and deep short-axon cell; AAC, anterior limb of the anterior commissure; LOT, lateral olfactory tract.

receive, respectively, endings from the superficial and deep short-axon cells, and centrifugal inhibitory and excitatory fibers, mainly through the lateral olfactory tract (LOT) and the anterior limb of the anterior commissure (AAC).

LeMagnen (1971b) pointed out the importance of a centrifugal control of the bulb activity and proposed the hypothesis of a selective integration of the olfactory information adjusted to the internal state of the behaving animal; he emphasized the central character of palatability.

A significant feature of the organization of the third-order level of the olfactory pathway is the generalized presence of feedback loops from the third-order to the second-order neurons. Therefore, the olfactory pathways may be schematically represented as an ascending linear chain of neurons, from the mucosa to the thalamic and neocortical projections (Motokizawa, 1974; Giachetti and MacLeod, 1975; Tanabe *et al.*, 1975b), provided with internal feedback loops. Moreover, output circuits are branched at several stages, where exchanges of information occur with the rest of the brain. The functions and interrelations of the internal and external controls are still largely speculative. It may be that the internal feedback loops concern the qualitative properties of odors, while the external circuits relate more specifically to the incentive properties.

Centripetal information and centrifugal internal (olfactory) and external (modulatory) influences overlap as peripherally as the mitral cells. The structural organization and function of this layer are more thoroughly investigated than those of the upper olfactory stages (Phillips *et al.*, 1961, 1963; Shepherd, 1963a,b; Nicoll, 1969, 1971a,b; McLennan, 1971; Guselnikova *et al.,* 1971; Westecker, 1971; Mathews, 1972; Kauer, 1974; Freeman, 1974; Getchell and Shepherd, 1975; Mori and Takagi, 1975). Mitral cells constitute a target zone, the functional study of which is likely to bring us information on the analysis of biologically meaningful odors. An apparatus designed to record chronic mitral cell multiunit activity was used in the rabbit by Moulton (1963), who noted the great influence of centrifugal fibers on this activity. In the rat, the same author (1968) observed modifications of mitral cell multiunit responses during olfactory discrimination learning, according to correct or incorrect choice of the animal.

Moulton's recording system has been used by Giachetti *et al.* (1970) with hungry (22 hr fasting) and satiated rats (end of single 2-hr daily meals) receiving amyl acetate, or stimulated by the odor of their mashed usual food. The authors have observed variations of the ratio of positive and negative responses to these odors in the two nutritional states. Positive or negative responses are defined as a significant increase or decrease of the summated mitral cell activity during the odor stimuli. It appears that the positive responses to food odor are more numerous, and that negative responses are scarcer in hungry than in satiated animals. In either nutritional state, the patterns of responses to amyl acetate are statistically identical to the pattern of responses to food odor in satiated rats. This phenomenon was interpreted as the removal of a centrifugal inhibition, released by specific hunger arousal, which selectively facilitates transfer of food odor information.

Insulin injections have been observed, by chronic recording, to duplicate the food deprivation effects on the multiunit mitral cell responses of rats subjected to food odor and citral stimulation (Pager *et al.*, 1972). In anesthetized rats, an insulin treatment has been reported to increase the positive responses of mitral cell units to food odor and to amyl acetate in the same manner, supposedly as a result of provoking general arousal (Cain, 1975). Thus the degree of selectivity of the nutritional modulation of mitral cell responses to food and nonfood odors, and its functional meaning with regard to the olfactory analysis of food odors, require further examination.

When the mitral cell multiunit responses to identical controlled stimuli are chronically collected in freely moving rats under strict conditions of food deprivation (H) or satiation (S), the variability of the results depends either on the coding mechanisms, or on modifications of the animal's internal state. The two processes will now be examined successively, although any distinction between them may be artificial.

CODING RELATED VARIABILITY

Qualitative discrimination between odors sharpens along with increasing order of neurons in the olfactory pathway (Giachetti and MacLeod, 1973; MacLeod, 1973; Tanabe *et al.*, 1975a). From a strictly chemical point of view, it may be supposed that food odor coding obeys the same rules as any other odor. Accordingly, the multiunit mitral cell responses may be expected to reveal some food/nonfood discrimination, if it does exist. The summated multiunit responses of mitral cells are collected chronically in hungry (H) and satiated free-moving rats bearing bipolar electrodes, and receiving 10 successive puffs of the same odor. Each animal thus yields patterns of 10 H and S positive, negative, and null responses, according to the criteria of Giachetti *et al.* (1970). The patterns of multiunit mitral cell responses to three different food odors (Sorolabo biscuits, unflavored pellets, eucalyptol-flavored pellets) are given in Fig. 2; those of responses to two nonfood odors (amyl acetate and eucalyptol) appear in Fig. 3.

The patterns of positive, negative, and null responses to the five stimulating odors (Table I) are distinct both for the H and the S situations ($\chi_8{}^2H = 306.17$; $p < 0.001$; $\chi_1{}^2S = 275.31$; $p < 0.001$). When compared, the summated patterns of response to the nonfood odors, namely, amyl acetate and eucalyptol, display significant differences in either nutritional state ($\chi_2{}^2H = 6.02$; $0.02 < 0.05$; $\chi_2{}^2S = 7.12$; $0.02 < p < 0.05$). This could be interpreted as a result of the discriminating mechanisms detected at the mitral cell level. The patterns corresponding to the three food odors do

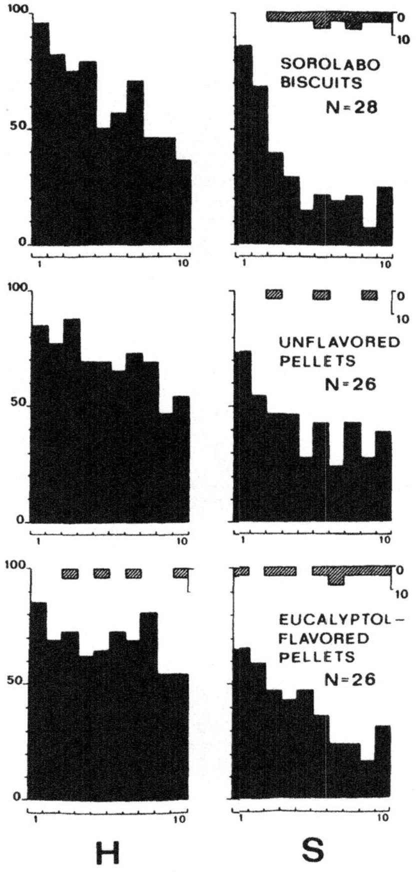

Fig. 2. Patterns of multiunit mitral cell responses to the odors of Sorolabo biscuits, unflavored pellets, and eucalyptol-flavored pellets in hungry (H) and satiated (S) rats. Abscissa: rank order of the stimulus; ordinate: percent positive (black areas) and negative (striped areas) responses; N: number of animals.

The rats had been accustomed to receiving a single 2-hr daily meal at stated hour. They have been used just before or after it (H or S situations). They were bearing one stainless steel bipolar electrode in each olfactory bulb (thread diameter 100 μm). The mitral cell layer was determined by means of cytological and stereotaxic techniques. Its characteristic high-amplitude activity may be easily recognized, visually and auditorily. The recorded sites were localized at random in the ventral part of the layer. During the recording sessions, the rats were freely moving in a carefully washed Plexiglass cage, designed to deliver the same stimulation in any position of the animal.

Two stimulating odors are delivered alternately every 2 min, up to a series of 10 identical 10-sec puffs. The stimulating odors were those of mashed foods, or of saturated vapor of pure chemicals, diluted 600 times by a permanent airflow. The olfactometer was constituted of glass and Teflon devices, and of solenoid valves.

The multiunit mitral cell activity (amplitude: 30–120 microvolts) was amplified, filtered in a 300–3000 Hz bandpass, fed into an oscilloscope, a loudspeaker, or a summating device (time constant: 0.5 sec), and paper recorded. The stimulating and recording conditions remained constant throughout the experiment.

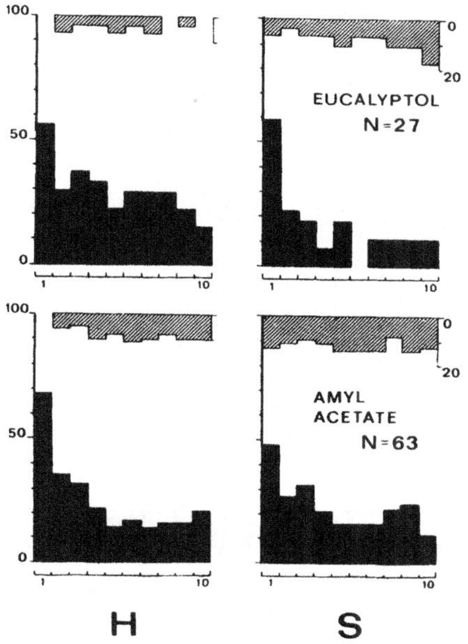

Fig. 3. Patterns of multiunit mitral cell responses to the odors of eucalyptol and amyl acetate in hungry (H) and satiated (S) rats. The data are obtained under the same conditions as those of Fig. 2, and are represented with the same conventions. Abscissa: rank order of the stimulus; ordinate: percent positive responses (black areas) and negative responses (striped areas); N: number of animals.

not, however, differ significantly one from the other (χ_2^2H = 2.38; χ_2^2S = 5.03). *The degree of similarity of the patterns of responses elicited by food and nonfood odors may therefore help to distinguish one group from the other.*

Since the differences between the food odor patterns are no more significant when their separation into 10 successive blocks is taken into account (χ_{18}^2H = 6.48; χ_{18}^2S = 15.4), grouping all of them into a common pattern seems to be justified. When this composite pattern is compared to the eucalyptol pattern, significant differences are evident in the H and S animals (χ_1^2H = 111.16; $p < 0.001$; χ_2^2S = 79.27; $p < 0.001$). The same observation is made with the amyl acetate pattern compared to the food odor pattern (χ_2^2H = 265.59; $p < 0.001$; χ_2^2S = 101.03; $p < 0.001$). In any case, the response patterns to nonfood odors display more negative responses and fewer positive responses than do those for food odors. Furthermore, the difference observed with the eucalyptol–amyl acetate pair is qualitatively similar to that characterizing the food odor–eucalyptol pair, although to a less significant extent ($0.02 < p < 0.05$ against $p < 0.001$).

TABLE I

Patterns of Positive, Negative, and Null Mitral Cell Multiunit Responses to Three Food Odors and Two Nonfood Odors in Hungry (H) and Satiated (S) Rats

Odor	No. of responses	H			S		
		+	0	−	+	0	−
Sorolabo biscuits	280	179	101	0	92	186	2
Unflavored pellets	260	181	79	0	109	149	2
Eucalyptol-flavored pellets	260	178	80	2	100	157	3
Eucalyptol	270	82	178	10	46	199	25
Amyl acetate	630	162	420	48	146	407	77

Thus *in any nutritional state, the food odors elicit a relatively high characteristic percentage of positive responses, and nearly no negative responses*. The nonfood odors elicit relatively low but variable percentages of positive responses, and rather variable percentages of negative responses. This observation has been extended to menthol (Pager, 1974) and citral (Pager *et al.*, 1972).

The results percentages of positive and negative responses may be referred to as *activation* (AL) *and inhibition level* (IL). The inhibition level, a main feature of the food/nonfood odor discrimination, is measured in mitral cell populations of several dozen cells. A negative multiunit response is the result of a predominance of inhibited units in the population. The scarcity of multiunit negative responses corresponds to a weak probability for inhibited units, and indicates that food odor stimulations provoke a massive bulb activation. Negative responses of bulb units, in curarized rats, are generally considered an element of qualitative information coding (Mathews, 1972; Giachetti and MacLeod, 1973). Coding could result in negative areas surrounding activated ones in the olfactory pathway (Haberly and Shepherd, 1973).

That the odors of unflavored pellets and eucalyptol pellets could involve some common coding mechanism is not surprising, since the presence or absence of eucalyptol is the only feature distinguishing the two complex mixtures of starch, casein, dextrose, cellulose, peanut oil, minerals, and vitamins. But Sorolabo biscuits present quite a different composition, and the three odors are clearly discriminated by human subjects in the concentration used in the experiments. The response patterns for food odors at the multiunit level reflect only approximately the coding of their chemical characteristics.

A common feature of the three food odors is probably their olfactory complexity, which is linked to the chemically composite character of the foods. Their coding should then require the activation of a large part of the bulb units, making the detection of a discrimination very improbable at the multiunit level. By stimulating fishes with food extracts or by L-leucine, Hara and Law (1972) observed that adapted responses of the olfactory bulbs present a higher plateau in the case of the first stimulus. The authors consider that food odor is chemically more complex and stimulates a greater number of receptors. Preliminary data of Cattarelli *et al.* (1976) obtained in chronically recorded rats indicate that the activation level reached by mitral cell populations is even higher in the case of grouped rat odor (a rewarding biological stimulus) than with food odors as olfactory stimulus.

The chemical complexity of food odors is not sufficient to explain the differences between food and nonfood response patterns. Figure 4 represents the response patterns to eucalyptol in H and S rats fed on the eucalyptol-flavored pellets since weaning. When these patterns are compared to those for eucalyptol from control groups (Fig. 2), a significant difference appears irrespective of the nutritional state ($\chi_1^2H = 20.36$; $\chi_1^2S = 15.64$; $p < 0.001$). This may be explained by the lack of negative responses to the food-associated eucalyptol, and by the higher percentage of positive responses elicited by this odor. After nutritional learning, the eucalyptol odor responses display features similar to those for food odor. When the H and S patterns are compared to those of food odors, a

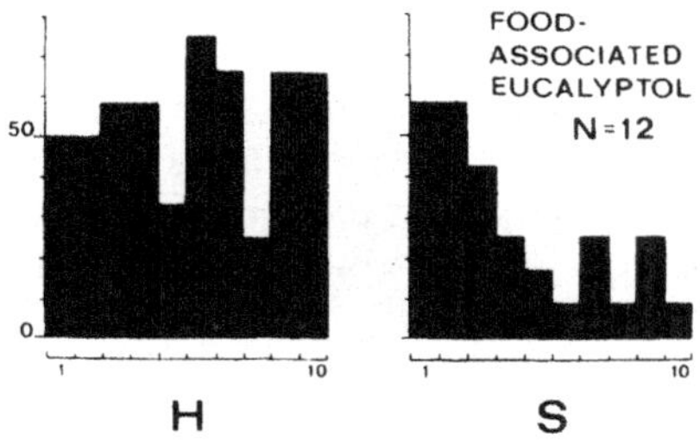

Fig. 4. Patterns of multiunit mitral cell responses to the odor of food-associated eucalyptol in hungry (H) and satiated (S) rats. Abscissa: rank-order of the stimulus; ordinate: percent positive responses (no negative response was obtained in this case); N: number of animals.

Animals were recorded as described in Fig. 2. They received serial stimulations with eucalyptol odor (1/600 diluted saturated vapor).

The 12 male rats of this nutritional learning experiment were selected at random from two litters, the mothers of which were fed eucalyptol-flavored pellets. The pups themselves were fed the same diet from weaning to adult age, when they were recorded.

One control group received only the unflavored pellets, and another received the eucalyptol-flavored pellets for 1 month when adult. The response patterns for eucalyptol were not significantly modified for any of them.

significant difference appears nevertheless ($\chi_1^2 H = 6.41$; $\chi_1^2 S = 4.19$; $0.02 < p < 0.05$). This difference is attributable to the slightly greater activation level elicited by the food odors. Consequently, if one excludes the hypothesis of common chemical properties for eucalyptol and food odors, it must be admitted that *the main part of the food odor elicited activation level is related to the nutritional meaning of these odors, that is, their biological interpretation, rather than to their chemical features.*

It must be recalled, however, that the eucalyptol patterns are more similar to those of food odors than the amyl acetate patterns are. One may therefore ask whether amyl acetate could become interpreted as a food odor under those experimental conditions. It may be that some odors are potentially interpretable as food odors because of the high activation and low inhibition levels associated with them. A discussion about the learned or unlearned nutritional meaning of odors is beyond our subject; the problem calls for extensive electrophysiological investigation.

VARIABILITY RELATED TO INTERNAL STATE

Since the behavioral guiding ability of food odors is modulated according to the nutritional state, the response patterns have been compared in the H and S situations for each stimulating odor. The details of the statistical analysis are given in Table II.

In any case, the difference between H and S patterns is a consequence of the striking increase in the activation level in hungry animals exposed to food odors. The increase is less marked in the case of eucalyptol, and not significant with amyl acetate. According to electrophysiological criteria, eucalyptol appears once more to be intermediate between food odors and amyl acetate, in spite of its neutral character according to the criterion of food intake behavior (LeMagnen, 1956). A clear discrimination between

TABLE II

Comparison between the Patterns of Responses to Five Stimulating Odors in the H and S Situations: Statistical Analysis

Odor	χ_1^2 Value	p
Sololabo biscuits	56.46	< 0.001
Unflavored pellets	37.23	< 0.001
Eucalyptol-flavored pellets	43.42	< 0.001
Eucalyptol	14.58	< 0.001
Amyl acetate	0.55	> 0.05

food odors and nonfood odors is, however, established by the fact that food odors display an even higher AL and a lower IL in the S situation than nonfood odors in the H situation (amyl acetate: $\chi_2^2 = 58.72$; $p < 0.001$; eucalyptol: $\chi_2^2 = 4.31$; $0.02 < p < 0.05$). The following classification may be established *in terms of activation level: H ×food odor > S ×food odor > H × nonfood odor > S × nonfood odor.*

An opposite classification is obtained in terms of inhibition level. Complementary variations of AL and IL may be observed from the first to the tenth response to the same stimulus. The IL tends to increase slightly as the stimuli are repeated, at the same time that the AL decreases. Since olfactory habituation, described by Adrian (1950), Moulton (1963), and Døving (1964), may be expected to be manifested in either kind of response, the recorded pattern is most likely the result of two overlapping processes. One is related to the chemical coding and yields positive, null, and negative responses. The other is an activation wave temporarily masking the coding pattern, more developed in the case of food odors, and amplified by food deprivation, especially for food odors.

Keeping in mind the various centrifugal modulatory systems described previously, unilateral sections of various pathways have been performed to test the hypothesis of the centrifugal nature of the food intake-linked activation of the bulb. Sectioning the lateral olfactory tract (LOT) does not modify the H/S response modulation, but does increase the activation level on the side of the lesion (Pager, 1976). Lesions of the olfactory peduncle (OP), and of the anterior limb of the anterior commissure (AAC), suppress the activation wave observed at the mitral cell level (Fig. 5 and

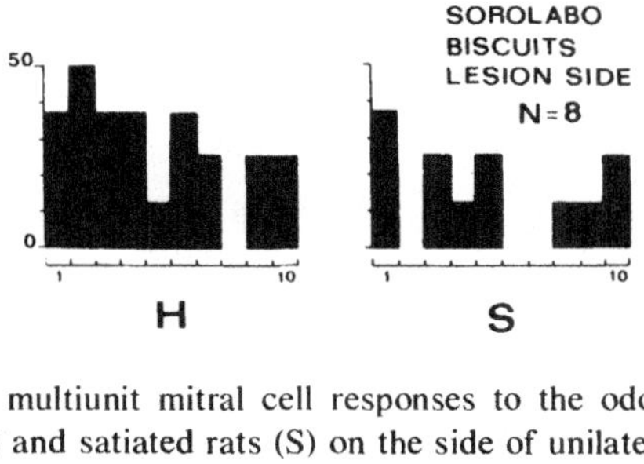

Fig. 5. Patterns of multiunit mitral cell responses to the odor of Sorolabo biscuits, recorded in hungry (H) and satiated rats (S) on the side of unilateral lesions of centrifugal pathways.

Abscissa: rank order of the stimulus; ordinate: percent positive responses; (no negative response was recorded in this group); N: number of animals.

The animals were recorded under the same conditions as described in Fig. 2. Four male adult rats were submitted to a transverse section of the olfactory peduncle at the level of the foramen opticum; four others had the anterior limb of the anterior commissure sectioned in the 9-mm anterior plane. Statistical tests allow all the data to be grouped into the same category.

Pager, 1974b). On the lesion side, no significant difference appears between the response patterns to Sorolabo biscuit odor in the H and S nutritional states ($\chi_1^2 = 3.22$). The patterns for the lesion-side bulbs differ significantly from those for food odors in the control bulb (χ_1^2H $= 44.99$; $p < 0.001$; χ_1^2S $= 5.46$; $0.02 < p < 0.05$). The activation level after the lesions is below the expected level, especially in the H rats. No negative response appears, however, indicating that their scarcity is a centripetal, coding related phenomenon. An oversimplified yet clarifying interpretation is that the centrifugal bulb activation is related to modulatory mechanisms external to the olfactory system, while LOT section effects result from internal modulatory mechanisms.

The amplitude of the gap between the H and S patterns, determined by centrifugal processes, may be used as an index to discriminate food odors from other odors. In a given group of mitral cell sites, the ability of an odor to yield a greater activation level in the H than in the S situation is called its *differential activating power (DAP)*. It is calculated by the following formula: $DAP = (D - A)/N$, where D is the number of responses which are positive in the H situation and nonpositive in the S situation and where, conversely, A is the number of responses which are nonpositive in the H situation and positive in the S situation. The theoretical limits of *DAP* may be calculated assuming that all the responses to an odor are positive in the H state and nonpositive in the S state, or, inversely, one obtains $-1 \leq DAP \leq +1$.

The *DAP* values have been calculated for all the odors previously mentioned and are collected in Table III, which may be referred to for a comparative survey of the main results described in this chapter. Besides the *DAP* values, the AL and IL are indicated in hungry and satiated animals. These data are extracted from experiments performed on 41 groups of rats (Pager, 1976). The patterns of these groups display a significant positive correlation between the HAL and SAL values ($r = 0.736$; $t_{35} = 6.793$; $p < 0.001$), which is interpreted as an indication that coding processes are not modified by nutritional state. As the *DAP* is not significantly correlated to the SAL values ($r = 0.238$; $t_{39} = 1.539$), it seems doubtful that the coded activating power of a stimulus could directly bring information on its *DAP*. The HAL values, on the contrary, are positively correlated to the *DAP* values ($r = 0.807$; $t_{39} = 8.547$; $p < 0.001$); both result from the same activating processes characterizing the internal state of specific arousal evoked by food deprivation.

Two kinds of information drawn from the multiunit recording of mitral cell activity may suffice to characterize odors with respect to their nutritional meaning: the percentage of one type of response and the differential activating power. The distribution of the stimulating odors mentioned in

TABLE III

Values of Indexes Characterizing Multiunit Mitral Cell Responses to Various Food and Nonfood Odors: Activation and Inhibition Level (AL and IL) in H and S Animals, and Differential Activating Power (*DAP* × 1000)

Odors	HAL (%)	SAL (%)	HIL (%)	SIL (%)	*DAP* × 1000
Sorolabo biscuits	63	32	0	3	300
Sorolabo biscuits after lesions	28	15	0	0	112
Unflavored pellets	69	41	0	1	257
Eucalyptol-flavored pellets	68	38	1	3	296
Eucalyptol-flavored pellets after aversive learning	54	52	0	0	42
Eucalyptol	27	17	3	9	133
Eucalyptol after nutritional learning	55	27	0	0	300
Amyl acetate	25	23	7	12	17
Menthol	38	57	7	6	−133
Citral	18	28	46	13	− 75

this chapter, in the AL × *DAP* plane, is represented in Fig. 6 for the H and S situations. As could be predicted from behavioral data, discrimination between food and nonfood odors appears sharper in the H situation. The odors of the three foods and of nutrition-associated eucalyptol are closely grouped. The nonfood odors of menthol, citral, amyl acetate, and control eucalyptol constitute a more scattered group, which also includes the odor of Sorolabo biscuits, after cutting the centrifugal pathways indicated in Fig. 5. In this group, amyl acetate appears to be an almost ideal neutral odor with regard to nutritional state.

An isolated point in the AL × *DAP* plane relates to the odor of eucalyptol-flavored pellets, when the first intake of this food is followed by a nauseating apomorphine injection (Pager and Royet, 1976). In this case, the *DAP* value has dropped to the area of the nonfood odors, the difference between the H and S patterns not being significant (Fig. 6). A general arousal coming from aversion brings the percentage of positive responses up to the same level in the H and S situations, confirming that an *AL increase specific only to food deprivation may characterize food odors*. This criterion, corresponding to the *DAP* index, seems to be more efficient in discriminating food odors than the AL itself, which has a poor discriminating power for separating food odors from other biologically meaningful odors displaying efficient activating properties, in the S and H situations.

The statement that food odors may be discriminated as such at the mitral cell level, mainly by means of a centrifugal activation amplified by

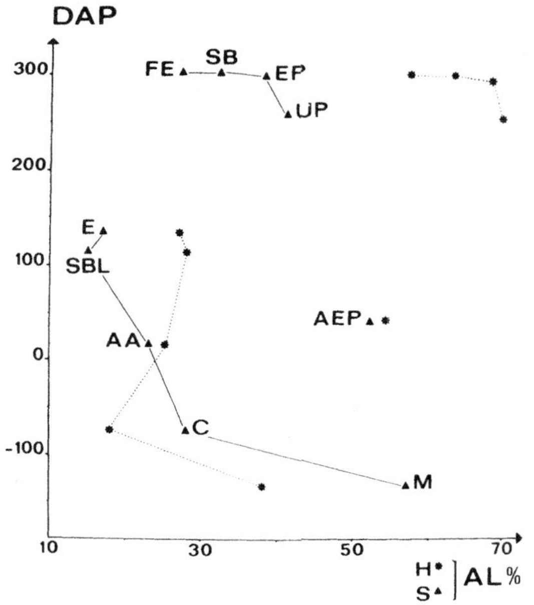

Fig. 6. Distribution of the stimulating odors in the AL × *DAP* plane. Abscissa: activation level (%). The two abscissae related to each odor respectively correspond to hunger activation level (HAL: stars) and to satiation activation level (SAL: triangles). Ordinate: differential activating power (×1000). AA: amyl acetate; AEP: aversive eucalyptol-flavored pellets; C: citral; E: eucalyptol; EP: eucalyptol-flavored pellets; FE: food-associated eucalyptol; M: menthol; SB: Sorolabo biscuits; SBL: Sorolabo biscuits, after AAC or OP lesions; UP: unflavored pellets.

food deprivation, raises the question of the biological significance of this phenomenon. The *DAP* index has been defined in parallel with palatability taken as the capacity of a sensory cue to release and maintain food intake in food-deprived animals. Up to now, *DAP* and palatability have always been observed to vary in the same direction. The time course of the rate of food intake during a 30-min period in food deprived rats, which measures palatability according to LeMagnen (1971b), reproduces the time course of the mitral cell AL observed under similar conditions, after a few minutes of induction. Both *DAP* and palatability have a central origin, and in addition, the rewarding property of bulb activating stimulation should be emphasized (Phillips and Mogenson, 1969; Phillips, 1970). Palatability, at the behavioral level, and *DAP*, at the bulb electrophysiological level, seem to be determined by nutritionally reinforcing characters of the stimulus. These functional analogies could lead us to investigate whether mitral cell activation corresponds to a neuronal evaluation of palatability.

REFERENCES

Adrian, E. D. (1950). The electrical activity of the mammalian olfactory bulb. *Electroencephalogr. Clin. Neurophysiol.* **2,** 377.

Allison, A. C. (1953). The structure of the olfactory bulb and its relationship to the olfactory pathways in the rabbit and the rat. *J. Comp. Neurol.* **98,** 309.

Bojsen-Møller, F., and Fahrenkrug, J. (1971). Nasal swell-bodies and cyclic changes in the air passage of the rat and rabbit nose. *J. Anat.* **110,** 25.

Cabanac, M. (1971). Physiological role of pleasure. *Science* **173,** 1103.

Cain, D. P. (1975). Effects of insulin injection on responses of olfactory bulb and amygdala single units to odors. *Brain Res.* **99,** 69.

Cattarelli, M., Pager, J., and Chanel, J. (1976). In preparation.

Dennis, B. J., and Kerr, D. I. B. (1975). Olfactory bulb connections with basal rhinencephalon in the ferret: An evoked potential and neuroanatomical study. *J. Comp. Neurol.* **159,** 129.

Døving, K. B. (1964). Studies on relation between the frog's EOG and single unit activity in the olfactory bulb. *Acta Physiol. Scand.* **60,** 150.

Freeman, W. J. (1974). Average transmission distance from mitral-tufted to granule cells in olfactory bulb. *Electroencephalogr. Clin. Neurophysiol.* **36,** 609.

Getchell, T. V., and Shepherd, G. M. (1975). Synaptic actions on mitral and tufted cells elicited by olfactory nerve volleys in the rabbit. *J. Physiol. (London)* **251,** 497.

Giachetti, I., and MacLeod, P. (1973). Supériorité du pouvoir discriminateur des cellules mitrales comparé à celui des récepteurs olfactifs. *J. Physiol. (Paris)* **66,** 399.

Giachetti, I., and MacLeod, P. (1975). Cortical neuron responses to odours in the rat. *Olfaction Taste, Proc. Int. Symp., 5th, 1974* p. 303.

Giachetti, I., MacLeod, P., and LeMagnen, J. (1970). Contrôle centrifuge des afférences olfactives en fonction des états de faim et de satiétéchezle rat. *C. R. Seances Soc. Biol. Ses Fil.* **164,** 841.

Guselnikova, K., Guselnikov, V., Tsitolovsky, L., and Engovatov, V. (1971). Studies of recurrent excitation in the olfactory bulb of the frog. *J. Higher Nerv. Activ. I.P. Pavlov* **21,** 191.

Haberly, L. B., and Shepherd, G. M. (1973). Current density analysis of summed evoked potentials in opossum prepyriform cortex. *J. Neurophysiol.* **36,** 789.

Hara, T. J., and Law, Y. M. C. (1972). Adaptation of the olfactory bulbar responses in fish. *Brain Res.* **47,** 259.

Holley, A., Duchamp, A., Revial, M. F., and Juge, A. (1974). Qualitative and quantitative discrimination in the frog olfactory receptors: Analysis from electrophysiological data. *Ann. N.Y. Acad. Sci.* **237,** 102.

Kauer, J. S. (1974). Response patterns of amphibian olfactory bulb neurones to odour stimulation. *J. Physiol. (London)* **243,** 695.

LeMagnen, J. (1956). Rôle de l'odeur ajoutée au régime dans la régulation quantitative à court terme de la prise alimentaire chez le rat blanc. *C. R. Seances Soc. Biol. Ses Fil.* **150,** 136.

LeMagnen, J. (1971a). Advances in studies on the physiological control and regulation of food intake. *Prog. Physiol. Psychol.* **4,** 203.

LeMagnen, J. (1971b). Un développement de la neurophysiologie des comportements: Les mécanismes de la régulation alimentaire du bilan d'énergie. *Arch. Ital. Biol.* **111,** 591.

Leveteau, J., and MacLeod, P. (1969). Reciprocal inhibition at glomerular level during bilateral olfactory stimulation. *Olfaction Taste, Proc. Int. Symp., 3rd, 1968* p. 212.

McLennan, H. (1971). The pharmacology of inhibition of mitral cells in the olfactory bulb. *Brain Res.* **29,** 177.

MacLeod, P. (1971). Structure and function of higher olfactory centers. *Handb. Sens. Physiol.* **4,** 182.

MacLeod, P. (1973). Quality coding through neural response patterns. Transduction mechanisms in chemoreception. *Proc. Symp. ECRO,* p. 295.

Mathews, D. F. (1972). Response patterns of single units in the olfactory bulb of the rat to odors. *Brain Res.* **47,** 389.

Mori, K., and Takagi, S. F. (1975). Spike generation in the mitral cell dendrite of the rabbit olfactory bulb. *Brain Res.* **100,** 685.

Motokizawa, F. (1974). Olfactory input to the thalamus: Electrophysiological evidence. *Brain Res.* **67,** 334.

Moulton, D. G. (1963). Electrical activity in the olfactory system of rabbits with indwelling electrodes. *Olfaction Taste, Proc. Int. Symp., 1st,* p. 71.

Moulton, D. G. (1968). Electrophysiological and behavioral responses to odor stimulation and their correlation. *Olfactologia* **9,** 69.

Nicoll, R. A. (1969). Inhibitory mechanisms in the rabbit olfactory bulb: dendrodendritic mechanisms. *Brain Res.* **14,** 157.

Nicoll, R. A. (1971a). Pharmacological evidence for gaba as the transmitter in granule cell inhibition of the olfactory bulb. *Brain Res.* **35,** 137.

Nicoll, R. A. (1971b). Recurrent excitation of secondary olfactory neurons: A possible mechanism for signal amplification. *Science* **171,** 824.

Pager, J. (1974a). A selective modulation of the olfactory bulb electrical activity in relation to the learning of palatability in hungry and satiated rats. *Physiol. Behav.* **12,** 189.

Pager, J. (1974b). A selective modulation of olfactory input suppressed by lesions of the anterior limb of the anterior commissure. *Physiol. Behav.* **13,** 523.

Pager, J. (1976). Contrôle centrifuge des réponses électriques du bulbe olfactif: Étude chez le rat en fonction de l'état nutritionnel. Thèse de doctorat d'état, Lyon, 1976.

Pager, J., and Royet, J. P. (1976). Some effects of conditioned aversion on food intake and olfactory bulb electrical responses in the rat. *J. Comp. Physiol. Psychol.* **90,** 67.

Pager, J., Giachetti, I., Holley, A., and LeMagnen, J. (1972). A selective control of olfactory bulb electrical activity in relation to food deprivation and satiety in rats. *Physiol. Behav.* **9,** 573.

Phillips, A. G. (1970). Enhancement and inhibition of olfactory bulb self-stimulation by odours. *Physiol. Behav.* **5,** 1127.

Phillips, A. G., and Mogenson, G. J. (1969). Self-stimulation of the olfactory bulb. *Physiol. Behav.* **4,** 195.

Phillips, C. G., Powell, T. P. S., and Shepherd, G. M. (1961). The mitral cells of the rabbit's olfactory bulb. *J. Physiol. (London)* **156,** 26P.

Phillips, C. G., Powell, T. P. S., and Shepherd, G. M. (1963). Responses of mitral cells to stimulation of the lateral olfactory tract in the rabbit. *J. Physiol. (London)* **168,** 65.

Pinching, A. J. (1971). Spatial aspects of the neuronal connexions in the rat olfactory bulb. *Olfaction Taste, Proc. Int. Symp., 4th,* p. 40.

Price, J. L., and Powell, T. P. S. (1970). The afferent connexions of the nucleus of the horizontal limb of the diagonal band. *J. Anat.* **107,** 239.

Schiffman, S. S. (1974). Physico-chemical correlates of olfactory quality. *Science* **185,** 112.

Schultz, E. F., and Tapp, J. T. (1973). Olfactory control of behavior in rodents. *Psychol. Bull.* **79,** 21.

Shepherd, G. M. (1963a). Responses of mitral cells to olfactory nerve volleys in the rabbit. *J. Physiol. (London)* **168,** 89.

Shepherd, G. M. (1963b). Neuronal systems controlling mitral cell excitability. *J. Physiol. (London)* **168,** 101.
Shepherd, G. M. (1972). Synaptic organization of the mammalian olfactory bulb. *Physiol. Rev.* **52,** 864.
Tanabe, T., Iino, M., and Takagi, S. F. (1975a). Discrimination of odors in olfactory bulb, pyriform-amygdaloid areas, and orbito-frontal cortex of the monkey. *J. Neurophysiol.* **38,** 1284.
Tanabe, T., Yarita, H., Iino, M., Ooshima, Y., and Takagi, S. F. (1975b). An olfactory projection area in orbito-frontal cortex of the monkey. *J. Neurophysiol.* **38,** 1269.
Tucker, D., and Beidler, L. M. (1956). Autonomic nervous system influence on olfactory receptors. *Am. J. Physiol.* **187,** 637.
Westecker, M. E. (1971). Reciprocal activation of two evoked potential components in the olfactory bulb. *Pfluegers Arch.* **324,** 297.

DISCUSSION

Cagan: You have presented some very significant results in demonstrating differences, at the level of the mitral cells, between the hungry and satiated, and I compliment you on that. I find it a refreshing change from many of the structure/activity studies of electrophysiological responses of the olfactory system. I think your approach is physiologically meaningful. Is this the first level in the olfactory system at which you see this kind of difference? Have you made any peripheral recordings as well?

Pager: I obtained a few records from the axons of the first-order neurons. I could not see any modulation of the responses, but they are rather difficult to obtain, and I had few of them.

Cagan: There are many implications regarding this feedback, particularly whether it extends out to the periphery.

Pager: Many mechanisms may be supposed to modulate this activity: sympathetic nervous activity, general arousal, even humoral mechanisms may not be excluded. Of course, those centrifugal mechanisms seem to be the main ones, but several others may be superimposed on them.

Doty: I am wondering if a dimension that you are looking at, other than food versus nonfood odors, is one of complexity of the volatile mixture reaching the epithelium, so that when you compare so-called food odor you are looking at a complex odor, and when you are using a so-called nonfood odor you are using a more simple mixture.

Pager: No, the dimension of complexity has been eliminated by the experiments with eucalyptol, which is sometimes a food odor and sometimes a nonfood odor, while the complexity, chemically speaking, remains the same. The effective learning experiment has been done with eucalyptol. It had been chosen initially as a control odor, but eucalyptol seems to be an intermediate between amyl acetate and food odors in several respects. I am not sure whether amyl acetate would have allowed positive learning like eucalyptol did. Maybe the genetically determined learning program for food odor accepts eucalyptol and excludes amyl acetate. I do not know.

DeWys: Did you look at any different concentrations or doses of the odor, and might there be a change from a positive to a negative, for example, if you get too high a concentration?

Pager: It also is a point remaining to be investigated, which will be done in the next years. The dynamics of the system are a function of concentration and also a function of the duration of food deprivation.

DeWys: You had only one electrode in the bulb? Did you try a series and see if there are distributions within the bulb?

Pager: As I am using freely moving and behaving rats, I have used only one electrode in each bulb, not to have too big a plug on the head of the animal. In acute experiments I recorded with several electrodes and, of course, the responses for the same odor are positive, nil, or negative according to the localization of the electrode in the mitral cell layer.

Nicolaidis: What is the correlation between general, nonspecific eating channels?

Pager: It is a point which is under investigation.

Senseman: Do you ever see changes in the activity of the mitral cells when you are presenting pure odors with no food odors?

Pager: Generally, the activity and responses remain steady for several weeks in the same site.

Senseman: You are using rather large microelectrodes. How many units do you measure using this particular technique?

Pager: I suppose there are several dozen. It is a reasonable estimate. It is not the EEG-dimension but it is a population, in fact a zone of the bulb.

SENSORY QUALITY AND INGESTION

Chapter 4

Intensity and Hedonic Functions for Chemosensory Stimuli

HOWARD R. MOSKOWITZ

Human observers can detect and discriminate between chemosensory stimuli at varying physical concentrations. Direct magnitude estimation (ratio scaling) has shown that the odorous intensity of olfactory stimuli of varying concentration can be described by power functions ($S = kC^n$) whose exponents are less than 1.0. In contrast, gustatory stimuli such as glucose are governed by power functions with exponents greater than 1.0 and sourness by exponents less than 1.0. The present set of experiments investigated a variety of different gustatory and olfactory stimuli, and required the observers to evaluate both perceived intensity and estimated hedonic tone (degree of liking/disliking). Hedonics and sensory intensity are highly correlated for some chemosensory stimuli (especially those which are principally unpleasant), and only moderately correlated for others (those stimuli which reach a peak pleasantness in the mid range of concentrations, and then decline). These two sets of correlations suggest that hedonic tone of some stimuli (viz., unpleasant ones) may completely depend on sensory intensity, whereas the hedonic tone of other stimuli (e.g., some pleasant ones) may only partially depend upon sensory intensity.

INTRODUCTION

Traditionally, psychophysical evaluations in the chemical senses have concerned model systems, whether they be tastants diluted in aqueous solutions, or simple chemical odorants presented either from a liquid solution or from an air dilution olfactometer. The aim of many of these studies has been to determine how the sensory systems work. By presenting to the observer such model systems of known composition, and by instructing the observer as to the attributes to evaluate (e.g., quality, intensity, hedonics), experimenters have obtained the following:

(a) Quality reports: evaluations of the different qualities evoked by simple chemical stimuli [e.g., Meiselman and Dzendolet (1967) for various salts, or Schiffman and Dackis (1975) for various vitamins and amino acids].

(b) Intensity reports: evaluations of the perceived intensity of these stimuli as a function of known variations in concentration [e.g., evaluation of different taste materials, as reviewed by Meiselman (1972), and evaluation of different odor materials, as reviewed by Berglund *et al.* (1971) and by Cain and Moskowitz (1974)].

(c) Hedonic reports: evaluation of the liking/disliking of stimuli, either as a function of the type of stimulus, or for single chemicals, as a function of concentration [e.g., evaluation of different sweeteners by Moskowitz (1971a); evaluation of different odorants by Doty (1975) and Moskowitz *et al.* (1976a)].

This paper concerns relations between physical concentration of stimuli (both tastants and odorants), perceived sensory intensity, and perceived hedonic value. It differs from many previous approaches in that it considers a continuum of stimuli, ranging from simple chemical stimuli which can be adequately measured in the laboratory and which comprise only one component, to simple foods which are systematically varied in one or two ingredients.

PSYCHOPHYSICAL MEASUREMENT METHODS

Traditional approaches in psychophysical assessment have used one or another variety of rating procedure, in which the observer estimates the intensity of a sensation or hedonic value. The two most widely used approaches today are category (or interval scaling), preferred by researchers in the food industry, and magnitude estimation (or ratio scaling), developed by psychophysicists (Stevens, 1953), and used principally by psychologists.

Category Scaling

With this method of assessment the experimenter provides the observer with a scale comprising a fixed number of categories (e.g., 1–9), and assumes that the perceptual difference between adjacent categories (e.g.,

1 and 2) equals that between other adjacent categories (e.g., 2 and 3), etc. When applied to gradations of physical concentration (e.g., tastants) the category scale is curvilinearly related to concentration. Quite often the curvature in the function straightens out when the experimenter plots the category rating versus log concentration [or log (concentration + k), where k is an additive constant chosen to maximize linearity)]. Category scaling is easily performed, even with naive observers, but is subject to numerous biases:

(a) Observers shy away from using the end points of the scale for fear of "running out of numbers."

(b) Observers do not use the scale as an equal interval scale, so that the perceptual difference between an 8 and a 9 may be much larger than the perceptual difference between a 5 and a 6 (in the middle of the scale).

(c) The form of the function relating stimulus concentration to category response varies with the spacing of the stimuli along the physical continuum, as well as the number of categories selected by the experimenter. Thus, reproducible equations from one laboratory to another are difficult to obtain, and quite often experimenters are content to settle for simple correlations between instrumental (or objective) assessments of concentration and category ratings made by observers.

Ratio Scaling

With this method of assessment the experimenter permits the observer to use a freely selected range of numbers in order to reflect perceived magnitude (or hedonic tone). The principal requirements are that the numbers which the observer provides should match the perceived intensities (or hedonic tones), and that ratios of numbers should reflect ratios of intensity. Ratio scaling (or magnitude estimation) usually yields power functions of the form $S = kC^n$ (S = magnitude estimate, C = concentration of the chemosensory stimulus, n = exponent, k = intercept). In contrast to category scaling, ratio scaling eliminates the following biases:

(a) There are no end points to the scale, so that observers do not have to worry about "running out of numbers."

(b) The form of the psychophysical function (viz., a power function for intensity evaluation of simple physical continua, such as acoustic energy rated for loudness) is relatively invariant with respect to the spacing of the stimulus. However, the slope of the function may vary somewhat if an initial standard stimulus is selected to be either the very lowest stimulus in the set (i.e., all other responses are rated relative to that lowest stimulus), the highest stimulus in the set, or the middle stimulus (Stevens, 1974).

In both types of measurements as used by psychophysicists, the aim is to develop functional relations or equations that relate perceived sensory intensity (or hedonic tone) to physically measured magnitude. Such an approach differs substantially from traditional approaches to sensory measurement in which interest is focused upon threshold values, and in which changes in threshold values are presumed to reflect changes in the sensory capacities of the individual [cf. studies by Fischer (1971) and by Henkin (1967)]. Threshold levels mark the concentration of stimuli where perception just begins, and represent benchmarks which are the lowest points on the sensory scale. With an assessment of the entire range of concentrations one can obtain a clearer picture of the characteristics of the taste and smell systems unavailable from measurements of just one point.

INTENSITY MEASUREMENT IN THE CHEMICAL SENSES

Experimenters interested in taste and smell have developed numerous scales for perceived sensory intensity versus concentration. The following results have been reported.

Taste

Sweetness

Sugars Sweetness (versus Concentration). When evaluated by the method of magnitude estimation this sweetness often yields power functions. The exponents of these power functions take on two different sets of values, depending upon the method of presentation. If the stimulus is presented in paper cups, sipped, expectorated, and rated (cf. Moskowitz, 1971a), then the exponents exceed 1.0, and average about 1.3. This means that perceived sweetness = $k(\text{concentration})^{1.3}$, indicating that sweetness grows more rapidly than concentration (viz., as an accelerating function). On the other hand, other experimenters have flowed stimuli over the tongue when the tongue is extended from the mouth (cf. Meiselman, 1972). For that procedure the power function exponent obtained is less than 1.0 (often around 0.5–0.7), suggesting that perceived sweetness grows less rapidly than concentration (viz., as a decelerating function). Substantial controversy exists about the most appropriate exponent for perceived sugar sweetness (Meiselman, 1972).

Artificial Sweeteners (e.g., Saccharin, Cyclamate). When evaluated by magnitude estimation, they are usually governed by power functions with exponents less than 1.0, but only in the mid ranges of concentration. At high concentrations (the values for "high" are idiosyncratic to each sweetener) the taste becomes bitter as well as sweet, and quite often sweetness peaks, and then diminishes, with increasing concentration (Moskowitz, 1970; Moskowitz and Klarman, 1975a). Recent studies with the new dipeptide sweetener, aspartyl phenylalanine methyl ester (aspartame), suggest that this sweetener may be governed by a power function whose exponent more nearly approximates the sucrose exponent (Cloninger and Baldwin, 1970), perhaps because aspartame does not have a pronounced bitter taste at moderate and high concentrations. A strong bitter taste would suppress sweetness and lead to a peak sweetness level at an intermediate range of concentration.

Saltiness

Most of the data on perceived saltiness as a function of NaCl concentration suggest that saltiness grows as an accelerating function of concentration (exponent exceeding 1.0) when the NaCl solution is sipped (as in normal eating and drinking), but that saltiness grows as a decelerating function of concentration (exponent less than 1.0) when NaCl solutions are flowed over the tongue (Meiselman, 1972).

Sourness and Bitterness

Both of these tastes are governed by power functions for taste intensity with exponents either equal to 1.0 (Beebe-Center, 1949) or exponents less than 1.0 (Meiselman, 1972; Moskowitz, 1968).

Smell

In virtually all experiments thus far reported, the concensus of findings is that the perceived intensity of odorants grows as a decelerating function of concentration. The exact value of the power function exponent varies, depending upon the physicochemical properties of the odorant (Cain, 1969; Moskowitz *et al.*, 1976a), and upon the mode of stimulus preparation and presentation. Odorants dissolved in a liquid medium such as water or diethyl phthalate (cf. Engen, 1965) yield lower exponents than similar odorants presented by a dynamic air dilution olfactometer (Cain, 1969; Moskowitz *et al.*, 1974b, 1976a).

PLEASANTNESS (HEDONIC) MEASUREMENT IN THE CHEMICAL SENSES

In contrast to the relatively large number of studies pertaining to taste and smell intensity which have appeared during the past decade and a half, relatively few studies in that period have systematically considered perceived pleasantness/unpleasantness of pure tastants, pure odorants, or even foods. To some extent the hiatus in the literature can be traced to a lag in the measurement procedure—traditional psychophysical measurement of magnitude lends itself quite well to the assessment of taste and smell intensity (whose evaluations are direct extensions of early studies of brightness and loudness) (Stevens, 1953). In contrast, the appropriate assessment of liking/disliking is hindered by two major problems. First, how does one go about assessing a psychophysical continuum such as hedonics, knowing that there are stimuli which are both liked and disliked? Are liking and disliking to be treated as one single continuum, with low numbers representing stimuli which are disliked, and high numbers (on the same numerical continuum) representing stimuli which are liked (e.g., Henion, 1971)? Or is the experimenter to set up two different scales, representing liking on one scale and disliking on the other (e.g., Doty, 1975; Moskowitz *et al.*, 1976a)? Second, the preponderance of individual differences in hedonic reactions to tastes and to odors often seems to have discouraged any substantial investigation of hedonic tone, especially by the method of magnitude estimation.

Taste

The principal findings for taste hedonics can be summarized as follows.

Sweetness

Observers report that the pleasantness of sweet solutions of sugars (either sucrose or other sugars) first increases with concentration, peaks (usually at a sweetness equal to that of 1.0 *M* glucose = 18% w/v, or 9% sucrose, which is equally sweet), and then subsequently diminishes with increasing concentration beyond the peak point (Engel, 1928; Kocher and Fisher, 1969; Moskowitz, 1971a; Moskowitz *et al.*, 1974a; Moskowitz *et al.*, 1976b). For artificial sweeteners which provoke more than sweetness alone, the simple function found for sugars (viz., one appearing to be an inverted U or L) no longer holds. Saccharin pleasantness may diminish with concentration at a different sweetness level than does sucrose pleasantness (Moskowitz and Klarman, 1975b).

Saltiness

Observers report that the pleasantness of the salt taste may conform to one of three functions. Either saltiness becomes increasingly pleasant with concentration, decreasingly pleasant (both monotonic), or conforms to an inverted U- or L-shaped function (showing a breakpoint) (Ekman and Åkesson, 1964; Kocher and Fisher, 1969; Moskowitz *et al.*, 1975c; Pangborn, 1970).

Sourness and Bitterness

Observers from western countries report that the pleasantness of sourness decreases with concentration (any increase in pleasantness with concentration occurs at a very low concentration of acid, which is barely perceptible; cf. Engel, 1928). On the other hand, for observers who have traditionally grown up with sour foods all their lives, and who are used to the sour taste (e.g., they may eat tamarind fruit for a confection, as do Karnataka laborers in India), perceived sourness becomes increasingly pleasant with concentration and sensory intensity, until an exceptionally strong sourness is reached. Only afterward does sourness become less pleasing with increasing levels (Moskowitz *et al.,* 1975c).

Smell

For smell, controversies surround many of the experimental findings. Psychophysical studies with pure odors show that some odorants become increasingly pleasant with concentration, whereas in the other studies, using the same procedures but with a different sample of observers, the same odorant will increase in pleasantness, reach a peak, and then decrease (Doty, 1975; Moskowitz *et al.,* 1976a). However, the following general findings deserve mention.

Unpleasant Odorants

Those which are clearly unpleasant (e.g., pyridine) continue to be unpleasant at higher concentrations, and become even more unpleasant.

Neutral Odorants

Those which are neutral tend to remain either neutral or become increasingly unpleasant with higher concentrations (e.g., some of the aliphatic alcohols; Moskowitz, 1974).

Pleasant Odorants

Those which are pleasant often level off in pleasantness, and fail to increase in pleasantness with concentration. Occasionally, the odorant may reach a peak pleasantness and diminish in pleasantness (sometimes passing neutrality and becoming definitely unpleasant). The point at which the odorant peaks in pleasantness is not fixed, either in terms of concentration (e.g., log dilution from saturation or ppm in air), or even in terms of perceived odor intensity (Moskowitz *et al.,* 1976a). This failure for pleasantness to "turn around" and diminish at a constant sensory level contrasts with the invariance of sweetness of sugars. At a sugar sweetness of

1.0 *M* (equivalent to glucose), perceived pleasantness of aqueous solutions maximizes for group data.

PROBLEMS WITH SIMPLE SYSTEMS

As discussed above, much of the current literature pertaining to intensity and hedonic aspects of chemosensory stimuli concerns model systems (aqueous solutions, dilutions of odorants, chemicals in air or in a liquid medium). In the "real world" such stimuli are rare for the following reasons:

(a) Most foods and odorants are complex stimuli comprising many different chemicals and embedded in a nonhomogeneous base. Foods contain many dozens of volatiles and taste components, so that knowledge gleaned from the analysis of simple model systems *may not* necessarily apply to real foods (although there is some suggestion that the sweetness–pleasantness function does reappear in food systems which are sweetened with various levels of sucrose; Moskowitz *et al.*, 1974a).

(b) We have expectations about food, in addition to our sensory inputs. These expectations may influence sensory responses (e.g., evaluation of perceived sweetness of wines vary in the presence of different coloring agents which simulate white versus rose wines; Pangborn *et al.*, 1963), but they may more powerfully influence hedonic responses. Some foods are expected to be sour and bitter, and may be rejected if they fail to meet the expected level of sourness (even though sourness is a taste which by itself is unpleasant).

(c) We evaluate foods according to a variety of criteria, only one of which is actual sensory property of the food. The others may stem from the momentary state of hunger versus satiety. Hunger may potentiate affect and make stimuli which are normally displeasing actually pleasing (as is witnessed by cannibalism among starving individuals). On the other hand, satiety may render the most acceptable foods displeasing. As Cabanac (1971) and his co-workers (Duclaux *et al.*, 1973) have shown, stimuli (whether they be simple solutions of glucose, the aroma of orange oil, or of other food-related materials) lose their acceptability to the observer when he is satiated. They call this effect "alliesthesia," or the regulation of external affective responses to external stimuli by the current state of the "milieu interieur."

THE APPROACH OF THE PRESENT PAPER

This paper concerns the psychophysical evaluation of chemosensory stimuli, using both assessments of intensity and assessments of hedonics.

It deals with responses to stimuli as functions of the following variables: (a) type of stimulus used (taste versus smell); (b) real world significance of stimuli (chemical versus food); and (c) type of response requested (intensity, affect). Because many of the studies reported here were performed over a 7-year period, the scaling procedures and the observers were not necessarily the same. Nonetheless, the major thrust of the approach throughout the 7 years was to obtain functions relating intensity and affect to known stimulus variations.

Study Series I: Taste of Simple Systems

In this series of studies, observers were presented with a variety of simple chemical stimuli and instructed to rate both the taste intensity and the taste pleasantness of the stimuli. Where magnitude estimation is indicated, the observers gave ratio-scale estimates of perceived taste intensity, and gave ratio-scale estimates of liking/disliking. Observers assigned positive numbers to reflect liking, 0 to reflect disliking, and negative numbers to reflect disliking.

Experiment 1: The Sweetness and Affect of Glucose Stimuli

In this study observers evaluated five concentrations of reagent-grade glucose for both liking and disliking (bipolar scales), as well as for perceived taste intensity. Each observer of a group of 75 gave magnitude estimates for intensity of sweetness and for the degree of liking/disliking. Figure 1 shows the relation between molarity and estimates of sweetness

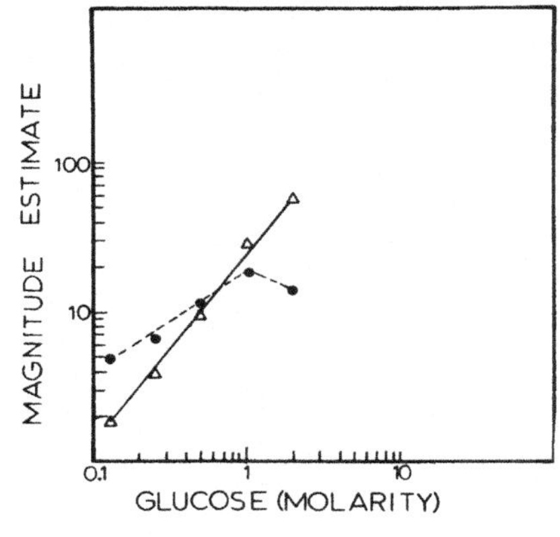

Fig. 1. Relation between the molar concentration of glucose and the perceived sweetness and pleasantness of the taste. The coordinates are log–log in which power functions appear as straight lines. The data are based upon median ratings. Triangles, sweetness; circles, pleasantness.

and estimates of liking/disliking, respectively. Note that the studies from which Fig. 1 was derived were performed over a period of several months, and that some observers are represented slightly more often than others.

The important thing to note about Fig. 1 is the remarkable difference between ratings of taste sweetness and taste liking for this group of Massachusetts observers (ages 17–43). Whereas the sweetness range is expanded, in comparison to the concentration range (i.e., sweetness grows as a power function, with exponent exceeding 1.0, which means that we expand the physical range of molarities to a larger subjective range of sweetness) the range of hedonic values is contracted for the same set of stimuli. In addition, one should note that rarely, on the average, are simple glucose solutions found to be unpleasant when aqueous systems are evaluated. In other contexts in which glucose would be evaluated against actual sweet beverages, one might expect other effects (e.g., glucose being negatively rated, as an unpleasant stimulus; B. P. Halpern, personal communication).

An analysis of the variability of the ratings for taste intensity versus concentration and for taste pleasantness can be performed for each concentration, in order to determine whether increasing concentrations of glucose concentration produce equally variable sets of magnitude estimations. Such an analysis could also reveal whether there are two or more clusters of responses, especially at high concentrations of glucose, where observers rate pleasantness (which would suggest that there are some individuals who note an increase in pleasantness beyond the sweetness of 1.0 *M*, and others who note a decrease in pleasantness). Table I presents the summary statistics for variability of the magnitude estimates versus concentration. Note that the measure of variability is the interquartile

TABLE I

Glucose Taste[a]

Concentration (*M*)	Sweetness[b] ($N = 790$)	Interquartile ratio	Pleasantness ($N = 770$)	Interquartile ratio
0.125	1.90	4.74	4.57	5.11
0.25	3.83	5.07	6.78	4.73
0.5	9.79	3.66	10.62	4.21
1.0	29.61	3.12	19.03	6.18
2.0	56.23	3.37	13.80	12.79

[a] The magnitude estimates were obtained from a base population of 75 individuals, most most of whom give at least eight replicate judgments per concentration.

[b] Log Sweetness = 1.27 log (concentration) + 1.40 $R = 0.997$.

range of the log ratings (or the interquartile ratio). There is an indication that at high glucose concentrations the variability of the ratings increases faster for estimates of hedonics than for estimates of perceived sweetness. This means that potentially there may be several different groups of respondents (or segments of the population), each with its own unique hedonic function for the sweet taste, but all exhibiting similar responses to glucose sweetness.

Experiment 2: The Taste and Affect of Artificial Sweeteners

In this study observers evaluated seven concentrations of reagent-grade Na saccharin, and seven concentrations of reagent-grade Na cyclamate. The same groups of observers participated as had participated in experiment 1. In some of the sessions the observers rated the artificial sweeteners as well as glucose. Altogether, the data obtained for these sweeteners are based upon 120 replicate judgments per concentration.

In contrast to the function for glucose sweetness, which is accelerating (versus concentration), the sweetness ratings for both Na cyclamate and Na saccharin are decelerating versus concentration. Figure 2 shows a log–log plot of perceived sweetness versus concentration, and a semilog

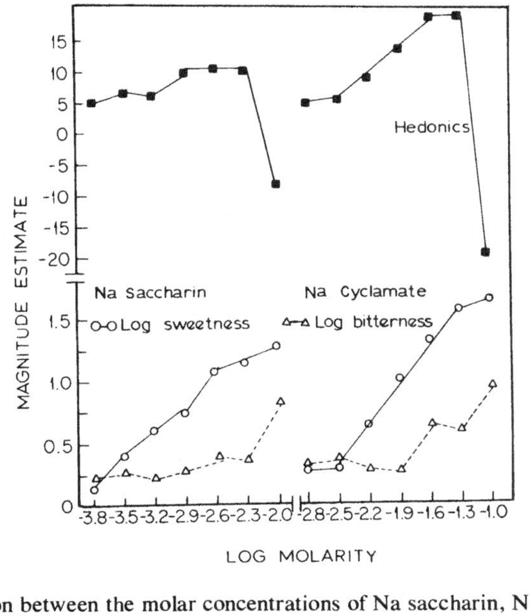

Fig. 2. Relation between the molar concentrations of Na saccharin, Na cyclamate, and perceived sweetness, perceived bitterness, and pleasantness/unpleasantness. The concentration scale is logarithmic, as is the sweetness scale. The pleasantness/unpleasantness scale is linear.

plot of perceived pleasantness (+) or perceived unpleasantness (−) versus concentration.

Both substances show decelerating functions for sweetness versus concentration [which in log–log coordinates corresponds to a straight line between 0.003 *M* and 0.025 *M* for Na cyclamate (slope = 0.82) and a straight line for Na saccharin between 0.003 *M* and 0.05 *M* (slope = 0.64)]. The pleasantness functions are much more curvilinear, and are almost quadratic functions. At high concentrations of both artificial sweeteners the magnitude estimates turn negative, indicating that the tastes become predominantly unpleasant. The region of the maximum pleasantness appears to be 0.00125–0.005 *M* for saccharin (corresponding to a sweetness level of 8–15, which equals the sweetness imparted by approximately 0.5 *M* glucose), and 0.025–0.05 *M* for cyclamate (corresponding to a sweetness level of 30, which equals the sweetness imparted by 1.0 *M* glucose). Both artificial sweeteners become unpleasant at higher concentrations. Cyclamate crosses neutrality at approximately 0.065 *M* (corresponding to a sweetness level of approximately 36, which equals the sweetness imparted by 1.3 *M* glucose), and saccharin crosses neutrality at approximately 0.006 *M* (corresponding to a sweetness level of 12, which equals the sweetness imparted by 0.6 *M* glucose).

One can conclude from the above comparisons that:

1. Functions relating sweetness to concentration differ from one sweetener to another, so that apparently the taste system must make different transformations relating concentration and sweetness, depending upon the sweetener (e.g., glucose versus cyclamate versus saccharin). One possibility is that there is internal masking of sweetness and bitterness, which appears more evident for cyclamate and saccharin than for glucose. Such internal masking would correspond to a simultaneous evocation of both sweetness and bitterness by the same molecule, which is known to occur for cyclamate and for saccharin (Cohn, 1914; Moskowitz and Klarman, 1975a; Rada *et al.*, 1967). Since the tongue sends simultaneous messages to the brain signaling both bitter and sweet, and since the brain does not necessarily recognize these as coming from a single source (viz., one chemical), the outcome is a naturally occurring "mixture experiment," such as might be obtained were the experimenter to mix quinine sulfate (bitter) and sucrose (sweet) at one fixed ratio, and then dilute that starting mixture with varying amounts of water. Such mixtures produce a diminished perceived sweetness and bitterness (McBurney and Bartoshuk, 1973; Moskowitz, 1972; Pangborn, 1960; Pilgrim, 1962).

2. The invariant "sweetness–pleasantness" breakpoint (which occurs for hedonic ratings at a subjective sugar sweetness corresponding to that produced by 1.0 *M* glucose = 18% w/v = 9% w/v sucrose) reoccurs here, but only for glucose and cyclamate and not for saccharin. Quite

possibly the presence of a secondary taste besides sweetness eliminates this invariance [cf. Moskowitz and Klarman (1975b) who also failed to find the invariant breakpoint for Na and Ca cyclamate and for Na and Ca saccharin].

3. The invariant "sweetness–pleasantness" relation occurs for group data, but there may be individual segments of the population that respond differently (especially at the higher concentrations, where some observers find the taste to be noxious, whereas others find it to be pleasant).

Experiment 3: The Taste and Effect of an Acid (Cyclamic Acid)

In this study, observers evaluated seven concentrations of cyclamic acid, an intensely sour tasting acid whose salts are sodium and calcium cyclamate. In one session observers rated perceived sourness, and in the other session observers rated liking/disliking. Note that the data were collected from several smaller experiments, in which different groups of observers from a pool of 75 individuals participated.

Figure 3 shows the relation between log concentration and log response (median perceived sourness, median perceived unpleasantness).

The results from experiment 3 suggest that for the sour taste, an entirely different intensity–pleasantness relation holds. First, in contrast to the sweetness functions for glucose (which conform to power functions with exponents exceeding 1.0), the perceived sourness of cyclamic acid grows as a decelerating power function which is governed by an exponent around 0.7–0.8. This value accords with several previous estimates for the exponent for the sour taste (cf. Moskowitz, 1971b), and suggests that the taste

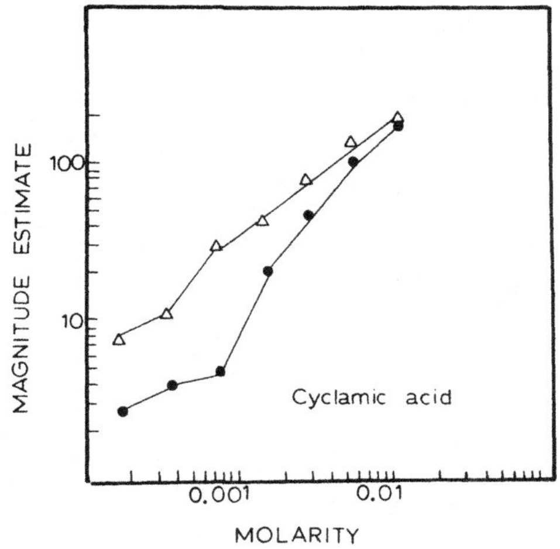

Fig. 3. Relation between the molar concentration of cyclamic acid and both perceived sourness (triangles) and perceived unpleasantness (circles). The coordinates are log–log.

system processes information from the H^+ molecule in a profoundly different way from the manner in which it processes information from the glucose molecule. The other important finding is that for these observers perceived unpleasantness can be considered as a power function of concentration, with an exponent slightly higher than the exponent for perceived sourness. In a sense, the hedonic tone associated with sourness is governed by the same rule for magnitude as governs the growth of perceived sourness. It is quite possible that we find it difficult to distinguish between the two things (viz., sourness and unpleasantness of sourness), whereas we can easily distinguish between sweetness and the pleasantness of sweetness.

Similar parallels between intensity functions and hedonic functions were found by Moskowitz *et al.* (1976b), who used category scales to evaluate the taste intensity and taste pleasantness/unpleasantness of simple solutions of citric acid and quinine sulfate. That study employed Indian medical students at St. John's medical college, and suggested that these populations of U.S. military men and scientists as well as Indian medical students perceive the hedonic tone of acid in the same way. Both groups find it difficult to distinguish, at least according to their psychophysical functions, between taste intensity and taste hedonics.

Study Series II: Odor of Simple Systems

In this series of studies, observers made psychophysical estimates of intensity and pleasantness for a variety of different odor stimuli. Some of these were relatively pure chemicals (e.g., Baker Grade), whereas others were actual aromas and flavors previously prepared by flavor and fragrance supply companies for use in foods. All of the stimuli were diluted in diethyl phthalate, an odorless solvent. The dilution factor was 4/1, which produced the following range of concentrations: 1, 1/4, 1/16, 1/64, and 1/256.

In any "experiment," an observer was presented with 30 different bottles, coded by letters. Those 30 bottles comprised six dilution series, one for each of six stimuli. The stimuli were presented to the observer in random order, from small, wide-mouthed scintillation bottles in which the stimulus solution lay at the bottle. A cotton wick protruded inside the bottle upward from the base, and increased the surface area from which the stimulus could evaporate to produce a cloud of stimulus odorant.

Observers rated the intensity of each odorant using magnitude estimation and the degree to which they liked or disliked the odorant, using a modified category scale (0 = dislike extremely, 1–4 were gradations of disliking, 5 = neutral, 6–9 were gradations of liking, with 9 = like extremely). Each observer from a set of 97 individuals evaluated every experimental series only once. Since there were many such experimental sets of 30 odorants, approximately 30–45 individuals evaluated each set. All observers were paid to evaluate the odorants and were drawn from college age students in the Framingham–Natick area of Massachusetts.

Figure 4 provides the summary functions relating the concentration of the odorant stimuli (in logarithmic coordinates) to the arithmetic mean of the hedonic ratings (in linear coordinates). Table II provides the parameters of power functions for odor intensity versus concentration, and logarithmic functions for odor hedonics versus concentration.

The important results of study series II are summarized as follows:

1. Odorants grow in perceived odor intensity according to power functions whose exponents are always less than 1.0. This means that the sensory system shrinks the range of perceived odor intensity as compared to the physical range of concentrations.

2. Complex combinations of chemicals used to develop flavors obey similar psychophysical laws—they also grow in odor intensity according to power functions, again with exponents less than 1.0.

3. There appear to be two, and possibly three, different types of hedonic functions. For chemicals that do not resemble foods, the odor hedonic function may either be flat (no change with concentration) or more typically slope downward with increasing concentration.

The former type of hedonic function (flat) is best described as either a random fluctuation of the hedonic ratings around a mid level or an inverted U- or L-shaped function (more rare). This hedonic function characterizes many of the flavors which have strong character impact, and can be recognized as foods or foodlike aromas. One may speculate here that for such chemicals the observers did not truly show gradations in hedonic tone with changes in odor intensity, but rather their judgments randomly varied about a single hedonic level.

The latter function (downward sloping) suggests that for those substances the observer provides magnitude estimates in accordance with intensity as well. The breakpoint noticeable for sugar sweetness is absent. The downward sloping functions appear to be similar to the function encountered for cyclamic acid, in which unpleasantness is highly correlated with sourness.

Study Series III: Hedonics of Taste Mixtures

In this series of experiments an attempt was made to develop and evaluate a more natural taste stimulus and yet maintain laboratory control over the stimulus constituents. Most taste stimuli in the environment comprise mixtures of stimuli (often with a dozen or more sapid chemicals present at varying concentrations). Therefore, the stimuli for this experiment were

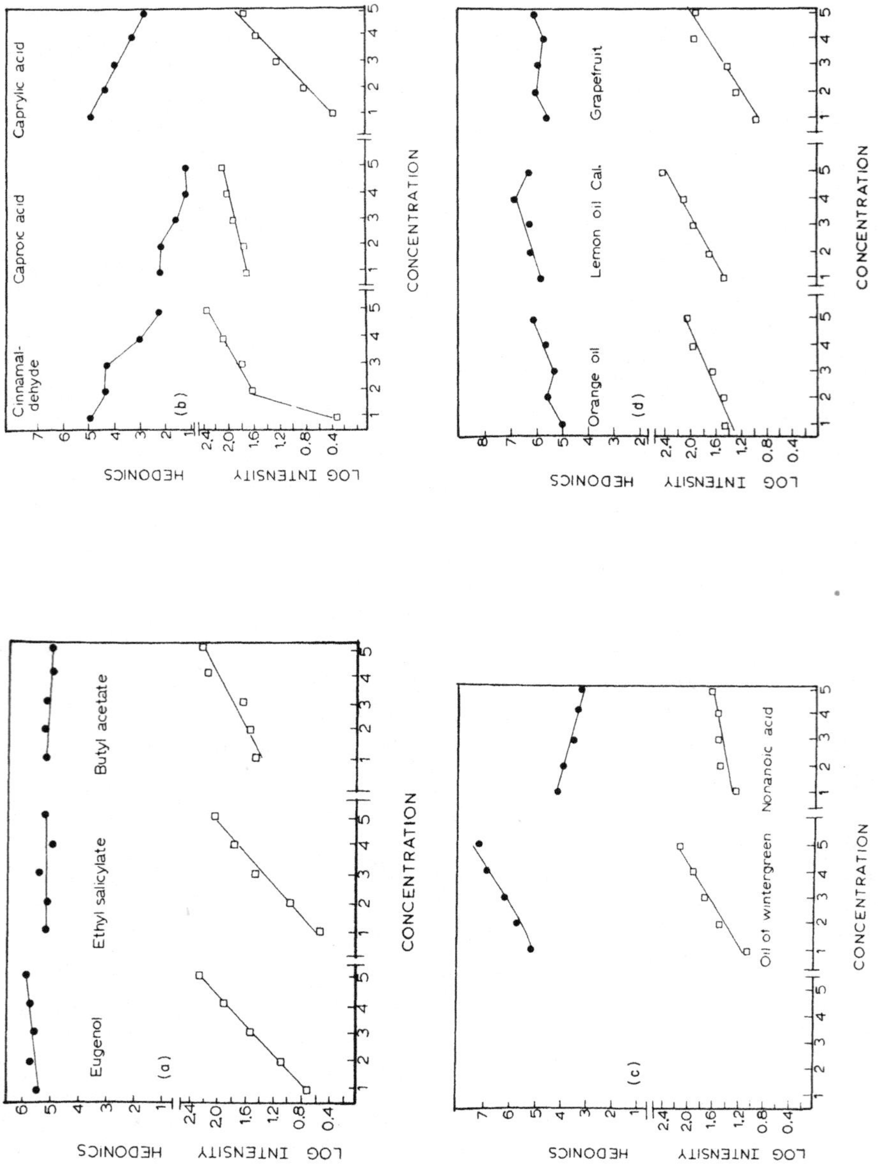
(a)
Eugenol
Ethyl salicylate
Butyl acetate
(b)
Cinnamaldehyde
Caproic acid
Caprylic acid
(c)
Oil of wintergreen
Nonanoic acid
(d)
Orange oil
Lemon oil Cal.
Grapefruit
HEDONICS
LOG INTENSITY
CONCENTRATION

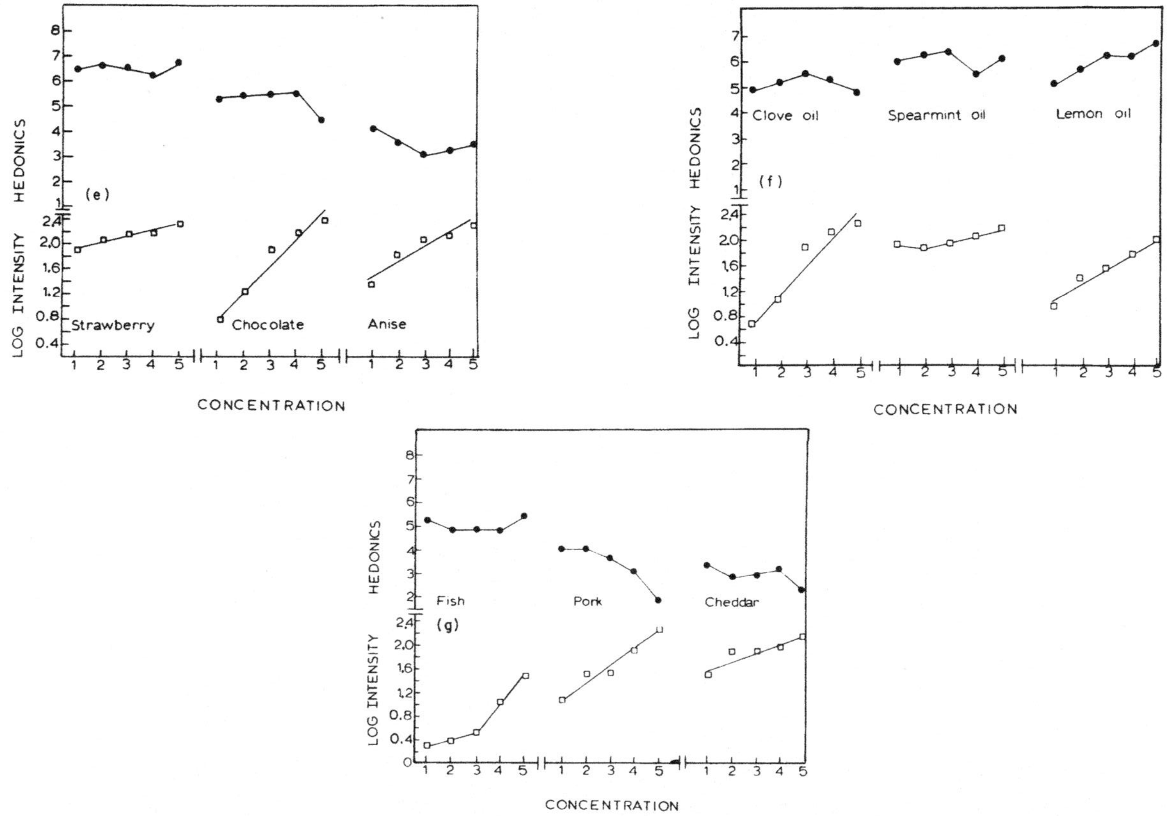

Fig. 4. (a–g) Relation between the concentration of an odorant in diethyl phthalate (abscissa), and both perceived odor intensity (bottom) and perceived pleasantness/unpleasantness (top). The concentration scale and the intensity scale are logarithmic, whereas the pleasantness/unpleasantness scale is linear. For reference purposes, a 9 on the pleasantness/unpleasantness scale represents "like extremely," a 5 represents "neutrality," and a 0 represents "dislike extremely." Cal., California.

TABLE II

Intensity and Hedonics of Odors[a]

1. Eugenol ($N = 41$ individuals)—cloves, spicy
   $\log I = 0.64(\log C) + 2.26$ $R = 0.99$
2. Ethyl salicylate ($N = 43$ individuals)—wintergreen
   $\log I = 0.63(\log C) + 2.09$ $R = 0.99$
3. Butyl acetate($N = 47$ individuals)—banana, nail polish remover
   $\log I = 0.35(\log C) + 2.22$ $R = 0.93$
4. *Trans*-cinnamaldehyde($N = 43$ individuals)—cinnamon, spicy
   $\log I = 0.78(\log C) + 2.72$ $R = 0.92$
   $H = -1.09(\log C) + 2.44$ $R = 0.95$
5. *N* Caproic acid($N = 40$ individuals)—fatty, waxy, oily, rancid
   $\log I = 0.17(\log C) + 2.23$ $R = 0.97$
   $H = 0.48(\log C) + 1.00$ $R = 0.96$
6. Caprylic acid($N = 37$ individuals)—fatty, waxy, oil, rancid
   $\log I = 0.60(\log C) + 2.01$ $R = 0.99$
   $H = -0.89(\log C) + 2.06$ $R = 0.99$
7. Clove oil($N = 40$ individuals)—like eugenol
   $\log I = 0.67(\log C) + 2.42$ $R = 0.96$
8. Spearmint oil($N = 42$ individuals)—spearmint
   $\log I = 0.16(\log C) + 2.16$ $R = 0.97$
9. Lemon oil($N = 39$ individuals)—lemon
   $\log I = 0.42(\log C) + 2.03$ $R = 0.98$
   $H = 0.56(\log C) + 6.66$ $R = 0.93$
10. California orange($N = 34$ individuals)
    $\log I = 0.52(\log C) + 2.16$ $R = 0.89$
    $H = 0.34(\log C) + 5.92$ $R = 0.85$
11. Lemon oil, California($N = 38$ individuals)
    $\log I = 0.35(\log C) + 2.37$ $R = 0.99$
12. Grapefruit($N = 44$ individuals)
    $\log I = 0.44(\log C) + 2.07$ $R = 0.97$
13. Nonanoic acid($N = 40$ individuals)—fatty, oily, fruity in dilution
    $\log I = 0.20(\log C) + 1.78$ $R = 0.93$
    $H = 0.28(\log C) + 3.53$ $R = 0.97$
14. Anise($N = 38$ individuals)—licorice
    $\log I = 0.37(\log C) + 2.20$ $R = 0.96$
15. Strawberry($N = 40$ individuals)
    $\log I = 0.16(\log C) + 2.09$ $R = 0.93$
16. Chocolate($N = 44$ individuals)
    $\log I = 0.66(\log C) + 2.29$ $R = 0.97$
17. Fish extract($N = 37$ individuals)
    $\log I = 0.42(\log C) + 1.28$ $R = 0.77$
18. Pork($N = 37$ individuals)
    $\log I = 0.46(\log C) + 2.26$ $R = 0.97$
    $H = -0.86(\log C) + 2.29$ $R = 0.90$
19. Imitation cheddar cheese ($N = 41$ individuals)
    $\log I = 0.20(\log C) + 2.13$ $R = 0.88$
20. Pork flavor ($N = 36$ individuals)
    $\log I = 0.46(\log C) + 2.26$ $R = 0.97$
    $H = -0.86(\log C) + 2.29$ $R = 0.90$
21. Wintergreen ($N = 36$ individuals)
    $\log I = 0.42(\log C) + 2.16$ $R = 0.98$
    $H = 0.79(\log C) + 7.15$ $R = 0.99$

[a] I = intensity (magnitude estimates); H = Hedonic rating (0–9); C = concentration (% in diethyl phthalate).

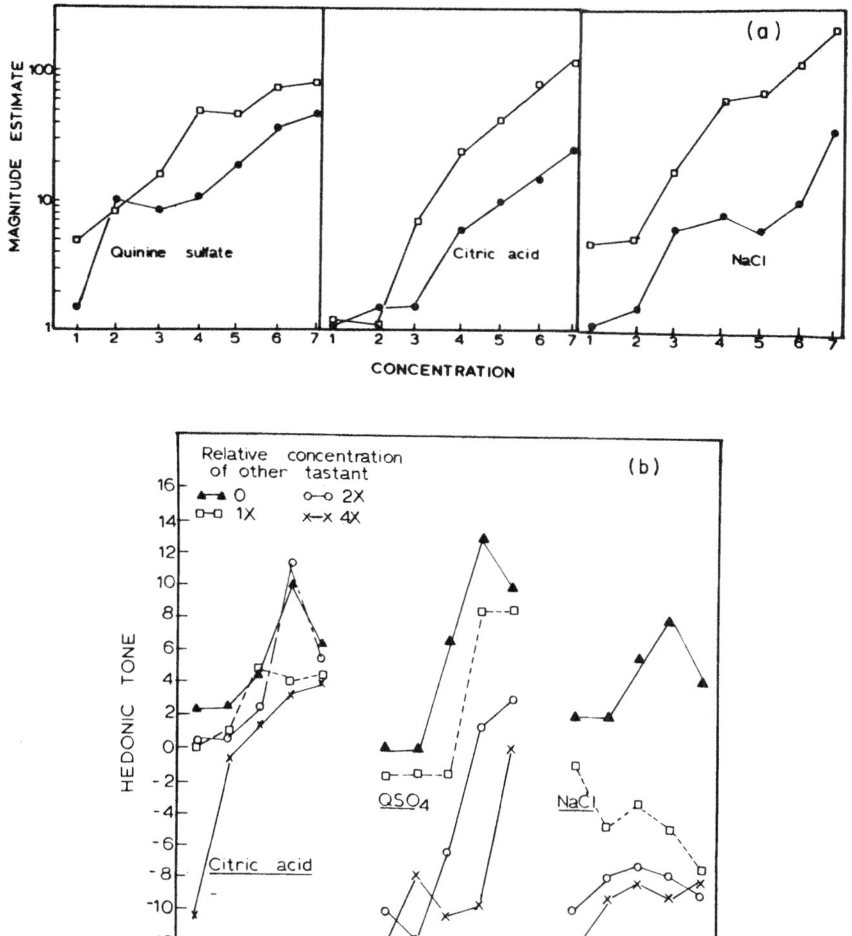

Fig. 5. (a) Relation between the concentration of NaCl, quinine sulfate, and citric acid, respectively, and perceived unpleasantness. The coordinates are log–log. For comparison purposes the taste intensity ratings are plotted as well. Circles, hedonics; squares, intensity. (b) Relation between the concentration of glucose (abscissa), in logarithmic coordinates, and the perceived pleasantness/unpleasantness of taste. Different concentrations of citric acid, quinine sulfate (QSO_4), and sodium chloride were added to the glucose solutions. A 0 reflects neutrality, positive numbers reflect liking, and negative numbers reflect disliking.

binary mixtures comprising a sweet substance (glucose at 2.0, 1.0, 0.5, 0.25, 0.5) which is usually judged to be pleasant, and a nonsweet substance (sodium chloride at 0.5, 0.25, 0.12 5 *M*, or a citric acid at 0.005, 0.0025, 0.00125 *M*, or quinine sulfate at 0.0005, 0.00025, 0.000125 *M*) which is judged to be unpleasant.

In three different experiments, groups of 16 observers selected from scientific staff and military volunteers at the U.S. Army Natick Laboratories evaluated the pleasantness (+ magnitude estimates) or the unpleasantness (− magnitude estimates) of varying mixtures.

Figure 5a shows the hedonic function of the components. Figure 5b shows how perceived hedonic tone changes in a mixture as a function of the constituents. The numbers shown in Fig. 5a are the median magnitude estimates from the group of 16 judgments per combination.

The major finding derived from the three sets of experiments are

(a) Glucose pleasantness functions again exhibit the breakpoint at 1.0 *M* concentration of glucose if no other taste is present.

(b) NaCl is primarily unpleasant, as is citric acid and quinine sulfate. The degree of unpleasantness of the three is monotonic with concentration, within this concentration range.

(c) The hedonic tone of mixtures is oftentimes intermediate to the hedonic tone of the components, although in some instances (e.g., glucose and citric acid) a mixture is more pleasant after the addition of an unpleasant tasting substance than would be expected on the basis of pure averaging of hedonic tones of components.

The implications of this set of mixture experiments are that chemosensory hedonics should be looked at from several points of view:

(a) The observation that some chemicals are innately pleasant tasting and some are unpleasant tasting may not predict mixture hedonics.

(b) Mixtures themselves may provide an unpleasant tasting substance, but the fact that the stimulus is itself complex and presents a pattern of sensory stimulation may be an influential factor in determining hedonics.

Study Series IV: Taste, Aroma, and Hedonics of a Simple Food System

In this series of experiments an attempt was made to determine the hedonic responses and the sensory impressions of individuals to a simple food system, fruit flavored drink (noncarbonated), which varied in two ways:

1. Concentration of flavoring material (actual cherry flavor + citric acid), which for each flavor varied in the ratio 0.25 normal, 0.5 normal, normal, 2× normal, 4× normal.

2. Concentration of sucrose (the recommended sweetener which the user would normally add), which for each flavor and for each level of flavoring material varied in the ratio 0.25 normal, 0.5 normal, normal, 2× normal, 4× normal.

The stimuli were made up in colorless solutions from material provided by the General Foods Corporation. The stimuli were frozen until use and then thawed and remixed before the experiment. They were served at room temperature. Observers were instructed in the method of magnitude estimation, and all observers (a total of 15 women, ages 18 through 43) had had previous experience in assigning numbers to both tastes and smells.

In the actual experiment each observer received approximately 8–10 ml of the colorless solution and was asked to sip as much of it as she wished, and then to rate the following attributes: sweetness and pleasantness of sweetness (two attributes); flavor intensity and pleasantness of flavor (two attributes); total intensity of the impression (one attribute).

The experiment required two mornings for the evaluation of all 25 samples which were presented in random order of concentrations within a single flavor type. Observers were instructed to use the same scale in all evaluations for two replicates per day. Each observer was paid for the completion of the study.

As in the previous studies with hedonics, positive numbers were to reflect degrees of liking, negative numbers were to reflect degrees of

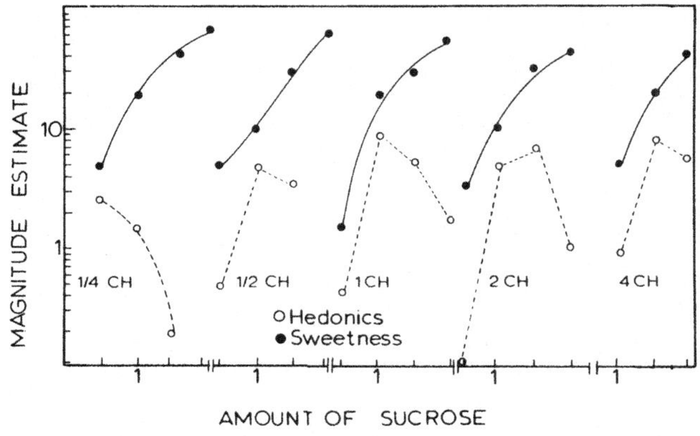

Fig. 6. Relation between the concentration of sucrose (abscissa) in % w/v, and both the perceived sweetness of a cherry-flavored drink and the perceived liking of the sweetness of the drink. The coordinates for both concentration and response variables are logarithmic. Each curve in this figure represents varying concentrations of sucrose and a fixed level of flavor (denoted by ¼ · · · 4, which reflect relative concentration, 1 = normal amount of flavor). Neutral and unpleasantness ratings are not shown. CH, fixed, relative amount of cherry flavoring.

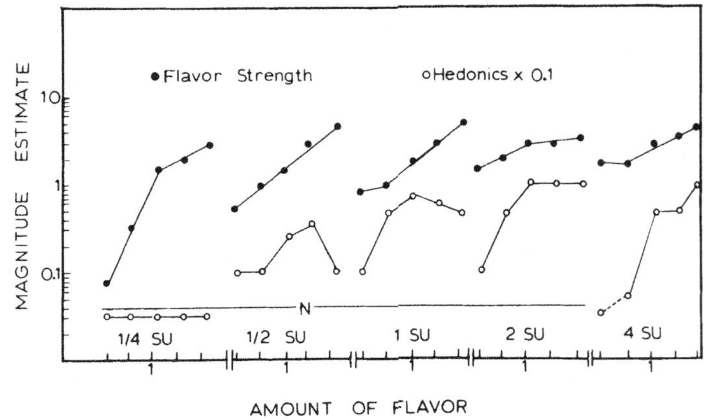

Fig. 7. Relation between the amount of flavoring (abscissa) in terms of ratio to "normality" (1 = normal), and both perceived flavor intensity and the perceived liking of the flavor. The coordinates are all logarithmic. Each curve in this figure represents a fixed level of sucrose (denoted by 1/4–4.0, which reflect relative concentration, 1 = normal amount of sucrose = approximately 10% w/v). SU, fixed, relative amount of sucrose.

disliking, and zero was to reflect neutrality (neither like nor dislike). The size of the numbers was left to the discretion of the observer.

Figure 6 shows the relation between sucrose concentration, perceived sweetness, and perceived pleasantness of sweetness. The concentration of flavor has been left as a parameter, fixed at a constant level for each curve. Figure 7 shows the relation between flavor concentration, perceived flavor level, and perceived pleasantness of flavor. Note that the observers in this experimenter were instructed to estimate the taste and flavor impressions, as well as the component hedonic reactions to taste and flavor. The data shown are the medians of 60 magnitude estimates (four from each of 15 individuals) and are plotted in log–log coordinates.

The principal results from study series IV can be summarized as follows:

1. With increasing sugar concentration perceived sweetness increases for all flavor levels (which are held constant), indicating the ability of the observer to track the change in concentration.

2. With increasing sugar concentration the liking/disliking of the sweet taste exhibits breakpoints, but these breakpoints vary as a function of the level and type of flavor. The breakpoint found for simple aqueous systems in which only sucrose or glucose is present is affirmed, but the actual level may vary somewhat (cf. Moskowitz *et al.*, 1974a).

3. With increasing flavor concentration the observers gave increasing estimates of flavor intensity.

4. With increasing flavoring concentration the observers often gave flat or inverted U- or L-shaped functions for liking/disliking.

DISCUSSION

The foregoing series of studies points to two aspects of hedonics in chemosensory systems which deserve extensive investigation. These are (1) the nature of hedonic functions for the two senses (taste versus smell), and (2) the nature of hedonic functions for simple systems that have little or no relation to food, as compared to simple systems which may resemble food.

Hedonic Functions for the Two Senses

Observers in psychophysical studies can reliably assign numbers to reflect both the degree of perceived sensory intensity and the degree of liking/disliking. This ability to differentiate hedonics and intensity is not surprising for the assessment of sensory intensity, since it has been previously shown over the past half-century by a variety of methods (cf. Engel, 1928). What is surprising is the nature and repeatability of the hedonic functions, even for such simple systems as glucose, cyclamic acid, caproic acid, etc.

Type I functions are those in which the mean estimates of liking/disliking are highly correlated with perceived sensory intensity. The correlation may be either positive or negative, but the important thing is that these type I hedonic functions are somehow functionally indistinguishable from the corresponding sensory functions. That is, were an investigator unable to determine what question was being asked (viz., how much does one like or dislike the taste/smell versus how strong does the taste/smell seem to be), then the function which emerges would be of no help in disentangling the two. Type I functions appear to be those which are generated when the observer evaluates unpleasant stimuli, such as the taste of cyclamic acid, the odor intensity of fatty acids, the taste of bitter quinine sulfate, etc.

Type II functions are those in which the mean estimates of liking/disliking are correlated with perceived sensory intensity, but somewhere in the mid range of sensory intensities a maximum hedonic point is reached. Beyond that maximum point hedonic tone either diminishes toward unpleasantness or to neutrality, and the previous correlation with sensory intensity is reversed. Type II functions appear to govern those taste stimuli which are sweet (e.g., glucose, artificial sweeteners, etc.).

An entirely different picture can be seen from considering olfactory hedonics, especially with regard to odorants derived from actual foods or flavoring. Many of these stimuli show linear functional relations between perceived sensory intensity and known physical concentration (in liquid dilution). Yet, the hedonic functions for many of these foodlike aromas are often randomly distributed about a middle range of acceptability. Such a hedonic function may reflect yet a third type *(type III)*, defined as no hedonic variation with concentration and sensory intensity.

It is quite possible that different processes underlie each type of hedonic function. For example, one could speculate that type I and type III functions may reflect the inability of observers to differentiate between hedonics and sensory intensity. Such an inability would produce one of two different appearing functions: (a) a completely flat hedonic function (no change in hedonic tone with changing concentration and sensory intensity), or (b) a hedonic function which is parallel to or perpendicular to the intensity function.

For case (b) the observer might "misinterpret" or "mislabel" sensory intensity as hedonic value, but judge them in the same way.

The data reported here imply that unpleasant-tasting materials exhibit type I hedonic functions, in which the unpleasantness ratings parallel or are perpendicular to the intensity ratings. One can conclude, for instance, that a 10 fold increase in the sourness of an acid would produce approximately a 10-fold increase in the unpleasantness of the same acid. Similar parallels have been noted for

(a) The taste and hedonics of citric acid and quinine sulfate for Indian medical students (Moskowitz *et al.*, 1976b).

(b) The unpleasantness of noise versus the noisiness or loudness of noise when the noise is generated by airplanes (Berglund *et al.*, 1973).

(c) The door intensity and odor unpleasantness of L-butanol, scaled by both magnitude estimation and category estimation (Moskowitz *et al.*, 1974b).

(d) The odor intensity and odor unpleasantness of a variety of odorants, in which odor intensity was scaled by means of equal intensity matches to butanol (and subsequently converted to derived magnitude estimates). Odor unpleasantness was scaled by a line matching procedure and converted to derived magnitude estimates (Moskowitz *et al.*, 1976a). The unpleasant odorants were pyridine, thiophene, cyclohexylamine, etc.

Comparison of Taste versus Smell Hedonics

It has been traditionally assumed that hedonic tone is a prepotent aspect of smell, and to a lesser extent of taste as well. Yoshida (1964) discussed this prepotency and noted, as did others (Schiffman, 1974; Woskow, 1964), that the first dimension which is extracted by multidimensional scaling of odor dissimilarity is almost universally a hedonic one.

Based upon this propotency, one would have expected that odor hedonics would show a large change with odor intensity, especially for pleasant odors. It does not, a result which is quite opposite to the changes shown when observers evaluate glucose pleasantness and unpleasantness.

Pleasant odors show little hedonic change with concentration, and in many cases the changes which are found are randomly distributed about a flat line (0 slope). For unpleasant odorants (e.g., caproic acid) unpleasantness increases monotonically with odor intensity, as is the case for unpleasant tastes (e.g., cyclamic acid and citric acid).

As a first approximation, therefore, unpleasant odorants and tastants obey similar hedonic laws—the change in degree of hedonic tone mirrors the change in sensory intensity. There is a strong possibility that the rate of change in hedonic tone is really nothing more than a mislabeled estimate of the rate of change of sensory intensity. This possibility can be further investigated by studies which require the observer to scale hedonics and intensity by the same procedure (either category scaling or magnitude estimation). Pleasant odorants, in contrast to unpleasant ones, show little or no change in hedonic tone with sensory intensity, whereas pleasant tastes show a systematic increase, a maximum point of hedonics, and then a decrease as concentration is varied.

Hedonics of Mixtures and Simple Food Systems

Mixtures of sensory impressions, whether of pairs of tastes (study series II) or of tastes + smell (study series IV), show behavior which differs from the rules determined for simpler systems. The maximum point of pleasantness with varying sweetness is a function of both concentration of the sugar and of the other taste substance present (e.g., citric acid, quinine sulfate, sodium chloride). There may or may not be a maximum point, depending upon which of the taste substances is added. In addition, the presence of a second taste material often reduces the hedonic level of the sweet taste, but in some cases may enhance it (e.g., citric acid and glucose). The reason for the enhancement and the reduction is not clear, but several hypotheses deserve mention:

1. Some tastes "clash with each other" (e.g., sweet and salt) and produce an unpleasant taste. At any pair of component concentrations the resultant taste is unpleasant (as it appears to be for mixtures of NaCl and glucose). Moskowitz (1972) quantified the degree of "clashing" between tastes, and reported that it was maximal for glucose–NaCl and fructose–NaCl mixtures.

2. We usually taste mixtures of sweet and sour, or sweet and bitter. As Beebe-Center (1949) pointed out in his demonstration of the gust scale applied to food evaluation, no food (including honey) elicits one taste quality alone. Therefore, it should come as no surprise that mixtures, being more abundant than pure taste materials, might be better accepted as well.

3. There is a form of hedonic averaging or intermediacy (Klitzner, 1975) in which the addition of an unpleasant-tasting substance to a pleas-

ant tasting one often produces an intermediately pleasant-tasting or unpleasant-tasting stimulus.

When observers evaluate cherry-flavored beverages (an example of a simple food system comprising both tastants and odorants), the rules noted for pure tastes are slightly altered. For example:

1. The breakpoint in the sweetness–pleasantness function may shift to different sweetness levels, depending upon the concentration of the flavoring which is maintained constant. In aqueous solution where there are no other flavoring ingredients this breakpoint occurs at 1.0 *M* glucose or its sweetness equivalent. As the flavoring level in the cherry-flavored fruit beverage increases (from ¼ normal through 4× normal) the breakpoint in terms of the perceived sweetness level rises slightly.

2. For the evaluation of an actual flavor acceptability (i.e., the cherry flavor) apart from the sweetness, there is no systematic hedonic functional relation. Occasionally (especially at low fixed sucrose levels), observers find the flavor intensity to be neutral. At intermediate fixed sucrose levels (0.5 normal and normal) the hedonic functions for flavor level exhibit a breakpoint (but at different perceived flavor levels depending upon the fixed sugar level). Finally, at higher fixed levels of sucrose, flavor hedonic functions either level off or continue to increase monotonically with flavor level.

AN OVERVIEW

The foregoing set of experiments points toward the use of psychophysical scaling procedures as potentially powerful tools in the investigation of taste and smell hedonics. The use of functional relations (or dose–response curves) for sensory intensity and for hedonic tone can produce insights about how we respond to the sensory versus the hedonic properties of stimuli. The use of two sensory modalities reveals differences in hedonic responses. Finally, the use of a continuum of stimuli, from pure chemicals to simple binary mixtures to simple foods, reveals the stages at which laws and rules found for one domain operate or fail to operate at a higher one.

ACKNOWLEDGMENTS

The author wishes to acknowledge the assistance of General Foods in providing the colorless Kool-Aid. The author also wishes to thank Eugene Zanella, James Ryckman, Clifford Gerbers, and Cynthia DuBose, who aided in the series of experiments. Dr. Bruce Halpern provided many helpful insights about discerning relations between model systems and real foods.

REFERENCES

Abramson, R., and Moskowitz, H. R. (1976). Effect of obesity and renal dialysis on taste intensity and taste preference. Paper presented to the Eastern Psychol. Assoc., New York.

Beebe-Center, J. G. (1949). Standards for use of the Gust scale. *J. Psychol.* **28,** 411–419.

Berglund, B., Berglund, U., Ekman, G., and Engen, T. (1971). Individual psychophysical functions for 28 odorants. *Percept. Psychophys.* **9,** 379–384.

Berglund, B., Berglund, U., and Lindvall, T. (1973). "Scaling Loudness, Noisiness and Annoyance of Aircraft Noise," Report no. 393. Psychol. Lab., University of Stockholm, Stockholm, Sweden.

Cabanac, M. (1971). Physiological role of pleasure. *Science* **173,** 1103–1107.

Cain, W. S. (1969). Odor intensity: Differences in the exponent of the psychophysical function. *Percept. Psychophys.* **6,** 349–354.

Cain, W. S., and Moskowitz, H. R. (1974). Psychophysical scaling of odor. *In* "Human Responses to Environmental Odors" (A. Turk, J. W. Johnston, Jr., and D. G. Moulton, eds.), pp. 2–28. Academic Press, New York.

Cloninger, M. R., and Baldwin, R. E. (1970). Aspartylphenylalanine methyl ester: A low calorie sweetener. *Science* **170,** 81–82.

Cohn, G. (1914). "Die Organischen Geschmackstoffe." Siemenroth, Berlin.

Doty, R. L. (1975). An examination of the relationship between the pleasantness, intensity, and concentration of 10 odorous stimuli. *Percept. Psychophys.* **17,** 492–496.

Duclaux, R., Feisthauer, J., and Cabanac, M. (1973). Effets du repas sur l'agrement d'odeurs alimentaires et nonalimentaires chez l'homme. *Physiol. Behav.* **10,** 1029–1033.

Ekman, G., and Åkesson, C. (1964). "Saltiness, Sweetness and Preference: A Study of Quantitative Relations in Individual Subjects," Report no. 177. Psychol. Lab., University of Stockholm, Stockholm, Sweden.

Engel, R. (1928). Experimentelle Untersuchungen über die Abhangigkeit der Lust und Unlust von der Reizstarke beim Geschmacksinn. *Pfluegers Arch. Gesamte Physiol. Menschen Tiere* **64,** 1–36.

Engen, T. (1965). Psychophysical analysis of the odor intensity of homologous alcohols. *J. Exp. Psychol.* **70,** 611–616.

Fischer, R. (1971). Gustatory, behavioral and pharmacological manifestations of chemoreception in man. *In* "Gustation and Olfaction" (G. Ohloff and A. F. Thomas, eds.), p. 187. Academic Press, New York.

Henion, K. E. (1971). Odor pleasantness and intensity: A single dimension? *J. Exp. Psychol.* **80,** 275–279.

Henkin, R. I. (1967). Abnormalities of taste and olfaction in various disease states. *In* "Chemical Senses and Nutrition" (M. R. Kare and O. Maller, eds.), pp. 95–113. Johns Hopkins Press, Baltimore, Maryland.

Klitzner, M. (1975). Hedonic summation: Test of a linear model. *Percept. Psychophys.* **18,** 49–54.

Kocher, E. C., and Fisher, G. L. (1969). Subjective intensity and taste preference. *Percept. Mot. Skills* **28,** 735–740.

McBurney, D. H., and Bartoshuk, L. M. (1973). Interactions between compounds with different taste qualities. *Physiol. Behav.* **10,** 1101–1106.

Meiselman, H. L. (1972). Human taste perception. *CRC Crit. Rev. Food Technol.* **3,** 89–119.

Meiselman, H. L., and Dzendolet, E. (1967). Variability in gustatory quality identification. *Percept. Psychophys.* **2,** 496–498.

Moskowitz, H. R. (1968). Scales of intensity of single and compound tastes. Doctoral Dissertation, Harvard University, Cambridge, Massachusetts (unpublished).

Moskowitz, H. R. (1970). Ratio scales of sugar sweetness. *Percept. Psychophys.* **7,** 315–320.

Moskowitz, H. R. (1971a). The sweetness and pleasantness of sugars. *Am. J. Psychol.* **84,** 387–405.

Moskowitz, H. R. (1971b). Ratio scales of acid sourness. *Percept. Psychophys.* **9,** 371–374.

Moskowitz, H. R. (1972). Perceptual changes in taste mixtures. *Percept. Psychophys.* **11,** 257–262.

Moskowitz, H. R. (1974). Odor intensity and pleasantness of aliphatic alcohols via an air dilution olfactometer. *Bull. Psychon. Soc.* **4,** 244 (abstr.).

Moskowitz, H. R., and Klarman, L. A. (1975a). The tastes of artificial sweeteners and their mixtures. *Chem. Senses Flavor* **1,** 411–422.

Moskowitz, H. R., and Klarman, L. (1975b). The hedonic tones of artificial sweeteners and their mixtures. *Chem. Senses Flavor* **1,** 423–429.

Moskowitz, H. R., Kluter, R. A., Westerling, J., and Jacobs, H. L. (1974a). Sugar sweetness and pleasantness: Evidence for different psychological laws. *Science* **184,** 583–585.

Moskowitz, H. R., Dravnieks, A. L., and Gerbers, C. (1974b). Odor intensity and pleasantness of butanol. *J. Exp. Psychol.* **103,** 216–223.

Moskowitz, H. R., Kumraiah, V., Sharma, K. N., Jacobs, H. L., and Sharma, S. D. (1975c). Cross-cultural differences in simple taste preferences. *Science* **90,** 1217–1218.

Moskowitz, H. R., Dravnieks, A. L., and Klarman, L. (1976a). Odor intensity and pleasantness for a diverse set of odorants. *Percept. Psychophys.* **19,** 122–128.

Moskowitz, H. R., Kumraiah, V., Sharma, K. N., Jacobs, H. L., and Sharma, S. D. (1976). Effects of hunger, satiety and glucose load upon taste intensity and taste hedonics. *Physiol. Behav.* **16,** 471–475.

Pangborn, R. M. (1960). Taste interrelationships. *Food Res.* **25,** 245–256.

Pangborn, R. M. (1970). Individual variations in affective responses to taste stimuli. *Psychon. Sci.* **21,** 125–128.

Pangborn, R. M., Berg, H. W., and Hansen, B. (1963). The influence of color on discrimination of sweetness in dry table wine. *Am. J. Psychol.* **76,** 492–495.

Pilgrim, F. J. (1962). Interactions of suprathreshold taste stimuli. *In* "The Physiological and Behavioral Aspects of Taste" (M. R. Kare and B. P. Halpern, eds.), pp. 67–78. Univ. of Chicago Press, Chicago, Illinois.

Rada, C. P., Tihanyi, S. G., and Zienty, F. B. (1967). A study of the true taste of Saccharin. *J. Food Sci.* **32,** 357–360.

Rodin, J., Moskowitz, H. R., Fleming, B., and Bray, G. A. (1977). Relationship between obesity, weight loss and taste responsiveness. *Physiol. Behav.,* in press.

Schiffman, S. S. (1974). Physiochemical correlates of odor quality. *Science* **185,** 112–117.

Schiffman, S. S., and Dackis, C. (1975). Taste of nutrients: Amino acids, vitamins, and fatty acids. *Percept. Psychophys.* **17,** 140–146.

Stevens, S. S. (1953). On the brightness of lights and the loudness of sounds. *Science* **118,** 576.

Stevens, S. S. (1974). "Psychophysics, Introduction to Its Perceptual, Neural and Social Prospects." Wiley, New York.

Thompson, D., Moskowitz, H. R., and Campbell, R. (1976). Effects of body weight and food intake on pleasantness ratings for a sweet stimulus. *J. Appl. Physiol.* **41,** 77–83.

Woskow, M. H. (1964). Multidimensional scaling of odors. Doctoral Dissertation, University of California, Los Angeles (unpublished).

Yoshida, M. (1964). Studies of psychometric classification of odors (4, 5). *Jpn. Psychol. Res.* **6,** 111–224 and 145–155.

DISCUSSION

Dethier: You made a great point of the hedonic curve of sweetness having a break in it and not being monotonic with concentration. If you ran your sodium chloride down to much lower concentrations, would you not then get a break?

Moskowitz: I think that there is a break in the salt function. In a recent paper (Moskowitz *et al.*, 1976b) we found a beautiful break for sodium chloride, but the break is at a lower level than the levels we have investigated here.

Dethier: But does this get at anything fundamental? Is it not simply that if we have different effective concentration ranges for different compounds, then we can somehow or other equate them in your ordinary curves?

Moskowitz: There are two ways of answering your question. One is by looking at some of the taste data that Engel (1928) published. Engel had observers classify whether they liked or disliked acid, quinine, sugar, and salt. He found that all of those taste materials showed an inverted U-shaped function. Acid and quinine sulfate are bitter and sour, respectively, and principally unpleasant. However, the region at which people find these to be pleasant is so low that it might be either (a) that they are hardly tasting the material, or (b) that they are probably tasting more than sour or bitter. I think the region that we are studying here is a region that evokes a moderately strong salt taste. Secondly, for odor as in the case of orange oil and grapefruit (and the other pleasant odors which were diluted over a 256/1 range), the odor intensities were quite different and the hedonic responses were virtually unchanged. Again, the concentration functions are quite different; yet there is hardly any change whatsoever. To answer your question, then, there seem to be "hedonic laws" which depend *only in part* on taste or smell intensity. The remaining portion depends upon quality of the taste (or smell). Thus, were we to attempt to coalesce the sweetness hedonic function and the salt hedonic function, we might, in fact, find them to be only partially superimposable. One way of testing this conjecture is to plot hedonic tone versus sensory intensity, and try to superimpose the hedonic functions for sweet and salt. I think that the curves in that plot would appear quite different, and discontinuous.

Pfaffmann: Are you going to include, beside Kool-Aid, some of the aperitifs described as bitter and sour, yet widely used by some people?

Moskowitz: All of this is done without any cognitive input. I think once you put cognitive expectations on it, something like Cynar would be quite acceptable. Yet, if you were to give the same materials, bitter and sweet, to a drinker without the coloring, without the bottle, and without telling him it was artichoke liquer, it might be absolutely repugnant. The same thing would probably happen with martinis. The average young drinker of a martini who has not learned to expect the pleasures, social or postingestional, of a martini may find it to be repulsively bitter. The mature drinker does not. We are not sure whether he (a) finds the taste to be changed, or (b) associates the taste with some pleasurable things.

Maller: Of what validity are these measures of hedonics in terms of predicting how much of something the people will actually consume. Do you have any data on it?

Moskowitz: No. I do not know of any good literature on how much one is going to drink versus how palatable the drink is. I did one experiment on sweetening cola beverage and found that the most liked level turned out to be about 15% sugar rather than 12%. I think people realize that what you taste on a short-term exposure and what you get accustomed to may not be the same things.

Cagan: For the odorants you examined, do you have any idea what relative contributions to pleasantness/unpleasantness occur through the trigeminal compared with olfaction? Have you made such an analysis?

Moskowitz: Not directly. However, knowing the pleasantness, we can infer some things. There is very little change in pleasantness with increasing or decreasing concentrations for pleasant odorants. Therefore, you can continue to diminish the concentration of the odorant until it no longer excites anything but a bare smell and still have a pleasing odor. At this low level most of this input is smell (olfactory) and not trigeminal.

Cagan: I wondered if you had an opportunity to evaluate hedonic ratings with anosmic patients who retain trigeminal function.

Moskowitz: No, none at all.

Cantor: I am curious about the materials that you used in that sugar work. First with respect to the Kool-Aid. Was that the commercial product and, if so, did it not contain some glucose to which you added sucrose?

Moskowitz: No, this was unsweetened Kool-Aid provided by General Foods. It was colorless and unsweetened. All it had was the flavoring and citric acid and preservatives. There was no sweetener in it whatsoever.

Cantor: Let me say I am not clear when you say sugar whether you mean sucrose.

Moskowitz: Sucrose.

Cantor: The other question is in relationship to the origin of the glucose that you were using. You know glucose may be a variable product that is the α-hydrate. It depends on how it was made whether it has taste-detectable quantities of side reaction products. These materials have strong flavors. In addition to that, you have a quite different situation in a glucose solution than in a sucrose solution. Sucrose is one identity; in a glucose situation, you have two identities—α-glucose and β-glucose—and certainly they differ in rate of solubility. I am curious as to whether (a) your glucose contains any of these materials and (b) whether you get a reaction from the isomeric forms.

Moskowitz: We used monohydrate glucose, principally from Merck. Solutions were maintained for 2 days under refrigeration. We had on one or another occasion made up glucose solutions from Eastman and so I cannot tell you the exact purity.

Cantor: I can tell you that both of those companies buy glucose commercially and do not do very much in purifying it.

Moskowitz: We waited at least 48–72 hr under refrigeration before we ever used any of the stimuli so anything that would have been used would have already come to equilibrium. In addition, the material was all hydrate (MW = 198).

Cantor: Is there any work in which there have been distinctions made between alpha and beta?

Moskowitz: Pangborn's data showed differences in the alpha and beta forms.

Kinnard: You said that (in the Kool-Aid experiment) you had 25 different variations of combinations of flavor and sugar, is that right? And you said that the flavored component had essentially no sugar.

Moskowitz: The product we used comprised the Kool-Aid cherry flavor and citric acid and preservatives but no sugars. We dry weighed the sugar and the cherry flavoring material and then we mixed up a liter of total solution.

Kinnard: How much variation in sugar do you have to impose before the subjects could detect the difference in sweetness.

Moskowitz: I believe the JND is about 10%, but we are talking about 400%, 200%, 100%, 50%, and 25%. We have an overall 16 : 1 variation. In contrast, the 10% change is 1.1 : 1. We used a wide range of sugar and a wide range of flavor, not within the micro area of distinctions.

Pangborn: Relative to measuring hedonic responses in a laboratory situation, I think one has to take certain precautions because one is also measuring influences of the contrast and convergence effects of the samples that preceded and followed those in the test situation. There is the artificiality of the environment. What is there to like and dislike about 50 ml of sodium chloride solutions in a small booth under controlled illumination and conditions of presentation? Also, the mental set, the instructions given the subject of the instructions that he himself imposes on the test situation. Do you like oil of wintergreen? For what? For mouthwash, for chewing gum, for moustache wax—whatever the mental set is. The individuality of single responses can vary considerably within the same individual from time to time and from one subject to another. In fact, they can be contradictory. Some show an increase and others show a decrease to where an average would be meaningless. Finally, it can be totally unrelated to actual intake and consumption, a good point that Maller brought forward. To be asked how much one likes 1 g of horseradish in the laboratory may not have meaning. Horseradish may be very offensive unless it is served on prime rib. And to say that the amount you like is going to influence the amount that you consume is, in that case, ridiculous, because we do not consume a large amount of horseradish even though we like it very much as compared, for example, to rice.

Moskowitz: Those are a very, very good set of points. I want to add one thing. If you cannot go out to the field, the next best thing to do is to build comparative conditions in these laboratory experiments. Even if you cannot generalize outside, you can see how much the people like Kool-Aid versus how much they like sugar solutions of the same sweetness level. If you cannot make one major bridge, you can at least start making several other types of bridges. If you can get enough bridges, enough sets of connections, possibly you can flesh out the structure.

Pangborn: As long as we do not extrapolate beyond the confines of the experiment.

Cantor: Is there in fact an incremental effect of additional flavoring on acceptance. You said that in the laboratory you can compare the Kool-Aid of a particular sweetness with the sweetness of a pure sugar solution and compare those hedonic readings. I was asking how the Kool-Aid, which is more "natural" for someone to drink, compares with the ratings on the pure sugar solution.

Moskowitz: We cannot compare the actual numbers in this experiment. We can compare the shapes of the curves and equivalent relative magnitude estimates. Kool-Aid exhibits the same kind of increased hedonic function and break (viz., inverted U) across all the different flavor levels. The break or peak occurs as the pleasantness of Kool-Aid maximizes at the sweetness produced at about 10% sucrose in that Kool-Aid base. The break would be about 9 or 8 or 7% if it were pure sucrose, because the flavor base tends to mask the sweetness. I think there is some comparability; I do not think that you can generalize all the way.

Pfaffmann: You have mentioned work on the postingestional phase. Do you find that breakpoint shifting as the person becomes depleted?

Moskowitz: No. However, if a person is sated, the breakpoint is eliminated. For example, you can change the breakpoint by loading a person with glucose solution, but the breakpoint does not change after meals alone or after overnight fasting (Moskowitz *et al.*, 1976b). Fat people have a higher breakpoint than normal weight people (Rodin *et al.*, 1977; Thompson *et al.*, 1976). They turn off not at about 1 *M*, but at about 2–2.5 *M*. I think in the case of salt preference during renal dialysis there is a slight shift as well (Abramson and Moskowitz, 1976). I think it is a very important thing that is not only of interest in and of itself but also as a dependent variable for other analyses.

Chapter 5

Influence of Internal Factors on the Perceived Intensity and Pleasantness of Gustatory and Olfactory Stimuli

GEORGE D. MOWER, ROBERT G. MAIR,
AND TRYGG ENGEN

In the chemical senses, as in other sensory modalities, perceived intensity and pleasantness appear to be described by different psychophysical functions. The perceived intensity of olfactory and gustatory stimuli is related monotonically to stimulus concentration (Engen, 1971; Stevens, 1960). The perceived pleasantness of gustatory stimuli, on the other hand, shows a rise and then a decline with increasing stimulus concentration (Ekman and Åkesson, 1965; Moskowitz, 1971; Moskowitz *et al.*, 1974). Although the relation is not linear, data presented by Moskowitz *et al.* (1974) suggest that the perceived hedonic value of a taste stimulus is based on the perceived intensity of that stimulus. Some odorants show a montonic relationship between pleasantness and intensity, while others show a rise and then a fall (Henion, 1971; Doty, 1975).

Earlier accounts of hedonic experience (Young, 1959; Pfaffmann, 1960) have stressed stimulus qualities as the major determinant of pleasantness. While this may be the case in homeostatically balanced subjects, it has also been shown that internal physiological state can modify the perceived pleasantness of external stimuli (Cabanac, 1971). In both olfaction (Duclaux and Cabanac, 1970; Duclaux *et al.*, 1973) and gustation (Cabanac *et al.*, 1968; Cabanac and Duclaux, 1970), a given food-related stimulus which is perceived as pleasant by a fasted subject becomes unpleasant following the ingestion of a satiating load. Nonfood related stimuli are unaffected by such manipulations (Duclaux *et al.*, 1973). For the internally stressed subject, therefore, hedonic experience is not strictly stimulus bound. These results indicate that perceived pleasantness of a given external stimulus depends on internal receptors, which signal the homeostatic usefulness of the external stimulus, as well as external receptors, which signal the nature and intensity of the stimulus.

It has been suggested that while hedonic judgments are affected by manipulations of internal state, the perceived intensity of external stimuli is independent of such manipulations (Cabanac, 1971). Data indicating the independence of perceived intensity and internal state have been presented for thermal stimuli (Mower, 1976) and for orange odor (Duclaux and Cabanac, 1970). No data have been presented that compare perceived intensity for olfactory and gustatory stimuli in both stressed and nonstressed states. The present study was undertaken to determine the effect of internal state on both the perceived intensity and hedonic value of olfactory and gustatory stimuli and on the relationship between the two types of judgments. Subjects were tested in three internal states (fasted, normal, and satiated) so that hedonic and intensity judgments could be compared under stressed and nonstressed conditions.

METHOD

Subjects

The subjects were ten men and seven women between the ages of 20 and 35. All were of normal body weight.

Stimulus Materials

Taste stimuli consisted of two series of different concentrations of sucrose solutions. In the first series, the concentrations were 0.1 *M*, 0.4 *M*, 0.7 *M*, 1.0 *M*, 1.3 *M*, 1.6 *M*, and 1.9 *M*. In the second series, the concentrations were 0.2 *M*, 0.6 *M*, 1.0 *M*, 1.4 *M*, and 1.8 *M*. Taste solutions were stored in a refrigerator when not in use.

Four food-related olfactory stimuli were used: Anethol (a synthetic licorice smell), orange, maple, and chocolate. Two series of different concentrations of Anethol (Fritzsche Brothers) were used. In the first series, Anethol was diluted in diethyl phthalate in concentrations of 100%, 50%, 25%, 12.5%, 6.3%, 3.1%, and 1.6% by volume. In the second series, anethol was diluted in paraffin oil in steps of 100%, 33%, 11%, 3.7%, and 1.2% by volume. Florida orange oil (Polak's Fruital Works, Inc.) was diluted in paraffin oil in steps of 100%, 33%, 11%, 3.7%, and 1.2% by volume. Imitation maple (John Wagner and Sons, Inc.) and chocolate (John Wagner and Sons, Inc.) extracts were diluted in distilled water in steps of 100%, 25%, 6.3%, 1.6%, and 0.39% by volume. Odorants were stored in sealed glass test tubes.

Procedure

The study was divided into three experiments. In the first experiment, six subjects participated and the seven-step concentration series of both sucrose and anethol stimuli was used. In the second experiment, four different subjects participated and the five-step concentration series of both sucrose and anethol was used. Another group of seven subjects participated in the third experiment and the orange, maple, and chocolate olfactory stimuli were used.

In all parts of the study, subjects were tested under three conditions of nutritive state: fasted, normal, and satiated. In the first experiment, subjects were food deprived for 18 hr prior to testing in the fasted condition. In the normal condition, testing sessions began 1½ hr after the subject ate his normal meal. In the satiated condition, subjects ingested 500 ml of a 1.22*M* glucose solution 1 hr after a meal, and trials were begun 20 min later. Subjects were tested at 1 PM on separate days for each of the conditions. The order of internal state conditions was counterbalanced between subjects.

Under each of these conditions, two tasks were required of the subjects: judgments of the intensity of the stimuli and judgments of the hedonic value of the stimuli. Subjects registered their intensity judgments by positioning a pointer on a dial with 15 equally spaced marks. Hedonic judgments were made in a similar manner, also using a 15-point scale. Only verbal labels appeared on the scales, and these labels were converted to numbers for analysis of results as follows: for intensity 2 = very low, 5 = low, 8 = moderate, 11 = high, 14 = very high; for hedonics 2 = very unpleasant, 5 = unpleasant, 8 = neutral, 11 = pleasant, 14 = very pleasant.

Experimental sessions were broken into blocks of hedonic and intensity trials. During each block, subjects made either an intensity or a hedonic judgment of each of the stimuli. The blocks of hedonic and intensity judgments were counterbalanced. In total, each subject made four intensity and four hedonic judgments of each stimulus under each condition of internal state.

For taste stimuli, 5 ml of sucrose solution was taken into the mouth, retained for approximately 3 sec, and expectorated. The subject then registered his judgment and rinsed with distilled water. All test stimuli were presented at room temperature as was the rinse water. Olfactory stimuli were presented on a cotton swab attached to a glass rod which was stored in a glass test tube between trials (cf., Engen, 1965). The various concentrations were presented randomly, and gustatory and olfactory stimuli were presented alternately throughout the experiment.

In the second experiment, several procedural changes were made to compare results under different methodologies. Fewer and more distinct test stimulus concentrations were used, and the food deprivation period was lengthened. Also, subjects were run in all three conditions on the same day to help maintain more consistent judgmental criteria between rating sessions. During each testing session, subjects made four intensity and four hedonic judgments of each of the olfactory and gustatory stimuli. The rating procedure and manner of stimulus presentation were the same as that described above.

The first test was administered after the subjects had fasted for 22 hr. Following this session, subjects ate their normal meal and were tested 1 hr later. This was defined as the normal condition. When this session ended, subjects drank a 500-ml solution of 1.22*M* glucose, waited for 20 min, and then were tested in the satiated condition. This procedure precluded counterbalancing, but it is unlikely that order effects had a strong influence on the results. In experiments similar to the present one it has been shown that repeated presentations of tastes and odors do not affect the perceived hedonic value of the stimuli (Cabanac *et al.,* 1968; Duclaux and Cabanac, 1970). Additionally, in the present experiment a rest period of about 1 hr was interposed between rating sessions.

In the third experiment, only olfactory stimuli (orange, maple, and chocolate) were used. Fasted, normal, and satiated conditions were defined as in the second experiment. Trials in the fasted and satiated conditions were run in the same day; trials in the normal condition were run on a subsequent day. During each testing session, subjects made four intensity and four hedonic judgments of each concentration of each odorant. The rating procedure and manner of stimulus presentation were the same as that described above. The odorants were presented successively in the order of orange, maple, and then chocolate.

RESULTS

Comparison of hedonic ratings of the food-related olfactory and gustatory stimuli under the three conditions of internal state suggested that the subjects be divided into two groups. Some subjects (shifters) showed changes in hedonic ratings across the three experimental conditions; others (nonshifters) did not. To distinguish shifters and nonshifters, a two-way analysis of variance was performed on individual data for each odor and taste. Shifters were defined as those subjects who showed a significant change in hedonic ratings ($p < 0.05$) across conditions.

A total of 10 subjects rated the intensity and pleasantness of sucrose and Anethol stimuli. For taste stimuli, three subjects were classified as shifters and three as nonshifters with the procedure of experiment one. Examples of individual data from a shifter and a nonshifter for sucrose are given in Table I. With the procedure of experiment two, there were three shifters and one nonshifter. Otherwise, the results under the two procedures were similar and are not considered separately.

Figures 1 and 2 show group data from experiment one for both intensity and hedonics for shifters and nonshifters, respectively. None of the shifters showed a significant change in hedonic ratings between the fasted and normal conditions. In each of these conditions, the taste stimuli (sucrose) were rated as neutral or slightly pleasant with stimuli in the middle concentration range being most preferred. In the satiated condition, all shif-

TABLE I

Mean Hedonic Ratings for Taste Stimuli of Two Subjects[a]

		Stimulus concentration (*M*)						
Subject	Condition	0.1	0.4	0.7	1.0	1.3	1.6	1.9
J. H. (shifter)	Fasted	8.0	8.3	10.3	9.5	7.5	5.5	5.0
hedonic data	Normal	8.0	8.0	9.1	7.5	7.0	7.8	7.8
	Satiated	8.0	6.0	5.3	3.3	3.5	3.3	2.0
M. R. (nonshifter)	Fasted	13.0	6.0	4.9	3.5	3.0	1.5	1.0
hedonic data	Normal	14.3	7.5	5.9	5.6	4.3	3.8	2.0
	Satiated	14.8	8.3	6.3	5.8	3.5	4.8	1.8

[a] Higher ratings are more pleasant. Subject J. H. is typical of shifters in showing decreased preference in the satiated condition and no change between fasted and normal conditions. Subject M. R. is typical of nonshifters in showing monotonically decreasing functions for perceived hedonic value with increasing intensity in all conditions.

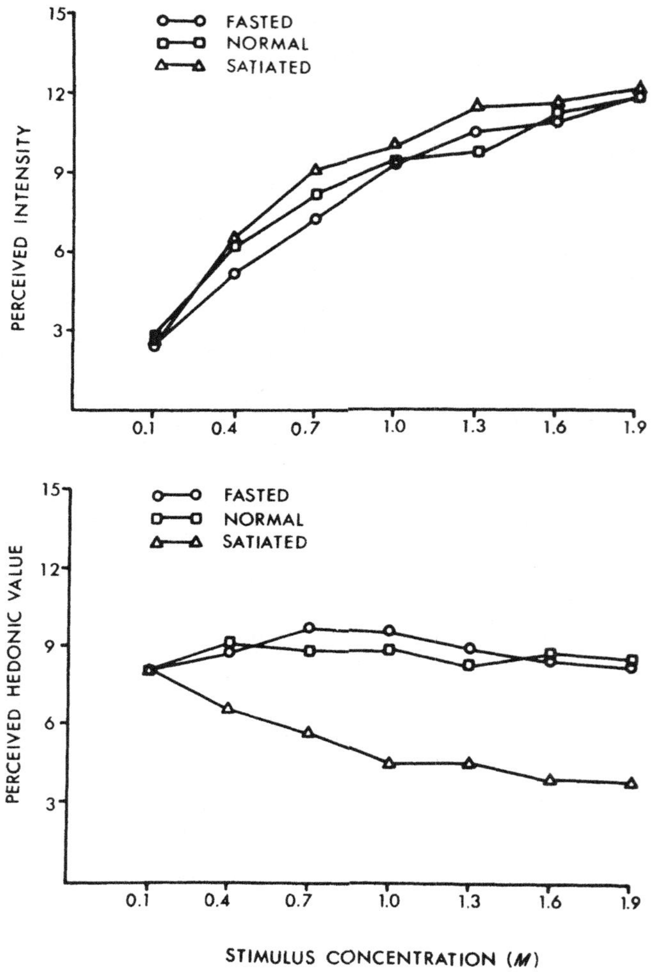

Fig. 1. Group mean category judgments of perceived pleasantness and perceived intensity of sucrose solutions by shifters ($N = 3$) under three conditions of internal state.

ters showed a monotonic decrease in perceived pleasantness with increasing stimulus concentration.

Unlike shifters, nonshifters did not show a rise and fall of perceived hedonic value with increasing stimulus concentration in the normal condition. Rather, nonshifters rated the sucrose solutions increasingly unpleasant with increasing concentration in all internal state conditions.

Figures 1 and 2 also show perceived intensity of sucrose solutions as a function of stimulus concentration for the shifters and the nonshifters. In all conditions, perceived intensity was a monotonically increasing function of stimulus concentration. In contrast to the hedonic ratings, perceived intensity was unaffected by internal state for all subjects. No subject showed consistent changes in ratings of intensity among the three conditions.

None of the 10 subjects tested showed a significant change in either hedonic or intensity ratings of Anethol among internal state conditions. In all conditions, perceived intensity was a monotonically increasing function

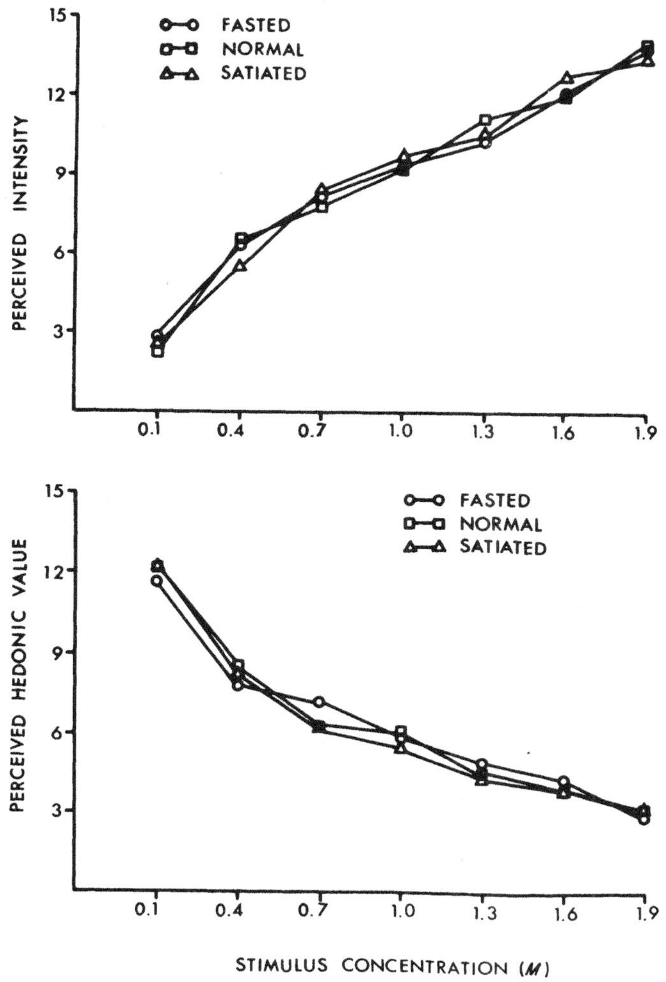

Fig. 2. Group mean category judgments of perceived pleasantness and perceived intensity of sucrose solutions by nonshifters ($N = 3$) under three conditions of internal state.

of stimulus concentration. Anethol was rated as neutral or pleasant at all stimulus concentrations, and there was a tendency for hedonic ratings to be slightly more positive with increasing stimulus concentrations in all conditions.

For the orange, maple, and chocolate odorants, subjects again could be classified as shifters or nonshifters. Most subjects were classified as shifters for some odorants and as nonshifters for others. Table II presents data from a single subject. This subject was a shifter for orange and maple and a nonshifter for chocolate.

Group data for shifters are presented in Fig. 3. For all odorants, shifters gave lower hedonic ratings in the satiated condition than in the fasted condition. All shifters for orange gave clearly different ratings between the fasted, normal, and satiated conditions. Maple odor was similar to sucrose taste in that hedonic ratings in the fasted and normal conditions did not differ. There were only two shifters for the chocolate odor. Both of these showed a clear difference between the fasted and normal condition; however, only one gave different ratings between the normal and satiated conditions. In the nonstressed condition, shifters tended to show a rise and

TABLE II

Mean Hedonic Ratings by a Single Subject of the Pleasantness of Various Concentrations of Orange, Maple, and Chocolate Olfactory Stimuli under Three Conditions of Internal State[a]

Odorant	Internal state condition	Stimulus concentration				
		1.2%	3.7%	11%	33%	100%
Orange	Fasted	9.3	10.5	11.0	11.0	11.5
	Normal	10.3	10.0	10.6	7.8	6.8
	Satiated	9.8	8.3	8.4	8.0	5.5
		0.39%	1.6%	6.3%	25%	100%
Maple	Fasted	8.1	9.5	9.5	10.0	11.3
	Normal	8.5	9.0	9.8	9.8	9.9
	Satiated	8.0	7.8	8.0	7.8	5.0
Chocolate	Fasted	8.0	8.4	8.6	7.0	7.0
	Normal	8.3	8.0	9.8	8.0	9.4
	Satiated	8.0	7.8	7.5	7.6	5.3

[a] This subject was a shifter for orange and maple odors and a nonshifter for chocolate.

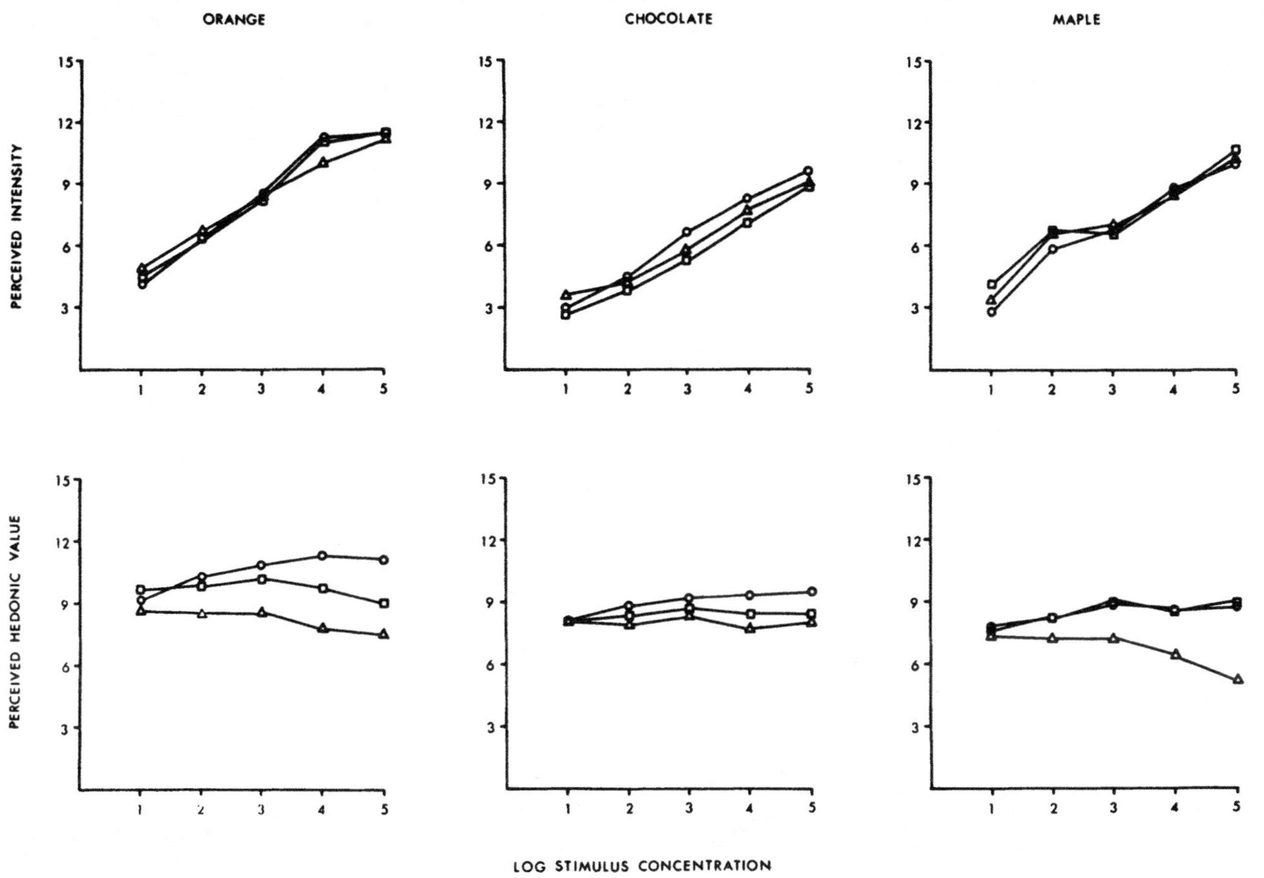

Fig. 3. Group mean category judgments of perceived pleasantness and perceived intensity by shifters for orange ($N = 5$), maple ($N = 4$), and chocolate ($N = 2$) olfactory stimuli under three conditions of internal state. These conditions were fasted, circles; satiated, triangles; and nonstressed, boxes.

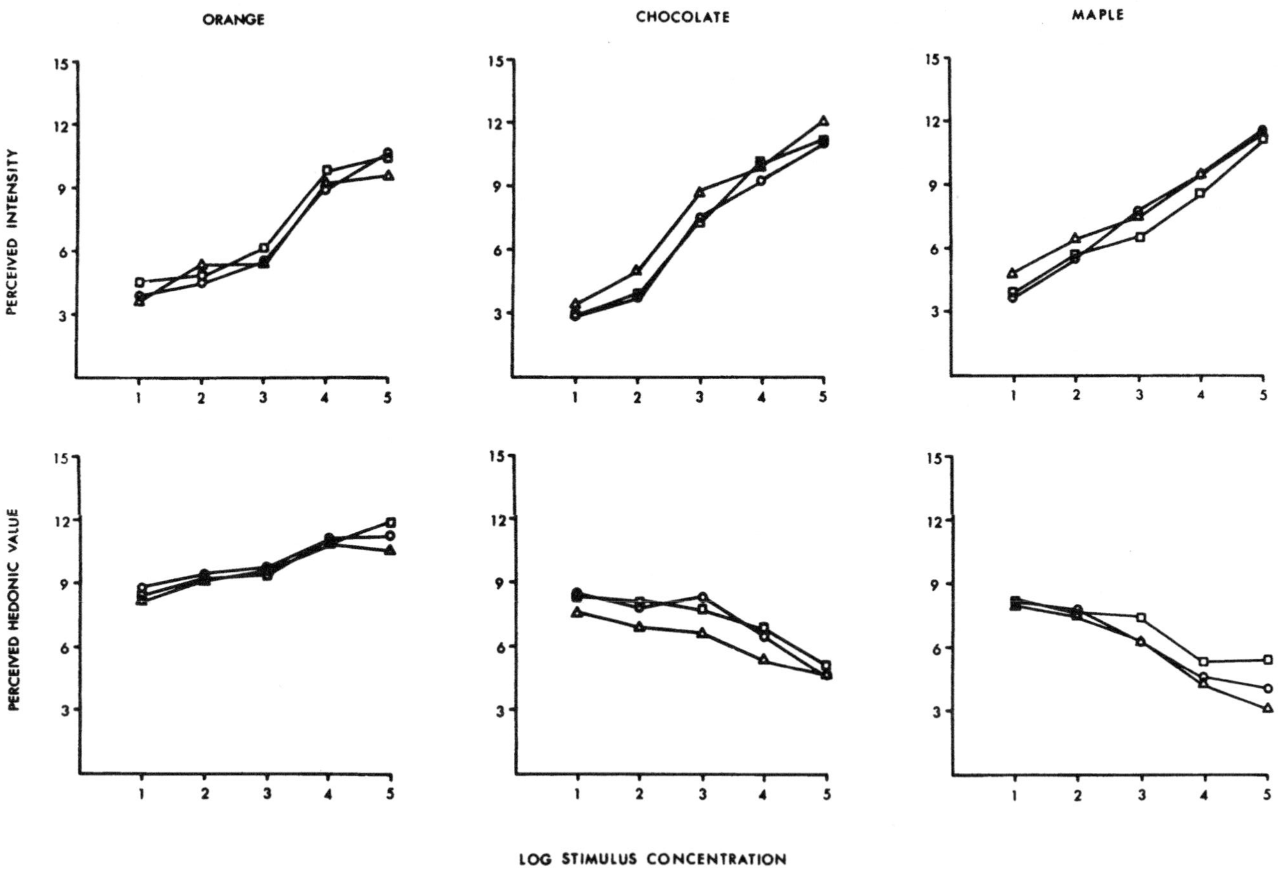

Fig. 4. Group mean category judgments of perceived pleasantness and perceived intensity by nonshifters for orange ($N = 2$), maple ($N = 3$), and chocolate ($N = 5$) olfactory stimuli under three conditions of internal state. These conditions were fasted, circles; satiated, triangles; and nonstressed, boxes.

TABLE III

Spearman Rank Correlation Coefficients Relating Hedonic Value and Perceived Intensity for Sucrose Taste and Orange, Maple, and Chocolate Odors for Shifters and Nonshifters[a]

	Shifters				Nonshifters			
Condition	Sucrose	Orange	Maple	Chocolate	Sucrose	Orange	Maple	Chocolate
Fasted	−0.04	0.90#	−0.30	1.00#	−1.00	0.87	−0.90	−1.00
Normal	−0.19	−0.43	0.22	0.50	−1.00	1.00	−1.00	−0.90
Satiated	−0.96#	−0.93#	−0.93#	−0.08	−1.00	1.00	−1.00	−1.00

[a] The critical value for significance ($p \leq 0.05$) for the orange, maple, and chocolate odors is 0.90 and for the sucrose stimuli is 0.71 (Siegel, 1956). The symbol # designates cases where the perceived hedonic value of a stimulus in a stressed condition differs from the nonstressed condition (see text for explanation).

then a fall in rated hedonic value with increasing stimulus concentration. When internal stress altered the perceived pleasantness of an odorant for shifters (fasted state for orange and chocolate, satiated state for orange and maple), there was a monotonic trend in rated hedonic value with increasing concentration. In the fasted state, this trend was for an increase in hedonic value with increasing concentration, and in the satiated state for

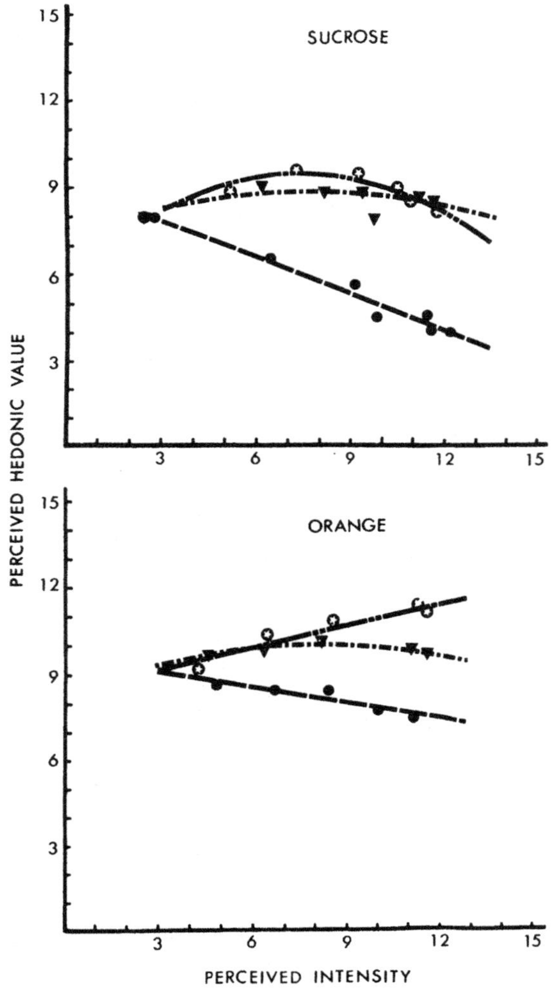

Fig. 5. Group data showing the relationship, for shifters, between perceived hedonic value and perceived intensity for sucrose taste ($N = 3$) and orange odor ($N = 5$) in all three internal state conditions. These conditions were fasted, stars; satiated, circles; and non-stressed, triangles. The least squares line of best fit appears where there was significant linear correlation (Spearman). In other cases, the best fit parabola appears.

a decrease in rated hedonic value with increasing stimulus concentration. In all conditions, perceived intensity of each odorant was a monotonically increasing function of stimulus concentration. Intensity ratings did not change in any consistent way with the changes in internal state.

Group data for nonshifters are presented in Fig. 4. As with shifters, perceived intensity of each odorant was unaffected by internal state manipulation. For maple and chocolate odorants, perceived pleasantness was a monotonically decreasing function of stimulus intensity in all three conditions. For orange odor, hedonic ratings increased monotonically with stimulus concentration in all three conditions. As with taste stimuli, nonshifters differed from shifters in that they did not show first ascending then descending ratings of perceived hedonic value with increasing concentration in the normal condition.

Table III shows the Spearman rank correlation coefficient for perceived intensity and pleasantness of sucrose taste and orange, maple, and chocolate odors for shifters and nonshifters. Nonshifters showed significant correlations in all three internal state conditions. Shifters only showed significant correlations in stressed conditions in which the perceived hedonic value was changed relative to the nonstressed condition.

Figure 5 shows the relationship between perceived intensity and hedonic value for shifters to sucrose taste and orange odor in all three internal state conditions. In the satiated condition, there is a negatively sloped linear relation between pleasantness and intensity for both types of stimuli. In the normal condition, this relationship takes the form of an inverted U. In the fasted condition, pleasantness of orange odor increased linearly with intensity, and pleasantness of sucrose was the same as in the normal condition.

DISCUSSION

All the results of these experiments showed that the perceived intensity of olfactory and gustatory stimuli was not affected by internal state. This result held in both sensory modalities for all subjects regardless of whether or not their hedonic judgments were affected. Intensity, therefore, appears to be largely independent of internal signals.

Perceived pleasantness of olfactory and gustatory stimuli, on the other hand, was markedly affected by manipulations of internal state in some cases. Shifters gave hedonic responses that were lower in the satiated than in the fasted condition. The perceived pleasantness of the stimuli, therefore, changed in a direction which would help restore homeostasis. This result agrees with that reported by Cabanac *et al.* (1968). Anethol was exceptional in that no subjects could be classified as shifters. Duclaux *et al.* (1973) have shown that nonfood-related odors have similar hedonic value in fasted and satiated conditions, and it is thus questionable whether Anethol was treated as a food-related odor.

Perceived pleasantness in stressed conditions (fasted and satiated) did not always differ from that in nonstressed conditions. Sugar taste and maple odor were perceived as equally pleasant in fasted and normal conditions by all shifters for these stimuli. Comparable results, suggesting no difference between deprived and nondeprived states, have been reported in experiments measuring rats' intake of sweet solutions (Ernits and Corbit, 1973).

Cabanac (1971) has argued that the perceived pleasantness of food-related olfactory and gustatory stimuli is not stimulus bound, but rather depends on the homeostatic usefulness of the stimulus as determined by internal signals. The present results suggest that this hypothesis should be modified. For all stimuli tested there were shifters, but there were also nonshifters for whom the perceived pleasantness of the stimulus was not affected by internal state. For olfactory stimuli, subjects usually shifted for some stimuli and not for others. The existence of nonshifters cannot be explained as an overall insensitivity to internal signals as has been suggested by Cabanac and Duclaux (1970), rather individual preferences must be considered. It appears that for a particular observer the hedonic value of some stimuli depends on internal state while for other stimuli it does not.

Whether or not a subject will show a shift in the hedonic value of a particular stimulus with changes in internal state appears to depend on the manner in which he perceives the pleasantness of that stimulus in the normal state. All nonshifters showed a strongly linear relationship between perceived hedonic value and intensity in the nonstressed condition. In the majority of these cases, the stimuli were judged increasingly unpleasant with increasing intensity. It appears that for nonshifters hedonic value is determined more by external sensory factors (e.g., intensity) than by internal factors. It is also possible that in these cases the subject did not perceive the stimulus as food related.

Shifters tended to judge the stimuli as neutral or pleasant in the normal state, and produced a nonlinear relationship between the ratings of pleasantness and intensity. This relationship showed a rise and fall in hedonic ratings with increased stimulus intensity, similar to that previously reported as typical for various tastes and smells in nonstressed conditions (Ekman and Åkesson, 1965; Moskowitz, 1971; Moskowitz *et al.*, 1974; Doty, 1975). But whenever internal stress affected the hedonic value of a stimulus, the relationship between perceived hedonic value and intensity became linear. In the case of food deprivation the slope of this function was positive and in the case of satiety it was negative.

These findings suggest that perceived intensity and perceived hedonic value may be related in two ways. One way is exemplified by shifters in the nonstressed state. In this condition the relationship takes the form of an inverted U. The second way in which perceived pleasantness and intensity are related is observed when perceived pleasantness is changed by internal stress. In this case, the degree of pleasantness or unpleasantness appears to reflect an intensity judgment. The direction of affect, positive or negative, is determined by the subject's homeostatic needs. A mechanism of this sort is clearly adaptive for the homeostatically stressed organism. Stimuli which are effective in restoring homeostasis are perceived as increasingly pleasant with increasing intensity. As a result, the most effective stimuli for restoring homeostasis are the most actively sought out and the stimuli most disruptive to homeostasis are the most actively avoided. Although the reason is not clear, nonshifters rate odors and tastes as shifters do in stressed conditions.

The existence of shifters and nonshifters indicates that in some cases hedonic value is stimulus bound and in others it varies with internal state. What are the factors determining the dependence of hedonic value on internal signals? Our speculation is that there are differences in the way that food-related odors are perceived in the normal condition that determine whether or not an individual observer will change his hedonic ratings with changes in internal state. When the sensory qualities of a food-related stimulus are strongly liked, or more usually, disliked, observers will give hedonic ratings that are intensity bound and that will not change with internal stress. When the sensory qualities of a food-related stimulus are not strongly liked or disliked in the normal state, observers will give hedonic ratings that are not linearly related to intensity. When these observers' ratings are changed by internal stress, hedonic value will become a linear function of intensity.

SUMMARY

Seventeen adult subjects used category scales to rate the pleasantness and intensity of food-related olfactory and gustatory stimuli under three conditions of internal state: fasted, normal, and satiated. The results showed that judgments of intensity were unaffected by manipulations of internal state while judgments of hedonic value were often affected by these manipulations. Most subjects showed changes in hedonic value depending on internal state for some stimuli but not for others. Whether the perceived pleasantness of a particular stimulus was changed for an individual appeared to depend on the perceived pleasantness of that stimulus in the normal condition. When changes in hedonic value did occur, they were always in a direction which would restore homeostasis. These changes were reflected in a shift from a nonlinear to a linear relationship between perceived intensity and perceived hedonic value.

REFERENCES

Cabanac, M. (1971). Physiological role of pleasure. *Science* **173,** 1103–1107.

Cabanac, M., and Duclaux, R. (1970). Specificity of internal signals in producing satiety for taste stimuli. *Nature (London)* **227,** 966–967.

Cabanac, M., Minaire, Y., and Adair, E. (1968). Influence of internal factors on the pleasantness of a gustative sweet sensation. *Commun. Behav. Biol, Part A* **1,** 77–82.

Doty, R. L. (1975). An examination of relationships between the pleasantness, intensity, and concentration of 10 odorous stimuli. *Percept. Psychophys.* **17,** 492–496.

Duclaux, R., Feisthauer, J., and Cabanac, M. (1973). Effets du repas sur l'agrement d-odeurs perception d'un stimulus olfactif alimentaire. *C.R. Hebd. Seances Acad. Sci., Ser. D* **270,** 1006–1009.

Duclaux, R., Feisthauer, J., and Cabanac, M. (1973). Effets du repas sur l'agrement d'odeurs alimentaires et nonalimentaires chez l'homme. *Physiol. Behav.* **10,** 1029–1033.

Ekman, G., and Åkesson, C. (1965). Saltness, sweetness, and preference. *Scand. J. Psychol.* **6,** 241–253.

Engen, T. (1965). Psychophysical analysis of the odor intensity of homologous alcohols. *J. Exp. Psychol.* **70,** 611–616.

Engen, T. (1971). Olfactory psychophysics. *Hand. Sens. Physiol.* **4,** Part 1, 216–244.

Ernits, T., and Corbit, J. D. (1973). Taste as a dipsogenic stimulus. *J. Comp. Physiol. Psychol.* **83,** 27–31.

Henion, K. E. (1971). Odor pleasantness and intensity: A single dimension? *J. Exp. Psychol.* **90,** 275–279.

Moskowitz, H. (1971). The sweetness and pleasantness of sugars. *Am. J. Psychol.* **84,** 387–406.

Moskowitz, H., Kluter, R., Westerling, J., and Jacobs, H. (1974). Sugar sweetness and pleasantness: Evidence for different psychophysical laws. *Science* **184,** 583–585.

Mower, G. (1976). Perceived intensity of peripheral thermal stimuli is independent of internal body temperature. *J. Comp. Physiol. Psychol.* **90,** 1152–1155.

Pfaffmann, C. (1960). The pleasures of sensation. *Psychol. Rev.* **67,** 253–68.

Siegel, S. (1956). "Nonparametric Statistics for the Behavioral Sciences." McGraw-Hill, New York.

Stevens, S. S. (1960). On the new psychophysics. *Scand. J. Psychol.* **1,** 27–35.

Young, P. T. (1959). The role of affective processes in learning and motivation. *Psychol. Rev.* **66,** 104–125.

DISCUSSION

Nicolaidis: You are talking about a hypothesis. There is no hypothesis. Cabanac expressed scientifically what everybody from the beginning of time has known and anyone could tell you: that when one is "stuffed" he no longer likes a nutrient. This mechanism is the one which is responsible for homeostatic achievements, there is no doubt. If you found some people switching from pleasantness to unpleasantness, according to their instinct, and some that did not, something must be wrong.

Engen: We did three experiments to try various things to make sure that we did not do something wrong. We have a total of 17 subjects and there is a possibility that there could be something strange about the subjects but we are not inclined to that explanation, and that is why we decided to present the data here.

Jordan: I am intrigued to hear this presentation because some three or four years ago we looked at the Cabanac phenomenon, too. So did Grinker at Rockefeller. We found that all the subjects did not show it either; to further confound the issue, we discovered that if you repeat these tests over and over again on the same subject, you find that some days the subject shifts and some days the subject does not shift. Some days you get a very profound effect and other days not.

Engen: Another aspect of that which I think might be quite interesting is that shifting versus nonshifting depends upon a person's perception of a stimulus, such as the orange flavor, in the normal condition. Those who shifted may have perceived the orange odor as a food-related odor, and those who did not may have thought of it as perfume. We have no evidence for this, but perhaps this applied to your subjects also. Recognition and identification of odors are very poor, and it is possible for a person to think that orange is food one day and perfume the next.

Jordan: We were just doing it with taste.

Miller: How many of these subjects were repeated, what were the intervals between repeats, and do you have any information about them other than the fact that they were available?

Engen: In the first experiment the person was tested on three separate days. He was fasted for one day, satiated for another, normal for a third. In the second experiment, we did them all on the same day. That means that we could not counterbalance the order of presentation; the subjects were always tested first in the fasting condition. They came in 22 hours hungry and did the taste and odor judgment, and then they had the regular meal and an hour and a half later the same day they were tested under the so-called normal condition. After that, they drank the glucose solution and then waited for 20 min to have that take an effect, and then we tested them in the satiated condition. This difference in the experiments did not make any difference as far as we could observe.

Lepkovsky: I followed Cabanac's research and I think you left out what I consider some of his most interesting work. Where you reduced the subject to 80% of his normal weight and compared him with those that are obese, Cabanac found that after satiation with glucose in the stomach, it then became unpleasant. However, when they reduce the weight to 80% of the normals, they are then given intragastric glucose load and it would taste pleasant. In other words, the weight had a lot to do with it. Then he took those that are obese and he found that the obese found the taste of glucose pleasant even after an intragastric load; thereby, he showed that the loss of weight, underweight, and obesity have something in common, especially with respect to taste.

Engen: Our subjects were all normal weight. We were trying to avoid that problem. We were aware of it but did not have enough time to investigate it, but it certainly is very important.

Engen: Various trends were not elaborated upon for statistical reasons. The effects I discussed are those we have statistical support for, but I would be quite willing to speculate about others.

Senseman: Well, that was my question too, except why is maple odor hedonically aversive in higher concentrations for the normal person?

Engen: These were imitation odors, of course, and they became aversive with increased concentrations. I do not find that surprising.

Senseman: Well, you said if they are satiated they do not become aversive. That is what your data says, is it not?

Engen: No. When we satiated the shifters, then their judgment suggested that it became less and less pleasant the stronger the concentration. The nonshifters were not affected by either the satiation or by fasting.

Hegsted: There is an assumption here that because they prefer it, they take more given the opportunity. I am reminded of the survey that was in the Sunday paper on the characteristics that people want in a presidential candidate, and they listed experience in Washington, experience in international affairs—they listed half a dozen things that obviously are not true.

Garcia: Even if you take psychology students and you ask them about such things as visual aftereffects, about half the people do not see them for a long time until you train them to report on that phenomenon. For example, you can see cosmic particles and for a long time the astronauts did not report this. The cosmonauts tended to report perceptual things. Then all of a sudden the American astronauts started to report them, too. It seems it was almost permissible.

Engen: I certainly see that as a possibility in our experiments. However, we were of course careful to do what we could to avoid that and we find no reason to explain the difference between the shifters and the nonshifters on that basis, but for all we know it is possible. We did not get inside their heads!

Pfaffmann: What about the correlation of shifters and nonshifters on taste and odor. Did the same people shift on the odor and the taste?

Engen: They are dissociated as far as we can tell. That is, those who shifted did not shift for the same odors. Some shift for some and not for others. Individual differences were definitely observed. It would be interesting to know the subject's perception of an odor, what he thinks of it, what he judges it to be, together with information about his experience with that odor and what it is associated with, food or nonfood and so on.

Pfaffmann: I think that latter point is very important and I wonder whether the difference in the pleasantness is not only dependent on nutritional status. Have they had any candies or sugar stimulation during a period or have they not had any, resulting in sensory deprivation?

Jordan: I think the other thing that seems apparent to us is the degree of sophistication of the subject in making a hedonic judgment. Some people just do not do it very well. When we did it with ourselves as subjects, we got pretty good at making these kinds of judgments. You bring a naive subject in and it is different. Going back to Cabanac's original work, a lot of work that was done used the subjects who were members of that laboratory. As far as I know, many of these people were not naive subjects.

Cagan: There is a striking difference in your data and I wonder about its significance. The group that you call the shifters, when hedonic rating is plotted against concentration, they were quite constant, both in taste and in smell, whereas the nonshifters were already dropping in hedonic rating as you increased concentration. Therefore, there was no further change due to satiety. It was marked in the taste data and, although not as striking for the olfactory data, there was a similar trend. From these data I wonder if you should call them ones who care and ones who do not care, because they were very constant in hedonic rating whatever the concentration.

Doty: You are probably quite aware of the Rosenthal and Ohne effects; that is, that demand characteristics of the experimental situation. A person comes into a situation with a hypothesis as to what the purpose of the experiment is and tends to respond to the situation in terms of his own perceptions. How important do you think such factors were in affecting your results? There are some ways of finding this out by questioning the subject afterward. Did you do that?

Engen: There is no way that this would influence these results because we had too many stimuli for the subjects to keep track of, and their judgments of them were separated in such a way that they could not manage that. However, I certainly appreciate that point. It is an important one just like Garcia's and must be considered, but I am convinced that the difference we observed has to do with a person's experience with these tastes which have somehow become meaningful in different ways to different people.

Chapter 6

Sensory–Neuroendocrine Reflexes and Their Anticipatory and Optimizing Role on Metabolism

STYLIANOS NICOLAIDIS

It is generally admitted that nutrients enhance appropriate metabolic responses by means of their mass action at the level of the inner milieu. The classic example is that of glucose which can elicit, after iv injection, all endocrine responses insuring its metabolism. More recently, the validity of this view has been questioned; the endocrine responses elicited by an iv load of glucose and its subsequent metabolism will not be as appropriate as those encountered when the same load of glucose is absorbed orally.

The physiology of the sensory equipment surrounding the apparatus of ingestion must be better understood today, since accumulating data extend its role further than its obvious functions of detection of edible foods and their mechano-enzymatic processing which contribute to making the greatest portion of food absorbable. The role of the sensory cephalic input is extended beyond the intestinal barrier and acts at the postabsorptive stage as well. This theory extends to the cephalic level the well-known intestinal postabsorptive together with the preabsorptive metabolic action on ingestants. Though from the intestinal level, a great part of the messages seems to be endocrine (such as gastrin or cholecystokinin); at the oral level the information seems to consist mainly of nervous discharges to which the CNS responds with neuroendocrine secretions. All these cephalic sensory–vegetative reflexes bear some common characteristics, such as rapidity of action and functional anticipation in the sense that these reflexes mimic at the preabsorptive stage the postabsorptive effect of the stimulating substances. This anticipatory effect is accompanied by an optimizing action on the further utilization of the absorbed substances. For example, orally ingested glucose will elicit a smaller (less diabetoid) and shorter hyperglycemia than that of the same amount of glucose intubated in the stomach or given intravenously. The metabolic "comfort" resulting from such preparatory actions of the sensory input is obvious. Orocephalic input seems to modulate the tonus of the responsive organs in such a way that the absorbed substrates elicit responses in these organs (directly or through brain receptors) which correspond to the optimum metabolism of the absorbed substrates. This is the reason permanently infused solutions with all necessary ready-to-use molecules are not as metabolically effective as their raw, nonprocessed but orally absorbed precursors (see later part of this paper).

The role of the anticipatory reflexes could be important in another function: that of eliciting the behavioral anticipatory response of early satiety. The preabsorptive metabolic changes could be the necessary link between the sensory cues and the postabsorptive cues which, at the level of inner receptors, generate the alternation of hunger and satiety. Available electrophysiological data favor the idea of convergent action of oral input upon central nervous system formations usually activated by metabolic changes of the inner milieu (Nicolaidis, 1969a).

In this review we will study sensory–neuroendocrine reflexes which bring about metabolic changes in relation with the regulation of energy homeostasis. We will not describe similar reflexes which result in changes of hydromineral metabolism. They have been briefly reviewed elsewhere (Nicolaidis, 1975).

OROGLYCEMIC RESPONSES

As early as 1927, Ciaccio and Racchinsa observed the action of saccharin on glycemia. The distinction between the oral and the systemic action of glucose loads has been known since 1954 (Scow and Cornfield, 1954). Several investigators using more or less aggressive techniques of sampling tried to understand the early responses to glucose given orally or intragastrically. Hyperglycemic as well as hypoglycemic responses have been described, depending on the precocity of sampling, on species, on experimental conditions, etc. (see Nicolaidis, 1969b).

The earliest glycemic responses in anaesthetized rats that had their oral receptors stimulated by sweet (sucrose or saccharin) solutions were first investigated in 1966 using a continuous aspiration through an iv canula and continuous essay by means of an "on line" Autoanalyzer (Technicon) (Nicolaidis *et al.*, 1966; Nicolaidis, 1969b). The main early response is hyperglycemic and can start as soon as 1 min after the beginning of the oral stimulation (Fig. 1). This response is present in animals deprived of food for more than 5 hr and increases with the degree of deprivation. The duration of the hyperglycemic response is more than 10 min and cannot be accounted for by postabsorptive effects, not only because of the esophageal ligature, but because of the similar effect of saccharine stimulation. The hyperglycemic response does not occur when basal glycemia is high. Control water stimulation ruled out the nonspecific (stress for example) origin of the phenomenon. Frequently, particularly when animals are not deprived, a hypoglycemic response of several minutes (1 to 8) precedes the hyperglycemic one. This early hypoglycemic response has been recently emphasized and considered as the only preabsorptive response in nonanaesthetized rats (Louis-Sylvestre, 1976). However, this finding has not been confirmed by Steffens (1970, 1976), who studied early glycemic responses of chronically catheterized, freely moving and eating rats in a scheduled meal situation as well as during meals eaten *ad libitum*. It is noteworthy that water stimulation of the tongue in the rat (Nicolaidis, 1969b) or water drinking in man (Vinik *et al.*, 1973) often results in a hypoglycemic response which is abolished by vagotomy.

In fact, early glycemic responses are the expression of the effect of

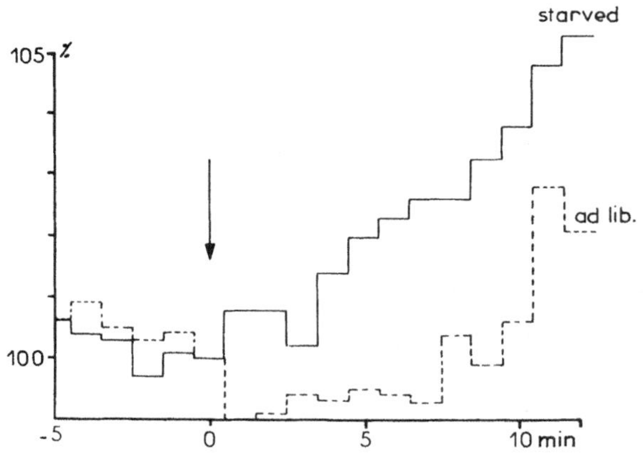

Fig. 1. Effect of oral stimulation with 30% sucrose solution on the blood glucose level of rats fed *ad libitum* (dashed line) or starved from 5 to 24 hr (solid line). Blood glucose level recorded continuously. Results expressed as mean percent of blood glucose level at zero time ($N = 11$).

preabsorptive, reflex-elicited release of both glucagon and insulin almost immediately after the beginning of a meal. The relative importance of the preabsorptive release of these hormones, together with other possible glucogenetic and glucolytic reactions in different metabolic states, is not well known. It seems that during food deprivation the anticipatory processes of glucogenesis prevail (early hyperglycemia), while during *ad libitum* feeding, processes of glucose storage are rather predominant (early hypoglycemia).

In most studies available in the literature, the effect of nutrients at the cephalic level has not been separated from the effect at the level of lower located receptors along the digestive tract. Even use of saccharine solutions as a stimulus restricted to the gustatory receptors is probably not the proper way of excluding gastrointestinal afferences, since we know little about perception of "sweet" and sweetlike substances at this level. Since Pavlov, sham feeding has been extensively used for studies of exocrine pancreatic secretion (for review, see Preshaw *et al.,* 1966). In the dog, oral stimulation by nutrients elicits at least gastric and pancreatic exocrine secretions. In the next section, we will deal with data suggesting that oral sensing also leads to endocrine secretion of the pancreatic islets.

EARLY INSULIN AND GLUCAGON RELEASE DURING INGESTION

Whether innate or conditioned, the early glycemic responses elicited by sensory stimulation of foods suggested oropancreatic reflexes similar to those of exocrine gastropancreatic secretions. Direct evidence for insulin and, more recently, glucagon release as soon as a meal starts has been awaiting the development of radioimmunoassay techniques.

To date most of the data concern insulin release. It is now well established that the amount of insulin released in response to oral administration of glucose is higher than in response to the same stimulus given intragastrically. Also, insulin release from intragastric administration of glucose is higher than from the same amount administered iv. These observations were made in humans (Balsano *et al.,* 1964; Elrick *et al.,* 1964; MacIntyre *et al.,* 1965; Perley and Kipnis, 1967) as well as in rats (Assal *et al.,* 1970; Steffens, 1970) and rabbits (Nijjar and Rerry, 1970).

When glucose is absorbed orally and follows its normal absorptive process, insulin release displays a bimodal profile. An almost immediate phase of discharge is followed, some 10 min later, by a second one, which is more sustained, lasting until the end of the postabsorptive hyperglycemia. Only the early phase of insulin release appears during a sham feeding of nonmetabolizable edulcorated bulk (Strubbe and Steffens, 1975; Hommel *et al.*, 1972). On the other hand, when nutrients are administered intragastrically or intravenously, only the second phase of insulin secretion appears. Such a dissociation was also described in human subjects who drank lemon juice sweetened with saccharin and showed only the early insulin response. In animals bearing esophageal fistula, insulin secretion occurs in spite of the absence of any food reaching the stomach (Basset, 1974; Hommel *et al.*, 1972).

Again, as in the case of early glycemic responses (Nicolaidis, 1969b), the oropancreatic reflexes are dependent on the depletion–repletion stage. In depleted rats, early insulin release is strongly attenuated (Strubbe and Steffens, 1975), resulting in a predominantly hyperglycemic early response.

In the oropancreatic hypoglycemic reflex, it is difficult to know what the sensory stimulation is that generates the insulin response. The gustatory input is clearly important since lingual application of saccharin is sufficient to elicit insulin release. However, the gustatory input is not necessary since ingestion of unflavored bulk can also initiate the early insulin release. Indeed, a number of sensory stimulations have been shown to be able to evoke insulin discharges (Parra-Covarrubias *et al.*, 1971). An important confounding variable here is the apparent facility of the insulin response to be classically conditioned to various stimuli (Woods *et al.*, 1970; Woods and Kûlkosky, 1976; Reynier-Rebuffel *et al.*, 1974). The presence of so easily established conditioned responses makes studies of sensory input difficult. Therefore, triggering insulin release by means of nonsweetened sham feeding does not imply that the gustatory sweet-detecting buds are not the receptors triggering the unconditioned reflex, since, previously in the animal's life, bulk always accompanied carbohydrate nutrients. The same situation holds true for the gustatory afferences which can trigger responses of β cells by means of a conditioned process reinforced by the insulinogenic action of the stimulating nutrients at the postabsorptive stage. However, if conditioning were the only mechanism responsible for the early systemic responses, the temporal sequence of events would be better followed; i.e., the conditioned sensory stimulus would be effective by the time or just before the unconditioned stimulus was effective. If eating a carbohydrate meal resulted in a postabsorptive systemic action on insulin release 10 min after the beginning of eating, the suppression of the systemic action, by means of a gastric fistula for example, would permit the insulinogenic effect of the meal to persist, but it would start after the same delay of 9 or 10 min displayed by the unconditioned process. In addition, the early metabolic responses persist as if they were unconditioned.

There is little known of the pathways and the central formations involved in the sensory pancreatic reflexes. The mesencephalic relay was studied by Brugsch *et al.* (1921), but he did not establish whether the structures localized were included in the afferent or efferent pathway of the reflex. The afferent pathways seem to project into the hypothalamic glucoregulatory areas and, particularly, into the lateral insulin secretory area (Kuzuya, 1962; Steffens *et al.*, 1972). The relationship of peripheral input to the tonic insulin inhibitory ventromedial hypothalamic area is not known, but there is at least a reciprocal activity between the lateral and medial glucoregulatory areas (Anand *et al.*, 1964; Oomura *et al.*, 1969).

There is more information in the literature on the descending efferences. Not only is it well known that an autonomic control of the endocrine pancreas exists, but also that there is an activation of the pancreatic parasympathetic innervation during a meal, as shown by experiments in which animals bearing esophageal fistulas sham eat. In these experiments the exclusive cephalic stimulation by nutrients elicits an insulin release (Basset, 1974; Hommel *et al.*, 1972). Section of the vagus suppresses the early insulin releasing response as does atropinization. However, action of the sympathetic nerves was also suggested in the early sensory–endocrine pancreatic response. Such a double innervation, reciprocal or synergic, is not known, but there is at least a reciprocal activity between the lateral and pancreatic islets the control might not only be double but probably multiple. Superimposed upon the substrate and nervous autonomic controllers, there are hormonal modifiers of islets' secretion. This indirect releasing action of sensory stimulation by nutrients is particularly suggestive in experiments with medullectomized animals which show diminished early insulin secretion. The hormonal elicitation of the early pancreatic endocrine (and exocrine) secretion is quite extensive. Growth hormone, cortisol, thyroxine, estrogens, and androgens seem to regulate the tone of α and β cells' responsiveness. However, these hormones do not seem to be important in the process of release of pancreatic hormones. They act rather as tonic amplifiers (positive or negative) of the secreting process, responding primarily to substrate controllers. An additional group of controllers acts upon the release process. This group includes hormones secreted from the gastrointestinal tract during ingestion. The already known factors as gastrin, secretin, cholecystokinin, pancreozymin, and enteroglucagon, and other factors such as the gastrointestinal inhibitory polypeptide (GIP). Secretion of these factors in response to orocephalic nutritive-bound stimulation has not been demonstrated, but since Pavlov's experiments there has been ample evidence for the existence of chain reactions such as sensory–gastric secretions, followed by duodenal and intestinal entrainment of endocrine and exocrine production, leading finally to hormonal release from the islets' cells.

We know that β cells possess two insulin compartments. Only the first one is depleted rapidly in response to sudden demand, such as to the triggering of neurohormonal controllers by sensory stimulation. The second compartment releases insulin progressively and in a prolonged manner in response to substrate and to long-acting endocrine factors.

The more recent assays of glucagon have also demonstrated the existence of an explosive release of this hormone as soon as a meal starts (Porte *et al.*, 1976; Steffens, 1976). Steffens has shown that this early glucagon release is due, at least partly, to oral stimulation, followed by gastrointestinal and systemic ones. Vagotomy does not abolish this early response; neither does atropine, suggesting that the essential nervous control of α cells of the islets of Langerhans is sympathetic (though not exclusively, since vagal stimulation is able to bring about glucagon secretion).

The early release of these two metabolic hormones makes the organism more able to face the "stress" of sudden nutritional influx. The anticipatory preparation of the inner processors of nutrients allows a smoother assimilation and storage of the ingestants and probably a better adaptation to the relative composition of the battery of metabolic hormones, and particularly to the insulin to glucagon ratio. In addition, a better adapted and more appropriate preabsorptive satiety might be the result of preabsorptive metabolic hormones known to act on hunger and satiety. In this respect, an alternative or complementary mechanism of preabsorptive hyperglycemia and satiety could be the secretion of adrenaline during food ingestion from intrahepatic chromaffin cells, recently demonstrated in the portal spaces of liver parenchyma (Martinez *et al.*, 1974; Racotta *et al.*, 1972).

PREABSORPTIVE METABOLIC CHANGES

Since both substrates and cofactors of their utilization are reflex-released in the inner milieu, one can expect preabsorptive metabolic consequences at the level of the body as a whole. Such reflex-elicited, rapid metabolic changes have been described (Nicolaidis, 1969b). They concern both the absolute level of metabolism (i.e., oxygen consumption) and the qualitative aspect of it (i.e., switch of fat metabolism of fasting to carbohydrate consumption).

It is well known that fasting brings about breakdown of fat stores resulting in a respiratory quotient [RQ = CO_2/O_2] close to 0.7. Feeding provides the inner milieu with new carbohydrates, switching the catabolism from lipid utilization to that of glucosides, the RQ of which is equal to 1. Another metabolic consequence of fasting is a hypometabolic state expressed by low 0_2 consumption. As feeding restores the metabolizable substrates in abundance, it also ends the hypometabolic state and switches the metabolic rate to a higher level.

Classically, both RQ and metabolic rate changes occur only as a consequence of postabsorptive invasion of the inner milieu by ingestants. In fact, these changes appear very early and seem to be reflex triggered as a consequence of external stimulation of afferences of the orodigestive tract.

For this study an open circuit, metabolic device was used. Rats were placed in a small air-conditioned chamber which limited movement and the expired air was continuously analyzed by O_2 and CO_2 analyzers. The main findings were as follows.

Early Changes of Metabolic Rate

From the very start of eating, the hypometabolic state of fasting was replaced by a remarkable increase in O_2 consumption, only partially due to the muscular activity associated with ingestion. Since it is difficult to quantify the fraction of energy spent by ingestive muscular activity, the reverse experiment was done. That is, the nonmuscular participation in the reflex-elicited switch of metabolic rate was investigated using paralyzed rats (Nicolaidis, 1969b). In 24-hr fasting, curarized rats, the metabolic rate increased in less than a minute after oral stimulation with sucrose and remained high for a period of more than 10 min, while control water stimulation was ineffective (Fig. 2).

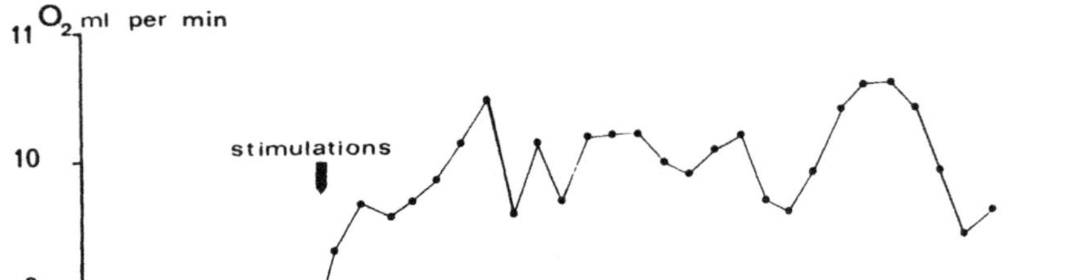

Fig. 2. Effect on O_2 consumption of starved, curarized, and artificially ventilated rats of a 50% sucrose solution applied on the tongue, the esophagus being ligated.

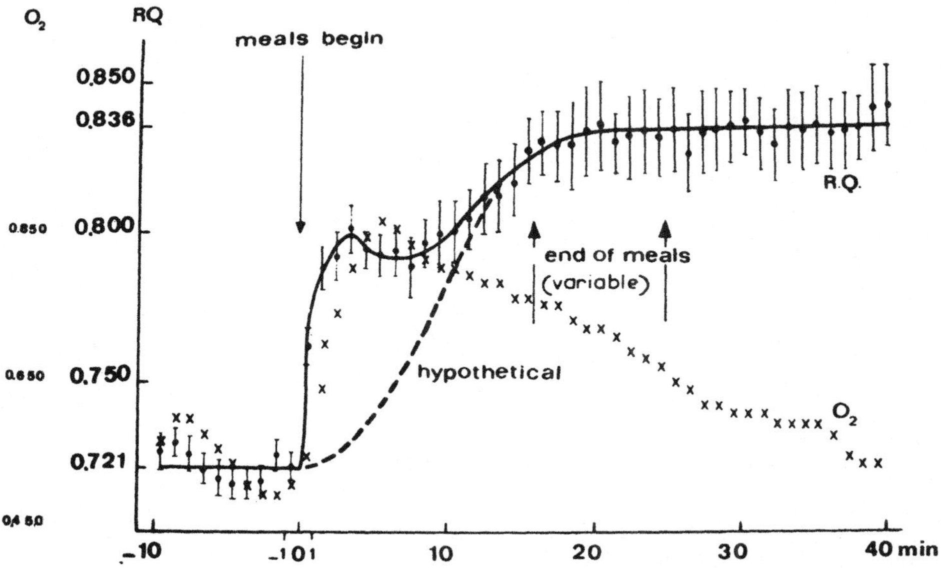

Fig. 3. Effect of meal ingestion on O_2 consumption (ml/min) and RQ (N = 17). Values are integrated over 1 min. The hypothetical dashed line interpolated toward the origin eliminates a possible epiphenomenon.

Early Switch from Lipolysis to Glycolysis

The increase of RQ was also dramatic and early in response to oral feeding. This increase again is partially nonspecific, due to muscular movement and hyperventilation (Fig. 3). As for the O_2 consumption, it is difficult to evaluate the proportion of RQ rise in response to muscular movements versus that of the specific phenomenon. The existence of a specific early rise in RQ is supported by several arguments, i.e. (a) RQ increases as a function of previous depletion (fasting), though rate of ingestion, and therefore muscular movements, do not; (b) replacement of the carbohydrate diet by a carbohydrate-free fat diet results in a comparable change in O_2 consumption, but in a substantially smaller increase in RQ. This minimal RQ rise is presumably associated only with the nonspecific ingestive muscular activity (Fig. 4). The absence of an early change in RQ in a carbohydrate-free diet is not due to the absence of sweet taste, since edulcoration of the fat diet by saccharin did not change the phenomenon. Neither is it a learned response, since it appeared in rats that had never tasted any diet other than their carbohydrate-rich powder. Furthermore, repeated ingestion of the same fat diet did not modify the shape and the magnitude of RQ changes. This suggests the existence of a specific gusta-

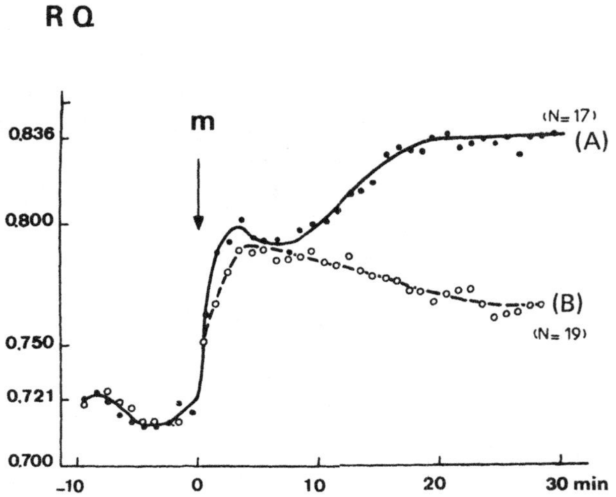

Fig. 4. Effect of ingestion of 100% fat (B) versus standard chow (A) meals on RQ of 24-hr starved rats.

tory or trigeminal (or both) sensory mechanism which initiates the absorption of "lipid" substrates and therefore prevents any specific rise of RQ (other than the nonspecific one) during the early stage. So the early metabolic events mimic the postabsorptive ones (low RQ) when exclusive lipid meals are consumed.

All these early responses show that cephalic (and presumably gastrointestinal) afferences are capable of triggering metabolic changes. These responses are anticipatory in the sense that they precede and are in the same direction as the metabolic changes which will be produced by the stimulating nutrients (either carbohydrates or lipids) after their intestinal absorption.

METABOLIC CONSEQUENCES OF THE SUPPRESSION OF OROGASTROINTESTINAL INFORMATION

The optimizing role of sensory endocrine reflexes fully appears when nutrients short-circuit entirely the afferences of the cephalic area. In an extensive study of the metabolic consequences of long-term nutritional iv infusions on rats, it was shown that utilization of perfusates was inadequate (Nicolaidis, 1971; Nicolaidis and Meile, 1972; Rowland *et al.*, 1973; Nicolaidis and Rowland, 1976). Rats were permanently attached to an infusing device which did not restrain their movements nor interfere with normal feeding and growth when saline was continuously infused. When these rats received nutritional infusions (from the simplest unbalanced glucose solutions to the most balanced ones) covering all or part of their energy needs, they did not show an inhibition of oral consumption equal to the iv administration. The reduction of oral feeding is of the order of 0.6 cal per infused calorie. So infused nutrients inhibit feeding only to the extent that they are utilized (i.e., their utilization was 60%). Similar observations of only partial reduction of oral feeding in response to shorter infusions of glucose solutions were made by Adair *et al.* (1968). All cases of greater inhibition of oral feeding associated with iv continuous nutritional infusions are due to a nonspecific suppressant effect of infusates, as is accounted for by loss of body weight (see also Nicolaidis and Rowland, 1976).

The insufficiency of infusates in inhibiting oral feeding is due to their incomplete utilization. This incomplete utilization is the consequence of inadequate release of insulin when sensory afferences of the orodigestive tract are eliminated. Such a mechanism could be suspected from data previously discussed in this review and is strongly supported by a number of experiments we shall mention briefly.

(a) It is known that time varying (high derivative) infusions are more efficient stimuli for insulin secretion than are their continuous (zero derivative) counterparts (Curry, 1971). Discontinuous infusions were shown to be more efficient as suppressive factors of oral feeding (Nicolaidis and Rowland, 1976). This suppressive effect of square wave infusions increases even more when given in the portal vein instead of the right atrium (M. Mather and S. Nicolaidis, unpublished).

(b) When the missing fraction of insulin is artificially provided by means of a parallel iv infusion, oral feeding is inhibited in a strictly proportional fashion, i.e., the reduction ratio of feeding approaches 1 cal per infused calorie (Rowland *et al.,* 1973).

(c) The same reduction ratio of almost 1 cal of food to one infused calorie is observed when the supplement of insulin is no longer administered but elicited endogenously. To demonstrate this, Rowland *et al.* (1975) used rats with lesions of the ventromedial hypothalamuc. After a control period of iv infusions which brought about a reduction of oral feeding equivalent to 60% of their caloric value, lesions of the ventromedial hypothalamic area were placed through implanted electrodes. On subsequent days, rats became hyperphagic, and iv infusions resulted in the theoretically expected ratio of reduction of oral feeding of 1 cal per infused calorie.

(d) Improvement of utilization of infusates by sensory stimulations. Another result of misutilization of continuous infusates in rats is the loss of body weight when, in the absence of oral food, the number of iv-administered calories is equal to what the same animals would ingest orally in order to maintain their normal body weight (BW). Previously described data indicate that one factor of misutilization of infusates is the inadequate secretion of insulin due to the short-circuit of sensory afferences. Therefore, restoration of sensory stimulations evoking nutritional ingestion should improve the utilization of iv infusates and their use for better maintenance of BW.

The following description of unpublished data of S. Nicolaidis, M. J. Meile, and N. Rowland demonstrates the beneficial effect of cephalic, nutrient-evoking stimulations upon the anabolic utilization of long-term iv infusions. Rats were implanted with indwelling jugular and atrial catheters attached to the skull (Nicolaidis *et al.*, 1974). Their caloric needs were determined under control, saline, permanent infusions. When, in the absence of oral food, an equivalent amount of nutrients was given iv (a commercially available solution, Trophisan-Egic), BW decreased (Fig. 5). After 2 days of such a negative balance of BW the same infusions were continued for another 2 days but, now, gustatory stimulations were added. In order to avoid providing the rat with postabsorptive nutrients, a saccharine solution was given orally by means of a revolving disk introducing into the cage of the animal small containers filled with the stimulating

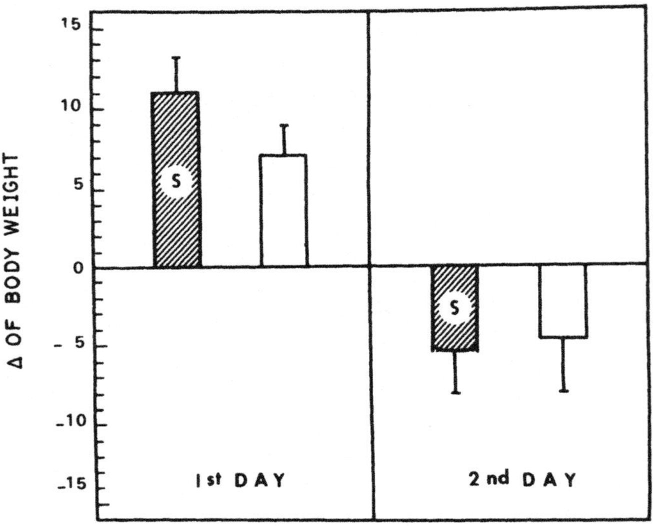

Fig. 5. Body weight balance (mg) of rats receiving exclusive iv feeding associated, during 2 consecutive days, with availability of saccharine solution (S) to drink or not at will ($N = 6$).

solution. During the first day, associating oral sweet solutions with iv infusions, BW increased in a statistically significant way (Fig. 5). However, this effect on BW balance seems temporary since, on the subsequent day, BW decreased again in spite of delivery of the same amount of oral saccharine solution. Finally, a new control, 2-day period followed with continuation of the same amount of infusates but, as during the first 2 days, without oral sweet stimulation. During this last period of exclusive parenteral alimentation, BW balance was negative again as during the first 2-day period of exclusive parenteral alimentation. During these experiments, water was always available *ad libitum*.

The beneficial effect on BW balance of oral stimulations associated with iv infusions can be obtained when a minute amount of powdered chow replaces the saccharine solution in the cups of the revolving disk. This paradigm provides the rats with a complex multifactorial stimulus of olfactory, gustatory, and gastrointestinal afferences.

Olfactory stimulations also seem to have a beneficial effect on BW balance under parenteral alimentation. Pilot experiments using air puffs directed in the cage of the animal through a container of powder chow show an improvement of BW balance, though this effect seems to be tenuous and has to be confirmed.

The nature and the number of sensory–vegetative reactions involved in these experiments of orodigestive stimulations are not known. It is possible that in addition to neuroendocrine responses, cephalic and gastrointestinal stimulations bring about strictly neurovegetative responses, adding their own metabolic effect to that of reflex-released hormones. Such purely sensory–vegetative effects have been described (Nicolaidis, 1969b).

NEUROPHYSIOLOGICAL MECHANISM OF SENSORY METABOLIC ANTICIPATORY RESPONSES

From the above experimental data, it is clear that chemosensory stimulations can trigger and mimic metabolic effects which normally occur when the stimulating chemicals are absorbed from the intestine and act systemically. This anticipatory action of chemosensory stimulations extends, even more effectively, to the global behavioral responses such as drinking and feeding. For example, not only is thirst quenched immediately after drinking but is exaggerated immediately after eating a salty morsel of fish. None of these responses can be accounted for by postabsorptive effects of the stimuli. Even the rapid absorptive phenomena described in the literature (Kare, 1976) cannot account for the anticipatory responses, since the minute amounts that reach the inner chemoreceptors in this manner cannot sustain the dose-dependent responses. It is necessary that some information about the chemical nature of the stimulus projects from the cephalic receptors onto the central units, the activity of which is basically modulated by the chemical composition of the inner milieu. This projection could be innate or wired by learned processes (Garcia *et al.,* 1961; Nachman, 1963; Rozin and Kalat, 1971), though the latter mechanism does not seem compatible with some of the properties of the anticipatory responses as have been discussed earlier in this paper.

Specific gustatory projections on chemosensitive hypothalamic units were first described by Nicolaidis (1968, 1969a,b) (Fig. 6). The conceptual difficulty of hypothalamic projections of sensory information was overcome by more recent anatomic studies (Norgren, 1976).

Today it is well established that the thalamocortical pathway of sensory input is not the only one, and that wide areas known to deal with vegetative neuroendocrine regulation are also affected by external stimulations (Emmers, 1973; Vincent *et al.,* 1972; Edinger, 1971; Rolls, 1976). The most explicit data of converging exteroceptive and interoceptive sensing concern osmoregulatory depletion–repletion information (Nicolaidis, 1969). In acute experiments of unit recordings on cats, osomosensitive units were first localized in the vicinity of classic osmosensitive areas by means of intracarotid (IC) hypertonic or water injections. These specific units (unre-

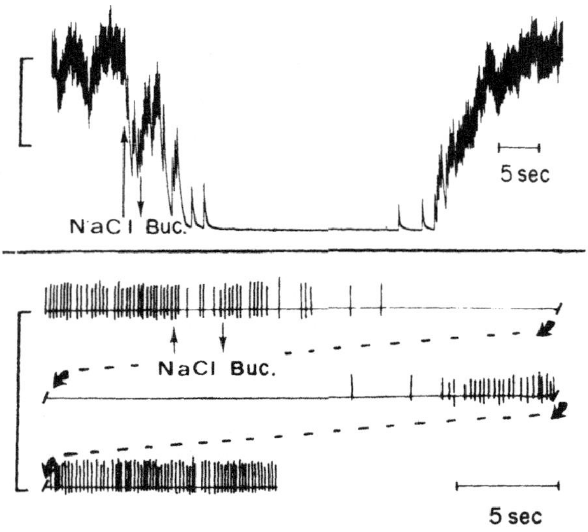

Fig. 6. Specific effect of lingual 1 *M* NaCl solution stimulation (arrows on and off) on activity of a lateral anterior hypothalamic osmosensitive unit of the cat. This activity is integrated on the upper recording. Buc., lingual.

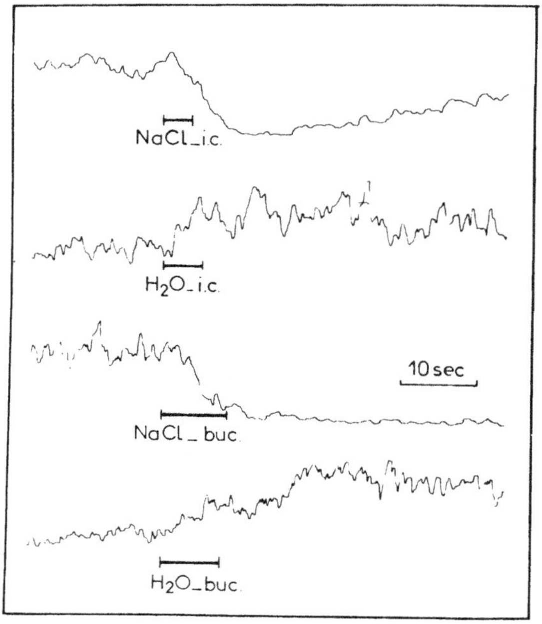

Fig. 7. Integrated activity of single units located in the postcommissural level of the lateral hypothalamus (area of the medial forebrain bundle). Hypertonic NaCl stimulations at either the intracarotid (i.c.) or lingual (buc.) level elicit very similar long-term change of activity of the same neuronal population (convergence). These units do not respond to nonspecific stimulations and reverse their responses when distilled water stimulation replaces the hypertonic one.

sponsive to general arousal) were then tested with lingual water and saline stimulations.

A summary of the essential findings follows: A small number of osmosensitive units respond to both IC and lingual specific stimulations. Some of the osmosensitive units of the lateral hypothalamus respond to IC injections of 1 *M* NaCl by decreasing their basal activity, and they do so in response to lingual stimulation by the same NaCl solution (Figs. 6 and 7). In distinct but intermingled lateral hypothalamic populations the reverse change of activity is also observed, i.e., both IC and lingual stimulations bring about an acceleration of previous activity. These responses are sensory specific and unaffected by noxious, photic, or acoustic stimulations. Similar convergent responses exist at the precommissural level (preoptic hypothalamus), but they tend to be less discriminatory for water and NaCl (Nicolaidis, 1971). More recently, it was shown on behaving monkeys that visual images evocative of repleting stimulations are able to elicit specific activity in LH units (Vincent *et al.,* 1972; Rolls, 1976).

Convergence on the same integrative population from both interoceptive and exteroceptive signals allows the systemically determined activity to be alleviated or aggravated by peripheral cues. The algebric summation of such messages causes the specific arousal canalized toward neuroendocrine release of (a) transmitters generating specific motivations and (b) endocrine factors controlling peripheral organs. So, announcement of future repletion not only allows anticipatory responses but, in addition, brings about a preparatory activation of secretory responsiveness and, as a consequence, potentialization and optimization of the postabsorptive effect of ingested metabolites.

Thus, recent physiological findings show the importance of "orality" and perhaps explain why man has always stressed its role, from the most ancient rituals to modern psychoanalytic concepts.

ACKNOWLEDGMENTS

A large part of our findings were obtained in collaboration with Mrs. M. J. Meile and Dr. N. Rowland. I thank Dr. M. Russek for criticizing this paper and Miss N. Duthoit for typing it.

REFERENCES

Adair, E. R., Miller, N. E., and Booth, D. A. (1968). Effects of continuous intravenous infusion of nutritive substances on consummatory behavior in rats. *Commun. Behav. Biol., Part A* **2,** 25–37.

Anand, B. K., Chhina, G. S., Sharma, K. N., Dua, S., and Singh, B. (1964). Activity on single neurons in the hypothalamic feeding centers. Effect of glucose. *Am. J. Physiol.* **207,** 1146–1154.

Assal, J. M., Levrat, R., and Stauffacher, W. (1970). Blood sugar and serum IRI after oral and intravenous administration of glucose in male Wistar rats with porto-caval anastomosis. *Diabetologia* **6,** 35.

Balsano, E., Pitucco, C., Masca, G., and Di Noto, V. (1964). New interpretation of oral glucose tolerance. *Lancet* **2,** 865–866.

Basset, J. M. (1974). Early changes in plasma insulin and growth hormone levels after feeding in lambs and adult sheep. *Aust. J. Biol. Sci.* **27,** 156–166.

Brugsch, T., Dresel, K., and Lewy, F. H. (1921). Zur Stoffwechselneurologie der Medulla Oblongata. *Z. Gesamte Exp. Med.* **25,** 262–270.

Ciaccio, C., and Racchinsa, S. (1927). The action of saccharine in various concentrations on hyperglycemia by dextrose. *Boll. Soc. Ital. Biol. Sper.* **2,** 309–316.

Cross, B. A., and Green, J. D. (1959). Activity of single neurons in the hypothalamus: Effect of somatic and other stimuli. *J. Physiol. (London)* **148,** 554–569.

Curry, D. L. (1971). Insulin secretion dynamics in response to slow rise and square wave stimuli. *Am. J. Physiol.* **221,** 324–328.

Edinger, H. M. (1971). Single unit firing patterns during drinking. *Proc. Int. Congr. Regul. Food Water Intake, 4th, 1971.*
Elrick, H., Stimmler, L., Hlad, C. J., and Arai, Y. (1964). Plasma insulin response to oral and intravenous glucose administration. *J. Clin. Endocrinol. Metab.* **24,** 1076–1082.
Emmers, R. (1973). Interaction of neural systems which control body water. *Brain Res.* **49,** 323–348.
Garcia, J., Kimeldorf, D. J., and Hunt, E. L. (1961). The use of ionizing radiation as a motivating stimulus. *Psychol. Rev.* **68,** 383–395.
Hommel, H., Fisher, U., Retzlaff, K., and Knofler, H. (1972). The mechanism of insulin secretion after oral glucose administration. II. Reflex insulin secretion in conscious dogs bearing fistulas of the digestive tract by sham-feeding of glucose and tap water. *Diabetologia* **8,** 111–116.
Kare, M. (1976). The function of taste. *Congr. E.C.R.O., 2nd, 1976.*
Kuzuya, T. (1962). Regulation of insulin secretion by the central nervous system. *J. Jpn. Soc. Intern. Med.* **51,** 65–74.
Louis-Sylvestre, J. (1976). Preabsorptive insulin release and hypoglycemia in rats. *Am. J. Physiol.* **230,** 56–60.
MacIntyre, N., Holdsworth, C. C., and Turner, D. S. (1965). Intestinal factors in the control of insulin secretion. *J. Clin. Endocrinol. Metab.* **25,** 1317–1324.
Martinez, I., Racotta, R., and Russek, M. (1974). Hepatic chromaffin cells. *Life Sci.* **15,** 267–271.
Nachman, N. (1963). Learned aversion to the taste of lithium chloride and generalization to other salts. *J. Comp. Physiol. Psychol.* **56,** 343–349.
Nicolaidis, S. (1968). Résponses des unités osmosensibles hypothalamiques aux stimulations salines et aqueuses de la langue. *C. R. Hebd. Seances Acad. Sci.* **267,** 2352–2355.
Nicolaidis, S. (1969a). Discriminatory responses of hypothalamic osmosensitive units to gustatory stimulation in cats. *Olfaction Taste, Proc. Int. Symp., 3rd, 1968* pp. 569–573.
Nicolaidis, S. (1969b). Early systemic responses to oro-gastric stimulation in the regulation of food and water balance. Functional and electrophysiological data. *Ann. N.Y. Acad. Sci.* **157,** 1176–1203.
Nicolaidis, S. (1970). The prandial calorigenic effect. *Nutr., Proc. Int. Congr., 8th, 1969* Excerpta Med. Found. Int. Congr. Ser. No. 213, pp. 216–219.
Nicolaidis, S. (1971). Réflexe poto-hidrotique et étude de son mécanisme neurophysiologique. *J. Physiol. (Paris)* **63,** 359–361.
Nicolaidis, S. (1975). Sensory-neuroendocrine relationships in the hydromineral balance. *Olfaction Taste, Proc. Int. Symp., 5th, 1974* pp. 271–275.
Nicolaidis, S., and Meile, M. J. (1972). Cartographie des lésions hypothamaliques supprimant la réponse alimentaire aux injections intracardiaques du 2-déoxy-d-glucose. *J. Physiol. (Paris)* **65,** 151A.
Nicolaidis, S., and Rowland, N. (1976). Metering of intravenous vs. oral nutrients and the regulation of energy balance. *Am. J. Physiol.* (in press).
Nicolaidis, S., Portet, R., and LeMagnen, J. (1966). Modification reflexe du quotient respiratoire chez le rat sous l'effet de stimulations alimentaires périphériques. *J. Physiol. (Paris)* **58,** 576.
Nicolaidis, S., Rowland, N., Meile, M. J., Marfaing-Jallat, P., and Pesez, A. (1974). A flexible technique for long-term infusions in unrestrained rats. *Pharmacol. Biochem. Behav.* **2,** 131–136.
Nijjar, M. S., and Rerry, W. F. (1970). Effects of intravenous and oral infusions of monosaccharides on serum insulin levels in rabbits. *Diabetes* **19,** 155–161.

Norgren, R. (1976). Taste pathways to hypothalamus and amygdala. *J. Comp. Neurol.* **166,** 17–30.

Oomura, Y., Ono, T., Ooyama, H., and Wayner, M. J. (1969). Glucose and osomosensitive neurons of the rat hypothalamus. *Nature (London)* **222,** 282–284.

Parra-Covarrubias, A., Rivera-Rodriguez, I., and Almaraz-Ulgalde, A. (1971). Cephalic phase of insulin release in obese adolescents. *Diabetes* **20,** 800–802.

Perley, M. J., and Kipnis, D. M. (1967). Plasma insulin responses to oral and intravenous glucose. Studies in normal and diabetic subjects. *J. Clin. Invest.* **46,** 1954–1962.

Porte, D., Halter, J., Kulkowski, P., Woods, S., Makous, W., and Chen, M. (1976). Afferent receptors, taste perception, and pancreatic endocrine function in normal and diabetic man. *Proc. Int. Symp. Food Intake Chem. Senses, 1976.*

Preshaw, R. M., Cooke, A. R., and Grossman, M. I. (1966). Sham-feeding and pancreatic secretion in the dog. *Gastroenterology* **50,** 171–178.

Racotta, R., Vega, C., and Russek, M. (1972). Liver catecholamines and preabsorptive satiation. *Fed. Proc., Fed. Am. Soc. Exp. Biol.* **31,** Abstr. 309.

Reynier-Rebuffel, A. M., Louis-Sylvestre, J., and LeMagnen, J. (1974). Conditioned insulin release: Effect on food intake. *Symp. Physiol. Food Fluid Intake, 26th,* p. 111.

Rolls, E. T. (1976). Neurophysiology of feeding. *Dalhem Workshop Appetite Food Intake, 1975* pp. 21–42.

Rowlan, N., Meile, M. J., and Nicolaidis, S. (1973). Action inadéquate des apports parentéraux sur l'insulino-sécrétion et sur le contrôle du comportement alimentaire chez le rat. *C. R. Hebd. Seances Acad. Sci.* **277,** 587–590.

Rowland, N., Meile, M. J., and Nicolaidis, S. (1975). Metering of intravenously infused nutrients in VMH lesioned rats. *Physiol. Behav.* **15,** 443–448.

Rozin, P., and Kalat, J. W. (1971). Specific hungers and poison avoidance as adaptive specialization of learning. *Psychol. Rev.* **78,** 459–485.

Scow, R. D., and Cornfield, J. (1954). Quantitative relation between the oral and intravenous glucose tolerance curves. *Am. J. Physiol.* **179,** 435–439.

Steffens, A. B. (1970). Plasma insulin content in relation to blood glucose level and meal pattern in the normal and hypothalamic rat. *Physiol. Behav.* **5,** 147–151.

Steffens, A. B. (1976). The influence of the hypothalamus on insulin and glucagon release in the rat. *Proc. Int. Symp. Food Intake Chem. Senses, 1976.*

Steffens, A. B., Mogenson, G. J., and Stevenson, J. A. F. (1972). Blood glucose insulin and FFA after stimulation and lesions of the hypothalamus. *Am. J. Physiol.* **222,** 1446–1452.

Strubbe, J. H., and Steffens, A. B. (1975). Rapid insulin release after ingestion of a meal in the unanesthetized rat. *Am. J. Physiol.* **229,** 4.

Vincent, J. D., Arnaud, E., and Bioulac, B. (1972). Activity of osmosensitive single cells in the hypothalamus of the behaving monkey during drinking. *Brain Res.* **44,** 371–384.

Vinik, A. I., Kalk, W. J., Keller, P., and Beumont, P. (1973). Overactivity of the entero-insular axis in maturity onset diabetes. *Lancet* **2,** 183–185.

Woods, S. C., and Kulkosky, P. J. (1976). Classically conditioned changes of blood glucose level. *Psychosom. Med.* **38,** 201–219.

Woods, S. C., Hutton, R. A., and Makous, W. (1970). Conditioned insulin secretion in the albino rat. *Proc. Soc. Exp. Biol. Med.* **133,** 964–968.

DISCUSSION

Bosma: Young infants may be satisfied by sham feeding on a blind nipple. An intensive period of suckling is followed by inaction and perhaps by sleep. The infant goes through the entire sequence of having been fed. After a suitable interval he will be hungry again and can be "cheated" again. Could this be related to what you would call feeding-related changes in metabolic state?

Nicolaidis: Before answering, let me make one more observation which is common in pathology. Diabetics treated with insulin often develop deep hypoglycemia and malaise. These patients carry a morsel of sugar and about 20 sec after they put it in their mouths they feel better. The effect is preabsorptive. If you calculate the systemic effect of 2–3 grains, you find a negligible increase of glycemia due to the postabsorptive effect. Obviously, the improvement is due to the reflex-released endogenous glucose which provides the emergency fuel until the intentional absorption of ingestants. If this influx of metabolites does not follow the sensory stimulation, the preabsorptive effect is dissipated and hunger (of the infants you are talking about) or malaise (or diabetics) reappears. The most probable mechanism of early satiety of infants sucking a blind nipple is the sensory-bound activation of central areas mimicking a state of repletion and enhanced by synergic neuroendocrinological responses.

Maller: Has anyone made use of the obvious application that when patients are fed intravenously, they should be given oral stimulation to make more efficient use of the calories?

Nicolaidis: In a recent series of 12 of my patients receiving exclusive iv alimentation, I tried several oral applications per day of sweet solutions and studied their body weight balance. A conclusion is difficult to draw because each patient had a different background and the variability of the results was too great. Only three patients had clearly improved during the days of oral sweet stimulation. My impression is that, though statistically not significant, these three patient's anabolic improvement was related to the oral stimulation.

Pfaffmann: You do not know whether it is just simply taste or could be due to all the oral stimulation, chewing, etc.?

Nicolaidis: I have no precise idea. It seems that more than one sensory modality acts on these reflexes.

Pfaffmann: Saccharin or sugar pills in the patients?

Nicolaidis: In coma cases I used saccharine solutions. I was giving them because there are some accidents due to sugar. You cannot use sweet solutions sweetened with sugar.

Pfaffmann: The reason I raise this question is that a related series of observations was made and reported in a recent *Psychology Review* paper, and the question dealt primarily with animals. The point was, if saccharin is used, and the animals have learned that this will start the anticipatory blood sugar change, but it does not get followed by the real sugar, extinction develops. Therefore, the treatment with saccharin might, in a sense, have unusual effects compared with the effect of the true sugar.

Nicolaidis: Yes. But it is possible that if you use sugar only as a gustatory stimulus not followed by intestinal absorption of it, responses to sugar will be the same as those to saccharin. Whether the oral stimulus is sugar or saccharin, if the experimental conditions result in some delayed repletion of the inner milieu the oral effect tends to extinguish. This is what happened to our rats fed intravenously and given 12 times a day a saccharine solution to consume. The anabolic improvement of the first day did not last to the subsequent period. There is a learned abolition of the oral effect of a stimulation when it is not followed by a systemic repletion.

Miller: I have a question regarding an infant sucking on a blind nipple. Several years ago, in order to get a technique in which we could hand-feed infant animals diets of various kinds, we developed a technique in which the animal's bellies were literally filled and then the animals were allowed to suck on nonlactating foster mothers. If you did not allow them to suckle, the ultimate utilization of the food decreased to a very low level and ultimately stomach clearance stopped and the animals died. They had to suckle. The reflexes involved in suckling these animals had little to do with how hungry the animals were. It had to do with other stimuli, part of which was how active they were at different times during the course of the day. It seems as though these were almost uncoupled events, the need of the animal for nutrient and the suckling process, associated but not directly related to each other. Coming back to Dr. Nicolaidis' comments, I would like to pursue this business about the animals receiving all their nutritional needs by infusion. You gave them a few calories in the form of sucrose and their weight gain increased for one day and then proceeded to drop along the same slope. Would you care to comment on why you think that occurs?

Nicolaidis: There is a learned abolition of the effect of oral stimulation when it is not followed by a systemic repletion. The fact that there is an innate, or salient, neural connection does not mean that the responses cannot be inhibited or reversed by a contradictory or nonconsequent systemic effect. In Bellow's experiments even water loses its quenching effect after several disconfirmations of rehydration due to the esophagostomy.

Unknown: What was the composition of the infusate and at what rate was it given during the course of the day?

Nicolaidis: The composition of infusates is given in Rowland *et al.* (1973). Rate of infusion usually was 60 ml per day of 1 kcal/ml solution. We tried several infusion schedules. Most of the data in this paper come from continuous, uniform 24-hr infusions. We also used infusions of one-third during the day, two-thirds during the night. These paradigms do not substantially improve the metabolic efficiency of infusates; on the contrary, square wave intermittent infusions increase it. This improvement of utilization of infusates is more obvious when one considers the extinction of oral food intake (Nicolaidis and Rowland, 1976) and reaches a maximum when infusions are triggered by meals and are delivered in the portal vein (M. Mather and S. Nicolaidis, unpublished data).

DeWys: Have you looked at the effect of concentration of the glucose or sucrose in the mouth? Could you get a dilute solution below the threshold? Is it just an osmotic effect? For the diabetic who puts the morsel of sugar in his mouth, there are osmotic effects as well as taste stimulation effects.

Nicolaidis: No, we did not. But hypotonic saccharine solutions have a similar effect.

DeWys: So you think it is a taste stimulation?

Nicolaidis: Yes, though not exclusively.

Gershoff: In these experiments you keep talking about efficiency of utilizations. Do you know that they are actually putting down tissue? It could also be water. Have you done nitrogen balance in your patients?

Nicolaidis: We have been considering this problem. Such an artifactual mechanism could account for short-term effects. Actually, there seems to be some water retention on the first day of iv infusions. But our data are based upon very long-term infusions, and hydrational states cannot account for statistically significant changes observed on the third or fourth day of infusion.

Lepkovsky: I am not clear as to your results of the RQ in the postabsorptive state. In the postabsorptive state you expect the RQ to go down instead of staying the same.

Nicolaidis: All metabolic states using lipids as catabolites (starvation or fat meals) bring about a low RQ around 0.7. Therefore, at the postabsorptive stage, carbohydrate or complete nutrients increase the RQ of a deprived animal, though fat nutrients do not. Our findings indicate that this postabsorptive effect is preceded by a similar preabsorptive one reflex-elicited by oral afferences. These afferences are sensory specific and, curiously enough, signal the lipid nature of nutrients.

Bosma: Margit Hamosh and Robert Scow have found that von Ebner's glands, at the base of the vallate papillae, secretes lipase. In the human infant, the composite of vallate papillae and these lipase-secreting glands approximate the bolus accumulation area at the junction of the oral palate and pharyngeal palate, or velum. Perhaps the action of suckling and the chemostimulation of milk at the vallate area elicits activity of these glands. Would the action of this lipase make the milk fat more accessible to taste? Another comment. Endocrine or endocrine-like functions of the oral mucosae might not be unexpected, in light of the origin of the anterior portion of the pituitary gland, as Rathke's pouch, from the embryonic stomodeum. As a neurosurgeon, you are aware that there may be vestiges of Rathke's pouch or of anterior pituitary tissue in the basal part of the sphenoid bone.

Nicolaidis: When we discover another function, we should again review pertinent developmental and comparative anatomy. In this perspective, I would point to the odd glandular relics that we call the vomarine eminences in the midline at the junction of oral and pharyngeal palate, whose function has not been explained.

Chapter 7

Taste Stimuli and Pancreatic Functions

MICHAEL NAIM AND MORLEY R. KARE

It is recognized that oral stimulation will modify the secretory processes along the gastrointestinal tract. Pavlov (1910) and his colleagues were the first investigators to study this phenomenon. They demonstrated that the chewing of palatable substances by the dog initiated gastric secretion. No such effect was seen when neutral tasting substances were offered. More recently, Nicolaidis (1969) extensively studied the role of oral stimulation in the regulation of metabolic processes. He suggested that the alimentary tract is not a passive canal in metabolic regulation. There are "gustatory–metabolic" reflexes which are initiated by food and fluids acting at a peripheral level. Examples of these anticipatory metabolic reflexes include (1) changes in respiratory quotient in response to eating; (2) hyperglycemic responses to oral stimulation with saccharin or sucrose in hungry rats; (3) the immediate (2–7 sec latency) increased sweating by dehydrated human subjects upon drinking fluids; and (4) in hyperhydrated rats, the large increase in diuresis on stimulation of the mouth with 1–3 ml of water, starting within 1 min. Conversely, stimulation with 5% NaCl was seen to give an immediate inhibition of diuresis. An inhibition of diuresis has also been reported by eating solid foods (Kakolewski *et al.*, 1968; Kakolewski and Valenstein, 1969).

In each of these investigations the physiological responses were extremely rapid, too rapid to be due to postingestional effects. Postingestional effects were ruled out on other grounds as well, including the use of esophageal ligatures and the knowledge that the time required for metabolism of ingested glucose to carbon dioxide is of the order of hours, not minutes (Wang *et al.*, 1962). The metabolic changes produced by the oral stimulation were anticipatory in the sense that they preceded but were in the same direction as the changes which had been produced by the substances as a result of postabsorptive systemic influences.

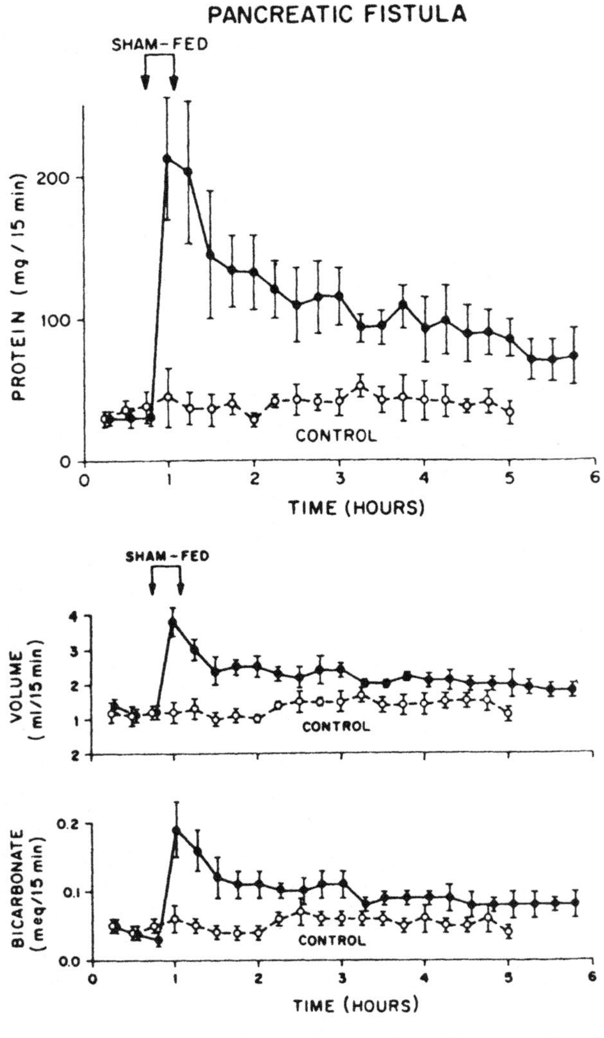

Fig. 1. Pancreatic volume flow and bicarbonate and protein output after sham feeding in dogs. Values are the mean ±SEM of six control experiments (open circles) and six sham-feeding experiments (filled circles) in three dogs. (From Preshaw *et al.*, 1966.)

The three phases of gastric secretion in response to stimuli have been defined as the cephalic, gastric, and intestinal phases. The cephalic phase refers to those stimuli acting in the region of the head, including anticipation, sight, taste, and smell of food. This phase is mediated entirely by the vagus nerve. The cephalic phase of secretion can be isolated and studied by sham feeding, in which ingested food is diverted to the exterior usually through an esophageal fistula (Konturek, 1974).

Stimulation of pancreatic secretion occurs in cephalic, gastric, and intestinal phases. While the cephalic phase of gastric secretion is known to be crucial, the pancreatic cephalic phase has been regarded as being of little consequence. However, sham feeding of men and dogs does produce pancreatic secretions (Preshaw *et al.*, 1966; Sarles *et al.*, 1968; Novis *et al.*, 1971). Preshaw *et al.* (1966) have shown that sham feeding of dogs causes a prolonged increase in protein and bicarbonate output (Fig. 1). Gregory (1962) concluded that sham-feeding in dogs causes mobilization and discharge of enzymes from the pancreas within 1–5 min after feeding and continues for approximately 30 min.

The action of sham feeding on the pancreas seems to be mediated mainly by vagal action on the gastric antrum, producing gastrin. The output of protein in response to sham feeding is almost totally inhibited by acidification of the antral pouch (Fig. 2), which is known to inhibit vagal release of gastrin (PeThein and Schofield, 1959).

Other gastrointestinal hormones—secretin and cholecystokinin—have been implicated in mediating pancreatic response. The hormonal action clearly depends on interaction with the parasympathetic nervous system. Anticholinergic drugs or surgical denervation will interrupt the hormonal effects (Henriksen, 1969; Moreland and Johnson, 1971; Preshaw, 1967; Wormsley, 1972). The pancreas has vagal innervation and there is evidence of direct neural stimulation of pancreatic secretion (Crittenden and Ivy, 1937). However, such a mechanism is probably unimportant compared to the vagally mediated hormonal response.

The neurohumoral mechanisms involved in control of pancreatic exocrine secretion may also operate in regulation of pancreatic endocrine function (Dupré, 1970). It has been noted that the delivery of glucose into the intestine in man is followed by the development of higher blood levels of immunoreactive insulin (IRI) than those obtained when similar levels of blood glucose are produced by intravenous infusions of glucose (McIntyre *et al.*, 1964; Elrick *et al.*, 1964; Perley and Kipnis, 1967).

Secretin, cholecystakinin, and gastrin were shown to cause a striking increase in insulin concentration in the pancreaticoduodenal venous plasma after their rapid endoportal injection in anesthetized dogs (Unger *et al.*, 1967). Fischer *et al.* (1972a) reported that insulin secretion during oral glucose tolerance tests in dogs occurs in several phases. Hommel *et al.* (1972) reported that an increase in circulating insulin at 5–10 min (Fig. 3) can be produced in dogs by sham feeding of glucose or tap water alone, without alteration of blood glucose levels. This phenomenon was absent after topical anesthesia of the oral mucosa (Fischer *et al.*, 1972b). Presumably, the initial release of insulin is stimulated via nervous receptors in the mouth.

While it has been shown that stimulation of taste receptors will affect pancreatic secretions, there has been limited consideration given to the specific effects which will result from specific taste stimuli. Behrman and Kare (1968) have shown that pancreatic flow and protein output can be affected to varying degrees by different taste substances. Water or sucrose solution mixed with a basal diet was found to produce a significantly greater secretory response than did the less appealing citric acid or quinine.

Although the cephalic phase seems to have a marked effect on internal

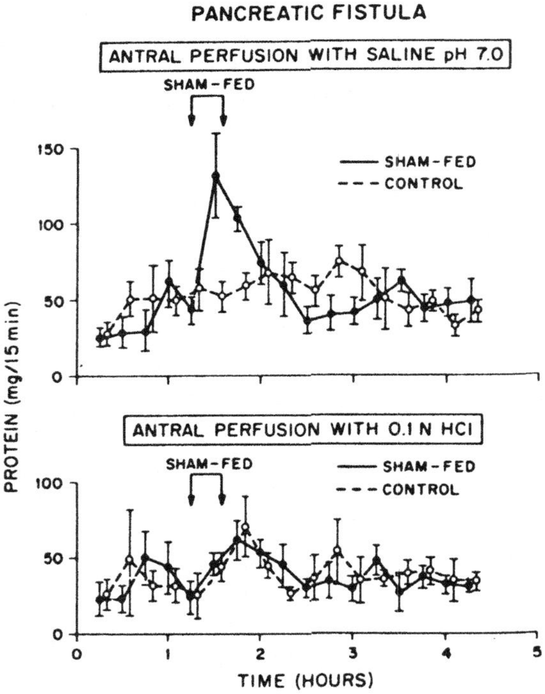

Fig. 2. Pancreatic protein output after sham feeding of dogs with antral perfusion of saline pH 7.0 or 0.1 *N* HCl. Values are the mean ±SEM of four to five experiments each with saline and acid perfusion in three dogs. (From Preshaw *et al.*, 1966.)

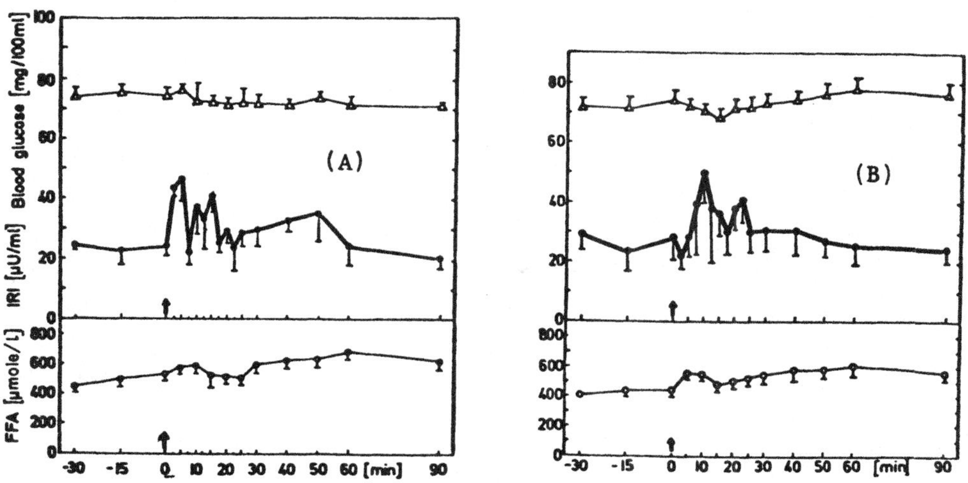

Fig. 3. Blood glucose levels (open triangles) as mg/100 ml, immunoreactive insulin (IRI) (filled circles) as μunits/ml, and free fatty acid (FFA) levels of serum (open circles) as μmole/liter over time before and after sham feeding of dogs. (A) Sham feeding of 1 g glucose-kg body weight in 50 ml of tap water (11 experiments in six animals); (B) sham feeding of tap water (nine experiments in five animals). The arrow at zero time indicates time of stimulus delivery. (From Hommel *et al.*, 1972.)

secretions, questions remain as to its overall significance and the impact of its modification on nutritional functioning. In collaboration with Dr. A. M. Merritt,* an experimental study was designed to further explore the relationships of specific chemical taste stimuli to pancreatic exocrine secretion. The preliminary experiments involved three dogs, each with a gastric and intestinal fistula (Thomas, 1941; Rudick and Dreiling, 1968). These pilot studies were used to develop a methodology for studying the effects of taste stimuli. The initial technique for stimulation of gustatory receptors was by application of swabs of aqueous solutions to the dogs tongues for a period of 6 min. Deionized water (at body temperature) was found to stimulate pancreatic secretion, whereas saliva had little or no effect. Saliva was thus used as a control stimulus.

In later experiments, a 40-min period of gastric and intestinal drainage prior to each stimulation provided a stable base line of pancreatic protein output (0.2–3.0 mg/5 min, as measured by absorbance at 280 nm). This base line was important for evaluation of differences found with various stimuli.

Next, employing four dogs with gastric and intestinal fistulae, three different solutions were used as stimuli: 0.29 *M* sucrose, 0.001 *M* quinine

* School of Veterinary Medicine, University of Pennsylvania.

sulfate, and 0.1 M citric acid. These solutions (applied to the dogs' tongues with swabs) were presented in a predetermined experimental order, the saliva control being used between each two stimuli. No more than one stimulation a day and three stimulations a week for each dog were applied. Stimulation with these substances at first produced a large increase of both the volume and protein content of pancreatic secretion (Figs. 4 and 5). However, after one or two trials with each stimulant with each dog, the increased output was no longer seen.

Although the initial taste stimulus was associated by the dogs with feeding, they learned quickly that food would not be forthcoming and no longer responded. This implies that the taste stimulus alone does not affect the pancreas and may, in fact, be a conditioned rather than a nonconditioned reflex. It is likely that the cephalic phase of pancreatic secretion is a complex function of higher processes than simple reflex stimulation, and that taste stimulation is only a tool of these processes. This interesting phenomenon of reduced response following repeated trials has not been reported with sham feeding. These data are compatible with those of Langlois *et al.* (1952). They suggested that the gastric response to orogastric reflexes can be delayed, and beside being excited by gustatory impulses, they are subject to inhibition by the hypothalamic areas.

Comparison of output following stimulation with the three solutions suggests that sucrose is a significantly more effective stimulant than quinine sulfate and citric acid (Table I).

Since swallowing is involved in the process of sham feeding, we next added 100 ml of the taste solution to 25 g of nonnutritive material (cellulose). Our assumption was that the swallowing action would trigger the correct feedback mechanism which would restore the pancreatic response. As it was technically difficult to gather adequate quantities of saliva, a mixture of water and cellulose was used for the control solution. Again, a predetermined experimental order was followed, with water stimulation used between each of the three taste stimulations. The same four dogs who had earlier been exposed to the swab stimuli were used. Pancreatic secretion was measured for volume and protein content for 90 min after each stimulus. After swallowing the mixtures, there were significantly less quantities ($p < 0.01$) of pancreatic juice and protein secreted in the first 40 min following stimulation by sucrose as compared to the initial swab stimulations (Fig. 6).

Presumably, this low level of secretion represents an aborted cephalic phase of secretion in dogs who have learned that these specific taste stimuli were not to be associated with feeding. This supports the conclusion that the response to swab stimulation was mediated through the

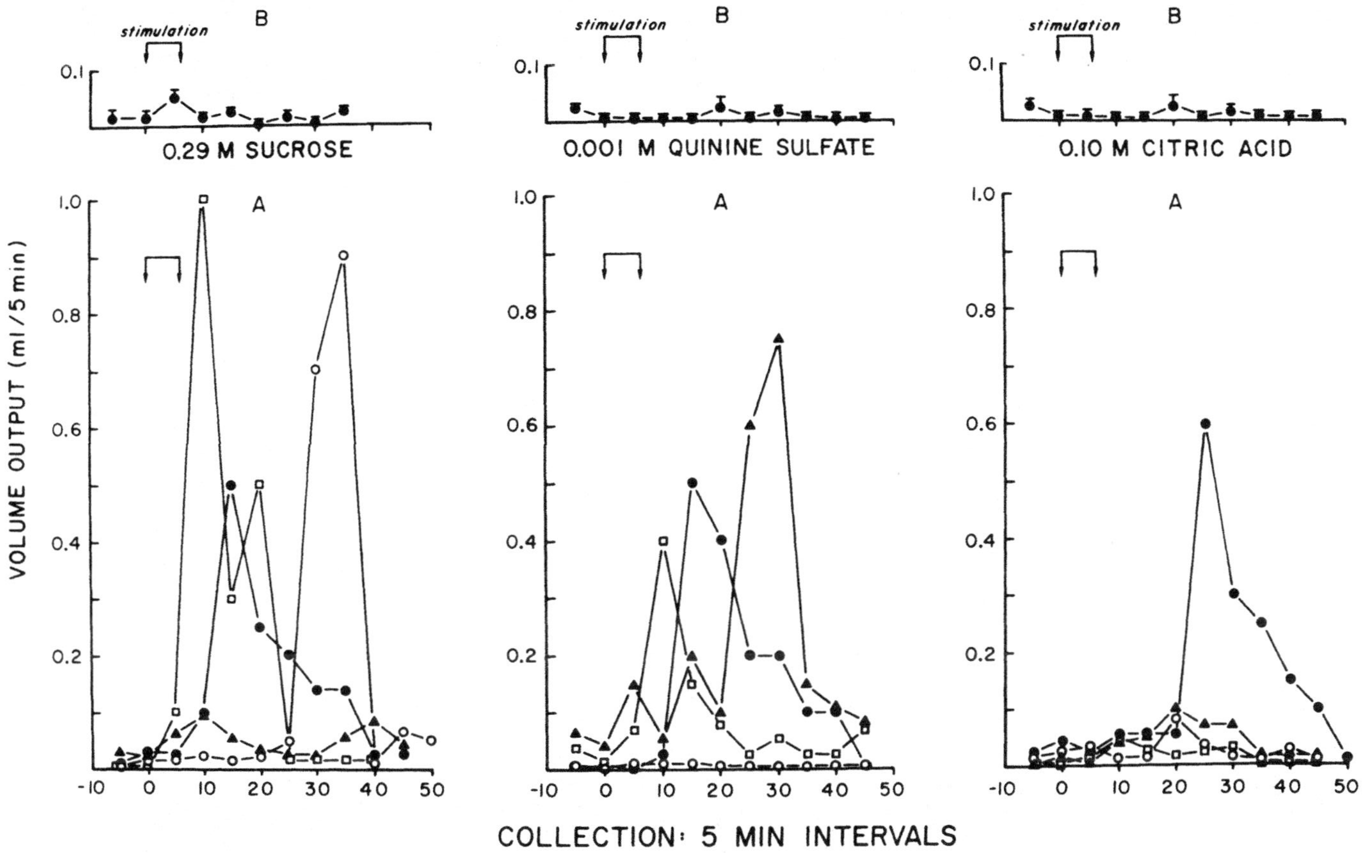

Fig. 4. Pancreatic volume flow after oral stimulation with three different taste stimuli using swab technique. (A) Initial response: values represent individual measurement on each of 4 dogs: filled circles, dog 1; squares, dog 2; open circles, dog 3; triangles, dog 4. (B) Later response: values are the mean ±SEM of four experiments for each stimulus with four dogs. M. Naim, M. R. Kare, and A. M. Merritt, unpublished data.

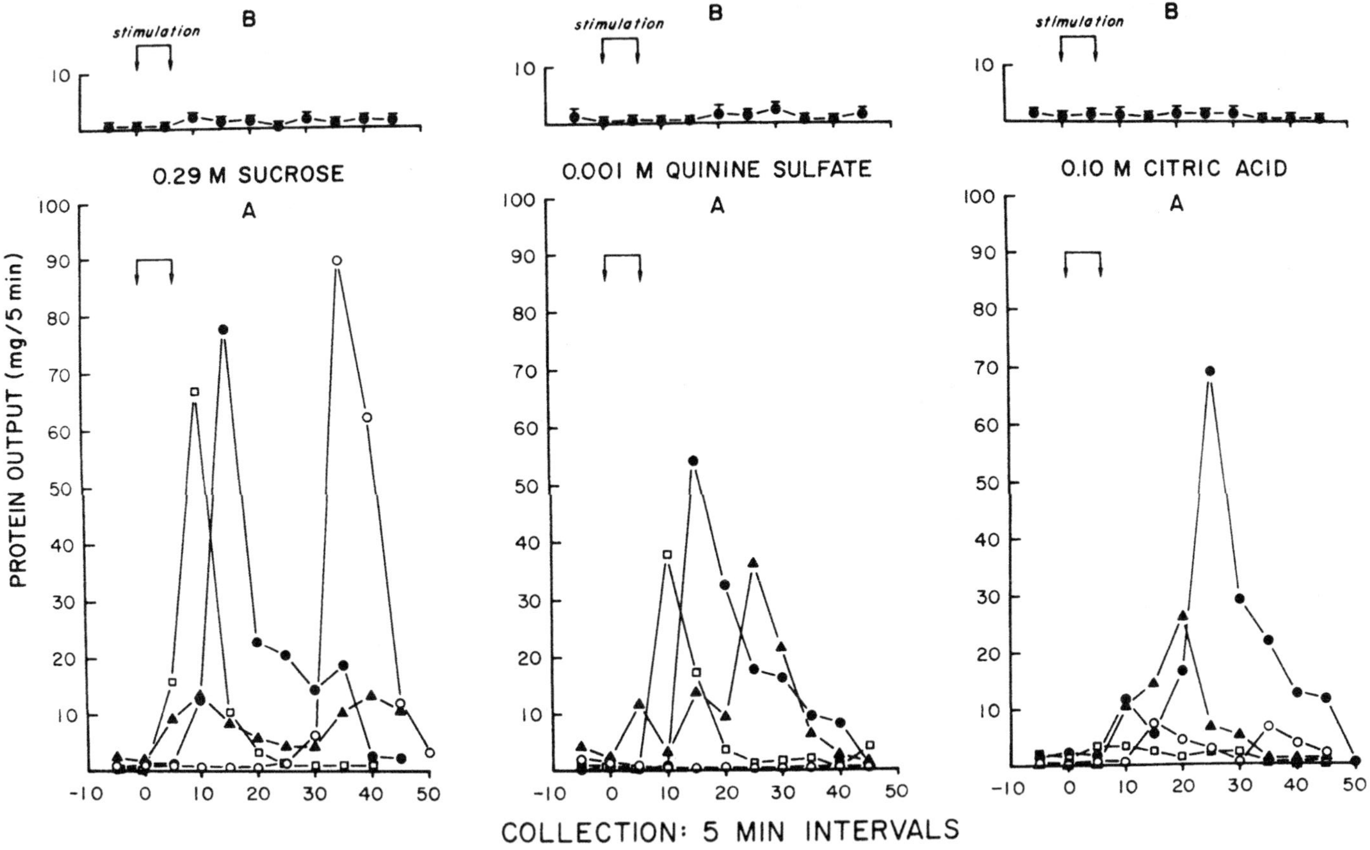

Fig. 5. Pancreatic protein output after oral stimulation with three different taste stimuli using swab technique. (A) Initial response: values represent individual measurement on each of 4 dogs: filled circles, dog 1; squares, dog 2; open circles, dog 3; triangles, dog 4. (B) Later response: values are the mean ±SEM of four experiments for each stimulus with four dogs. M. Naim, M. R. Kare, and A. M. Merritt, unpublished data.

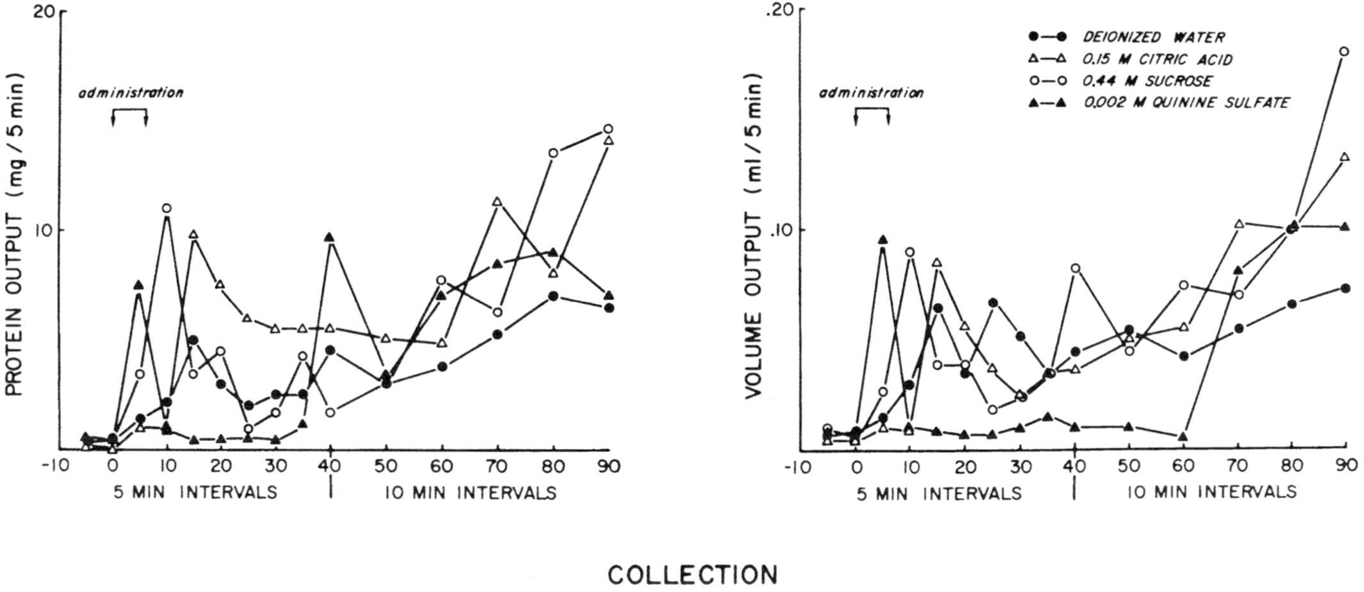

Fig. 6. Pancreatic protein and volume output after oral administration with four different stimuli using taste–cellulose mixtures. Values are mean of eight experiments for each stimulus with four dogs. M. Naim, M. R. Kare, and A. M. Merritt, unpublished data.

TABLE I

Pancreatic Exocrine Secretion of Dogs after Taste Stimulation by Swab[a]

	Total protein output (mg/45 min)		Total volume output (ml/45 min)	
Taste stimulus	Initial response	Later response	Initial response	Later response
0.29 *M* sucrose	a[b]	c	a	c
	132.36	10.86	1.45	0.18
	± 24.40	± 6.50	±0.34	±0.02
0.001 *M* quinine sulfate	b	c	ab	c
	82.42	10.10	1.18	0.10
	± 30.13	± 3.81	±0.46	±0.04
0.1 *M* citric acid	b	c	bc	c
	73.10	9.42	0.60	0.14
	± 37.22	± 3.43	±0.34	±0.04
Saliva (basal secretion)	c	c	c	c
	7.74	7.03	0.11	0.12
	± 2.70	± 2.82	±0.03	±0.04
	$F^{c}_{(7,28)} = 7.68$ $p < 0.001$		$F^{c}_{(7,28)} = 6.42$ $p < 0.001$	

[a] Values are the mean ± SEM of four experiments for each taste stimulus and six experiments for saliva with four dogs. M. Naim, M. R. Kare, and A. M. Merritt, unpublished data.

[b] All groups not designated with the same letter are statistically different at least at 5% level.

[c] Analysis of variance.

cephalic phase. (The values seen with swab stimulation were similar to those of sham feeding; Preshaw *et al.,* 1966.)

In this experimental system, we can not eliminate the possibility that the low values observed within 40 min of ingestion of the cellulose mixture could have been due to local internal stimulation. After 40 min, there was a peak of pancreatic secretion not seen with the swab stimulation. This probably was due to internal postingestional stimulation by the taste mixture, even though the gastric and the intestinal fistulae were kept open.

Since the learning process that occurred after repeated swab stimulation seems to have eliminated the cephalic phase of secretion that would otherwise have been seen with ingestion of the taste–cellulose mixture, future work in this experimental system might involve a study of what ingredients could be added to the cellulose to restore the cephalic phase of pancreatic secretion. That is, would a combination of some taste stimuli and/or odors restore this phase, or must complete food be added?*

The question arises as to whether alteration of pancreatic secretion has physiological or nutritional significance. To explore this, we chose an experimental model in which rats were fed raw and heated defatted soybeans as a protein source. In the raw form, soybean contains trypsin inhibitors and other substances that cause undesirable physiological response in various animal species (Kunitz, 1947; Birk, 1961; Lepkovsky *et al.*, 1965; Gertler *et al.*, 1967; Liener and Kakade, 1969; Rackis, 1974). By live steam treatment, a process referred to as toasting, the nutritional value of defatted soy can be raised to almost that of meat and milk. Both raw and processed soy protein products can have flavors objectionable to different groups of people (Rackis *et al.*, 1975).

The reason heating raw soybean improves its nutritional value is not fully understood. As heating raw soybean also improves its flavor, it is possible that the improved flavor enhances its nutritional value by affecting the digestive system through such factors as internal secretion and gastrointestinal motility. This, of course, is a complex preliminary question.

Initially, the protein efficiency ratio (PER) was compared in controls fed heated soy diets as opposed to similar diets with added aversive taste stimuli (Table II). The PER values were found to be the same. The aversive taste stimuli did not appear to inhibit gastric or pancreatic function to the extent of interfering with food utilization.

The explanation for this may be an overoptimum concentration of digestive enzymes in the small intestine. Inhibition of pancreatic enzyme secretion due to offensive taste, if it did occur, therefore, would not occur to a great enough extent to alter digestion. Bielorai and Bondi (1963) found incomplete digestion in the small intestine after feeding raw soy. Gastric evacuation also proceeded more slowly after feeding raw soy to rats, as reported recently by Lepkovsky *et al.* (1975). This suggests a study in which proteolytic enzyme inhibitors could be administered to exper-

* After this paper was presented, a further study was conducted to answer this question. Oral and intragastric administration of the same taste–cellulose mixtures to the same four dogs using the same experimental design was carried out. A full restoration ($p < 0.01$) of pancreatic response within 40 min after oral administration with sucrose–cellulose mixture was observed.

Pancreatic response to oral administration of quinine sulfate–cellulose or citric acid–cellulose was significantly lower, similar to the water–cellulose control. Pancreatic responses to intragastric administrations of the four taste–cellulose mixtures were low and not different from one another.

TABLE II

Effect of Aversive Taste Stimuli on the Protein Efficiency Ratio (PER) in Rats[a]

Diet[b] containing	Per	Total body gain (g)	Total food consumption (g)	% Pancreas from body weight
HS	3.01 a[d]	156 ab	288 ab	0.262 b
RS	2.19 b	109 c	276 ab	0.401 a
HS + 0.02%[c] quinine sulfate	2.94 a	142 b	267 b	0.259 b
HS + 2.0%[c] sucrose octaacetate	2.97 a	164 a	307 a	0.286 b

[a] M. Naim, M. R. Kare, and D. E. Ingle, *J. Nutr.*, in press; 21-day experiment, 12–13 animals/group.

[b] HS indicates heated defatted soybeans, RS indicates raw defatted soybeans. Commercial HS or RS (Ralston Purina Co.) was the source of protein (18%) in diet.

[c] Percent in diet.

[d] All groups not designated with the same letter are statistically different at 5% level.

imental animals until digestive enzymes become a limiting factor. Under these circumstances, the objectionable flavors in soy might theoretically influence nutritional value.

In trying to evaluate any avoidance behavior as a result of offensive flavors in soy diets, we administered a 30-min preference test to rats in which they could choose between heated soy (HS) and raw soy (RS) diets. As expected, there was persistent preference exhibited for the HS diet ($p < 0.01$). The question arises as to whether the preference of experimental animals for HS is due to taste, smell, texture, or postingestional factors. The influence of texture as a variable was tested by offering RS in unground form (flakes) (Fig. 7). This markedly improved the acceptability of the RS diet. As HS is preferred in the ground form, we can conclude that texture is not an important factor (Fig. 8).

Preference for a diet containing RS over that containing HS could be achieved by simultaneously adding an appealing taste (sodium saccharin) to unground RS and an aversive taste (quinine sulfate or sucrose octaacetate) to the diet containing ground HS. The animals were thus induced by flavoring to choose the less nutritious RS diet (Fig. 9).

Any postingestional factors favoring the more nutritious HS should be apparent in this experiment on long-term preference testing. Postingestional factors have been shown to modify the amount ingested and the shape of preference–aversion curves derived from preference tests (Jacobs 1962; Mook 1963). In fact, the animals did switch from the appeal-

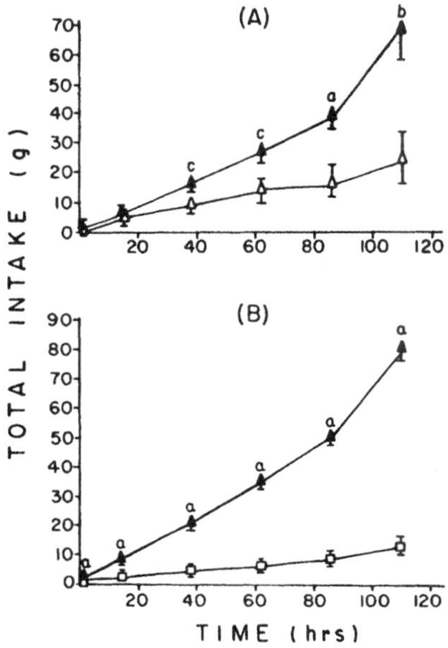

Fig. 7. Preference test between (A) a diet containing HS (filled triangles) versus a diet containing unground RS (open triangles); (B) a diet containing HS (filled triangles) versus a diet containing RS (squares). Values indicate the average accumulative food intake in grams ±SEM of 10–12 rats (a = $p < 0.001$; b = $p < 0.01$; c = $p < 0.05$). (M. Naim, M. R. Kare, and D. E. Ingle, *J. Nutr.*, in press.)

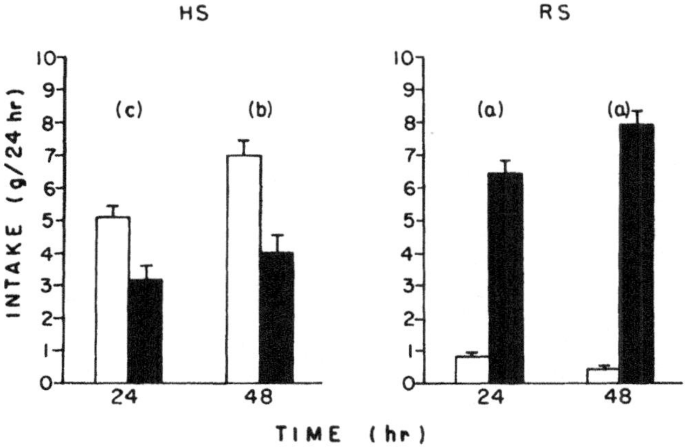

Fig. 8. Preference tests between diets containing ground defatted soybeans (light areas) versus diets containing unground defatted soybeans (dark areas). Values represent average food intake in grams ±SEM of 14 rats (a = $p < 0.001$; b = $p < 0.01$; c = $p < 0.05$). (M. Naim, M. R. Kare, and D. E. Ingle, *J. Nutr.*, in press.)

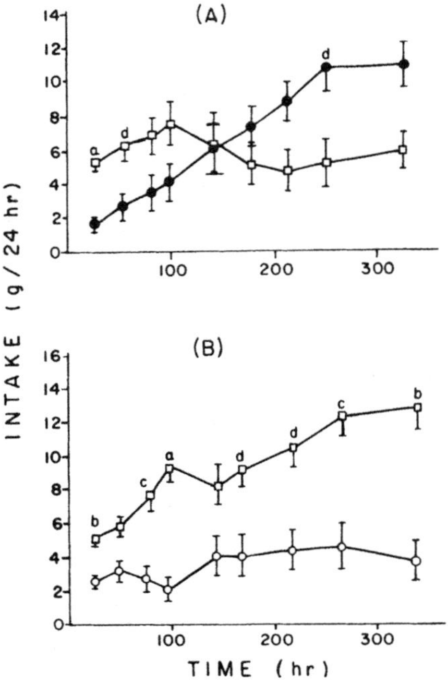

Fig. 9. Preference tests using diets containing unground RS and appealing taste versus diets containing HS and aversive taste. (A) HS + 2.0% sucrose octaacetate (filled circles) versus unground RS + 0.35% sodium saccharin (squares); (B) HS + 0.02% quinine sulfate (open circles) versus unground RS + 0.35% sodium saccharin (squares). Values represent average food intake in grams ±SEM of 12–14 rats (a = $p < 0.001$; b = $p < 0.01$; c = $p < 0.02$; d = $p < 0.05$). (M. Naim, M. R. Kare, and D. E. Ingle, *J. Nutr.*, in press.)

ing RS to the HS containing the aversive taste of sucrose octaacetate after 6–7 days. This could be explained either by habituation to the aversive taste of the sucrose octaacetate, or postingestional factors favoring the more nutritious diet. It is possible that the rats associated through learning the appealing taste with bad nutrition and aversive taste with good nutrition. However, no preference was exhibited for the HS containing quinine for the 2-week duration of the experiment. Either no habituation occurred to the aversive quinine taste or perhaps the quinine exerted a pharmacologic effect, leaving the animals with the postingestional message that quinine plus HS is poor nutrition. We can conclude, however, that at least for the first 6–7 days, postingestional information did not substantially influence these animals.

We next used vacuum treatment to determine the influence of objectionable odors in the preference for HS over RS (Table III). First, it was demonstrated that the animals (group 2) preferred vacuum-treated RS

TABLE III

Chronological Series of Overnight Preference Tests between Each Pair of Diet Treatments[a]

Group	HS(V)[b] vs	HS	RS(V)[b] vs	RS	HS + QSO_4[c] vs	HS	Producing of anosmia by 5% $ZnSO_4$	HS(V)[b] vs	HS	RS(V)[b] vs	RS	HS + QSO_4[c] vs	HS
1	5.8 a[d] ±0.6	7.7 ±0.6					+	6.1 a ±0.5	9.6 ±1.0				
2			7.7 d ±0.5	3.5 ±0.4			+			6.4 ns ±0.4	6.0 ±0.3		
3	5.1 c ±0.4	8.7 ±0.5					−	5.8 b ±0.5	8.2 ±0.6				
4			7.7 a ±0.3	6.6 ±0.4			−			7.7 b ±0.7	4.0 ±0.3		
5					3.2 d ±0.6	16.4 ± 0.8	+					3.7 c ±0.5	6.4 ±0.7

[a] M. Naim, M. R. Kare, and D. E. Ingle, *J. Nutr.*, in press; food intake g/animal ± SEM of 16–22 animals/group.

[b] (V) = subjection to vacuum treatment before diet was mixed.

[c] QSO_4 = 0.02% quinine sulfate.

[d] Comparisons were made between each pair of diet treatments for each group of animals using paired *t*-test. Letters to right of values designate significant difference a = $p < 0.05$; b = $p < 0.02$; c = $p < 0.01$; d = $p < 0.001$; ns = no significant preference.

over untreated RS. This preference was apparently due to removal by the vacuum treatment of an objectionable odor (odors). Next we found that the animals (group 1) exhibited a preference for HS as compared to vacuum-treated HS. We then rendered these animals anosmic using 5% zinc sulfate (Alberts and Galef, 1971) and repeated the preference tests 96 hr later. The preference for untreated HS persisted. Further, other anosmic animals could not distinguish between RS and vacuum-treated RS. Apparently, vacuum treatment of HS removes a positive taste stimulant.

We used five groups of animals in these experiments. Groups 3 and 4 were used as control animals in that they were exposed to the same preference tests as groups 1 and 2, respectively, but were not rendered anosmic. Since groups 3 and 4 duplicated their preference 1 week later, no habituation occurred. Group 5 was a control for anosmia in that the animals rejected the aversive taste of quinine both before and after anosmia.

Several conclusions can be drawn from our experimental data. First, we confirmed the work of Behrman and Kare that appealing taste stimulants have greater effect on pancreatic exocrine secretion than less appealing stimulants. Our experiments with dogs suggested further that this effect of the taste stimulus is part of a learning process in which taste stimuli are associated with feeding. The taste stimuli in themselves may not be enough to influence internal secretion.

We were unable to demonstrate an effect of an aversive taste stimulus mixed with the HS diet on the PER. Therefore, the question of the nutritional significance of taste stimuli remains unanswered. Further study is suggested in this area under conditions limiting the concentration of proteolytic enzymes in the small intestine (e.g., trypsin inhibitors). Our data do show that raw defatted soybean contains aversive odors and probably aversive tastes. These stimuli in combination with proteolytic enzyme inhibitors could theoretically be of nutritional significance and may help, in part, to explain why raw soy is so poorly utilized.

REFERENCES

Alberts, J. R., and Galef, B. G., Jr. (1971). Acute anosmia in the rat: A behavioral test of a peripherally-induced olfactory deficit. *Physiol. Behav.* **6,** 619–621.

Behrman, H. R., and Kare, M. R. (1968). Canine pancreatic secretion in response to acceptable and aversive taste stimuli. *Proc. Soc. Exp. Biol. Med.* **129,** 343–346.

Bielorai, R., and Bondi, A. (1963). Relationship between "antitryptic factors" of some plant protein feeds and products of proteolysis precipitable by trichloroacetic acid. *J. Sci. Food Agric.* **14,** 124–132.

Birk, Y. (1961). Purification and some properties of a highly active inhibitor of trypsin and α-chymotrypsin from soybeans. *Biochim. Biophys. Acta* **54,** 378–381.

Crittenden, P. J., and Ivy, A. C. (1937). The nervous control of pancreatic secretion in the dog. *Am. J. Physiol.* **119,** 724–733.

Dupré, J. (1970). Regulation of the secretions of the pancreas. *Annu. Rev. Med.* **21,** 299–316.

Elrick, H., Stimmler, L., Hlad, C. J., and Arai, Y. (1964). Plasma insulin response to oral and intravenous glucose administration. *J. Clin. Endocrinol. Metab.* **24,** 1076–1082.

Fischer, U., Ziegler, M., and Michael, R. (1972a). The mechanism of insulin secretion after oral glucose administration. I. Multiphasic course of insulin mobilization after oral glucose administration in conscious dogs. Differences to the behavior after intravenous administration. *Diabetologia* **8,** 104–110.

Fischer, U., Hommel, H., Ziegler, M., and Jutzi, E. (1972b). The mechanism of insulin secretion after oral glucose administration. III. Investigations on the mechanism of a reflectoric insulin mobilization after oral stimulation. *Diabetologia* **8,** 385–390.

Gertler, A., Birk, Y., and Bondi, A. (1967). A comparative study of the nutritional and physiological significance of pure soybean trypsin inhibitors and of ethanol-extracted soybean meals in chicks and rats. *J. Nutr.* **91,** 358–370.

Gregory, R. A. (1962). "Secretory Mechanisms of Gastro-intestinal Tract," pp. 143–151, Mongr. Physiol. Soc., No. 11. Arnold, London.

Henriksen, F. W. (1969). Effect of vagotomy or atropine on the canine pancreatic response to secretin and pancreozymin. *Scand. J. Gastroenterol.* **4,** 137–144.

Hommel, H., Fischer, U., Retzlaff, K., and Knöfler, H. (1972). The mechanism of insulin secretion after oral glucose administration. II. Reflex insulin secretion in conscious dogs bearing fistulas of the digestive tract by sham-feeding of glucose or tap water. *Diabetologia* **8,** 111–116.

Jacobs, H. L. (1962). Some physical, metabolic, and sensory components in the appetite for glucose. *Am. J. Physiol.* **203,** 645–659.

Kakolewski, J. W., and Valenstein, E. S. (1969). Antidiuresis associated with the ingestion of food substances. *Olfaction Taste, Proc. Int. Symp., 3rd, 1968* p. 593–600.

Kakolewski, J. W., Cox, V. C., and Valenstein, E. S. (1968). Short-latency antidiuresis following the initiation of food ingestion. *Science* **162,** 458–460.

Konturek, S. J. (1974). Gastric secretion. *In* "Gastrointestinal Physiology" (E. D. Jacobson and L. L. Shanbour, eds.), pp. 227–264. Univ. Park Press, Philadelphia, Pennsylvania.

Kunitz, M. (1947). Crystalline soybean trypsin inhibitor. II. General properties. *J. Gen. Physiol.* **30,** 291–310.

Langlóis, K. J., Lim, R. K., Rosiere, G., Stewart, D. I., and Stumpff, D. L. (1952). Unconditioned orogastric secretory reflex. *Fed. Proc. Fed. Am. Soc. Exp. Biol.* **11,** 88–89.

Lepkovsky, S., Furuta, F., Koike, T., Hasegawa, N., Dimick, M. K., Krause, K., and Barnes, F. J. (1965). The effect of raw soya beans upon the digestion of proteins and upon the function of the pancreas of intact chickens and of chickens with ileostomies. *Br. J. Nutr.* **19,** 41–56.

Lepkovsky, S., Dimick, M. K., Furuta, F., Feldman, S. E., and Park, R. (1975). Stomach and upper intestine of the rat in the regulation of food intake. *J. Nutr.* **105,** 1491–1499.

Liener, I. E., and Kakade, M. L. (1969). Protease inhibitors. *In* "Toxic Constituents of Plant Foodstuffs" (I. E. Liener, ed.), pp. 7–68. Academic Press, New York.

McIntyre, N., Hodsworth, C. D., and Turner, D. S. (1964). New interpretations of oral glucose tolerance. *Lancet* **2,** 20–21.

Mook, D. G. (1963). Oral and postingestional determinants of the intake of various solutions in rats with esophageal fistulas. *J. Comp. Physiol. Psychol.* **56,** 645–659.

Moreland, H. J., and Johnson, L. R. (1971). Effect of vagotomy on pancreatic secretion stimulated by endogenous and exogenous secretin. *Gastroenterology* **60,** 425–431.

Nicolaidis, S. (1969). Early systemic responses to orogastric stimulation in the regulation of food and water balance: Functional and electrophysiological data. *Ann. N.Y. Acad. Sci.* **157,** 1176–1203.

Novis, B. H., Banks, S., and Marks, I. N. (1971). The cephalic phase of pancreatic secretion in man. *Scand. J. Gastroenterol.* **6,** 417–421.

Pavlov, I. P. (1910). "The Work of the Digestive Glands" (transl. by W. H. Thompson), 2nd ed. Griffin, London.

Perley, M. J., and Kipnis, D. M. (1967). Plasma insulin responses to oral and intravenous glucose. *J. Clin. Invest.* **46,** 1954–1962.

PeThein, M., and Schofield, B. (1959). Release of gastrin from the pyloric antrum following vagal stimulation by sham feeding in dogs. *J. Physiol (London)* **148,** 291–305.

Preshaw, R. M. (1967). Integration of nervous and hormonal mechanisms for external pancreatic secretion. *Handb. Physiol., Sect. 6: Aliment. Canal.* **2,** 997–1005.

Preshaw, R. M., Cooke, A. R., and Grossman, M. I. (1966). Sham-feeding and pancreatic secretion in the dog. *Gastroenterology* **50,** 171–178.

Rackis, J. J. (1974). Biological and physiological factors in soybean. *J. Am. Oil. Chem. Soc.* **51,** 161A–172A.

Rackis, J. J., McGhee, J. E., and Honig, D. H. (1975). Processing soybeans into foods: Selected aspects of nutrition and flavor. *J. Am. Oil Chem. Soc.* **52,** 249A–253A.

Rudick, J., and Dreiling, D. A. (1968). A Thomas cannula adaptor for the collection of pancreatic secretion. *Surgery* **63,** 683–685.

Sarles, H., Dani, R., Prezelin, G., Souville, C., and Figarella, C. (1968). Cephalic phase of pancreatic secretion in man. *Gut* **9,** 214–221.

Thomas, J. E. (1941). An improved cannula for gastric and intestinal fistulas. *Proc. Soc. Exp. Biol. Med.* **46,** 260–261.

Unger, R. H., Ketterer, H., Dupré, J., and Eisentraut, A. M. (1967). The effects of secretin, pancreozymin, and gastrin on insulin and glucagon secretion in anesthetized dogs. *J. Clin. Invest.* **46,** 630–645.

Wang, C. H., Snipper, L. P., Bilen, O., and Hawthorne, B. (1962). Catabolism of glucose and gluconate in intact rats. *Proc. Soc. Exp. Biol. Med.* **111,** 93–97.

Wormsley, K. G. (1972). The effect of vagotomy on the human pancreatic response to direct and indirect stimulation. *Scand. J. Gastroenterol.,* **7,** 85–95.

DISCUSSION

Nicolaidis: On just the theoretical position you made, I disagree with your statement that there are some learning effects and no innate effects. If you do not reinforce any natural reflex which has been perfectly demonstrated as being innate, if you repeat several times, extinguish without reconfirming it, it disappears. You can superimpose the extinction of a natural reflex. The fact is that if you did it several times and you see a sort of depression of the response, it does not mean that you are dealing with a learned situation, but that just could mean that you have extinguished a preexisting natural nonconditioned reflex.

Naim: I am not sure that from the design of our experiment that we could answer if it is learning. However, with a few of the dogs we tried 1 or 2 months later and they still did not respond. Therefore, it appeared to be learning because it happened very quickly and remained for a long time.

Burghardt: I want to add a comment to Nicolaidis' point. You may be talking about what psychologists and others call habituation; that is, increase of an innate response which is itself considered by many people to be a most primitive or basic form of learning. Your point, I think, is well taken. It is questionable whether it is conditioning and the building up of a new response.

Naim: It does not seem to me to be habituation since they responded behaviorally. For example, the dogs recognized very well the aversive taste of quinine relative to the appealing taste of sucrose. I think the association that no feeding is taking place occurs particularly in the swab stimulation technique, resulting in the learning effect.

Nicolaidis: It is not very important because we know that the frontier between conditioned and innate is not so clear. In rats the first sweet stimulus is effective with no previous learning. These animals have never tasted a sweet taste before. This reflex persists under anesthesia which would not occur in a learned situation. If you try to extinguish the response it is very difficult. It is very clear that there is something innate or at least something appropriate or salient to elicit.

Naim: Using the cellulose technique, we had the impression that for the first stimulation with sucrose, the response was much lower than in later repeated experiments which probably started to break learning that occurred when the swab technique was used.

Lepkovsky: I want to raise a different question. Hawk showed that you can digest food without any stimulus at all if the basal secretion is adequate. If that is the case, why are there so many, many mechanisms to tremendously increase the flow of these digestive secretions. Even in our work on the chicken, we could remove the pancreas—almost all of it—and they get along alright. If the basal secretions are adequate, what is the meaning of all these stimuli?

Naim: I think that any stimulation which will increase gastric and pancreatic secretions will not have nutritional significance in processed food (not natural food). Processed food does not contain many undesirable factors like enzymes and growth inhibitors which can be found often in natural food, and can reduce digestibility. I would also like to remind you of the three phases that control pancreatic secretion: the cephalic phase, the stomach, and the intestinal. You can cause some distention of the stomach or apply food directly to the gastric fistula and get pancreatic response. However, it is difficult to stimulate one phase exclusively. Even Pavlov, 70 years ago, said during his experiments when he put food directly into the stomach, he was not sure whether the dogs did not realize by cephalic phase that they were eating. Therefore, these three phases could be overlapping.

Lepkovsky: Hawk worked on human beings. He messed up the food: charcoal, bread, butter, everything mixed in. One of the subjects, after eating, threw up this food; but those who did not digested their foods normally.

Nicolaidis: Lepkovsky's question is very important. What is the reason for all these reflexes since the basal level is sometimes sufficient to allow survival of the animal? I think these animals do survive in our laboratories, but under natural conditions they would not survive. All these quick responses to all of these emergency situations, which are nonsteady physiological states, are very important in natural conditions where fighting is everyday life. Probably all these reflexes are just what has permitted these animals to survive until now.

Miller: There are so many questions like this in biology. Many of these systems are just overactive because they can handle large loads all at once, and this is also true of the gastrointestinal tract. If things come in a slow, steady rate, the system can handle it at a low output. It is true for many animals that they have a burst of food appear and that is not true in the lab.

Garcia: One thing that has emerged from our work is that whenever we see conditioning in the gastrointestinal tract, the postingestional factors, while they are powerful unconditioned stimuli, are not very good for conditioning. The conditioning factors are almost all gustatory. We have induced animals to try to avoid poison with postingestional factors, and they just will not improve. If you mask the gustatory factors in any way, they cannot adapt to this situation and they will not survive. The minute you have any kind of a taste edge they immediately survive, so there is a big difference between the gustatory factors and the postingestional ones. The latter are reflexive and you cannot affect them.

NUTRITIVE STATE

Chapter 8

Nutritional State/Taste Interactions in Food Intake: Behavioral and Physiological Evidence for Gastric/Taste Modulation

K. N. SHARMA, H. L. JACOBS, V. GOPAL, AND S. DUA-SHARMA

INTRODUCTION

Taste has been implicated both in man and in animals in such adaptive functions as food selection, preferences, or rejection of noxious stimuli (Maller and Desor, 1974). Boring (1941), in historical perspective, describes how the receptors involved in savoring were identified and classification schemes evolved to describe and analyze these sensations. Over the years a considerable amount of data has accumulated, particularly in animals (Epstein, 1967), suggesting that such taste and olfactory perceptions may be useful in predicting selection and acceptance of nutrients. In spite of this growing interest and parallel advances in methodology to quantitatively measure sensitivity, discriminability, and intensity of oral stimuli related to taste and olfaction, very little systematic consideration has been given to the role of taste as part of the physiological system controlling food intake.

Animals must actively seek and select their food in their natural environment. To achieve this, naturally occurring food must have at least two different sets of properties (Le Magnen, 1971). Initially, the oral sensory appraisal of food leads to its acceptance or rejection and, when accepted, is eaten in definite amounts. "This sensory activity to foods is a critical determinant of innate or acquired feeding responses, insuring an oral selection and a metering of intakes. Through the second step of action of foods in the feeding process, these orally determined responses to food are "regulated." At the postabsorbtive and systemic level, food as a nutrient acts as a metabolic signal on regulatory centers and, through positive and negative feedback mechanisms, "modulates" oral feeding responses" (Le Magnen, 1971, p. 465).

Recent studies have shown that gastric and intestinal sensory mechanisms, operating at the intermediate level between oral and systemic factors, are concerned with the "sensory" appraisal of food including its texture, viscosity, volume, temperature, and other physicochemical properties of the diet. These alimentary signals play an important role in the peripheral mechanism of hunger and thirst and influence the central neural mechanisms regulating food intake (Sharma, 1967a,b). The gastrointestinal mechanism is a fast-acting system and shares in large measure the organizational control characteristics of the oral sensory system (Sharma *et al.,* 1975; Sharma, 1975). The flow of information from these oral and gastric receptors to the brain is not all in one direction, but various strategies are used so that the animal can react in strict accordance with the requirements of the external and internal environment. In the present chapter both behavioral and neurophysiological investigations related to taste and gastric mechanisms will be reviewed, particularly in the context of the physiological systems involved in food intake and energy exchange. This analysis of information transfer in peripheral loci brings forth the importance of centrifugal control systems and suggests that the interactions operate at multilevels as well as locally.

ENERGY BALANCE—INPUT/OUTPUT VARIABLES

Conventionally, output is recognized as coordinate with input in the analysis of energy balance. The energy expended (work done and heat produced) and the energy stored (tissue fat, carbohydrates, protein, and other forms, e.g., ATP), must equal energy input (food intake) for maintenance of energy balance. For the most part, consideration of output variables has been largely ignored, and the feeding side has been considered as an independent problem. Part of the reason for this is a general feeling, usually not explicit, that the correlations between input and output are linear, that individual differences are small, and that the effects of shifts in output are insignificant compared to the degrees of freedom for input change. For example, Brobeck (1967) concludes that "there is very little control of energy output, so if there is to be an energy balance, it will have to be controlled by way of energy intake." Classic nutritionists considering the problem of energy requirements, weight control, etc., also make the above assumptions and also tend to consider output as almost invariant. However, occasional observations on human subjects suggest the contrary. Hervey (1975) has pointed out that energy expenditures of a reasonably homogenous group of young men ranged from 2000 to 5500 cal/day. There are other examples that could be cited, and a few nutritionists are beginning to seriously consider the possibility that output changes may be an important factor in energy balance (Durin *et al.,* 1973; Hegsted, 1974).

Consideration of these issues has modified our own thinking about food intake and energy balance. In Fig. 1, we have separated two problems, *control of input* (top) and *regulation of energy level* (bottom). The upper section describes the classic way most of those interested in the role of the nervous system in the control of intake have attacked the problem. The available diet provides nutrient signals that the brain in turn compares to other signals coming from the energy–nutrient pool (vertical arrow), which provide information about the blood level of those nutrients. The brain

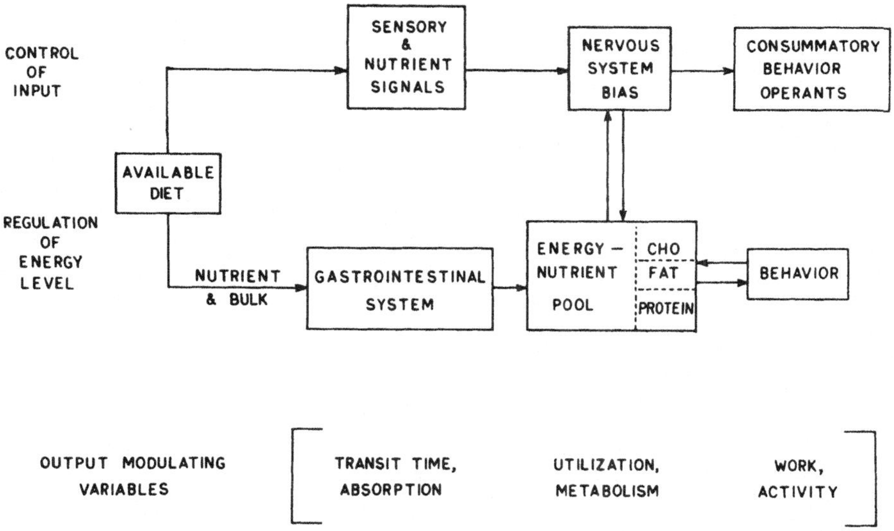

Fig. 1. Scheme of input control (top) and energy regulation (bottom) as interacting research problems. The brain modulates intake by continuous comparison of shortage or surplus in the energy–nutrient pool with sensory *and* nutrient signals in available diets. The energy–nutrient level determines the role of "taste" in food intake (vertical arrow pointing up), and conversely, "taste" signals can influence metabolic activity in the energy–nutrient pool (vertical arrow pointing down). Body weight is influenced by series of output modulating variables as well as by food intake. CHO, carbohydrate (Jacobs and Sharma, 1973).

then uses this information in modulating intake (consummatory behavior) or learned food-related behavior.

To consider the problem of energy exchange independently, scan the bottom section of Fig. 1. The available diet is usually ingested, but in this scheme it could be limited, as in a paired feeding experiment, or even intubated. Following across the bottom, the available diet provides nutrients and bulk going to the body (here called the energy–nutrient pool). From this point of view, the regulation of body weight can be effected by a series of output modulating variables, before, during, and after the food enters the energy–nutrient pool. It is easy to see that the combined action of gastric emptying time, intestinal absorption rate, efficiency of food utilization and metabolic activity (as in the classic "luxus consumption" concept, or in the lipogenic effect of spaced feeding), and the behavioral output of muscular work and general activity can be important in regulating the energy–nutrient pool. As the nutrient pool is modified, it can influence feeding itself by providing the appropriate signals as to its condition, which the nervous system uses in the control channel (vertical arrow).

Several years ago, we outlined a model of intake control which assumed that sensory as well as nutrient signals were part of the control system; the role of ingesta-produced signals (taste, aroma, etc.) in food intake was modulated by the state of energy balance (Jacobs and Sharma, 1969). Work by Nicolaidis (1969) on taste stimuli and RQ and blood glucose changes in hungry animals also suggested interaction between sensory signals (control channel) and the energy–nutrient pool (regulation channel). How would our model (Jacobs and Sharma, 1969) and the Nicolaidis scheme of taste-initiated metabolic responses fit into the scheme of Fig. 1? Very simply. *The addition* of sensory signals from the available diet in this figure, and the pair of vertical arrows on the right connecting the control channel with the regulation channel, provide a biological basis for the role of taste in energy exchange, as well as in food intake, by the various possibilities of interaction between these channels. Thus, in this scheme, body weight is not only influenced by food intake and changes in output, but by taste-induced appetite shifts from energy deficits (Jacobs and Sharma, 1969), or by taste-induced metabolic shifts in the energy–nutrient pool (Nicolaidis, 1969).

TASTE AND ENERGY DEFICIT—BEHAVIORAL ANALYSIS

Richter's views of need-related changes in palatability and taste sensitivity are well known (Richter, 1939, 1956). Our approach has been to vary hunger by food deprivation, insulin, or thyroxin injection. We then observed intake and preference shifts in response to "liked" items, e.g. fat, glucose, or saccharin, or "disliked" items, e.g., cellulose, NaCl, or quinine, which were added to stock diets or put into solution. Intact, adult rats and dogs, hypothalamic hyperphagic rats, and neonatal rats have been used as subjects (Jacobs and Sharma, 1969; Vasudev and Dua-Sharma, 1974). In our interpretation of hunger-mediated preference shifts, our logic has been identical to Richter's, but we have been interested in general energy deficit rather than specific nutrient needs. Our model assures that general hunger increases the importance of sensory signals in selection and intake, so that pleasant foods increase in pleasantness and unpleasant foods increase in unpleasantness (Jacobs and Sharma, 1969). We find that animals under energy deficit tend to be more discriminating and overreactive to taste. Similar observations were made by Nisbett (1972), who found that the food-deprived organism is highly discriminating, and food deprivation enhances the acceptance of good-tasting food. Hammer (1968) gave 30-min two-choice tests of sucrose and maltose at equal concentrations. The preference for the sweeter sucrose solution was greater at higher concentration in rats which were food deprived to a 20% body weight loss. Campbell (1958) put rats on a 40 cal/day deprivation schedule and presented them with a series of choices between two sucrose solutions in a 2-hr test. The preference threshold at each point was defined as that concentration which produced a 75% preference. Although all rats preferred the sweeter solution at all concentrations in the 2-hr test, hungry rats needed much less sweetness added to reach the preference threshold. One of the most precise animal studies from a psychophysical point of view has been that of Koh and Teitelbaum (1961), who used a tracking procedure where the rat is trained to lick a tube containing the test solution to avoid shock, or to obtain a pellet when hungry. The concentration was decreased until the animal failed to discriminate. Thresholds were determined for NaCl, QHCl, HCl, and sucrose. They found that the hungry, food-rewarded rats had significantly lower thresholds than the shock-motivated subjects ($p < 0.002$). This difference was greatest for sucrose, 47% for shocked rats, and 21% for hungry, food-rewarded rats. The value for hungry rats was quite close to the electrophysiological threshold obtained by Hagström and Pfaffmann (1959). Results such as those described above suggest that hunger does produce a change in taste sensitivity.

In our original series of diet dilution experiments, we found that adapted hungry rats ate approximately 20 g of stock diet during a 2-hr daily meal, increasing intake on a high-fat diet and decreasing intake on a high-cellulose diet (Jacobs and Sharma, 1969). This is contrary to classic observations in *ad libitum* fed rats, who adjust volume intake to caloric density, i.e., they "eat for calories." We interpreted this change in behavior as due to taste, i.e., a hunger-induced increase in fat palatability and a decrease in cellulose palatability. However, there is another equally acceptable explanation, a special case of Booth's argument that meal-time restrictions increase the importance of distinction as a limiting factor in intake (Booth, 1972). Thus, an alternative explanation of our data is based upon the fact that fat is more dense (less filling) than stock diet and cellulose is less dense (more filling) than stock diet. Our rats may be ignoring taste and calories as well, and merely eating to a constant stomach volume within the limited meal time.

We carried out a series of experiments to analyze this possibility. Figure 2 presents the results of an experiment using a graded series of dry cellulose diets ranging from stock (zero cellulose) to 100% cellulose. Each cellulose concentration was presented for 2 days in rats adapted to a 2-hr meal. If the animals were eating to maintain constancy of caloric intake, they should show an upward slope. If they were eating in proportion to gastric-filling capacity, they should decrease intake in proportion to the decrease in density with increasing amounts of the cellulose in the diet. Direct measures of density of the diets were obtained by weight–volume measures in a graduated cylinder. An empirical density gradient superimposed on this figure would show that stock diet is 0.71 g/ml, 40% dilution is

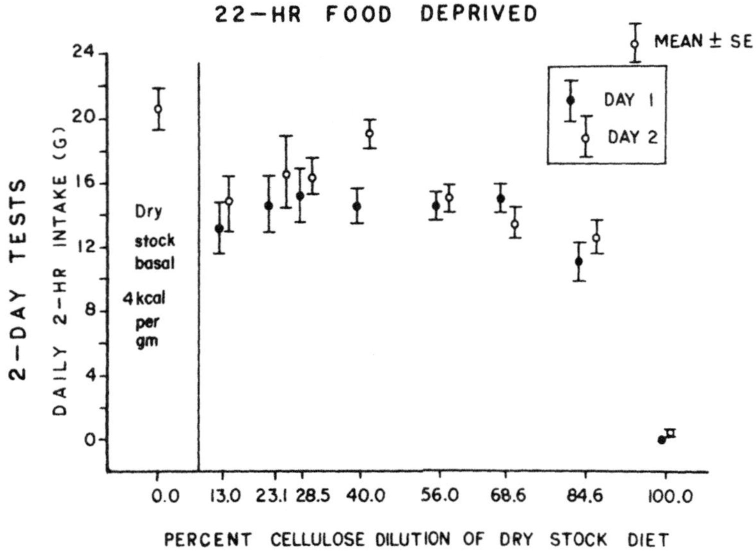

Fig. 2. Effect of increased cellulose dilution on the intake of stock diet in hungry rats in 2-day tests.

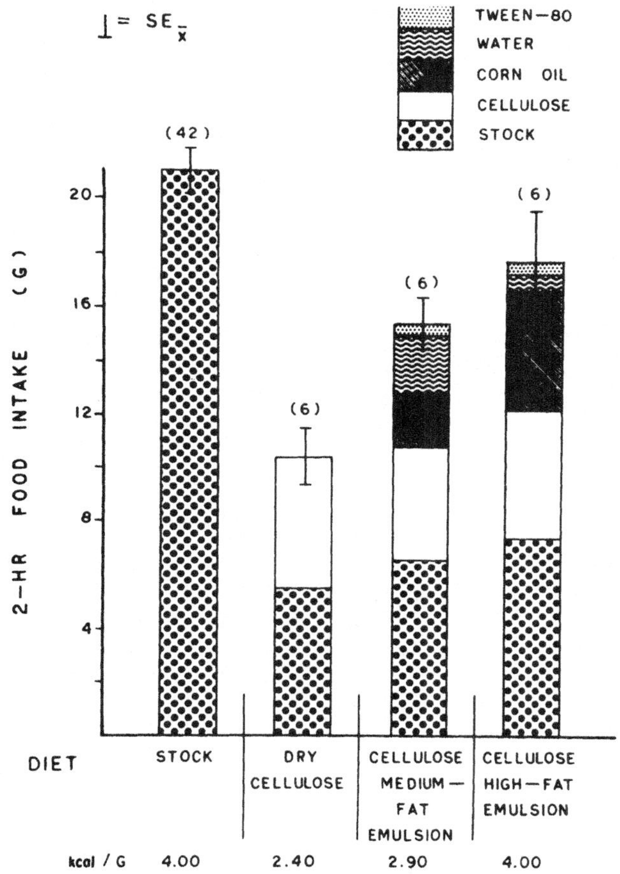

Fig. 3. Analysis of the proportion of each dietary component provided by the ingestion of various diluted diets in 22-hr deprived rats.

0.52 g/ml, 56% dilution is 0.46 g/ml, and 100% cellulose is 0.28 g/ml. The results do not support eating for caloric constancy (upward slope) or the hypothesis of eating to a constant stomach volume (downward slope). Accepting the palatability alternative for the moment, we would say that the rats "disliked" all of the dilutions equally, except for the case of pure cellulose, which we previously found to be completely ignored by hungry animals (Jacobs and Sharma, 1969). Figure 3 extends this argument by showing the actual components ingested with each diet. Diluting the stock diet with cellulose decreased total intake, and adding 13% corn oil (medium fat) or 25% corn oil (high fat) emulsion increased total intake. As we go from dry cellulose through the medium- and high-fat emulsion, the rats ingested *more* stock diet, plus *more* fat, plus the added water and

Tween-80 emulsifier in the two diets on the right. This suggests that the diets with the higher total intake do in fact fill the gastric cavity to a greater volume, presumably because taste potentiation overcomes distention inhibition.

Our final analysis of the distention argument proceeded as follows. If hungry rats on a 2-hr meal were eating in response to distention alone, we would expect a negatively accelerated intake curve in which the hungry animal would eat most rapidly at the beginning of the meal (when its stomach was empty) and slowly grind to a stop at the end of the 2 hr, as distention becomes more and more important. We began our empirical analysis of this point by dividing up the 2-hr meal of hungry rats into 30-min periods and presenting either a "palatable" high-fat diet or an "unpalatable" high-cellulose diet. Figure 4 shows the results of this experiment. The high-fat diet intakes follow the argument for the importance of distention in limiting intake. The rats ingest the palatable high-fat diet rapidly at the beginning, where distention is weakest, and slow down at the end of the meal, where taste is competing with distention. The unpalatable cellulose diet behaves quite differently. After the first 30-min session, its intake drops sharply and stays at that low level for the last three

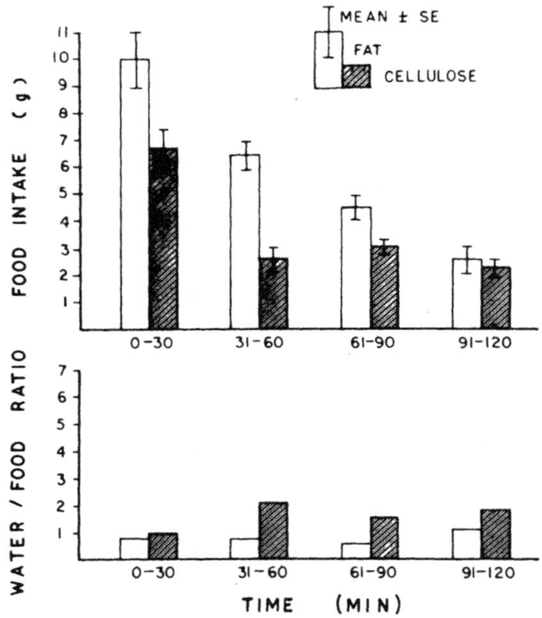

Fig. 4. *Top:* Intake of 25% fat diet (fat), or 40% cellulose diet (cellulose) in 22-hr food-deprived rats ($N = 14$). *Bottom:* Water/food ratios for same conditions ($N = 14$).

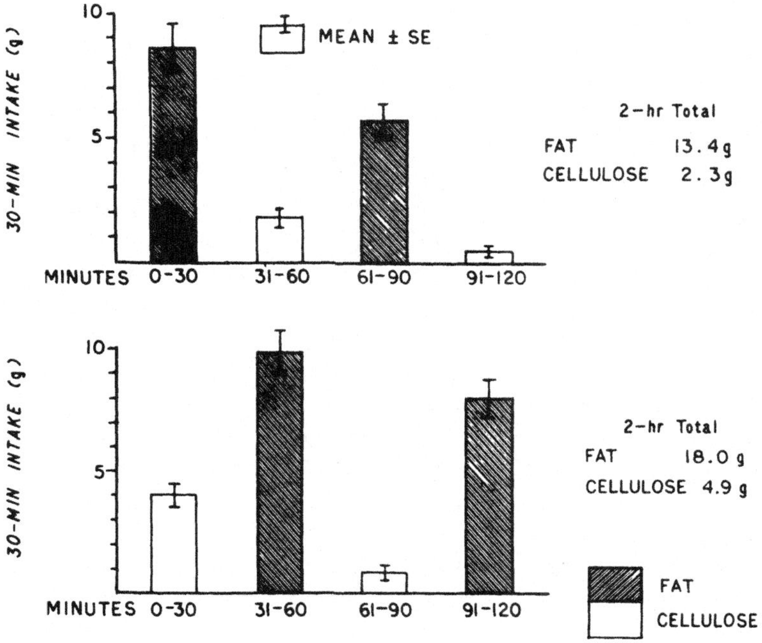

Fig. 5. Intake of 25% fat diet (fat) or 40% cellulose diet (cellulose), alternated at 30-min intervals in a 2-hr test ($N = 14$).

periods. This suggests that "bad taste" limits intake in the second and third 30-min periods, before distention reaches its maximum importance. The water/food ratios at the bottom of Fig. 4 are also independent of our presumed distention effect. As expected, high-cellulose induced more water intake than the high-fat diet.

We concluded our analysis of distention effects by alternating the high-fat and high-cellulose diets across the four 30-min periods. As shown in Fig. 5, we used two orders of presentation in alternating diets. Taste effects should be shown at the beginning of the 2-hr test to be overcome by the increasing importance of distention at the end of the test.

Let us look at the unpalatable cellulose first. When it was in the first and third sessions with taste maximized (bottom of figure), a total of 4.9 g were eaten. When it was in the second and fourth session (top of figure) where distention was maximized, total intake was reduced over 50% to 2.3 g. Thus, distention can easily overcome taste when the unpalatable high-cellulose diet is used. The results are quite different with the palatable high-fat diet. With taste maximized in the first and third session (top of figure) total fat diet intake was 13.4 g. When it was available in the second and fourth session, where distention was maximum, intake *increased* to 18.0 g. Thus the highly palatable high-fat diet easily overrides distention. In conclusion, this analysis of patterns of intake within the 2-hr time constraints shows that the level of palatability is more important than distention in determining maximum intake at the end of the daily meal.

It is also conceivable that oral cues and gastric distention interact in a fashion related to the sensory and the metabolic properties of the food, and the priority and course of the signalization are biased by the continuous monitoring of the external–internal factors (Jacobs and Sharma, 1969; Sharma *et al.,* 1967; Sharma, 1972). In hungry human subjects, Cabanac (1971) found that after ingestion of 200 ml of glucose solution, sucrose solution, or orange syrup, all three became relatively more unpleasant. If, however, the subjects reduce their body weight by 10% and kept it stable at this level for several weeks, the ingestion-induced change to unpleasantness disappeared. The return to control body weight restored the unpleasantness following ingestion.

We would interpret these results as follows. The loss in body weight acted much like increased energy deficit in our model (Jacobs and Sharma, 1969), increasing food pleasantness, delaying the ingestion-induced unpleasantness, and thus allowing the possibility of increased intake and the ultimate return to normal weight.

The behavioral work discussed in this section has an analog in more basic physiological analysis (Sharma, 1972; Sharma *et al.,* 1975). In the following sections, electrophysiological evidence is represented which shows that gustatory afferent discharge is influenced in a predictable fashion depending upon oral and gastric factors. Identical gastric signals under *ad libitum* feeding or chronic food deprivation bring about qualitatively dissimilar gustatory responses. This shows that the control system is complex and that nutritional status is an important factor biasing the way sensory signals are handled by the organism.

ENERGY DEFICIT—GASTRIC MODULATION OF TASTE

Electrophysiological Analysis

Gustatory receptors responding to various nutrient substances show distinguishing features not rigidly specific to a particular substance used. The quality and intensity coding of the gustatory message (Ganchrow and Erickson, 1970), the temperature and flow rate (Cohen *et al.,* 1955), and the spatial and temporal characteristics of application are all reflected in the gustatory discharge pattern. Pfaffmann (1970) pointed out that afferent taste units have multiple sensitivity and that fine discrimination is due to overlapping sensitivities rather than high receptor specificity. Studies on intracellular recordings from gustatory papillae (Sharma and Doss, 1973; Sharma *et al.*, 1973) show low membrane depolarization characteristics modulated by various forms of stimuli applied locally or elsewhere.

The resting membrane potential in typical gustatory receptors of frog varies between −10 and −25 mV. These cells are electrically excitable and 1–3 nA pulses of 4 msec duration are sufficient to depolarize most of these cells. The electronic potentials generated had characteristic exponential waveforms from which time constants were calculated for the rising phase. Membrane resistances and capacitances were also determined. Electrical stimulation and topical application of a chemical substance have additive effects and influence excitation level, the rate of rise of receptor potential, and other depolarization characteristics (Sharma and Doss, 1973). When the impaled receptor cell was stretched to varying degrees, membrane potential showed further depolarization changes. A stretch of 50 U produced 10–15 mV depolarization, and the depolarization level was virtually linear for increasing stretch up to 150 U. As in the case of electrical stimulation, the chemical application interacted with the state of receptor stretch, and the resultant change was guided by these interacting factors. These results suggest that the input signal is profoundly influenced by the state of the receptor stretch and the dynamic features of the receptor characteristics as influenced by local factors besides the nature, amount, and mode of application of the chemical substance.

Another important dimension modulating the receptor activity is the state of the nutritional status of the animal. In well-fed frogs (type I), the unit activity recorded through gustatory afferents on topical lingual application of glucose gradually increases and obtains peak values in 1–2 sec, after which the activity slowly returns to the base line within 2–3 sec. The magnitude and time course of glucose response obtained in a typical type I animal is shown in Fig. 6. With increase of glucose concentration, the magnitude of response increases but the overall response profile remains similar. In chronically food-deprived animals (type II), the magnitude of glucose response is higher as compared to type I animals (Fig. 6), and the peak frequency is obtained almost immediately. This increased activity in type II animals remains at peak level for about 500 msec before it gradually returns to the basal level in 3–4 sec. In the case of NaCl, though the magnitude of response is higher and may last longer in type II animals, the peak frequency level is obtained almost immediately as observed in type I animals. Amino acid responses also show shorter latency and higher magnitude of response as compared to type I, but the responses to quinine and

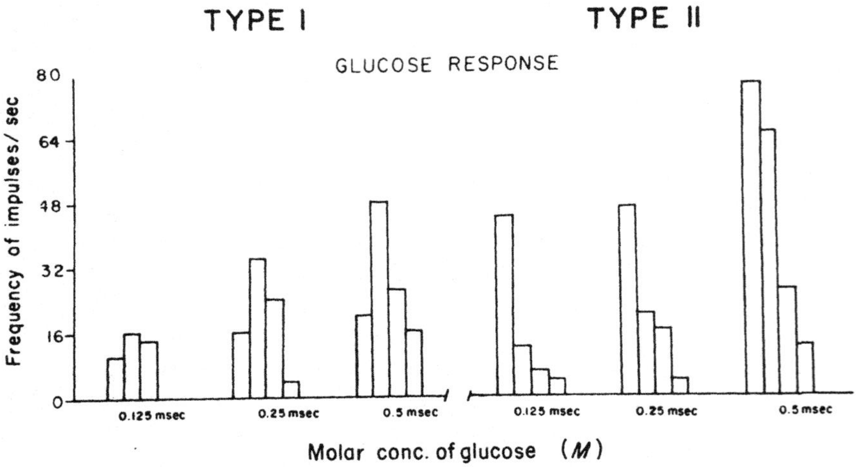

Fig. 6. Glucose responses obtained from the glossopharyngeal nerve in type I and type II animals. Each bar represents activity for 500 msec.

acetic acid have not shown a consistent pattern appreciably different from that observed in type I preparations.

Gastric distention inhibits the gustatory evoked activity obtained by lingual application of substances like glucose, NaCl, quinine, and amino acids (Dua-Sharma *et al.*, 1973). The amount and pattern of inhibition produced depends on the rate, degree, and duration of distention as well as the mode and temporal relation of gastric distention to gustatory evoked activity. In well-fed animals (type I), 1.0 ml distention may be sufficient to produce 15 to 30% inhibition of the evoked gustatory response. With 4–5 ml distention this may increase to 80–90% inhibition, and the gustatory discharges may be completely abolished with further increase of distention volumes to 6–8 ml in the majority of the cases. The inhibition is obtained within a minute of the gastric distention, and the inhibitory effects may continue as long as gastric distention is maintained; effects have been observed for over 40–50 min during maintained gastric distention. Similarly, a rapid rate of equal distention volumes produces higher magnitude of peak inhibitory response, which may later stabilize at values equivalent to that produced by a slower rate. Thus, the inhibitory changes follow a time course of dynamic and static phases after a variable latency of 5–10 sec to 1–2 min. Similar results are obtainable from intracellular recording at the receptive site (Sharma and Doss, 1973).

Figure 7 presents in graphic form the distention effects on glucose-evoked glossopharyngeal afferent activity in frogs under intact conditions and after gastric vagotomy. The peak frequency discharge of 36/sec obtained on glucose application within 30 sec is decreased to 30/sec when repeated with maintained gastric distention of 1.0 ml. After left gastric vagotomy, the glucose-evoked activity is seen enhanced to almost 54/sec, and remains elevated after initial decrement for over 5 sec. If gastric distention is now repeated with glucose application, the peak activity level reaches 48/sec frequency discharge, which is lower than that obtained by glucose alone in vagotomized condition but is significantly higher than that observed under intact conditions. Thus, the amount of inhibition produced on the gustatory response by gastric distention is considerably minimized after the left gastric vagotomy. This distention-evoked inhibitory influence on gustatory response is further diminished by bilateral gastric vagotomy. By contrast, cervical sympathectomy does not seem to influence the inhibitory effects of gastric distention on gustatory responses. It does, however, decrease the magnitude of basal gustatory response by 5–40%.

Distention effects on gustatory responses to NaCl application are similar to that seen for glucose, the inhibitory influence being potentially minimized after gastric vagotomy. Distention effects on gustatory responses to amino acids have also been generally similar to that observed for glucose and NaCl. The inhibitory effects, however, are less marked. For example, 1 ml distention produced 5–20% decrease in amino acid-evoked gustatory response. With 4–5 ml distention this inhibition may be increased to 30–45%. Coupling of gastric distention with lingual applica-

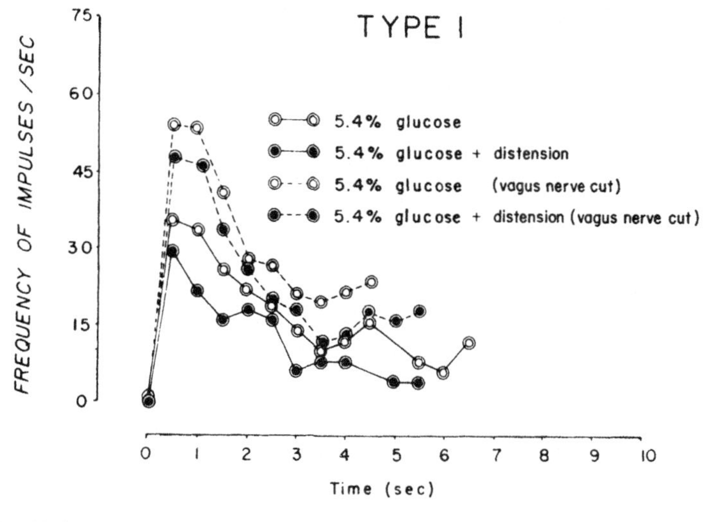

Fig. 7. Unit recordings from glossopharyngeal afferents in type I frogs on topical application of glucose to the tongue alone or coupled with gastric distention (1.0 ml) before and after vagotomy.

tion of quinine produced variable effects, the effects ranging from slight facilitation to moderate inhibition. Though the acetic acid-evoked gustatory response was also inhibited by gastric distention, repeated trials usually failed to show the effect consistently.

Intestinal distention, particularly jejunal distention, also showed qualitatively similar inhibitory effects on gustatory responses as observed on gastric distention. The latency of response varied from 10–30 sec to 2–3 min, and a stable detectable inhibition of 10–25% was obtained with 0.5–1 ml distention of the miniature balloon placed in the gut. Distention with higher volumes did not produce stable results, and often resulted in deterioration of the experimental preparation.

In chronically food-deprived (type II) frogs, contrary to the findings in type I animals, gastric distention facilitates rather than inhibits gustatory receptor discharge. Figure 8 compares the gustatory responses to glucose under conditions of empty and distended stomach before and after gastric vagotomy. In the example shown, there is a marked increase in the frequency of average glucose-evoked gustatory discharge from 29/sec to 85/sec when glucose is applied during maintained distention of 3.5 ml. The facilitatory effect persists for 2–3 sec before the activity gradually returns

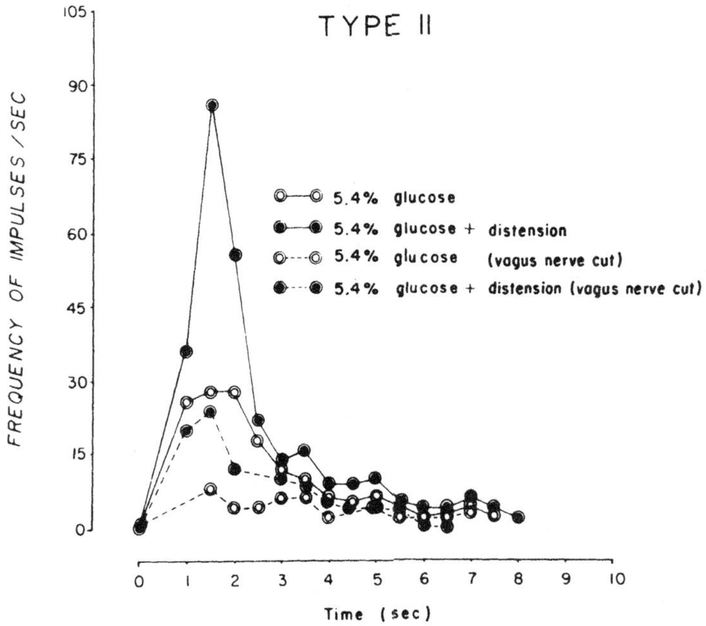

Fig. 8. Unit recordings from glossopharyngeal afferents in type II frogs on lingual application of glucose alone or coupled with gastric distention before and after vagotomy.

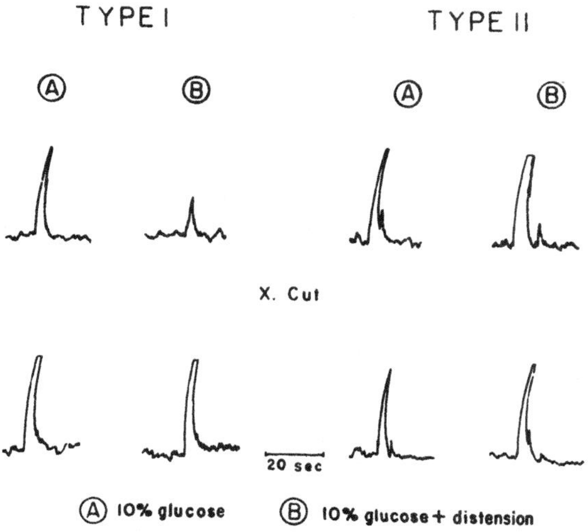

Fig. 9. Summated responses from glossopharyngeal afferents on lingual application of glucose and gastric distention in type I and type II frogs before and after gastric vagotomy. X. cut, 10th nerve was cut.

to basal level. Figure 9 shows the summated responses recorded from glossopharyngeal afferents in type I and type II animals on lingual application of glucose coupled with gastric distention before and after sectioning of the left gastric vagus. Clearly, gastric distention produces facilitation in type II and inhibition in type I animals.

If gastric distention is maintained in type II animals, a second application of glucose after an interval of 8–10 min may produce identical facilitation. Repeating the solution earlier shows less marked facilitation and may not show any response if applied within 2–3 min of first application. The amount of facilitation increases with higher volumes of distention, and maximal facilitation has been obtained with 6–8 ml distention. Beyond these volumes, there may not be further increase in facilitation. Thus, the rate and degree of distention and the temporal relation of gustatory stimulus to gastric distention influence the amount and pattern of facilitation observed. The glucose-evoked gustatory activity is decreased after gastric vagotomy, but as illustrated in Fig. 9, gastric distention (3.5 ml) after vagotomy does increase the average gustatory discharge from 8/sec to 24/sec. This 200% increase in evoked response is quite comparable to the distention-evoked facilitation obtained before vagotomy. Thus, while the gustatory evoked response by itself is considerably reduced after vagotomy, the magnitude of facilitatory change remains virtually similar.

Could this facilitation be mediated via extravagal fibers? In an earlier study Sharma *et al.* (1972) had observed that a dual efferent control of gastric receptors exists in which vagal influence is inhibitory and sympathetic influence is facilitatory to the gastric receptor discharge. It is likely that a similar dual control system operates at the oral level, which may explain the differential effects of gastric distention on gustatory discharge, facilitating the activity in chronically food-deprived animals and inhibiting it in well-fed animals. To test this, an electrical stimulation series was undertaken.

Electrical Stimulation

Stimulation of gastric vagal fibers with standard parameters (pulses of 1 msec duration, 50–60 cps, and 0.34 V passed for 10 sec) produces a 15–40% inhibition of the gustatory response recorded from glossopharyngeal (IX) afferents. Ipsilateral stimulation (e.g., stimulation of left gastric vagal fibers and recording from left IX afferents) produces more marked inhibition of the gustatory receptor discharge, and the bilateral stimulation effects are additive. The maximal inhibitory effects are obtained almost immediately on stimulation, and these inhibitory influences may last for 8–10 min with gradual diminution. Figure 10 shows that these changes occur at the receptor level. Stimulation of the gastric vagus brought about decreased excitability (hyperpolarization) in the receptor, the activity returning to the basal level in 8 to 10 min—a time course similar to that observed in gustatory unit discharges.

After unilateral gastric vagotomy, stimulation of the central stump produces inhibition of the gustatory response equivalent to the effects seen on stimulation of intact fibers. Stimulation of the peripheral cut end, however, does not seem to modulate gustatory receptor discharge. Central stimulation effects are present after bilateral gastric vagotomy and indicate that the inhibitory effects of gastric distention are mediated via central connections of gastric vagal fibers.

Sympathetic stimulation effects are facilitatory to the gustatory receptor discharge (Figure 10), and the resultant effects depend on the mode and parameters of stimulation used. Standard stimulation of left cervical ganglia may produce 60–180% facilitation of the gustatory response ipsilaterally, which may be less marked contralaterally. This facilitation may be further enhanced to about 300% if bilateral stimulation is done. The maximal facilitatory effects are produced almost immediately following the stimulation, and these facilitatory effects may last for 6–10 min. As in the case of the gastric vagus, stimulation of the central cut ends produces equivalent facilitation to that observed with intact sympathetic nerves.

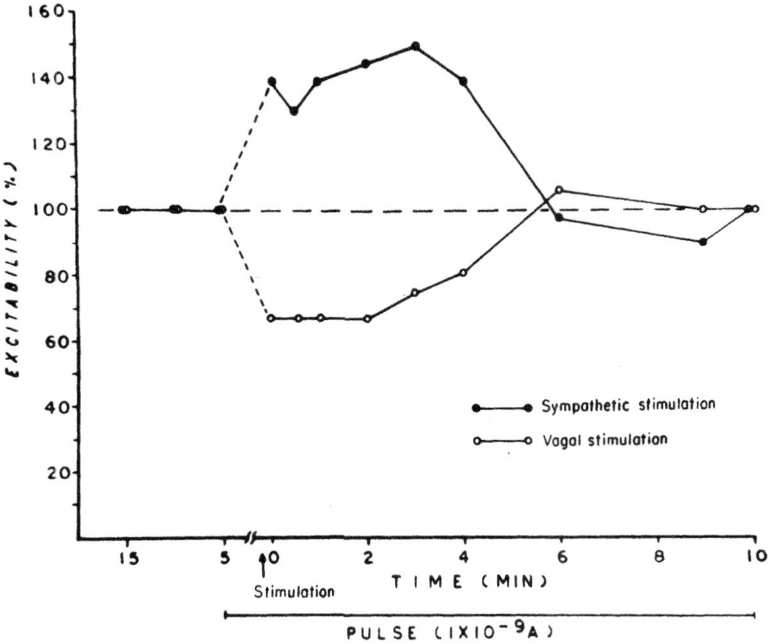

Fig. 10. The depolarization level obtained by single pulses (4 msec duration) is shown as 100% excitability. The dotted line indicates the extrapolated values for 100% excitability level. Vagal stimulation (pulses of 0.34 V, 50 cps, 1 msec duration for 10 sec) decreases the excitability while similar sympathetic stimulation increases it.

Stimulation of the peripheral cut ends also shows some facilitation (15–60%) of the evoked gustatory response. The peripheral cervical sympathetic stimulation effects thus differ from peripheral gastric vagal stimulation effects in that the peripheral vagal stimulation does not change the gustatory evoked response.

Thus, while the distention series allowed the effects of nutritional state on gustatory responses to be evaluated under more physiological conditions, the electrical stimulation series emphasized the nature and pathways of the control system. Gastric distention effects on gustatory discharge in rats show similar differential modulation in that gastric distention produces facilitation of the gustatory evoked activity in chronically food-deprived preparations and an inhibition in well-fed animals. However, the nature and pathways involved in these orogastric interactions in rat have not been worked out in detail. Hellekant (1971) recorded efferent activity in chorda tympani nerve of rat and found that gastric distention decreased the efferent activity. This inhibitory influence was diminished after sectioning vagal nerves, suggesting that the inhibitory impulses initiated at the gastric level pass via vagal fibers.

Besides gastric mechanoreceptors (Sharma, 1967b), the presence of gastric chemoreceptors has also been reported (Sudakov and Rogacheva, 1963; Sirotin, 1961). In our recent studies we have found chemoreceptors responding to sweet substances (glucose, sucrose, saccharin), amino acids (lysine, glycine, glutamic and aspartic acid), and salts existing in the stomach wall of the frog and presenting distinguishing discharge patterns related to the general class of the nutrient substances. These gastric afferent impulses project to the region of the fasciculus solitarius, and probit analysis shows that the responses are characteristic for each test solution in that the relative frequency among parallel neurons and the time course of the frequency could be recognized by means of a quality message, while the total amount of activity (i.e., overall height) among the neuronal population appeared to convey the stimulus concentration (Ramakrishna and Sharma, 1975). Since it is known that sensory input from glossopharyngeal and vagus nerves in amphibians converge in the fasciculus solitarius (Herrick, 1944), it is quite likely that the gastric chemoceptive impulses interact with gustatory impulses at the brainstem level and thus influence feeding reflexes.

SUMMARY

Chemosensory signals from food provide the sensory basis of the hedonic matrix that controls food acceptance, choice, and intake. Classically, this has been of interest to the sensory physiologist and sensory psychologist, food and flavor chemist, and, ultimately, the market researcher and food manufacturer (see Moskowitz, chapter 4).

Our view is that sensory and nutrient signals from food are also intimately involved as primary *biological* variables in food acceptability, choice, and intake, and should be of direct interest to the nutritionist and metabolic physiologist as well.

We have described a series of experiments which extends the behavioral work on our sensory/metabolic model of food intake (Jacobs and Sharma, 1969), and broadens our electrophysiological work on the role of nutritional state in autonomic modulation of the gastric afferent system (Sharma *et al.*, 1972) to include direct work in the taste system itself.

Although most of the work has been on animal models, and the bridge between the electrophysiological events and the food-related behavioral events is still very primitive, our analysis of the general problem of the role of sensory and nutrient signals, input control, and energy exchange (Fig.

1) is potentially as applicable to man as to the rat. What is needed is more direct work on man.

ACKNOWLEDGMENT

Supported in part by PL-480 Grant NIH 01-015-1.

REFERENCES

Booth, D. A. (1972). Caloric compensation in rats with continuous or intermittent access to food. *Physiol. Behav.* **8,** 891–899.

Boring, E. (1941). "Sensation and Perception in the History of Experimental Psychology." pp. 437–458. Appleton, New York.

Brobeck, J. R. (1967). Discussion. *In* "Chemical Senses and Nutrition" (M. R. Kare and O. Maller, eds.), p. 135. John Hopkins Press, Baltimore, Maryland.

Cabanac, M. (1971). Physiological role of pleasure. *Science* **173,** 1103–1107.

Campbell, B. A. (1958). Absolute and relative sucrose preference thresholds in hungry and satiated rats. *J. Comp. Physiol. Psychol.* **51,** 795–797.

Cohen, M. J., Hagiwara, S., and Zotterman, Y. (1955). The response spectrum of taste fibers in the cat: A single fiber analysis. *Acta Physiol. Scand.* **33,** 316–332.

Dua-Sharma, S., Sharma, K. N., and Jacobs, H. L. (1973). The effect of chronic hunger on gustatory responses in the frog. *Physiologist* **16,** 300.

Durnin, J. V. G.A., Edholm, O. G., Miller, D. S., and Waterlow, J. C. (1973). How much food does man require? *Nature (London)* **242,** 418.

Epstein, A. N. (1967). Oropharyngeal factors in feeding and drinking. *Handb. Physiol., Sect. 6: Aliment. Canal* **1,** 197–218.

Ganchrow, J. R., and Erickson, R. P. (1970). Neural correlates of gustatory intensity and quality. *J. Neurophysiol.* **33,** 768–783.

Hagström, E., and Pfaffmann, C. (1959). Relative taste effectiveness of different sugars for the rat. *J. Comp. Physiol. Psychol.* **52,** 259–262.

Hammer, L. (1968). Relation of reinforcement value to consummatory behavior. *J. Comp. Physiol. Psychol.* **66,** 667–672.

Hegsted, D. M. (1974). Energy needs and energy utilisation. *Nutr. Rev.* **32,** 33–38.

Hellekant, G. (1971). The effect of stomach distention on the efferent activity in the chorda tympani nerve of the rat. *Acta Physiol. Scand.* **83,** 527–531.

Herrick, C. J. (1944). The fasciculus solitarius and its connections in amphibians and fishes. *J. Comp. Neurol* **81,** 307–331.

Hervey, G. R. (1975). The problem of energy balance in light of control theory. In "Neural Integration of Physiological Mechanisms and Behavior" (G. J. Mogenson and F. R. Calaresu, eds.), pp. 109–127. Univ. of Toronto Press, Toronto.

Jacobs, H. L., and Sharma, K. N. (1969). Taste vs calories: Sensory and metabolic signals in the control of food intake. *Ann. N.Y. Acad. Sci.* **157,** 1084–1125.

Jacobs, H. L., and Sharma, K. N. (1973). Taste as a biological parameter in the control of food intake and the regulation of energy balance. Presented at a symposium in honor of J. A. F. Stevenson, University of Western Ontario, London, Canada, May, 1973 (unpublished).

Koh, S. D., and Teitelbaum, P. (1961). Absolute and behavioral taste thresholds in the rat. *J. Comp. Physiol. Psychol.* **54,** 223–229.

Le Magnen, J. (1971). Olfaction and nutrition. *Handb. Sens. Physiol.* **4,** Part I, 465–482.

Maller, O., and Desor, J. A. (1974). Effect of taste on ingestion by human newborns. *In* "Oral Sensation and Perception: Development in the Fetus and Infant" (J. Bosma, ed.), pp. 279–291. U.S. Govt. Printing Office, Washington, D.C.

Nicolaidis, S. (1969). Early systemic responses to orogastric stimulation in the regulation of food and water balance: Functional and electrophysiological data. *Ann. N.Y. Acad. Sci.* **157,** 1176–1203.

Nisbett, R. E. (1972). Hunger, obesity and the ventromedial hypothalamus. *Psychol. Rev.* **79,** 433–453.

Pfaffmann, C. (1970). Physiological and behavioural processes of the sense of taste. *Taste Smell Vertebr., Ciba Found. Symp., 1969* pp. 31–50.

Ramakrishna, T., and Sharma, K. N. (1975). Organization and characteristics of gastric chemoceptive neurons in frog brainstem. *Proc. Indian Acad. Sci., Sect. B* **82,** 1–24.

Richter, C. P. (1939). Salt-taste thresholds of normal and adrenalectomized rats. *Endocrinology,* **24,** 367–371.

Richter, C. P. (1956). Salt appetite in mammals: Its dependence on instinct and metabolism. *In* L'instinct dans le comportment des animaux et de l'homme" (P. P. Grassé, ed.), pp. 577–629. Masson, Paris.

Sharma, K. N. (1967a). Alimentary receptors and food intake regulation. *In* "Chemical Senses and Nutrition" (M. R. Kare and O. Maller, eds.), pp. 281–291. Johns Hopkins Press, Baltimore, Maryland.

Sharma, K. N. (1967b). Receptor mechanisms in the alimentary tract: Their excitation and functions. *Handb. Physiol., Sect. 6: Aliment. Canal* **1,** 225–237.

Sharma, K. N. (1972). Nutrition and neural controls in alimentary signalization: Periphery revisited. *Proc. Indian Sci. Congr., 59th* pp. 1–16.

Sharma, K. N. (1975). Ontogenetic and nutritional modulation of alimentary signalization. *In* "Growth and Development of the Brain" (M. A. B. Brazier, ed.), pp. 191–202. Raven Press, New York.

Sharma, K. N., and Doss, M. J. K. (1973). Excitation and control of gustatory chemoreceptors. *Proc. Int. Conf. Med. Biol. Eng., 10th, 1973* Vol. 3, p. 53.

Sharma, K. N., Dua-Sharma, S., and Gopal, V. (1967). Orogastric factors involved in the modulation of the sensory control of food ingestion. *Indian J. Physiol. Pharmacol.* **11,** 22, 1967.

Sharma, K. N., Jacobs, H. L., Gopal, V., and Dua-Sharma, S. (1972). Vagosympathetic modulation of gastric mechanoreceptors: Effect of distention and nutrition state. *J. Neural Trans.* **33,** 113–154.

Sharma, K. N., Dua-Sharma, S., Doss, M. J. K., and Jacobs, H. L. (1973). Distention-evoked differential modulation of membrane depolarisation characteristics of lingual chemoreceptors. *Ann. Conf. Soc. Neurosci.* Vol. 3, p. 174.

Sharma, K. N., Dua-Sharma, S., and Jacobs, H. L. (1975). Electrophysiological monitoring of multilevel signals related to food intake. *In* "Neural Integration of Physiological Mechanisms and Behavior" (G. J. Mogenson and F. R. Calaresu, eds.), pp. 194–212. Univ. of Toronto Press, Toronto.

Sirotin, B. Z. (1961). Electrophysiological study of reception from certain internal organs in man. Report I. Impulses from receptors of the resected stomach and small intestine. *Bull. Exp. Biol. Med. (Engl. Transl.)* **50,** 873–877.

Sudakov, K. Y., and Rogacheva, S. K. (1963). The afferent and efferent activity of the fiber of the vagus nerve during fasting and after taking food. *Fed. Proc., Fed. Am. Soc. Exp. Biol.* **22,** Trans. Suppl. 306–310.

Vasudev, R., and Dua-Sharma, S. (1974). Gustatory metabolic consequences of thyroxine treated rats. *Congr. Physiol. Sci.* [*Proc.*], *26th, 1974* Vol. 11, p. 232.

DISCUSSION

Senseman: Is it likely that there are circulating hormones modulating receptor input?

Sharma: Topical application of various substances as well as systemic administration has shown that the tactually evoked activity may be modified for 20 to 30 min or even more. The modulation is dependent on the locus and the amount, mode, and nature of the substance used. These types of studies would indicate that the local environment as well as systemic changes can bring about modulation of gastric receptor activity.

Nicolaidis: Do you cut the nerve so there is no centrifugal inhibition?

Sharma: It depends on the experiment. While recording from gustatory afferents, the gastric vagal fibers are intact.

Nicolaidis: Could there be centrifugal inhibition? What happens if the nerve is cut?

Sharma: If we record from gastric afferents, say from sympathetic nerve (S_1) in the frog, and cut the gastric vagal fibers unilaterally, the stimulation of the peripheral cut end of the gastric vagus does produce inhibition of the tactually evoked response recorded through S_1. Stimulation of the central cut ends of the gastric vagal fibers also produces a similar inhibitory response. If, however, the gastric vagal fibers are sectioned on the contralateral side as well, then the stimulation of the central gastric vagal fibers does not produce any inhibition in the tactually evoked response. It would, thus, appear that stimulation of the central cut end following unilateral gastric vagotomy was producing the inhibitory effect mediated through contralateral gastric efferent fibers, and when these contralateral fibers were also cut the inhibition was lost. In the case of the oral cavity, the story is somewhat different. If you cut the glassopharyngeal nerve and stimulate its peripheral cut end, it does produce an inhibition of the gustatory response. But if one stimulates the central cut end, it facilitates the response. This facilitation is routed, in frogs, via hypoglossal–sympathetic connective. In other words, the centrifugal influence is a dual one—inhibitory and facilitatory—and the efferent control is mediated through other cranial nerves as well.

Nicolaidis: Could you measure latencies in these responses?

Sharma: We have not done a systematic study on the latencies for the gustatory responses. But the preliminary data indicate that the responses are oligosynaptic and in some cases the latencies have been as low as 10 to 20 msec.

Nicolaidis: Usually electrophysiology defines what has been observed previously by behavioral studies. You seem to have used electrophysiological methods to define behavioral responses. When you apply glycine locally to the gastric mucosa, you observe a decrease in the tactually evoked response, and, curiously enough, when you give systemic glycine, you observe the opposite effect, i.e., the facilitation of the tactually evoked response.

Sharma: Most of these interactions are multilevel. Glycine applied locally did inhibit tactually evoked response, while the systemic application produced different results. It is quite likely that the primary side of action of systemic administration of glycine was different, which could have released the peripheral centrifugal inhibition thereby producing a facilitatory type of evoked response from the gastric receptors. It would be interesting to develop a working model and see how various priorities are established and interact at different levels of neuraxis.

Nicolaidis: Can you test frogs behaviorally after systemic injection of glycine? Could that increase food intake?

Sharma: Yes. This could be attempted, though we have not done it so far.

Burghardt: Were these acute experiments with the frogs? Did you actually get them to eat? I am wondering whether they actually ever showed a preference for sugar?

Sharma: No. The food deprivation is chronic over months, and primarily electrophysiological investigations have been done in such chronically food-deprived preparations. We did not study the food preference behavior in frogs but rather analyzed the electrical activity from orogastric afferents on giving various types of substances.

Burghardt: What did you feed them and what species of frogs were they?

Sharma: The studies were conducted in *Rana tigrina* and *Rana hexdectyla*, the two common varieties of South Indian frogs. Essentially, two groups of frogs were used. Type I comprised the frogs which were well fed and were maintained in 2 ways: (1) freshly collected frogs for immediate use; (2) others kept up to 3 months during which period their weight was maintained with daily manual feeding of 2–6 g minced meat. The second group of frogs (type II) were chronically food deprived for periods of up to 12 months.

Miller: At what temperature were you keeping the frogs?

Sharma: The frogs were maintained in tanks at 22° ± 2°C.

Pfaffmann: What is the weight difference between type I and type II in the actual testing situation?

Sharma: If the frog has been kept on food deprivation for, say about 6 to 8 months, the body weight is reduced by almost 50 to 60%. The weight is lost in a steplike fashion every 4 or 5 days.

Chapter 9

Influence of Protein Nutrition on the Olfactory Bulb

HANS FISHER

The importance of protein nutrition in brain metabolism has received added confirmation in recent years through the findings of Wurtman and associates. These workers demonstrated that rat brain synthesis of neurotransmitters depended on the availability of the amino acid precursors tryptophan (for serotonin) and tyrosine (for dopamine and norepinephrine); in turn, the level of precursors was dependent on the protein nutriture of the animal (Wurtman and Fernstrom, 1975). The recent discovery of a relatively high concentration of the histidine-containing dipeptide, carnosine, in the olfactory bulb of rats and mice has raised interesting questions regarding the relationship between the dietary supply of histidine and the role of imidazole peptides in olfaction.

Histidine, one of the basic amino acids, was first isolated and identified in 1896 (Greenstein and Winitz, 1961). Its indispensability in the diet of mice and rats was reported by Abderhalden in 1922 and confirmed by Rose and Cox in 1924.

Several histidine-containing dipeptides have been found in animal tissues. Carnosine (β-alanyl-L-histidine) was isolated in 1900 by Gulewitsch and Admiradzibi. Anserine (β-alanyl-L-1-methylhistidine) was first described by Ackermann *et al.* in 1929, and more recently ophidine (β-alanyl-L-3-methylhistidine) and homocarnosine (γ-aminobutyrylhistidine) have been isolated and identified (Crush, 1970). The chemical structure of these dipeptides is shown in Table I.

TABLE I

Chemical Structure of Histidine and Histidine-Containing Dipeptides

Compound	Structure
Histidine	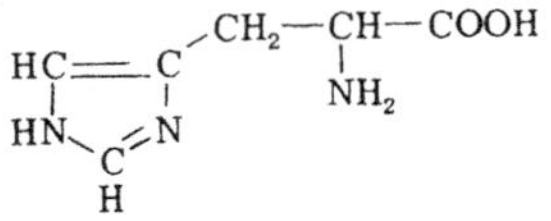
Carnosine	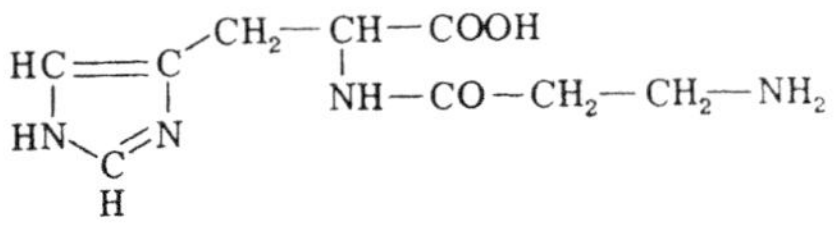
Homocarnosine	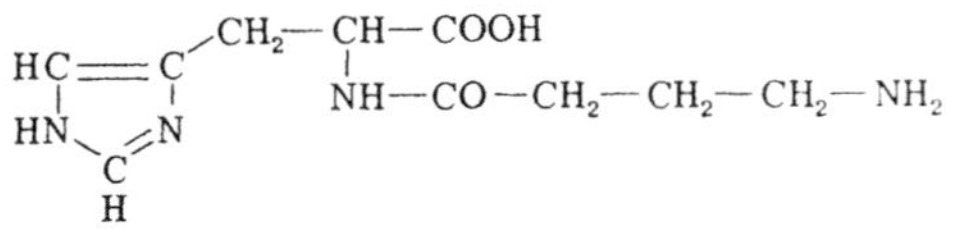
Anserine	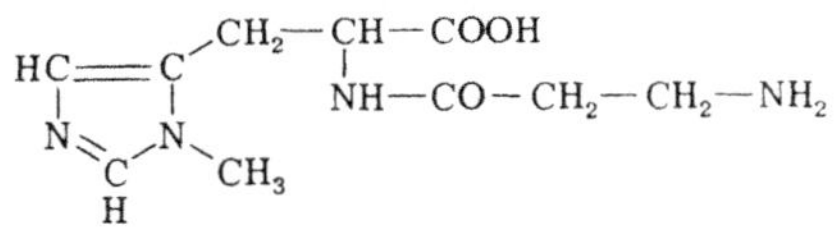
Ophidine	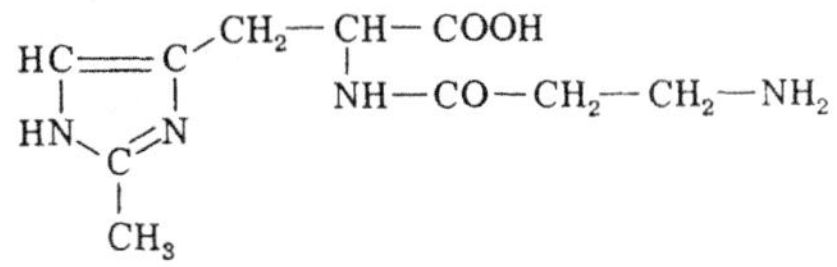

Until recently the study of histidine-containing dipeptides was severely handicapped by considerable difficulty in the methodology of separating and determining their concentration in animal tissues. Only through recent advances in column chromatography have better separations of these compounds been achieved with few of the disadvantages that plagued earlier methods (Crush, 1970).

Since the isolation and identification of carnosine at the turn of the present century it has been known that histidine-containing dipeptides were present in muscle tissue of various animal species. It was early demonstrated that they constituted the second most important nitrogenous constituent of muscle after creatine (du Vigneaud and Behrens, 1939). Many of the muscle tissue values for histidine-containing dipeptides (Table II) predate the latest chromatographic analytical techniques and must, therefore, be viewed with caution. In general, mammalian species contain both carnosine and anserine in muscle tissue with anserine often, but not always, present in larger amounts than carnosine. In all avian species studied to date, anserine is the predominant dipeptide with carnosine present at much lower muscle concentrations (Table II).

Ophidine has been reported as the quantitatively most important histidine-containing dipeptide in muscle tissue of the whale and dolphin, with smaller concentrations of carnosine and no detectable anserine (Table II). Snake muscle also contains ophidine; in fact, the name is derived from the Latin word for the zoological suborder for snakes.

In 1960, Hosein and Smart reported the presence of anserine and carnosine in rat brain tissue using column and paper chromatographic techniques. In 1962, Abraham *et al.* confirmed the presence of carnosine in brain tissue of several species but showed that homocarnosine was present in larger concentration than carnosine, and made no mention of finding anserine. Very recent observations from our laboratory indicate that anserine is generally not present in rat brain (Chung-Hwang *et al.*, 1976), although a trace was found on a single occasion (M. Quinn and H. Fisher, unpublished results). Table III gives brain homocarnosine concentrations of several animal species.

TABLE II

Muscle Concentration of Histidine-Containing Dipeptides of Various Animal Species[a]

Animal	Muscle	Carnosine (mg/100 g tissue)	Anserine (mg/100 g tissue)	Ophidine (mg/100 g tissue)
Agouti	Leg	160	550	—
Mouse	Mixed	15	25	—
Rat	Mixed	70	170	—
Cat	Mixed	220	400	—
Seal	Pectoral	200	60	
Rabbit	Gastrocnemius	80	502	
Frog	Mixed	120	—	—
Duckling	Breast	+	230	
Goose			120	
Chicken	Pectoral	278	983	
Pigeon	Pectoral	10	140	—
Fin whale	Longissimus dorsi	97		1082
Blue whale	Longissimus dorsi	9		1080
Dolphin		131		485

[a] Modified after Crush (1970).

TABLE III

Concentration of Brain Homocarnosine in Several Animal Species

Animal	Brain homocarnosine (nmole/mg)
Rat[a]	0.06
Mouse[b]	0.08
Hamster[c]	+
Guinea pig[a]	0.18
Rabbit[a]	0.18
Chicken[a]	0.03
Dog, cat[a]	Trace

[a] Abraham *et al.* (1962).
[b] Neidle and Kandera (1974).
[c] M. Quinn and H. Fisher (unpublished data).

Tsunoo *et al.* (1963a), reported the isolation and identification of free histidine and anserine in chicken brain and their inability to confirm the finding of homocarnosine in chicken brain as reported by Abraham and co-workers (1962). The same workers also isolated and identified anserine from the eye of the chicken (Tsunoo *et al.*, 1963b). This contrasts with the observation of Krause (1936), who reported finding carnosine, but no anserine, in ocular tissue of cattle.

The most interesting recent addition to our knowledge concerning the distribution of histidine-containing dipeptides has been the finding that in brain tissue of rat and mouse carnosine is highly concentrated in the olfactory bulb (Neidle and Kandera, 1974; Margolis, 1974).

Little is known concerning the function of histidine-containing dipeptides in the tissues in which they are found. Severin and associates (1963) have suggested a role in muscle contraction, but the exact nature of this function is still not known. These workers made the interesting observation that portions of frog muscle that were rich in nerve endings contained more carnosine than did the portions that were free of nerve tissue. Also, different sections of muscles that had a uniform distribution of nerve endings showed a similar carnosine content. This led the authors to suggest the possibility that carnosine participates in the act of transmission of the stimulus from nerve to muscle, in other words, as a neurotransmitter. More recently, Severin *et al.* (1966) have reported that carnosine and anserine stimulated acetylcholine synthesis when added to an appropriate enzyme preparation.

It has been reported by Nagai (1976) that carnosine and other histidine-containing dipeptides are effective in promoting wound healing. This author also reported a synergistic beneficial action in this regard of carnosine with cortisone that was far greater than the effect of cortisone alone.

Very recently, Margolis and co-workers (1974) were puzzled by their observation that peripheral deafferentation of the primary olfactory pathway led to a decrease in olfactory bulb weight and the disappearance of the specific olfactory marker protein from the bulb; yet they found no alterations in the activities of neurotransmitter-metabolizing enzymes or in high-affinity neurotransmitter uptake systems.

Similarly, the surgical removal of the olfactory bulb resulted in the rapid, specific decrease of the concentration of the marker protein without altering the levels of the neurotransmitter-metabolizing enzymes. These findings led Margolis to postulate that "perhaps the primary olfactory chemoreceptor neurons utilize a novel transmitter candidate at their synapses in the olfactory bulb" (Margolis *et al.*, 1974). The discovery of carnosine in relatively high concentrations in mouse olfactory bulbs and in nasal olfactory epithelium (Neidle and Kandera, 1974; Margolis, 1974), plus the marked decrease in carnosine concentration in olfactory bulbs after peripheral deafferentation (Ferriero and Margolis, 1975), suggested that carnosine might be the neurotransmitter responsible for olfaction.

The suggested interrelationship between carnosine and olfaction led Fisher and associates (Chung-Hwang *et al.*, 1976) to postulate the usefulness of nutrition intervention as a tool in further delineating this relationship. As early as 1924, Hunter (1925) reported that the carnosine content of white muscle of the cat could be decreased by fasting and increased by liberal feeding of meat. Fuller *et al.* (1947) reported that histidine deficiency in the rat led to a decreased carnosine content of muscle; muscle anserine concentrations were either not affected at all or only slightly. Similar observations were reported in 1958 for salmon fed a histidine-deficient diet (Lukton, 1958), and in 1960 by Ousterhout in the growing chick. In the same year, Leveille *et al.* (1960) reported from our laboratory that adult cockerels that were in nitrogen equilibrium on a histidine-free diet showed a decrease in breast muscle carnosine and anserine content. These workers pointed out that the decreased concentration of these two histidine-containing dipeptides needed to be considered in relation to functional changes. Until it is determined that a given concentration is associated with a specific behavioral or metabolic activity, it cannot be assumed that levels of histidine-containing compounds that fall below those found on histidine-rich diets are necessarily abnormal.

Abraham *et al.* (1961) found that the feeding to rats of γ-aminobutyric acid led to the detection of homocarnosine in both heart and skeletal muscle, where none was detected under normal conditions.

A cooperative research effort has been under way between the Monell Chemical Senses Center and the Department of Nutrition at Rutgers University to elucidate the effect of dietary histidine on the carnosine concentration of rat olfactory bulbs. In our initial studies (Chung-Hwang *et al.*, 1976), a maintenance mixture of indispensable amino acids was fed with or without histidine to rats weighing approximately 230 g for a period of 8 weeks. Brain olfactory bulbs and muscle tissue were removed and analyzed for free histidine and histidine-containing dipeptides using a modification of the analytical method of Christman (1971). Table IV shows the marked decrease in the carnosine concentration of both olfactory bulb and breast muscle of rats receiving a histidine-free diet. For rats receiving the complete, histidine-containing diet, the olfactory bulbs also showed a highly significant decrease to approximately half the initial level of carnosine. In the case of breast muscle, however, the final carnosine concentration on the complete diet was significantly greater than at the start.

Table V shows that the homocarnosine concentration of whole brain, excluding the olfactory bulb, decreased to a nondetectable level on the histidine-free diet; it remained essentially at the starting level on the histidine-containing diet.

The anserine concentration of breast muscle in rats receiving the histidine-free diet (Table VI) was significantly lower than that of rats re-

TABLE IV

Changes in the Carnosine Concentration of Olfactory Bulb and Breast Muscle of Histidine-Free group[a,b]

Weeks on diet	Olfactory bulb		Breast muscle	
	Complete	Histidine-free	Complete	Histidine-free
0	1.03 ± 0.18		1.36 ± 0.04	
2	1.21 ± 0.10	0.51 ± 0.06[d]	1.07 ± 0.04	0.61 ± 0.06[d]
4	1.06 ± 0.11	0.49 ± 0.05[d]	1.71 ± 0.30	0.60 ± 0.08[c]
6	0.20 ± 0.01	0.01 ± 0.004[d]	1.42 ± 0.09	0.50 ± 0.06[d]
8	0.56 ± 0.04	0.01 ± 0.003[d]	1.69 ± 0.16	0.37 ± 0.04[d]

[a] From Chung-Hwang *et al.* (1976).

[b] The results are expressed as means ± SEM and represent three pooled tissue samples from four rats each (12 rats per group). Values are in nmoles/mg wet weight of tissue.

[c] A significant difference with $p < 0.05$ in comparing the histidine-free values to their respective controls.

[d] A significant difference with $p < 0.01$ in comparing the histidine-free values to their respective controls.

TABLE V

Effect of Histidine Deprivation on the Homocarnosine Concentration of Whole Brain[a,b]

Weeks on diet	Homocarnosine[c] (nmole/mg wet weight of tissue)	
	Complete	Histidine-free
0	0.14 ± 0.02	
2	0.10 ± 0.03	N.D.[d,e]
4	0.13 ± 0.06	N.D.[e]
6	0.12 ± 0.05	N.D.[e]
8	0.17 ± 0.06	N.D.[e]

[a] From Chung-Hwang *et al.* (1976).

[b] The results are expressed as means ± SEM and represent three pooled tissue samples from four rats each (12 rats per group).

[c] Whole brain excluding olfactory bulb.

[d] Not detectable.

[e] A significant difference with $p < 0.01$ in comparing the histidine-free values to their respective controls.

TABLE VI

Effect of Histidine Deprivation on the Anserine Concentration of Breast Muscle[a,b]

Weeks on diet	Muscle Anserine (nmole/mg wet weight of tissue)	
	Complete	Histidine-free
0	2.39 ± 0.21	
2	2.70 ± 0.14	2.56 ± 0.20
4	4.63 ± 0.86	3.62 ± 0.63
6	4.42 ± 0.30	3.47 ± 0.21
8	4.94 ± 0.34	3.44 ± 0.11[c]

[a] From Chung-Hwang *et al.* (1976).

[b] The results are expressed as means ± SEM and represent three pooled tissue samples from four rats each (12 rats per group).

[c] A significant difference with $p < 0.05$ in comparing the histidine-free values to their respective controls.

TABLE VII

Effect of Histidine Deprivation on the Free Histidine Concentration of Olfactory Bulb, Whole Brain,[a] and Breast Muscle[b,c]

Weeks on diet	Olfactory bulb		Whole brain		Breast muscle	
	Complete	Histidine-free	Complete	Histidine-free	Complete	Histidine-free
0	0.27 ± 0.17		0.06 ± 0.006		0.35 ± 0.02	
2	0.08 ± 0.003	0.02 ± 0.003[e]	0.03 ± 0.003	0.01 ± 0[e]	0.14 ± 0.02	0.04 ± 0.01[d]
4	0.09 ± 0.01	0.02 ± 0.002[e]	0.07 ± 0.003	0.04 ± 0[e]	0.20 ± 0.03	0.03 ± 0.02[e]
6	0.07 ± 0.01	0.01 ± 0.003[e]	0.09 ± 0.01	0.003 ± 0.003[e]	0.22 ± 0.03	0.07 ± 0.02[e]
8	0.10 ± 0.005	0.01 ± 0.003[e]	0.10 ± 0.003	0.003 ± 0.003[e]	0.15 ± 0.08	0.03 ± 0.01[e]

[a] Whole brain excluding olfactory bulb.

[b] From Chung-Hwang *et al.* (1976).

[c] The results are expressed as means ± SEM and represent three pooled tissue samples from four rats each (12 rats per group). All values are in nmoles/mg wet weight of tissue.

[d] A significant difference with $p < 0.05$ in comparing the histidine-free values to their respective controls.

[e] A significant difference with $p < 0.01$ in comparing the histidine-free values to their respective controls.

ceiving the histidine-containing control diet; in both cases, however, there was a highly significant increase in anserine concentration after 8 weeks on the respective diets in comparison to the starting level.

The concentrations of free histidine in the olfactory bulb in whole brain excluding the olfactory bulb and in breast muscle is given in Table VII. On the histidine-free diet, the free histidine level decreased sharply in all three tissues. It also decreased on the complete diet in the olfactory bulb and in breast muscle but remained essentially unaltered or increased slightly on the complete diet for whole brain tissue. Lysine, which was also measured (not given), did not show any change in concentration on either the histidine-free or the histidine-containing diet for any of the tissues analyzed.

The question of repletion of carnosine in the depleted olfactory bulb has been raised. M. Quinn and H. Fisher (unpublished results) have shown that repletion is possible and rapid. Table VIII shows that the carnosine concentration of the olfactory bulb increased more than threefold in 11 days on a diet that provided only 0.07% L-histidine (the maintenance requirement). In this experiment the rats (a different strain from those used

TABLE VIII

Effect of Dietary Histidine Depletion and Repletion on Hemoglobin and the Carnosine Content of Rat Olfactory Bulb[a]

Time on experiment	Diet	Carnosine (nmole/mg wet tissue) (olfactory bulb)	Hemoglobin (g/100 ml)
0	Stock	0.5	17
2 weeks	Histidine-free	0.7	18
	Maintenance[b]	0.8	18
4 weeks	Histidine-free	1.5	16
	Maintenance	1.7	16
8 weeks	Histidine-free	0.7	14
	Maintenance	2.5	15
11 weeks	Histidine-free	0.5	12
	Maintenance	2.0	16
Repletion—11 days	Maintenance[c]	1.8	14
	Maintenance	2.2	16

[a] From M. Quinn and H. Fisher (unpublished data).

[b] Provided 0.07% L-histidine.

[c] Received the histidine-free diet for 11 weeks, followed by 11 days on the maintenance diet (0.07% L-histidine).

in the other study) maintained higher carnosine levels than before; apparently this occurs at the expense of hemoglobin, which is rich in histidine.

These findings raise the possibility of a direct means of testing the olfactory response of rats before and after carnosine depletion. Such studies are now in progress using a behavior technique in which the rat searches for an odorized food pellet, and the time elapsed until it finds it is monitored.

Finally, if it is shown that the carnosine concentration of the olfactory bulb influences the animal's olfactory response, the histidine requirement for growth and maintenance will have to be reassessed in this context. This may be particularly important for species such as the adult chicken or man, where histidine has been found dispensable for maintaining nitrogen balance over short time spans (Leveille *et al.*, 1960; Rose *et al.*, 1951).

REFERENCES

Abraham, D., Pisano, J. J., and Udenfriend, S. (1961). Synthesis of homocarnosine in muscle *in vivo*. *Biochim. Biophys. Acta* **50,** 570–572.

Abraham, D., Pisano, J. J., and Udenfriend, S. (1962). The distribution of homocarnosine in mammals. *Arch. Biochem. Biophys.* **99,** 210–213.

Ackermann, D., Timpe, O., and Poller, K. (1929). Über das Anserin, einen neuen Bestandteil der Vogelmuskulatur. *Hoppe-Seyler's Z. Physiol. Chem.* **183,** 1–10.

Christman, A. A. (1971). Determination of anserine, carnosine, and other histidine compounds in muscle extractives. *Anal. Biochem.* **30,** 181–187.

Chung-Hwang, E., Khurana, H., and Fisher, H. (1976). The effect of dietary histidine level on the carnosine concentration of rat olfactory bulbs. *J. Neurochem.* **26,** 1087–1091.

Crush, K. G. (1970). Carnosine and related substances in animal tissues. *Comp. Biochem. Physiol.* **34,** 3–30.

du Vigneaud, V., and Behrens, O. K. (1939). Carnosine and anserine. *Ergeb. Physiol., Biol. Chem. Exp. Pharmakol.* **41,** 917–973.

Ferriero, D., and Margolis, F. L. (1975). Denervation in the primary olfactory pathway of mice. II. Effects on carnosine and other amine compounds. *Brain Res.* **94,** 75–86.

Fuller, A. T., Neuberger, A., and Webster, T. A. (1947). Histidine deficiency in the rat and its effect on the carnosine and anserine content of muscle. *Biochem. J.* **41,** 11–19.

Greenstein, J. P., and Winitz, M. (1961). "Chemistry of the Amino Acids," Vol. 3. Wiley, New York.

Gulewitsch, W., and Admiradzibi, S. (1900). Über das Carnosin, eine neue organische Base des Fleischextraktes. *Ber. Dtsch. Chem. Ges.* **33,** 1902–1903.

Hosein, E. A., and Smart, M. (1960). The presence of anserine and carnosine in brain tissue. *Can. J. Biochem. Physiol.* **38,** 569–573.

Hunter, G. (1925). Carnosine of muscle and iminazole excretion in the urine. *Biochem. J.* **19,** 34–41.

Krause, A. C. (1936). Carnosine of the ocular tissues. *Arch. Ophthalmol.* **16,** 986–989.

Leveille, G. A., Shapiro, R., and Fisher, H. (1960). Amino acid requirements for maintenance in the adult rooster. IV. The requirements for methionine, cystine,

phenylalanine, tyrosine and tryptophan: The adequacy of the determined requirements. *J. Nutr.* **72,** 8–15.
Lukton, A. (1958). Effect of diet on imidazole compounds and creatine in Chinook salmon. *Nature (London)* **182,** 1019–1020.
Margolis, F. L. (1974). Carnosine in the primary olfactory pathway. *Science* **184,** 909–911.
Margolis, F. L., Roberts, N., Ferriero, D., and Feldman, J. (1974). Denervation in the primary olfactory pathway of mice: Biochemical and morphological effects. *Brain Res.* **81,** 469–483.
Nagai, K. (1976). "Studies of Carnosine and Related Chemicals on the Physiology of Wound Repair Mechanism." Department of Physiology, School of Dentistry, Nihon University, Tokyo.
Neidle, A., and Kandera, J. (1974). Carnosine—an olfactory bulb peptide. *Brain Res.* **80,** 359–364.
Ousterhout, L. E. (1960). Survival time and biochemical changes in chicks fed diets lacking different essential amino acids. *J. Nutr.* **70,** 226–234.
Rose, W. C., and Cox, G. J. (1924). The relation of arginine and histidine to growth. *J. Biol. Chem.* **61,** 747–772.
Rose, W. C., Haines, W. J., Warner, D. T., and Johnson, J. E. (1951). The amino acid requirements of man. II. The role of threonine and histidine. *J. Biol. Chem.* **188,** 49–58.
Severin, S. E., Bocharnikova, I. M., Vul'fson, P. L., Grigorovich, Yu. A., and Solov'eva, G. A. (1963). The biological role of carnosine. *Biokhimiya* **28,** 510–516.
Severin, S. E., Shu-Hsien, L., and Popova, I. A. (1966). Participation of imidazole, AMP, CoA, and their acetyl derivatives during synthesis and biosynthesis of acetylcholine. *Probl. Neirokhim.* pp. 169–180.
Tsunoo, V. S., Horisaka, K., Kawasumi, M., Aso, K., and Tokue, S. (1963a). Über die freien Aminosäuren im Hühnerhirn. *J. Biochem. (Tokyo)* **54,** 355–362.
Tsunoo, V. S., Horisaka, K., Sasaki, S., Aso, K., and Tokue, S. (1963b). Über die freien Aminosäuren im Auge des Huhnes. *J. Biochem. (Tokyo)* **54,** 363–368.
Wurtman, R. J., and Fernstrom, J. D. (1975). Control of brain monoamine synthesis by diet and plasma amino acids. *Am. J. Clin. Nutr.* **28,** 638–647.

DISCUSSION

Naim: Is it known if carnosine has a function in metabolism of muscle?

Fisher: It is not known. There are three possible implications. One is in muscle contraction, but nobody knows how and in what manner. The other one is the suggestion of Margolis that it is a neurotransmitter for olfaction; and finally there is some Japanese work that suggests that carnosine is particularly potent synergistically with cortisone in wound-healing. There are some very impressive data but nothing more specific than that. These are basically all end results, rather than modes of action.

Cagan: Would it be possible to find and use an inhibitor of carnosine synthetase?

Fisher: It is certainly worth looking at that as another possibility of interfering with carnosine metabolism. A question arises as to the other dipeptides. Do they play a similar role?

Cagan: Could you localize histochemically either the enzymes of synthesis or breakdown of carnosine? It might be especially interesting to examine this because of the highly ordered structure of the olfactory bulb.

Fisher: In principle I would think that it is feasible, but to translate that into practice, I am not sure.

Miller: You did show some data on total brain carnosine and then you had some data on that which is found in the olfactory bulb.

Fisher: The data for whole brain was for homocarnosine. There is virtually no carnosine detectable in whole brain after removal of the olfactory bulb, so that carnosine is highly concentrated only in the olfactory bulb.

Miller: During the course of these 8–11 weeks on a histidine-free diet, what was the effect on food intake and body weight?

Fisher: Food intake in the first series of studies was about 15% less than for the controls. Body weight decreased very slowly—a gram or 2 every other day—so that they ended up weighing somewhat less than they started. The other animals gained very slowly on the maintenance diet. There was a difference but it was not very large.

Quinn: In the second study, animals were pair-fed for that whole study; that is, the animals on maintenance were fed exactly what was eaten by the histidine-free animals. They all lost weight; however, the histidine-free animals lost more weight precipitously than the animals on maintenance.

Hegsted: What is the technique of measuring these dipeptides?

Fisher: This is a column chromatographic separation, with ion exchange resins available today that are better than we had before. The compounds are detected by ninhydrin color formation. It is essentially an amino acid column chromatographic separation which has been highly sensitized for these compounds. I might add one other observation. We looked at other free amino acids and they showed no change at all. We were concerned because the concentration of all amino acids in the so-called complete maintenance diet is much, much lower, of course, than in the stock diet, so we were wondering whether all concentrations of free amino acids would be exceedingly low, but lysine, for example, remained almost at exactly the same concentration in brain and in the olfactory bulb as it has been in animals on the stock diet. It did not change at all. Of course, histidine decreased tremendously.

Chan: Is there any effect of age on the concentration of carnosine in the olfactory bulb? Is there a critical age?

Fisher: We are just in the process of studying age as a variable. This is what seems to be one of the factors that one has to take into account in carrying out this type of experiment. Depending on the starting age of the animal—these were so-called adult rats—but whether they are 8 weeks old when you start or 3 months old makes an enormous difference in whether or not the carnosine concentration has already reached a peak and has leveled off. It is in this regard also that there seem to be strain differences among rats. The first strain at the same age had a much higher concentration than did the second strain we used.

Kinnard: Then there is a suggestion that variation between species other than a rat would even be greater than within species?

Fisher: Oh yes. One species we thought would be particularly interesting to study was the hamster, because of its need for reproductive purposes to use its sense of smell. Our first endeavors in this regard were shattered when early analyses showed no carnosine at all in the olfactory bulb of the hamster, but later on Mr. Quinn did find some carnosine. Again, it is just possible that another one of these dipeptides is fulfilling the same role. It may not have to be carnosine.

Pfaffmann: Was the amount in the hamster relatively small compared to the rat?

Quinn: It was of the same magnitude on a weight basis.

Chapter 10

The Role of Vitamins and Minerals in Taste

STANLEY N. GERSHOFF

Considering the relationship of the senses to the acceptance of foods by man and animals, it is surprising that experimental nutritionists have paid essentially no attention to the effect of nutritional status on the senses of taste and smell. The published material on the role of vitamins and minerals on taste and smell can be summarized in a few minutes. The lack of information does not appear to be caused by a lack of methodology. It is relatively easy to measure taste in man and animals. There is no difficulty in producing deficiencies of either vitamins or minerals in animals or in feeding excessive amounts of these nutrients to animals. To do such experiments in man presents ethical problems. However, with funding and some initiative it is possible to identify and study people malnourished with regard to one or more nutrients. We have considerable knowledge of the metabolic functions of vitamins and minerals, and it seems likely that deficiencies of many of these nutrients should be accompanied by alterations in taste and smell.

In recent years a considerable amount of interest has been engendered by reports that hypoguesia can be prevented or cured in some people by divalent cations, particularly zinc, copper, and nickel. Henkin and his colleagues have been associated with many of these studies.

Taste abnormalities in man responsive to micronutrient treatment have been observed under three sets of conditions. They may be caused by a disease, a treatment for a disease in which the therapeutic agent affects taste, or by a dietary deficiency.

In 1963 Henkin *et al.* reported that patients with adrenal–cortical insufficiency exhibited a greater sensitivity for each of the four modalities of taste: sweet, salt, bitter, and sour. Treatment with deoxycorticosterone acetate did not alter these lowered thresholds, whereas treatment with glucocorticoids (prednisolone) returned them to normal. In a later study similar effects were observed for olfactory thresholds (Henkin and Bartter, 1966).

There are some data which suggest that diabetics may have an increased threshold in their ability to taste phenylthiocarbamide (Rao and Sisodia, 1970).

Studies in Brazil (Azievédo *et al.,* 1965), Israel (Brand, 1963), and England (Sunderland and Cartwright, 1968), among others, indicate that iodine deficiency, particularly when severe enough to cause nodular goiters, is associated with a decreased ability to taste phenylthiocarbamide. In a recent study, McConnell *et al.* (1975) reported that 15 of 18 untreated patients with hypothyroidism showed decreased acuity for at least one of the four modalities of taste, and 14 of 17 of these patients showed decreased detection acuity for olfactory stimuli. In their subjects both detection and recognition thresholds were abnormal. After treatment with thyroid hormones the measured abnormalities in taste and smell either were corrected or improved for most of the patients. The same authors rendered rats hypothyroid with large doses of radioactive iodine. The rats developed taste abnormalities which could be corrected after 2 to 3 weeks of intraperitoneal injections of thyroxine. There are also some data suggesting an association between abnormalities of taste and pseudohypoparathyroidism and gonadal dysgenesis (Henkin *et al.,* 1969). It is interesting that in each of the examples of taste alterations just cited abnormal function of an endocrine gland was involved.

In 1967 Henkin *et al.* reported that in 4–6 weeks 23 of 73 patients treated with D-penicillamine for scleroderma, cystinuria, rheumatoid arthritis, and pulmonary fibrosis developed hypoguesia associated with a decrease in ceruloplasmin. This was reversed by providing oral copper supplements. $CuSO_4$ in the concentrations given has a bitter, metallic, unpleasant taste. The patients with hypoguesia did not complain until their taste returned to normal. Then they refused further copper therapy. Penicillamine is known to act as a vitamin B_6 antagonist. It inactivates pyridoxal phosphate by forming a thiazolidine derivative of it (Jaffe, 1972). Most of the patients in this study took 50 mg of pyridoxine daily.

Interestingly, very few patients with Wilson's disease when treated with penicillamine develop hypoguesia (Henkin *et al.,* 1969). Since penicillamine when used in Wilson's disease lowers serum and tissue levels toward normal while in other types of patients the use of the drug lowers their stores of copper to below the normal level, these observations have been interpreted as indicating that copper is involved in the regulation of taste acuity and that its depletion is associated with hypoguesia. Observations in rats consistent with these have been reported by Kare and Henkin (1969). They found that rats fed penicillamine increased their preference for NaCl solutions. They could correct this by removing penicillamine from the diet or by adding copper to it.

Oral D-penicillamine causes an increase in the urinary excretion not only for copper but also of zinc (McCall *et al.*, 1967). However, it also increases the absorption of these cations. Even so, the increase in copper excretion is of such a magnitude that people treated with the drug go into negative copper balance. This is not so for zinc. If the diet is adequate in zinc the increase in zinc absorption may be so much greater than the increase in excretion that patients treated with penicillamine are in positive zinc balance. As a consequence of this, penicillamine may decrease serum copper but not serum zinc. It has been suggested that the mechanisms by which penicillamine affects zinc and copper are different.

Penicillamine contains a thiol group. In rat studies, the use of another thiol-containing drug, 5-mercaptopyridoxal, has similar effects to those observed with penicillamine in altering NaCl preference. Although 5-mercaptopyridoxal did not affect serum or urine copper levels, oral copper administration reversed its effect on taste in rats (Henkin and Bradley, 1969). Recently (Erikssen *et al.*, 1975) loss of taste and smell in humans has been reported when thiocarbamides which contain thiol groups were used therapeutically. Since association constants of thiol groups with zinc and nickel are high, these cations were used to treat patients with hypoguesia. In one study (Henkin and Bradley, 1970) two patients with hypoguesia were used, the first with multiple myeloma of the rare G3 subclass which produces excessive thiols, and the other with cystinuria treated with penicillamine. Use of either nickel or zinc lowered their detection and recognition thresholds for representatives of each of the four taste qualities to normal.

In another report (Henkin *et al.*, 1971), 35 patients with complaints of idiopathic hypoguesia, dysguesia, hyposmia, and dysosmia were treated with zinc with considerable success. About half of these patients had had respiratory illness directly preceding or at the time hypoguesia symptoms developed.

A study (Solomons *et al.*, 1974) of six patients with regional enteritis or ileocolitis with low plasma zinc levels indicated that two had severely and two minimally impaired taste recognition. The use of zinc was partially effective in treating their taste problems. Recent reports indicate that taste dysfunction may be associated with impaired zinc absorption and decreased saliva levels. When 1–2 μCi of ^{65}Zn were administered orally to 13 patients with acrodermatitis enteropathica (Henkin and Aamodt, 1975), two absorbed 99 ± 1%, six absorbed 65 ± 4%, and five absorbed only 22 ± 5% of the labeled zinc. These five complained of dysfunction of taste and smell. Their serum zinc and leucocyte alkaline phosphatase values were below normal. In another study (Henkin *et al.*, 1975) 34 subjects with normal taste acuity showed parotid saliva zinc levels of 51 ppb ± 14 (SD), while 47 subjects with idiopathic hypoguesia had saliva levels of 10 ppb ± 6.

There may be some questions concerning the methodologies used in all of these studies, particularly those used in measuring taste and smell acuity. Nevertheless, it appears that there are significant numbers of people with abnormalities in their senses of taste and smell who can be helped by the oral administration of copper and zinc, and probably nickel. A question which has not been resolved is what is the mechanism of this protective action by cations. One possibility is that the divalent cations studied are associated with and necessary for biochemical reactions involved in taste and smell. If they are not available because of dietary restriction, metabolic disease, or because they are complexed by therapeutic agents, dysfunction of taste and smell may occur. An interesting hypothesis proposed by Henkin *et al.* (1969) is that taste buds contain pores and that the penetration by the tastant through the taste pore and its membrane is essential for normal taste. This portion of the taste process is influenced by thiol and metal concentrations. In ways which are not clearly understood the conformation of the protein molecules lining the pores of the taste bud are affected by thiol and cation concentrations. The alterations in these proteins, which may be visualized as folding and refolding, control the diameter and permeability of the pores and their membranes. It is claimed that support for this hypothesis has been obtained by scanning electron microscopy, which indicates that many more taste pores may be closed in penicillamine-treated rats than in controls.

While these observations surely have been of interest to clinicians and physiologists interested in taste and smell, they received little notice from nutritionists until Hambidge *et al.* (1972) reported hypoguesia and zinc deficiency in middle class Denver children. These workers measured the hair zinc of 338 apparently normal Caucasians from birth to 40 years of age. In neonates hair zinc averaged 174 ppm. Then at 3 months of age there was a marked drop in zinc which lasted until 4 years of age, during which period the average hair zinc level was 88 ppm. It then rose sharply and from 4–17 years of age it was 153 ppm, and from 17–40 years of age, 180 ppm. Ten children over 4 years of age had hair zinc levels less than 70 ppm. Seven of these had histories of poor appetite. Eight were under the tenth percentile for height, a size not explainable on a familial basis. Taste acuity was tested in six of these children. Five showed evidence of objective hypoguesia. When their diets were supplemented with zinc, taste acuity was normalized in each child, and their hair zinc increased. A number of important questions have been raised by this study. Is zinc deficiency common among children? Can it be diagnosed by taste tests? Is hair zinc a good index of zinc deficiency? The answers to these questions are not clear. Recently we did zinc analyses on 430 samples of hair from northern Thai village children 6 months to about 8 years of age with an average age of about 4½ years (Gershoff *et al.*, 1977). Twenty-eight percent of the boys and 36% of the girls had hair levels below 70 ppm, a level below which zinc deficiency exists according to Hambridge and his associates. Four percent of the boys and 10% of the girls had hair levels below 30 ppm. In comparing these boys and girls and children with hair zinc levels above 70 ppm, there were no significant differences, or even suggestions of differences, in morbidity, height, weight, or hemoglobin. It seems obvious that studies of the relationship of hair zinc levels to taste acuity in such a population ought to be done. It should be pointed out that hair levels of zinc may not be a reliable index of zinc nutriture. When experimental animals are fed diets which only vary in their zinc content, there is a very good correlation between hair and dietary zinc levels. However, in a recent study (Gershoff *et al.*, 1977) in our laboratory, eight groups of rats were fed diets varying in protein, carbohydrate, and vitamin but not mineral contents. There were large differences in hair zinc observed with a difference of 2½ times between the groups with the most and least amount of hair zinc, even though their diets contained the same amount of zinc. In these animals there was no apparent relationship between growth and hair zinc.

If one considers the many minerals known to be essential in the diet, it is apparent that little effort has been spent on studying the effects of mineral nutriture on taste. By comparison with the vitamin literature, however, it is voluminous. There are almost no studies of the effects of vitamins on taste.

Vitamin A is necessary for maintaining the integrity of epithelial tissue, and taste buds consist of modified epithelial cells. Thus, it is not surprising that Bernard *et al.* (1961) found an effect of vitamin A deficiency on taste preferences in rats. Normal rats reject 8 μM quinine sulfate solutions and prefer NaCl in low concentrations to water. These authors found that in vitamin A-deficient rats these preferences disappear. Repletion with vitamin A resulted in a return to normal for NaCl selection but not for quinine rejection.

It has also been reported that vitamin A-deficient rats lose their preference for sugar (Seifter, 1960). It is of interest that zinc appears to be essential in vitamin A metabolism. Smith *et al.* (1973) have reported that zinc is necessary to maintain normal concentrations of vitamin A in plasma. They took two groups of rats which had been fed a diet deficient in vitamin A and zinc for 28 days. One group was supplemented with vitamin A, the other with vitamin A and zinc for 2 weeks. They found that the group receiving the vitamin A alone showed liver levels of vitamin A 6 times greater but plasma levels only one-half as great as rats receiving both vitamin A and zinc. Recently, we have reported (Huber and Gershoff, 1975) data suggesting that zinc deficiency affects the metabolism of vitamin A by rat retinas and livers.

About 40 years ago, Richter and his associates published a series of studies having to do with cravings of rats for nutrients which were deficient in their diets. I believe that it is generally thought that most of their results reflected a learned response based on the beneficial effects they experienced in ingesting the needed nutrients. In one of their studies (Richter *et al.*, 1937), rats which had been on a diet deficient in B vitamins showed such an immediate and fierce preference for a thiamin solution, and to a lesser extent for a riboflavin solution, that it appeared that their appetites for the vitamins were based on taste rather than on the beneficial effects they conferred.

There appears to be a mythology which says that animals are capable of selecting those foods which are best for them, presumably by taste. Twenty years ago we were trying to develop palatability tests for cats. Our cats were given choices between two commercial pet foods. It turned out that one of them was deficient in vitamin B_1. However, it was so palatable that some of our cats ate it exclusively and developed a thiamin deficiency characterized by convulsions. Under many circumstances animals learn to select what is good for them. Adrenalectomized rats, for instance, will learn to drink saline instead of water (Fregley, 1958) and conversely, hypertensive rats will learn to avoid salt solutions (Fregley, 1959). However, it seems impossible to generalize about the learning experience of animals. One would expect that thiamin-deficient rats would shun sucrose solutions, yet Behrman and Maller (1965) have reported that their thiamin-deficient rats not only do not develop an aversion to sucrose solutions but may increase their consumption of them.

I have tried to present a summary of the information which I have found in the literature concerning minerals and vitamins and the sense of taste. I should now like to discuss in what I admit is a simplistic way the kind of studies which might provide further information on this subject. One could of course do screening studies and feed groups of rats or other animals a large number of diets, each deficient to varying degrees in a vitamin or mineral or even containing excess amounts of individual vitamins and minerals, and determine whether taste acuity was affected. The logistic problems involved and the fact that such a shotgun approach to a problem is inefficient argues against it. Since taste involves the tongue and other oral structures, as well as the nervous system, one might look to the literature and determine which vitamins and minerals have been associated with the maintenance of these tissues. The same might be done for micronutrients which affect the endocrine glands, since there appears to be evidence of taste dysfunction in endocrinopathies.

Vitamin A studies should be expanded since data already exist which indicate that vitamin A nutriture affects taste acuity. It would be desirable to extend these studies to better understand the way in which the vitamin exerts its effects and to include the effects of excess vitamin A.

Symptoms of niacin deficiency include glossitis and lesions of the gastrointestinal tract. Pellagra, which is caused by a deficiency of niacin imposed on otherwise imperfect diets, is associated with lesions in the mouth, esophagus, stomach, and colon. Microscopically, the papillae are atrophied. In addition, there is major pathology throughout the nervous system. Since pellagra remains a major health problem in many parts of the world, niacin studies would seem appropriate.

Thiamin deficiency remains a major public health problem in Asia. The deficiency results in lesions of the central nervous system but usually not of the peripheral nerves. It is usually not characterized by morphologic changes in tissues which are associated with taste. The few studies of relationships of thiamin and taste have not been very productive, and if I were setting up studies of micronutrients and taste, I would probably not give thiamin a high priority.

The same is true of riboflavin, although equivocal changes have been described in the nervous tissue of several species of animals fed riboflavin-deficient diets, and some oral pathology has been reported in man. However, "pure" riboflavin deficiency can be produced without any tongue defects.

Although there is only unpublished data (Morley Kare) that vitamin B_6-deficient rats have an altered sense of taste, I would put vitamin B_6 studies high on my list of priorities. A deficiency of vitamin B_6 is associated with marked neurological damage in a variety of animals including man. In experimental vitamin B_6 deficiency in man, changes of the mouth and tongue similar to those seen in pellagra have been reported. A deficiency of the vitamin has been associated with some endocrine abnormalities of the pituitary, pancreas, adrenals, thyroid, and gonads. Vitamin B_6 is particularly important in sulfur amino acid metabolism and may play a role in determining tissue levels of thiol groups. Currently, there is considerable interest in vitamin B_6. A variety of circumstances including a number of genetic diseases have been described in which it appears that vitamin B_6 requirements of people may not be met. A major concern has been whether contraceptive pills produce a vitamin B_6 deficiency state in women. Besides animal work, studies of such women, particularly in developing countries where vitamin B_6 intakes are ordinarily marginal, might be particularly interesting.

Studies of pantothenic acid and vitamin B_{12} might be of some value. Deficiencies of both vitamins may be accompanied by neurologic changes and lesions of the gastrointestinal tract. In addition, pantothenic acid-deficient animals show abnormal adrenal function. Similarly, previous deficiency studies suggest some basis for looking at vitamin C, biotin, and folic acid deficiencies. There is little reason to believe that studies of relationships of vitamins D, E, and K to taste would be fruitful.

In looking at minerals it is clear that further work is necessary to understand better the roles which zinc, copper, and nickel play in the physiology of taste and in their interrelations to each other. There already is a literature which indicates that high levels of zinc induce a copper deficiency. Zinc deficiency results in hyperkeratization of epithelial cells. In the zinc-deficient squirrel monkey, parakeratosis of the tongue has been reported. Zinc deficiency has been associated in man and animals with endocrine abnormalities involving the pituitary, pancreas, and gonads.

It would be interesting to study iron nutriture and taste. Iron deficiency results in histologic lesions of the gastrointestinal tract. In iron-deficient people, of which there are large numbers, lesions of the mouth and tongue similar to those encountered in some of the B vitamin deficiency states are often seen. In addition, it is claimed that pica is often observed in iron-deficient animals and people. The reported effects of hypothyroidism on taste indicate a need for more studies of iodine nutriture and taste.

The possibility that the feeding of excessive amounts of micronutrients might affect taste should not be ignored. Recently, there have been reports (Duffield, 1973; Himmelhoch and Hanin, 1974) that the therapeutic feeding of lithium carbonate results in dysguesia in about 5% of people. The discontinuance of treatment resulted in a return to normal taste acuity in 2 to 3 days.

A review of the literature concerning the biological effects of the other minerals which act as nutrients does not provide one with many reasons for suggesting that one or another should be examined in taste studies. In working with minerals, it should always be remembered that they are often biologically interrelated. Thus, manganese affects iron requirements, molybdenum affects copper, and cadmium affects zinc, copper, and iron.

One other caution is in order. Nutrition studies in which essential nutrients are deleted usually have profound effects on food intake. The mechanisms by which this occurs are not well understood. The deficiencies which develop are often accompanied by severe pathology, and it is reasonable to assume that the experimental animals involved may feel ill. Under these circumstances, the investigator may have difficulty in determining whether alterations in eating behavior and taste acuity are the result of specific effects of his experimental regimen or are secondary to other metabolic events set in action by the diets fed.

REFERENCES

Azievédo, E., Krieger, H., Mi, M. P., and Morton, N. E. (1965). PTC taste sensitivity and endemic goiter in Brazil. *Am. J. Hum. Genet.* **17,** 87.

Behrman, H. R., and Maller, O. (1965). Appetite for sugar and fat in the thiamine deficient rat. *Psychon. Sci.* **3,** 523.

Bernard, R. A., Halpern, B. P., and Kare, M. R. (1961). Effect of vitamin A deficiency on taste. *Proc. Soc. Exp. Biol. Med.* **108,** 784.

Brand, N. (1963). Taste sensitivity and endemic goiter in Israel. *Ann. Hum. Genet.* **26,** 321.

Duffield, J. E. (1973). Side effects of lithium carbonate. *Br. Med. J.* **1,** 491.

Erikssen, J., Seegaard, E., and Naess, K. (1975). Side-effect of thiocarbamides. *Lancet* **1,** 231.

Fregley, M. J. (1958). NaCl appetite of adrenalectomized rats. *Proc. Soc. Exp. Biol. Med.* **97,** 144.

Fregley, M. J. (1959). Specificity of NaCl aversion of hypertensive rats. *Am. J. Physiol.* **196,** 1326.

Gershoff, S. N., McGandy, R. B., Nondasuta, A., Pisolyabutra, U., and Tantiwongse, P. (1977). Nutrition studies in Thailand. 3. Trace minerals in human and rat hairs. *Am. J. Clin. Nutr.,* in press.

Hambidge, K. M., Hambidge, C., Jacobs, M., and Baum, J. D. (1972). Low levels of zinc in hair, anorexia, poor growth and hypoguesia in children. *Pediatr. Res.* **6,** 868.

Henkin, R. I., and Aamodt, R. L. (1975). Zinc absorption in acrodermatitis enteropathica and in hypogeusia and hyposmia. *Lancet* **1,** 1379.

Henkin, R. I., and Bartter, F. C. (1966). Studies on olfactory thresholds in normal man and in patients with adrenal cortical insufficiency: The role of adrenal cortical steroids and of serum sodium concentration. *J. Clin. Invest.* **45,** 1631.

Henkin, R. I., and Bradley, D. F. (1969). Regulation of taste acuity by thiols and metal ions. *Proc. Natl. Acad. Sci. U.S.A.* **62,** 30.

Henkin, R. I., and Bradley, D. F. (1970). Hypoguesia corrected by Ni^{++} and Zn^{++}. *Life Sci.* **9,** 701.

Henkin, R. I., Gill, J. R., Jr., and Bartter, F. C. (1963). Studies on taste thresholds in normal man and in patients with adrenal cortical insufficiency: The role of adrenal cortical steroids and of serum sodium concentration. *J. Clin. Invest.* **42,** 727.

Henkin, R. I., Keiser, H. R., Jaffe, I. A., Sternlieb, I., and Scheinberg, I. H. (1967). Decreased taste sensitivity after D-penicillamine reversed by copper administration. *Lancet* **2,** 1268.

Henkin, R. I., Graziadei, P. P. G., and Bradley, D. F. (1969). The molecular basis of taste and its disorders. *Ann. Intern. Med.* **71,** 791.

Henkin, R. I., Schechter, P. J., Hoye, R., and Mattern, C. F. T. (1971). Idiopathic hypoguesia with dysguesia, hyposmia and dysosmia. A new syndrome. *J. Am. Med. Assoc.* **217,** 434.

Henkin, R. I., Mueller, C. W., and Wolf, R. O. (1975). Estimation of zinc concentration of parotid saliva by flameless atomic absorption spectroscopy in normal subjects and in patients with idiopathic hypoguesia. *J. Lab. Clin. Med.* **86,** 175.

Himmelhoch, J. M., and Hanin, I. (1974). Side effects of lithium carbonate. *Br. Med. J.* **4,** 233.

Huber, A. M., and Gershoff, S. N. (1975). Effects of zinc deficiency on the oxidation of retinol and ethanol in rats. *J. Nutr.* **105,** 1486.

Jaffe, I. A. (1972). The antivitamin B_6 effect of penicillamine: Clinical and immunological implications. *Adv. Biochem. Psychopharmacol.* **4,** 217.

Kare, M. R., and Henkin, R. I. (1969). The effects of D-penicillamine on taste preference and volume intake of sodium chloride by the rat. *Proc. Soc. Exp. Biol. Med.* **131,** 559.

McCall, J. T., Goldstein, N. P., Randall, R. U., and Gross, J. B. (1967). Comparative metabolism of copper and zinc in patients with Wilson's disease (hepatolenticular degeneration). *Am. J. Med. Sci.* **254,** 13.

McConnell, R. J., Menendez, C. E., Smith, F. R., Henkin, R. I., and Rivlin, R. S. (1975). Defects of taste and smell in patients with hypothyroidism. *Am. J. Med.* **59,** 355.

Rao, G. N., and Sisodia, P. (1970). Diabetes and phenylthiocarbamide (PTC) tasting ability. *J. Assoc. Physicians India* **18,** 577.

Richter, C. P., Holt, L. E., Jr., and Barelare, B., Jr. (1937). Vitamin B_1 craving in rats. *Science* **86,** 354.

Seifter, E. (1960). *Chem. Eng. News* **38,** 81.

Smith, J. C., Jr., McDaniel, E. G., Fan, F. F., and Halsted, J. A. (1973). Zinc: A trace element essential in vitamin A metabolism. *Science* **181,** 954.

Solomons, N. W., Khactu, K. V., Sandstead, H. H., and Rosenberg, I. H. (1974). Zinc nutrition in regional enteritis (RE). *Clin. Res.* **22,** 582A.

Sunderland, E., and Cartwright, R. A. (1968). Iodine estimations, endemic goiter and phenylthiocarbamide (PTC) tasting ability. *Acta Genet. Stat. Med.* **18,** 593.

DISCUSSION

Moskowitz: Most of the work on disease or deficiency states with taste or smell is done at threshold. To study deficiency states and disease states, we should emphasize looking at the taste system as if it were an instrument and look at its entire range of function. How do people under these states respond to intensity variations of a number of materials of biological and nonbiological importance?

Gershoff: I agree, but we have many problems in doing nutrition studies. In the classic type of nutrition study, nutrients are often entirely removed from the diet. That does not happen in real life, where nutritionally-deficient people and animals usually receive diets somewhat deficient in a number of nutrients, with the degree of deficiency and number of nutrients involved possibly varying enormously. We know that the effects of a deficiency of a nutrient may vary considerably depending on how much of the nutrient is in the diet. The example I cite repeatedly is riboflavin deficiency. If you feed a dog a diet with no riboflavin after a period of time the animal, with little warning to the observer other than slight anorexia, goes into a coma and will die in 12 hr. At autopsy there is very little pathology that can be found in such an animal. If there had been a small amount of riboflavin in the diet, the dog would have lived weeks longer and would not become comatose, but would develop fatty livers and lose hair. A bit more riboflavin and the animal might develop cataracts. Thus, we come to the question of how much of the various nutrients should be in experimental diets designed to study the effects of vitamins and minerals on taste. If one considers the numbers of nutrients involved and decides to use graded levels, an enormous amount of work will have to be done. It seems that one of the things we ought to do is take a look at those nutrients which, on the basis of our previous experience, might be the most profitable to look at. One of the problems with doing a taste test is that one may feed two substances that taste very much alike such as lithium and sodium in equal concentrations, and neither animals nor people can discriminate between them. After an animal has become sick several times from ingesting lithium he then can tell the difference between them in 5 sec. However, if asked to discriminate between the two in another situation such as a shock situation, he may not be able to do it.

Miller: If one looks at marginal nutrient deficiency, particularly if you are looking at things like iron and vitamin C, the epithelia are probably among the tissues that respond most immediately to these deficiencies. For most other tissues, it takes a much greater extent of deficiency to cause similar effect, so your comments concerning marginal deficiencies seem particularly appropriate. I have a feeling that taste acuity changes, if one were to measure these things, may show up at a level of deficiency at which the gross changes we usually look for in deficiency studies have not yet occurred.

Gershoff: It would be nice if one could find that taste tests could be used as a diagnostic tool for nutrient deficiencies. One could use such tests in the field and not have difficulties with human study committees. There would also be less objections from the subjects than occur when blood or urine samples are needed.

Cagan: Your remarks about specific roles of zinc and thiols being involved in biochemical mechanisms in taste receptors should be made very cautiously. There has been much speculation about such possibilities, but no experimental data to support a role of either zinc or sulfhydryl groups as part of the peripheral receptor mechanisms.

Fisher: I want to continue with what Dr. Cagan said. You started out by suggesting that we not take a shotgun approach and yet it seems to me that you ended up suggesting precisely that again.

Gershoff: With a smaller shotgun.

Fisher: I would make a plea for doing more in the area of the chemistry, physiology, and biochemistry of the taste receptors to look for nutritional involvement. One of the striking observations is that most of the micronutrients are either stored to some extent or at least are present in appreciable quantity even at time of death in various tissues of the body, so that it will be very difficult to find an involvement unless it is really a very specific and very direct involvement of any one nutrient in taste reception.

Gershoff: Let us face the facts. It costs a lot of money to do the kind of work we have been discussing. If I had financial resources I would start looking at people who suffer from vitamin A deficiency, and I would take a look at vitamin B_6 because we have large numbers of people who, we have reason to think, have metabolic abnormalities related to these nutrients. If I could get some data that suggested that they had difficulties in taste related to these two vitamins, then I could do the kinds of studies you want to do. The problem with the shotgun approach is that usually you cannot approach research in this way. You have to have some kind of a basis for doing costly experiments. Frankly, it came as a shock to me in preparing my paper in this field to find out how little has been done all through the years on the relationships of vitamin and mineral nutrition to taste and smell. If I wanted to do studies in this field I would try to pick those nutrient deficiencies to study that seemed to be of public health significance.

Chapter 11

The Control of Food Intake: When and How Are Amino Acids Involved?

Q. R. ROGERS AND P. M. B. LEUNG

It has been known for several decades that dietary protein and amino acids in some way influenced food intake. Osborne and Mendel (1916) were among the first to be aware that when a poor quality protein was fed to animals, their growth was poor and they did not eat satisfactorily. It was not until the work of Rose (1931), however, in his studies on amino acid requirements, that it was suggested that an amino acid deficiency was synonymous with a depression of food intake. Although it is known that the percentage and quality of protein in the diet can vary considerably, and food intake will remain normal if the minimum requirements of the protein and amino acids are met, amino acids are thought to be involved more intricately than simply with amino acid deficiencies. In order to examine the overall question of "When and how are amino acids involved in the control of food intake?," we would like to ask and discuss several questions.

WHAT ARE THE FOOD INTAKE RESPONSES OF ANIMALS TO DIETARY PROTEIN AND AMINO ACIDS?

The responses of animals to proportions and qualities of protein and amino acids in the diet may be divided into four categories.

No Apparent Effect

As mentioned earlier, there appears to be a broad range of dietary protein level above the minimal requirements, but not excessively high, and a wide range of disproportionality possible among the dietary amino acids, which within reasonable ranges a normal animal may tolerate the excess, deficiency, or disproportionality with little or no effect on food intake. Thus, under many conditions the total energy intake may be controlled by factors unrelated to the quality of the protein or the proportionality among the amino acids.

Depression of Food Intake and Avoidance of a Diet Which Is Deficient (or Unbalanced) in One or More Indispensable Amino Acids

This includes the phenomena of amino acid deficiencies and amino acid imbalances. Amino acid imbalances have been studied extensively and result from the addition to a low-protein diet of one or more amino acids, other than the one that is growth limiting, in amounts that individually are not toxic at the level fed. Under these circumstances, the depression of food intake and growth resulting from ingestion of imbalanced diets are readily "corrected" by a supplement of the growth limiting amino acid(s) (Harper *et al.*, 1970). Thus, amino acid imbalance is a form of an amino acid deficiency caused by the addition of a mixture of indispensable amino acids lacking the most limiting one to a low-protein diet. This raises the requirement of the most limiting amino acid. Figure 1 shows the growth and food intake of animals fed an amino acid basal diet limiting in threonine or isoleucine or the same basal diet devoid completely of threonine or isoleucine. There is a long and continuous weight loss when animals are fed the threonine- or isoleucine-devoid diets. The food intake of the animals consuming the amino acid-devoid diet remained low at all times. With the amino acid imbalance, as has been routinely observed, the depression in weight is also the result of a depression in food intake (Harper *et al.*, 1970; Leung *et al.*, 1968a). Since an amino acid imbalance is a general phenomenon and has been repeatedly demonstrated with a mixture of amino acids devoid of a variety of growth limiting amino acids added to a low-protein diet, it has been used as a model to study the influence of amino acid balance on the control of food intake. In the present discussion, we will confine ourselves to the amino acid imbalance as an effect of a deficiency on food intake control rather than dwell on the concept of amino acid imbalance and its practical significance in nutrition.

Food intake depression is but one of the two major parameters of the

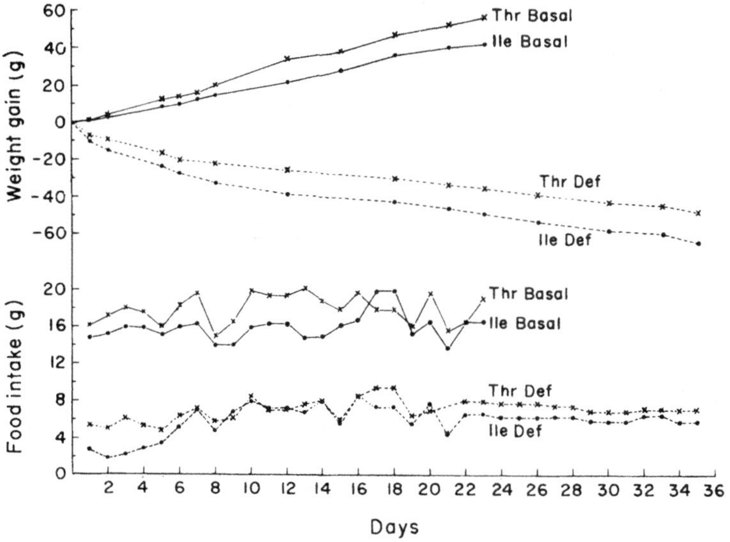

Fig. 1. Changes in body weight and food intake of rats fed basal amino acid diets low in threonine or isoleucine or the same diets completely devoid of threonine or isoleucine. The compositions of the threonine or isoleucine basal and devoid diets have previously been described in detail (Leung and Rogers, 1970). Def, devoid.

behavioral feeding response to an amino acid imbalance or deficiency. It also affects the dietary preference in animals as shown in Fig. 2. The animals, if offered a choice, will reject a diet with an amino acid imbalance and select a protein-free diet or one with a balanced pattern of amino acids (the low-protein control diet or the corrected diet). Also, the animals will select a diet containing a balanced pattern of amino acids over the protein-free diet. The selection of the protein-free diet is considered to be an adverse choice, since the animal cannot grow or survive on a protein-free diet, whereas a partially deficient diet, such as the one with the imbalance of amino acids, can support a slow rate of growth after the animals have been adapted to it. Thus, preference for a protein-free diet over a diet with disproportionate amounts of amino acids is a sensitive measure of dietary amino acid balance.

Depression of Food Intake and Avoidance of a Diet Containing an Excessive Quantity of Protein

If animals are prefed a low-protein diet and then are given a diet containing from a medium to a high level of protein as illustrated in Fig. 3 (Anderson *et al.*, 1968), there is an immediate depression in food intake. How-

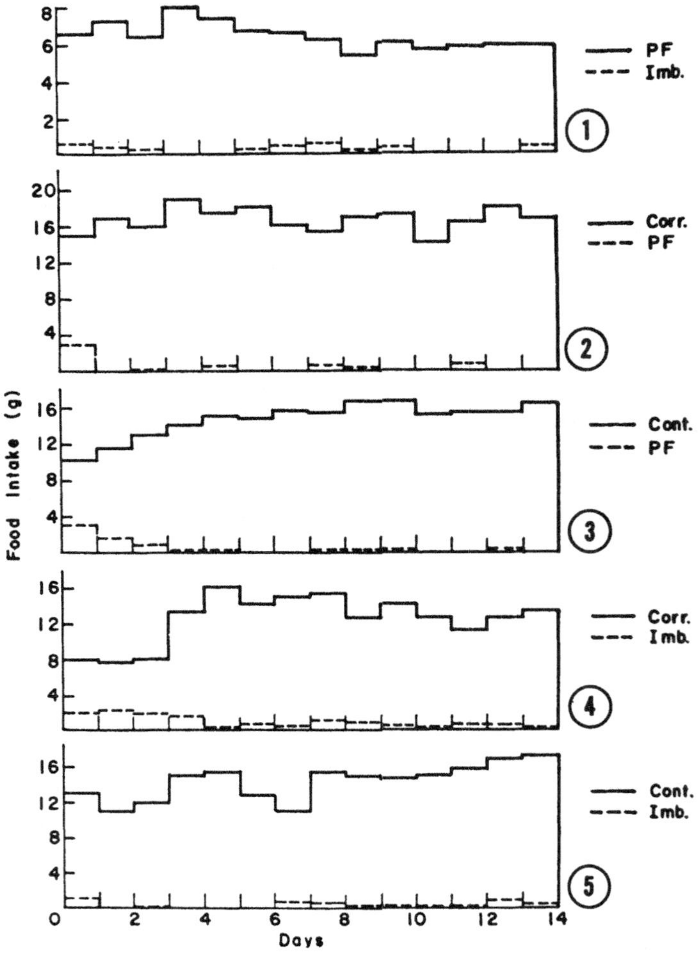

Fig. 2. (1) Food selection by rats fed an imbalanced diet low in threonine and protein-free (PF) diet; (2) food selection by rats fed the corrected (Corr.) diet and protein-free (PF) diet; (3) food selection by rats fed the control (Cont.) diet and protein-free (PF) diet; (4) food selection by rats fed imbalanced (Imb.) diet low in threonine and the corrected (Corr.) diet; (5) food selection by rats fed control (Cont.) diet and the threonine-imbalanced (Imb.) diet. Control diet: 6% casein plus 0.3% DL-methionine; imbalanced diet: control plus 5.4% amino acid mixture lacking threonine; corrected diet: imbalanced plus 0.45% DL-threonine and protein-free diet—same composition as the control diet except that protein and methionine were replaced by mixed carbohydrate (dextrin–starch 2 : 1). [Taken with permission from Leung *et al.* (1968b).]

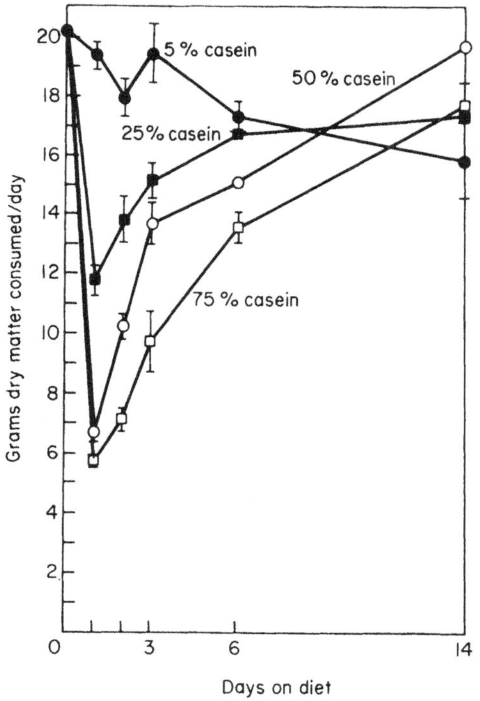

Fig. 3. Effect on food consumption of feeding 5, 25, 50, or 75% casein in rats previously fed 5% casein for 10 days. Dry matter consumed (grams) is plotted against days on the diet. Each point represents the average consumption per day of the remaining rats (94 rats on day 0, 22–25 on day 1, 4–5 rats on day 14). Filled circles, 5% casein; filled squares, 25% casein; open circles, 50% casein; open squares, 75% casein. Values are means ± SE. [Taken with permission from Anderson *et al.* (1968).]

ever, the animals resume eating normally and gain weight as they adapt to the higher level of dietary protein after a period of about 1–2 weeks. As mentioned earlier, animals will select a diet containing a balanced protein over a protein-free diet (Leung *et al.*, 1968b) whether it is low in protein, or even if it contains 30% or more protein (Fig. 4). However, if the animals are allowed a choice between a diet containing 25% protein versus one containing 50% protein, the animals will consume more of the diet containing 25% protein (Fig. 5) (Harper, 1967). Thus it appears that animals will select a level of protein in the diet somewhere between 25 and 50% if they are allowed to make a choice, and will avoid diets containing either high or low levels of protein. It is of interest to note that the animal will select 3 or perhaps 4 times its minimum dietary protein requirement in order to protect itself against an amino acid deficiency.

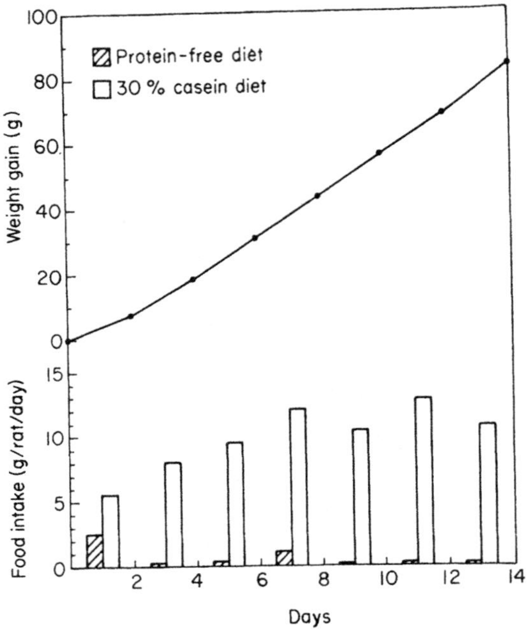

Fig. 4. Dietary choice of rats offered a protein-free diet and a 30% casein diet (10 rats/group). [Taken with permission from Rogers *et al.* (1967).]

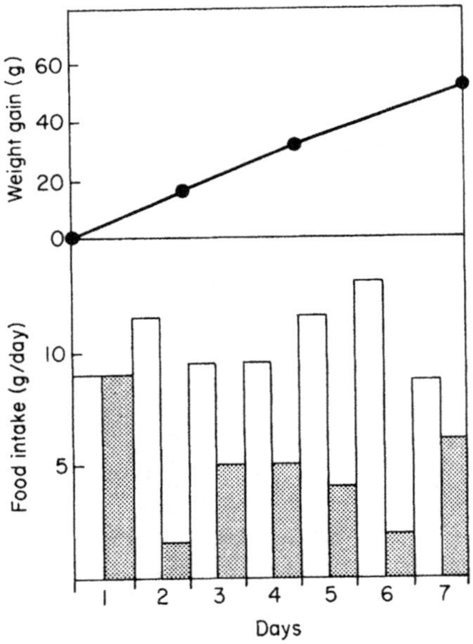

Fig. 5. Food intake of rats offered choice between 25 and 50% casein diet. Open areas, 25% casein; stipled areas, 50% casein. [Taken with permission from Harper (1967).]

Depression in Food Intake and Avoidance of a Diet Containing an Excess of a Single Amino Acid

This phenomena has been reported with many amino acids in high excess; among those most thoroughly studied include high tyrosine, methionine, leucine, and phenylalanine; each of the amino acids, if added at a high enough level, appears to depress food intake and growth of animals (Harper *et al.*, 1970; Sauberlich, 1961; Maramatsu *et al.*, 1971; Peng *et al.*, 1973). Figure 6 shows the effect of various levels of dietary leucine on growth and intake of rats (Spolter and Harper, 1961). The animals depressed their food intake in proportion to the leucine concentration in the diet, and thus the depression of growth is also proportional to dietary leucine. When the rat is offered a choice, it will reject a diet containing an excess of leucine and select a protein-free diet (Fig. 7) (Rogers *et al.*, 1967). The animal will also choose the high-leucine diet, in which isoleucine and valine have been supplemented to alleviate the

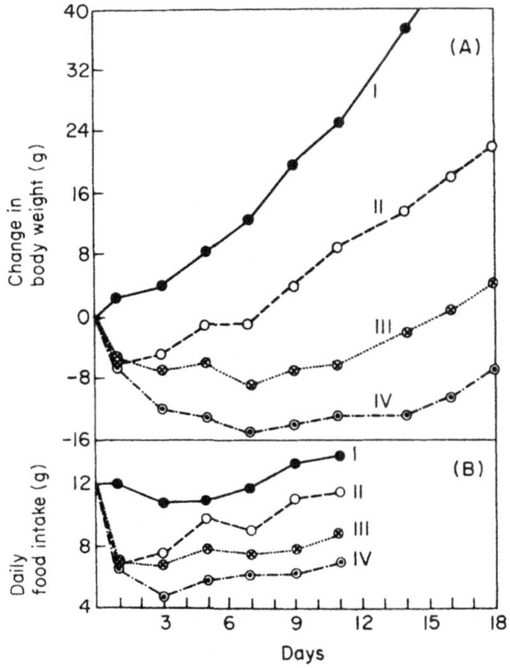

Fig. 6. Effect of various levels of dietary L-leucine on (A) growth and (B) food consumption of rats. Average initial weight of each group of six rats was 111 g. I, 9% casein (control); II, 9% casein plus 3% L-leucine; III, 9% casein plus 5% L-leucine; IV, 9% casein plus 7% L-leucine. [Taken with permission from Spolter and Harper (1961).]

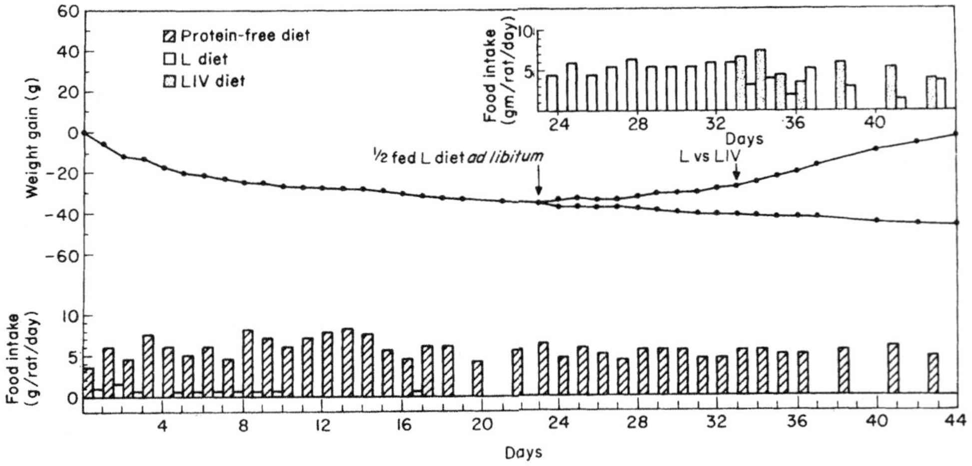

Fig. 7. Dietary choice of rats offered a 5% leucine diet (L) and a protein-free diet. LIV denotes the 5% leucine diet supplemented with 0.2% L-isoleucine and 0.2% L-valine (ten rats/group). [Taken with permission from Rogers *et al.* (1967).]

growth and food intake depression, to the unsupplemented diet containing the excessive leucine (Fig. 7).

WHAT ARE THE METABOLITES MEDIATING THE EFFECT OF AMINO ACIDS ON FOOD INTAKE?

The most commonly accepted idea concerning the metabolites mediating the amino acid effect are the amino acids themselves (Harper *et al.*, 1970; Rogers and Leung, 1973). For example, when a single meal of an amino acid-imbalanced diet is fed, the amino acids added to cause the imbalance increase in the blood plasma shortly after the ingestion of the imbalanced diet, whereas the limiting amino acid, threonine, actually decreases within 3 hr as compared to the basal diet, which contains the same amount of the limiting amino acid as the imbalanced diet (Fig. 8) (Leung *et al.*, 1968d). From these data alone, one cannot readily perceive whether the deviation of the level of the limiting amino acid from the control is a signal sufficient to cause a depression in food intake, or whether the excess of the amino acids in the plasma is the metabolic shift which is responsible, or whether both are involved. Either one or both events are probably rapid enough to influence the food intake depression which also occurred within 3 hr. Figure 9 illustrates the changes in the limiting amino acid threonine with time in the plasma and tissue (muscle) after a single meal of the amino acid-imbalanced diet. The growth limiting amino acid, threonine, falls within 3 hr after pair-feeding equal quantities of the imbalanced diet and remains low during the absorptive period (Fig. 9). There is a good correlation between the time period during which the food intake is being depressed (Leung *et al.*, 1968c) and the drop in plasma threonine. Also, it should be noted that the threonine concentration in the muscle is even more severely depressed in the imbalanced group (Fig. 9). The muscle

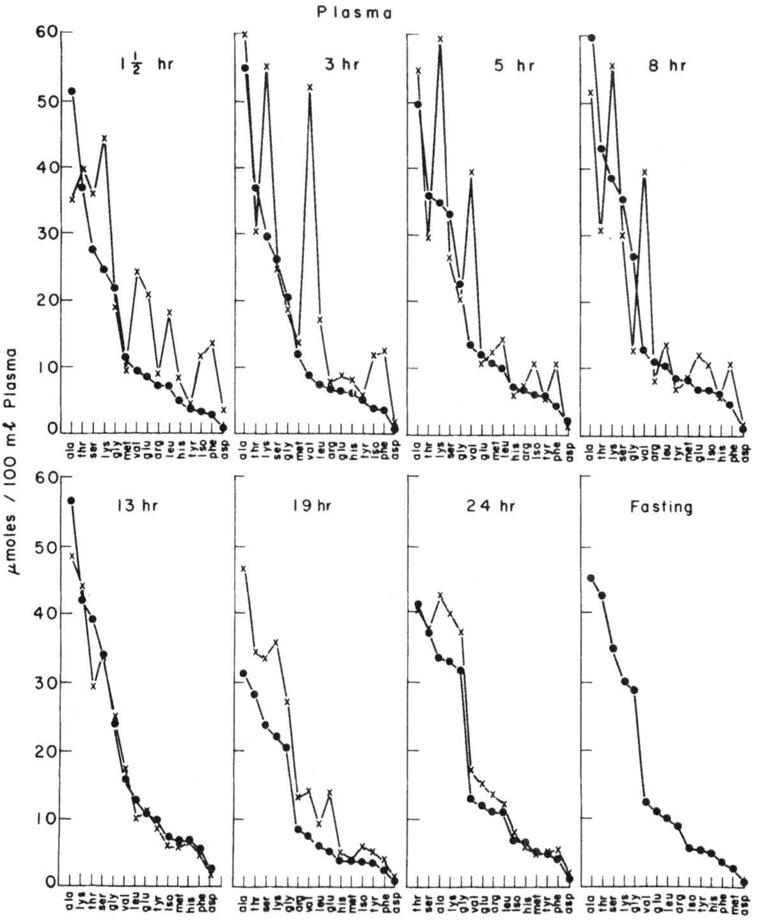

Fig. 8. Plasma aminograms of rats trained to eat a single 1-hr meal daily and fed the control diet or the imbalanced diet. Filled circles, control diet: 6% casein plus 0.3% DL-methionine; ×—×, imbalanced diet: 6% casein plus 0.3% DL-methionine and 5.4% amino acid mixture lacking threonine. Each point represents one pooled sample from five rats. [Taken with permission from Leung *et al.* (1968d).]

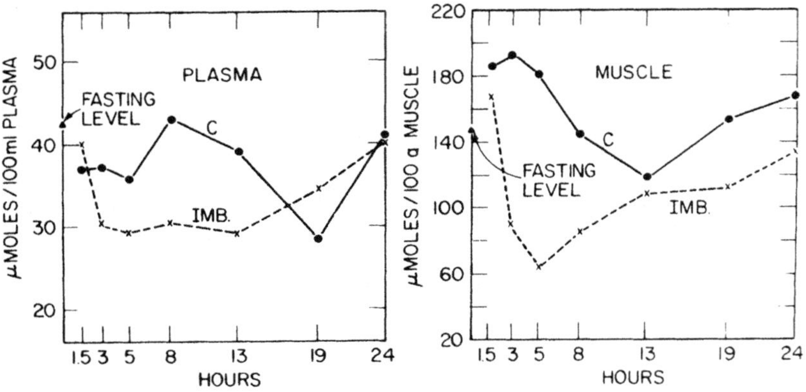

Fig. 9. Effect of amino acid imbalance on threonine concentrations in plasma and tissue (muscle) of rats fed the control diet (C) and the imbalanced diet (Imb.). Control diet: 6% casein plus 0.3% DL-methionine; imbalanced diet: 6% casein plus 0.3% DL-methionine and 5.4% amino acid mixture lacking threonine. [Taken with permission from Leung *et al.* (1968d).]

being 40% of the body weight represents a much greater body pool of the amino acids than the plasma, which represents only 3–4% of the body weight (Rogers and Harper, 1968). Thus, there is a significant drop in the tissue pool of threonine during the time the level of the amino acid falls in the plasma. Indeed, the plasma may well be a reflection of the changes occurring in the muscle (Rogers and Harper, 1968).

In examining the relationship between plasma and brain amino acid concentration, Harper and his associates (Peng *et al.,* 1972) have demonstrated that when rats were fed an imbalanced diet limiting in threonine, the concentration of the growth limiting amino acid, threonine, fell more rapidly and more drastically with time than that in the plasma, and that the amino acids added in excess to cause the imbalance were not elevated as much as those in the plasma (Fig. 10). Thus, it appears that the decrease in plasma and tissue of the limiting amino acids in rats fed the amino acid-imbalanced diet is associated with a depressed concentration of the growth limiting amino acid in the brain, and it has therefore been suggested that the amino acids per se in some way influence or mediate the dietary amino acid effects on the control of food intake.

The one common metabolite which may arise from all amino acids during their metabolism—ammonia—has also been suggested as a mediator of the effect of amino acids on the control of food intake. Thus, Russek (1971) and Noda (1975) and their co-workers have both suggested that ammonia, which arises during the catabolism of amino acids during excessive intake or under imbalanced situations, serves as a metabolite to limit food intake when there is an ingestion of disproportionate amounts of dietary amino acids. Figure 11 illustrates Noda's results, which he uses to argue that when animals are fed the amino acid imbalanced diet, blood ammonia may mediate the reduction in food intake. It should be pointed out, however, that when the "corrected" diet was fed (one that had been supplemented with the growth limiting amino acids to elevate the food intake and prevent the growth depression), the ammonia level was elevated to about the same degree as seen after feeding the imbalanced diet (Fig. 11). Thus, the elevated ammonia levels in the plasma of rats fed the amino acid-imbalanced diet were not unique, and there does not appear to be a good correlation between blood ammonia and food intake. We therefore do not think that ammonia is the mediator for signaling the effect of an amino acid deficiency on food intake depression.

Another piece of evidence arising from some infusion experiments lends support to the idea that the limiting amino acids per se, and not the am-

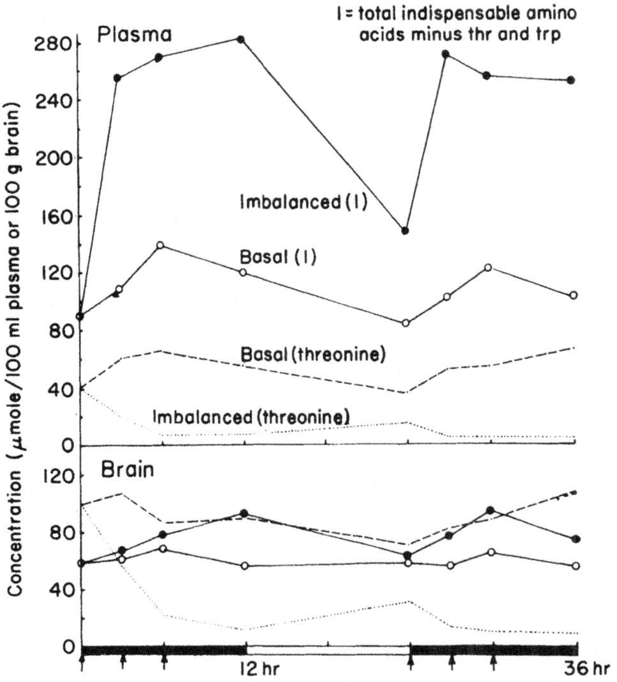

Fig. 10. Effect of force-feeding at times indicated by arrows of basal (6% casein plus 0.3% methionine and 0.2% threonine) or threonine-imbalanced (basal plus 9% amino acid mixture minus threonine) diet on concentrations of threonine and total indispensable amino acids minus threonine and tryptophan of plasma and cerebrum of rats. [Taken with permission from Peng *et al.* (1972).]

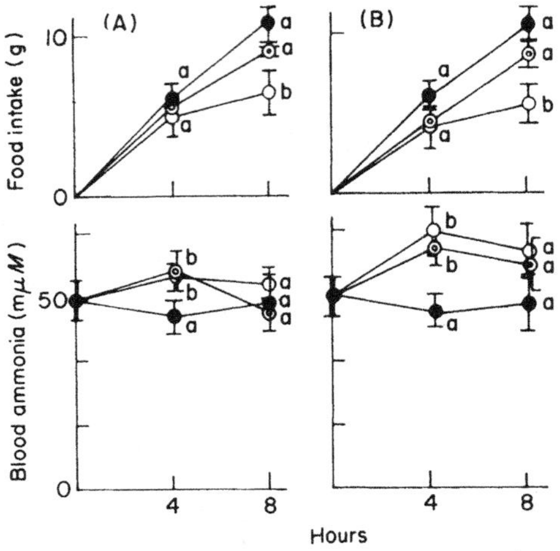

Fig. 11. Food intake and blood ammonia of rats after feeding the imbalanced or corrected diet. Rats were fed the basal diet and trained for 10 days to consume their required energy within 8 hr (9 AM to 5 PM). At 9 AM on day 11, rats were divided into five groups and were given the basal-, histidine-, or tryptophan-imbalanced diet or their respective corrected diets, and were killed after 4 or 8 hr to collect blood samples. (A) Feeding the histidine-imbalanced or its corrected diet; (B) feeding the tryptophan-imbalanced or its corrected diet. The histidine-imbalanced diet contained 6.5% essential amino acids lacking histidine, and the tryptophan-imbalanced diet contained 5% branched-chain amino acids (valine, leucine, and isoleucine). Their respective corrected diets were prepared by adding 0.4% histidine and 0.4% tryptophan to the imbalanced diet, respectively. Top figures, cumulative food intake; bottom figures, blood ammonia levels. Filled circles, basal diet; open circles, imbalanced diet; circles within circles, corrected diet. Each point represents the average of five rats. Vertical bars denote standard errors of the mean. Values followed by the same letter are not statistically significant. Differences between values followed by letters a and b are statistically significant ($p < 0.05$). [Taken with permission from Noda (1975).]

monia or the excess amino acids in the plasma, mediate the food intake depression when amino acid imbalanced or deficiency diets are fed. If one infuses a small amount of the limiting amino acid, be it isoleucine or threonine, into the carotid artery or the jugular vein of rats fed the imbalanced diets involving either isoleucine or threonine as the limiting amino acids (Figs. 12, 13, and 14) (Leung and Rogers, 1969), the animals receiving a small quantity of the limiting amino acids through the carotid artery maintained their food intake of the imbalanced or deficient diet, whereas those that receive the same amount of the limiting amino acid through the jugular vein did not. Also, the infusion of the limiting amino acid isoleucine through the portal vein did not prevent the food intake depression of rats fed the imbalanced diet limiting in isoleucine (Fig. 14). These results point to the possibility that the concentration of the limiting amino acid per se is mediating the food intake depression of animals fed the imbalanced or deficient diet through some receptor system in the brain.

It has not been shown whether ammonia levels increase sufficiently high, as a result of the intake of an excess of all or of a single dietary amino acid, to account for the food intake effects.

It has been suggested that serotonin- and catecholamine-containing brain neurons are under specific dietary control (Wurtman and Fernstrom, 1975). Wurtman and his associates have demonstrated that the rate of synthesis of neurotransmitters by monoaminergic neurons depends on the availability of the amino acid precursors tryptophan (for serotonin), and tyrosine (for dopamine and norepinephrine).

The consumption of a single meal of a protein-free diet elevates brain tryptophan levels and thus the levels of serotonin and its major metabolite, 5-hydroxyindoleacetic acid; whereas, the addition of protein to the diet supresses the increase in brain tryptophan and serotonin. The latter phenomena appears to occur as a result of the contribution of protein to plasma of large amounts of the other neutral amino acids which compete

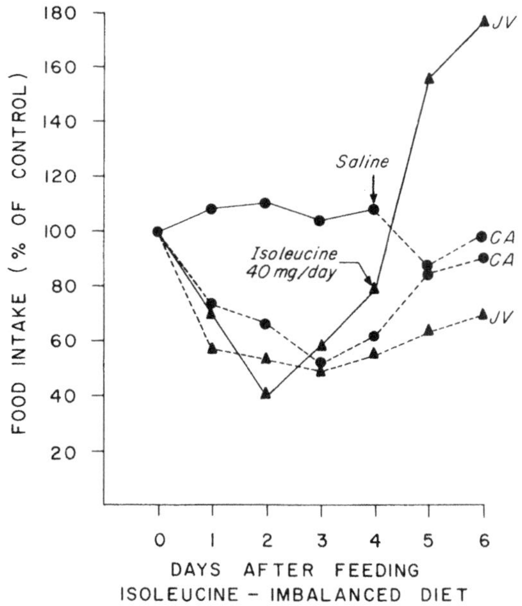

Fig. 12. Food intake of rats fed the isoleucine (Ile)-imbalanced diet and infused with saline or isoleucine through the carotid artery (CA) or the jugular vein (JV). Circles connected by solid line, CA, Ile (4 mg/day); triangles connected by solid line, JV, Ile (4 mg/day); circles connected by broken line, CA, saline; triangles connected by broken line, JV, saline. [Taken with permission from Leung and Rogers (1969).]

with tryptophan for uptake into the brain (Wurtman and Fernstrom, 1975). On the other hand, consumption of protein elevates brain tyrosine level and thus accelerates brain catecholamine synthesis (Wurtman and Fernstrom, 1975). Although tryptophan in plasma is distributed between two pools—free circulating tryptophan (accounting for 10–20%) and serum albumin-bound tryptophan (accounting for the rest)–the binding of serum tryptophan to albumin does not appear to limit the availability of the amino acid to the brain, since diet-induced changes in brain tryptophan do not correlate at all with what happens to serum-free tryptophan (Wurtman and Fernstrom, 1975). It is not known how the control of brain monoamine synthesis by diet and plasma amino acids relate to the control of food intake involving the ingestion of disproportionate amounts of dietary amino acids. However, in studies with amino acid imbalances limiting in an amino acid other than tryptophan, all amino acids are elevated, including tryptophan, in the plasma as a result of the excess indespensable amino acids other than the limiting one. Nevertheless, the food intake of animals ingesting the imbalanced diets remain depressed, despite

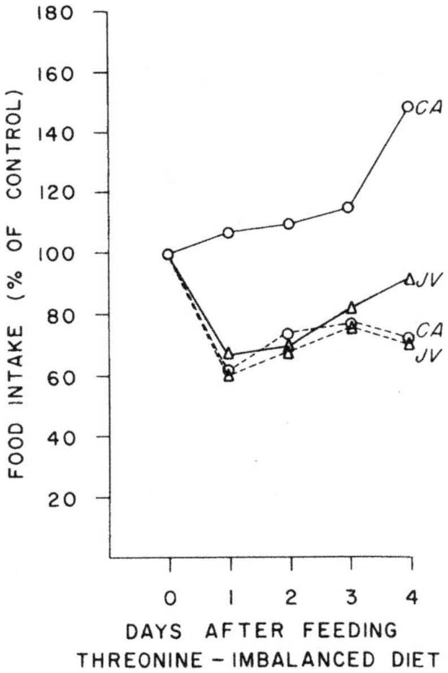

Fig. 13. Food intake of rats fed the threonine (Thr)-imbalanced diet and infused with saline or threonine through the carotid artery (CA) or the jugular vein (JV). Circles connected by solid line, CA, Thr (4 mg/day); circles connected by broken line, CA, saline; triangles connected by solid line, JV, Thr (4 mg/day); triangles connected by broken line, JV, saline. [Taken with permission from Leung and Rogers (1969).]

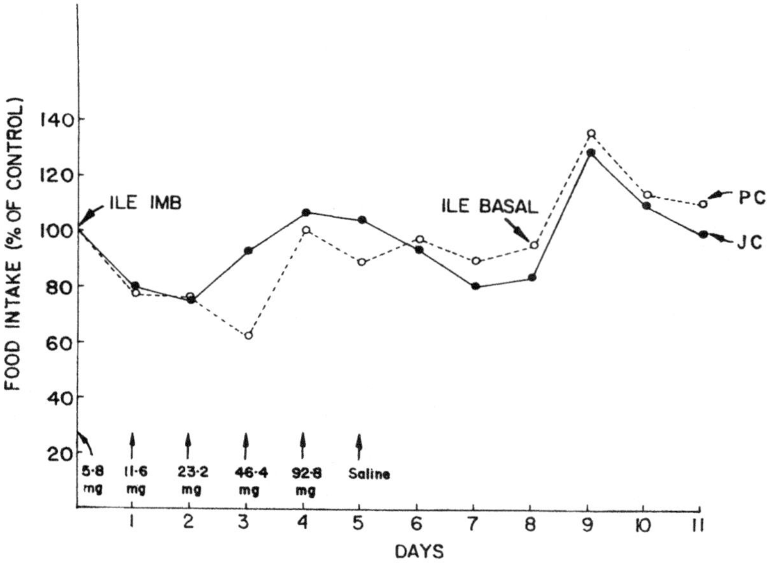

Fig. 14. Food intake patterns of portal(PC)- or jugular(JC)-cannulated rats fed an isoleucine-imbalanced (Ile imb) or isoleucine-basal (Ile basal) diet and infused with various concentrations of ammonia acetate or saline. For the compositions of the imbalanced or basal diet, see Leung and Rogers (1969).

the increased plasma, and presumably brain tryptophan. Also, tryptophan behaves as a typical limiting amino in the plasma when an amino acid imbalance is created involving tryptophan as the most limiting amino acid (Pant *et al.*, 1974). Although brain catecholamines have been proposed to influence food intake in some manner when the dopamine terminals in the brain are destroyed through lesions or other means (Striker and Zigmond, 1976), it is not clear whether the catecholamines in the brain are mediating the food intake of animals fed diets containing disproportionate amounts of amino acids.

WHERE IS THE RECEPTOR SYSTEM MEDIATING THE EFFECT OF DIETARY AMINO ACIDS ON THE CONTROL OF FOOD INTAKE?

There have been suggestions that the receptor system may be in the gastrointestinal tract, liver, and/or the brain. Sharma and Nasset (1962) have indicated that it is possible that the increased rate of mesenteric

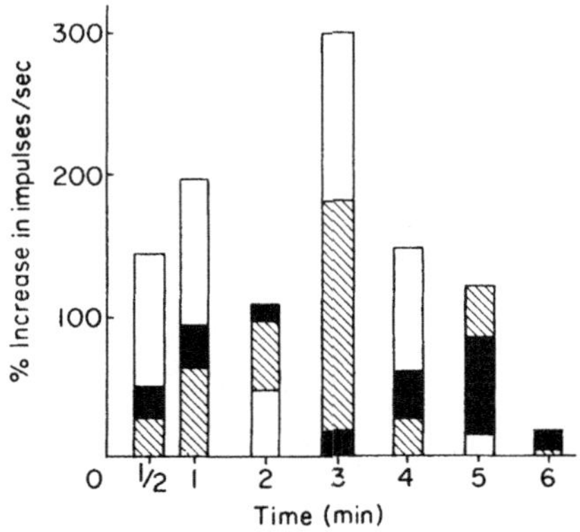

Fig. 15. Mean values (calculated for 5 sec) of percentage increase in activity of mesenteric nerves after Amigen perfusion. Black, shaded, and white areas indicate responses from small, medium, and large fiber groups, respectively. [Taken with permission from Sharma and Nasset (1962).]

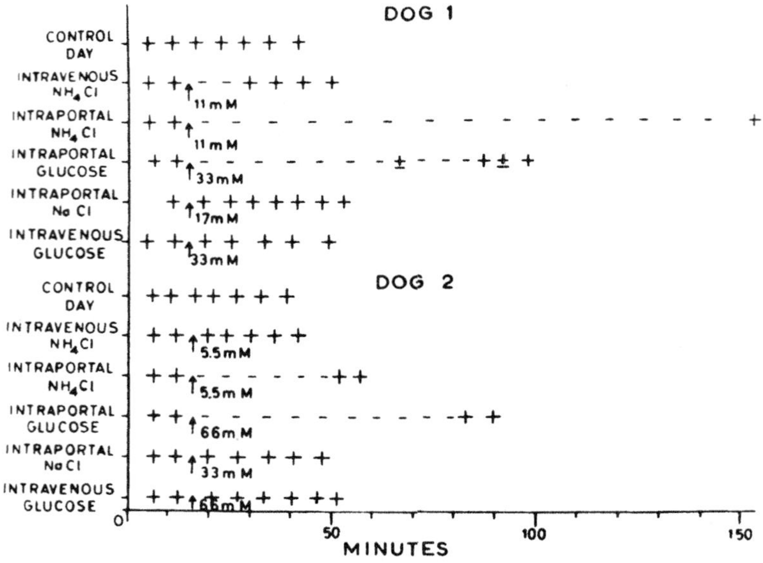

Fig. 16. The action of intraportal ammonium chloride (NH_4Cl) on food intake of 22-hr fasting dogs compared with the action of intraportal glucose. Abscissa, time in minutes; +, eats normally; ±, eats slowly and does not finish portion; −, does not eat. The arrows indicate the moment of injection and the number beside them the total in mmoles. Note that intraportal ammonium chloride is about five times more anorexigenic than glucose, compared on a molar basis. [Taken with permission from Russek (1971).]

nerve impulses evoked by perfusion of amino acids through the small intestine might constitute one of the mechanisms to signal the presence of nutrients ingested, and perhaps thereby to influence ingestion of food. Figure 15 illustrates their findings on the increase in electrical activity of mesenteric nerves after infusion of Amigen (an enzymatic casein hydrolyzate). However, no direct behavioral effect or direct linkage of this increased firing or electrical activity to food intake has been demonstrated.

Russek and co-workers have suggested that there exists what they term a "hepatic gluco-ammonium receptor" that monitors ammonia from the catabolism of excess amino acids from the portal circulation, and that it is this mechanism which accounts for the strong satiating effect of a high level of protein in the diet (Russek, 1971). Figure 16 shows the findings of Russek on the infusion of ammonium chloride (NH_4Cl) intravenously and intraportally to dogs and their respective effects on food intake of the animals. The intraportal infusion of ammonium chloride produced a complete and rather prolonged satiation (more than 2 hr). The same 22-hr fasting dogs given the same doses injected in the jugular vein had either no effect or produced a very short anorexia.

We have implanted chronic cannulas in either the portal vein, jugular vein, or the carotid artery of rats fed *ad libitum* a 6% casein (low-protein) diet and infused continuously over a 24-hr period of time with graded levels of ammonium acetate (Leung and Rogers, 1974) in gradually increasing quantities (Fig. 17). The carotid cannulated rats are the most sensitive to ammonium acetate infusion as compared to the jugular- or portal-cannulated animals. Infusion of 0.001 μmoles of ammonium acetate per day (in a total volume of 6 ml) reduced the food intake of the carotid- and jugular-cannulated rats by 50 and 10%, respectively. Ammonium acetate infusion did not affect the food intake of the portal-cannulated rats until 4.67 μmoles ammonium acetate per day were infused, where there was a reduction in food intake of only 5%. The same dose of ammonium acetate depressed the food intake of jugular-cannulated animals over 30% (Fig. 17). These differences in responses between the carotid-, jugular-, or portal-cannulated animals indicate that the liver is apparently not the site for sensing ammonia and thus causing the food intake depression of rats fed a low-protein diet. It would appear that the acute high dose of ammonium chloride administered by Russek (1971) would be sufficient enough to cause severe glycogenolysis in the liver via either a direct effect or possibly via glucagon secretion, thus causing sufficient glucose release to have a satiety effect. This effect of ammonium chloride could possibly create a pharmacological effect rather than a physiological effect on food

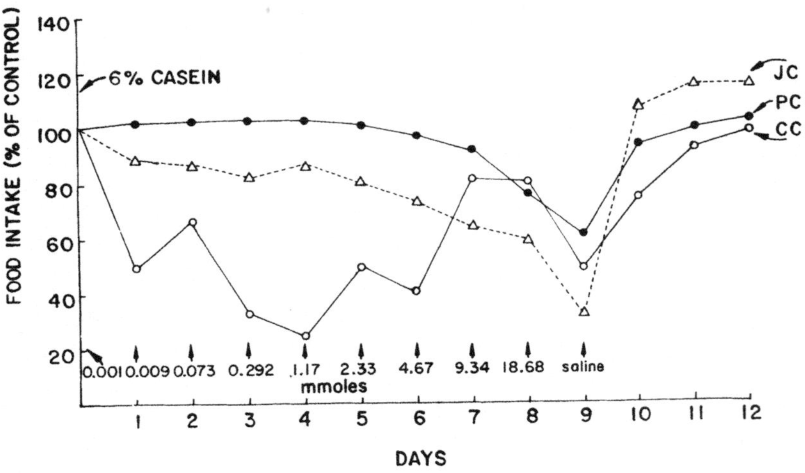

Fig. 17. Food intakes patterns of portal (PC), jugular (JC)-, or carotid (CC)-cannulated rats fed 6% casein diet and infused with various concentrations of ammonium acetate or saline.

intake. Our experiment (Fig. 17) comparing the levels of ammonia infused into the carotid-, jugular-, and portal-cannulated rats indicates that it takes from 100 to 1000 times as much as ammonia, on a continuous infusion basis, to depress the food intake of the portal cannulated rats as compared to the jugular- or especially the carotid-cannulated animals. It therefore appears that the receptor system for ammonia is in the central nervous system and not in the liver. It should be pointed out that it has not yet been shown whether ammonia concentration in the blood varies to such an extent as to cause a food intake depression as a result of a high-protein diet or as a result of any other amino acid effects.

As mentioned earlier, the infusion of a small amount of the limiting amino acid(s) (an amount which is not sufficient when present in the diet to prevent food intake depression of the imbalanced diet) into the carotid artery completely prevented the food intake depression of animals ingesting the imbalanced diets, whereas the same amount of the limiting amino acid infused into the jugular vein did not alleviate the food intake depression (Leung and Rogers, 1969). These results provide evidence for a food intake regulatory function of some portion of the brain which is sensitive to the concentration of the growth limiting amino acid in the blood when the animals are fed the amino acid-imbalanced or -deficient diet.

Panksepp and Booth (1971) have observed that micro injections of a balanced amino acid solution into the dorsolateral perifornical hypothalamic area, where consummatory behaviors are obtained with electrical stimulation, inhibit feeding in rats. As shown in Fig. 18, the injection of 1 μl of a mixture of amino acids with a balanced pattern depressed food intake of rats with cannulas implanted into the dorsolateral perifornical area, whereas saline or glucose would not. These results suggest the possible involvement of the lateral hypothalamus in responding to circulating amino acids, perhaps in excess. It is not clear what mechanism is involved under this circumstance in the control of food intake, or whether one of the amino acids is simply acting as a neurotransmitter. It would be interesting to see if other neural areas that have fiber connections with the dorsolateral perifornical hypothalamus would also respond in a similar manner to injected amino acids.

In an attempt to determine what neural structures might be involved in mediating the food intake depressing effect of animals fed disproportionate amounts of dietary amino acids, electrolytic lesions were placed in several areas in the central nervous system including the ventromedial hypothalamic nuclei, the limbic system, and other extrahypothalamic neural areas such as the amygdala, septum, and the prepyriform cortex (Rogers and Leung, 1973; Leung and Rogers, 1970, 1971, 1973). Only one area was found, in the anterior prepyriform cortex, which when destroyed by electrolytic lesions prevented the food intake depression of an imbalanced and deficient diet (one that is devoid of a single indispensable amino acid). Ablation of certain areas in the medial amygdala alleviated the food intake depression effect of an imbalanced diet (a milder form of deficiency) but not a diet devoid of a single indispensable amino acid (the most severe form of a deficiency). Figure 19 shows the food intake responses of some of the individual animals with prepyriform cortical lesions fed an amino acid-imbalanced or -deficient (devoid) diet. The animals bearing such lesions did not reduce their food intake of an amino acid-imbalanced diet limiting in threonine or an amino acid diet completely devoid of threonine. The food intake response of animals with prepyriform cortical lesions to an amino acid-devoid diet is truly amazing, since it has been well established that all animals prefed a low-protein diet without exception depress their food intake markedly within 1 day when fed an amino acid diet devoid of a single indispensable amino acid. It is also known that animals never adapt to such a deficiency by increasing their food intake with time (Rogers and Leung, 1973). With repeated studies in the prepyriform corti-

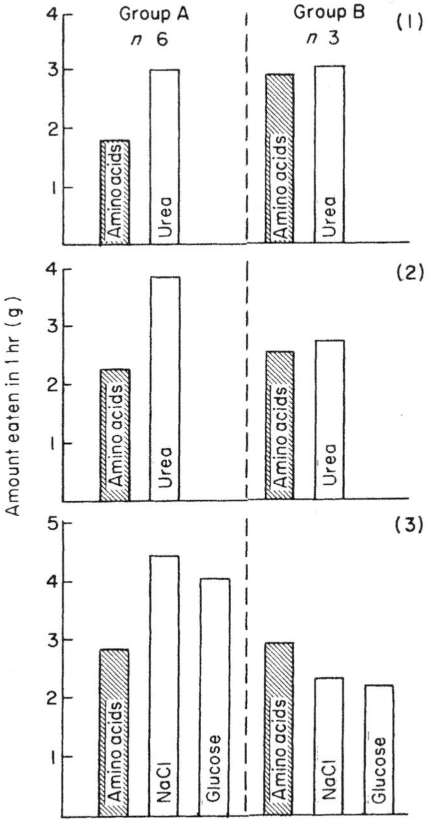

Fig. 18. Effect of intrahypothalamic injections of amino acids. Group A: cannulae tips of rats were located in dorsolateral perifornical hypothalamus, where consumatory behavior had been obtained with electrical stimulation. Group B: cannulae tips were located in the far lateral hypothalamus or in mid-lateral hypothalamus near the ventral surface of the brain. The effect of intrahypothalamic injections of amino acids were determined on three separate occasions (top, middle, and bottom figures). In the first two runs, the effects were compared with those of urea; in the third, they were compared with those of NaCl and D-glucose. Amino acids and glucose were injected as 10% w/v solutions; urea and NaCl were injected as 4.2% w/v and 2.0% w/v solutions (approximately equiosmotic to the amino acid solution). Each 10 g of the amino acid mixture contained 0.33 g histidine HCl, 0.79 g lysine HCl, 0.29 g tryptophan, 0.72 g methionine, 0.72 g phenylalanine, 0.91 g leucine, 1.14 g isoleucine, 0.82 g valine, 0.72 g threonine, 2.60 g sodium glutamate, 0.31 g glutamine, and 0.65 g glycine. L-Isomers were used. During all tests, intrahypothalamic injections (1 μl each) were given 1 hr, 30 min, and 5 min before the rats were given access to food. For the first run, food was removed 30 min before the start of the dark period and returned 1 hr later; for the second and third runs, food was removed 1 hr after the start of the light period and returned 6 hr later. [Taken with permission from Panksepp and Booth (1971).]

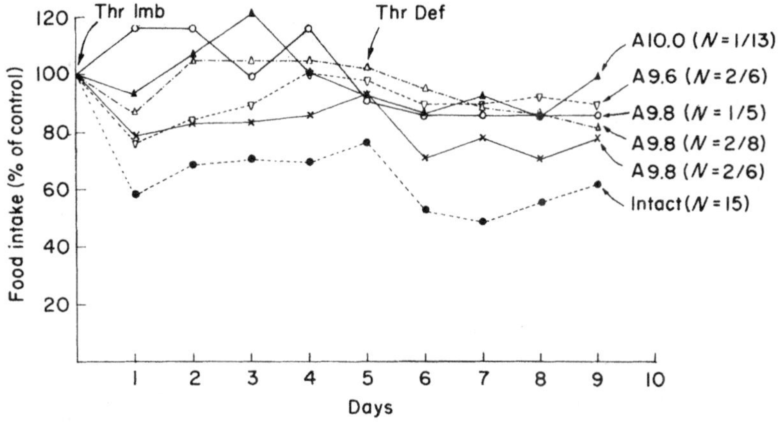

Fig. 19. Effect of threonine imbalance (Thr Imb) and deficiency (Thr Def) on food intake of selected individual or groups of rats with prepyriform lesions. Number of animals: filled circles, intact (nonoperated) = 15; inverted triangles, A9.6 = 2; open circles, A9.8 (lateral 4.0) = 1; open triangles, A9.8 (lateral 4.0) = 2; ×—×, A9.8 (lateral 3.0) = 2; filled triangles, A10.0 = 1. [Taken with permission from Leung and Rogers (1971).]

cal area, we have found that animals bearing such lesions reduced their food intake as do intact animals when fed diets containing an excess quantity of protein (75% casein), a single amino acid in great excess (6% casein plus 6% L-leucine) (Rogers and Leung, 1973), or a diet that is deficient in thiamine (authors' unpublished observations). Figure 20 shows a representative cross section of the brain with areas of the effective lesions in the anterior prepyriform cortex. Figure 21 illustrates a composite-superimposed overlay of the cross section of the effective lesions of 29 animals that showed little or no depression in food intake of the imbalanced or deficient diet. It is apparent that the lesions in various animals varied considerably; however, the minimum effective size of the lesions has not been determined. The anterior prepyriform cortical area, as defined above, therefore appears to be specifically part of the system which senses and causes a food intake depression of animals fed amino acid-imbalanced and -deficient diets. It is possible that this specific area causing the effect of the growth limiting amino acids on the control of food intake is the receptor system itself. The animals with effective lesions in the prepyriform cortex also altered their dietary choice as compared with the intact animals, selecting the imbalanced diet instead of the protein-free diet (Fig. 22). Thus, the animal with effective prepyriform lesions not only had his food intake depression abolished when fed the imbalanced diet, but also exhibited a preference response resembling a normal situation; that is, animals will normally choose a diet containing protein or amino acids over

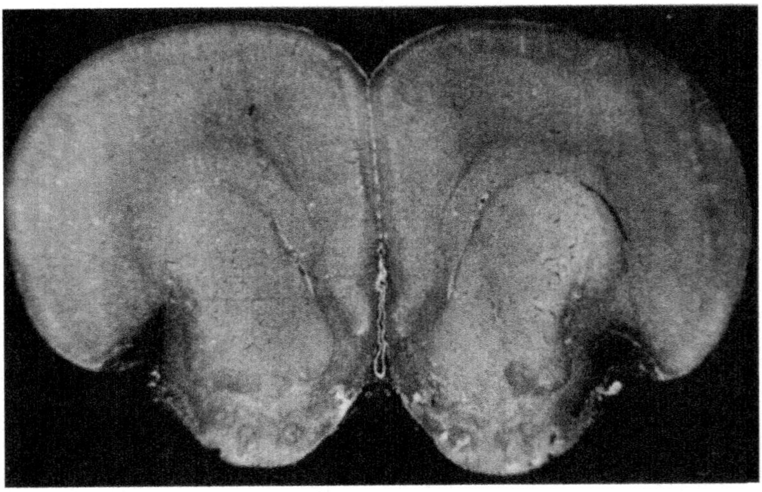

Fig. 20. Cross section of representative anterior prepyriform cortical lesions (A9.8) in animals that exhibit little or no depression in food intake of the imbalanced or deficient diets. [Taken with permission from Rogers and Leung (1973).]

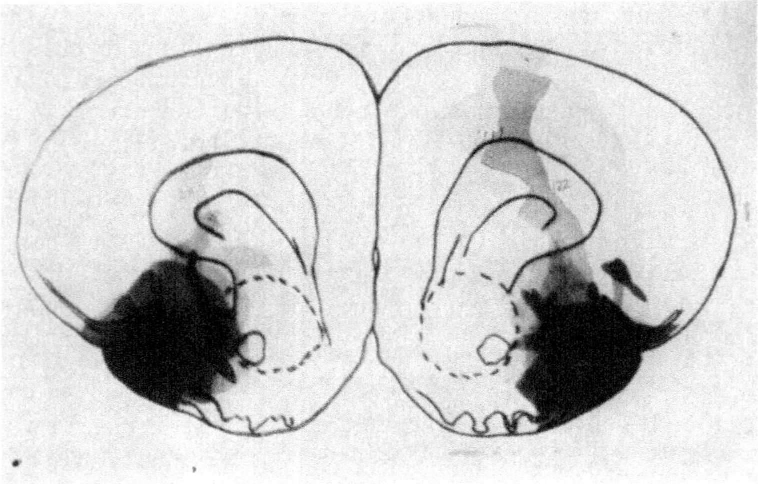

Fig. 21. Outline of cross section of anterior prepyriform cortex (A9.8) showing effective bilateral lesions (superimposed overlay of 29 animals). [Taken with permission from Rogers and Leung (1973).]

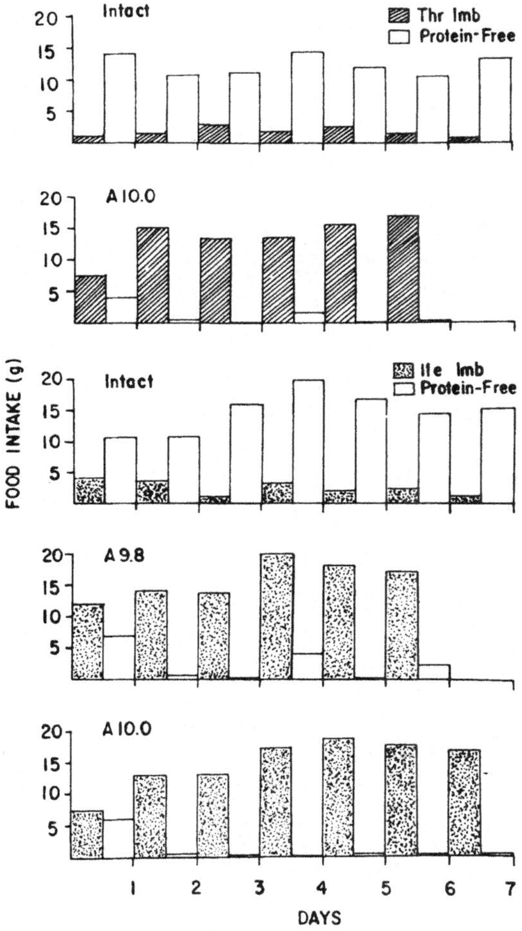

Fig. 22. Food selection of rats with prepyriform lesions fed isoleucine imbalanced (Ile Imb) and protein-free diet. Intact (nonoperated). $n = 5$; group A9.8 and A10.0, $n = 2$ and 4, respectively. [Taken with permission from Leung and Rogers (1971).]

a protein-free diet (Leung *et al.*, 1968b). The fact that the animals with effective prepyriform lesions did not consume, indiscriminately, the experimental diets with different taste and olfaction characteristics, such as the imbalanced diet versus the protein-free diet, indicates that the lack of response of the animals with lesions to the imbalanced or deficient diet is not a result of the abolishment of the sense of smell or taste or altered caloric needs, as occurs in cold-exposed rats. Animals exposed to a cold environment do not depress their food intake of the imbalanced diet be-

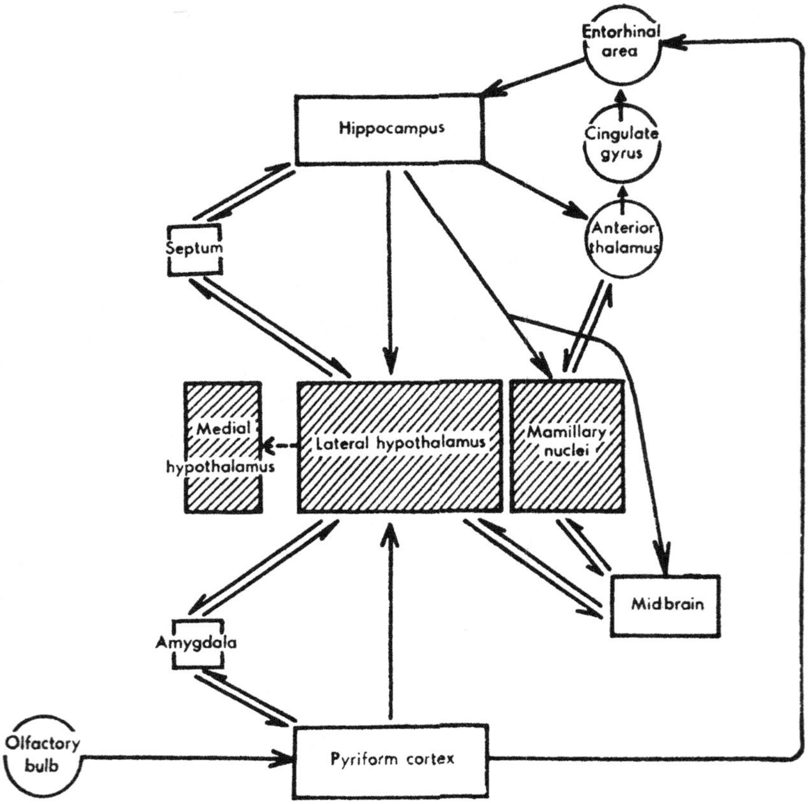

Fig. 23. A schematic diagram of the principal fiber connections of the hypothalamus. [Taken with permission from Raisman (1966).]

cause of increases in caloric needs, and yet they still prefer the protein-free diet over the imbalanced diet when allowed to make the choice (Harper and Rogers, 1966).

Figure 23 shows the neuroconnections to and from the lateral hypothalamus, the traditionally recognized feeding area (Raisman, 1966). Fiber connections have been found linking the olfactory bulb, prepyriform cortex, amygdala, septum, midbrain, hippocampus, and ventral medial hypothalamus to the lateral hypothalamus. It has been shown that lesions in some of these areas modify feeding behavior of animals ingesting disproportionate amounts of dietary amino acids in a variety of ways (Rogers and Leung, 1973; Leung and Rogers, 1970, 1971, 1973; Leung *et al.*, 1972). Some of these effects have implications for our next topic.

WHAT IS THE MECHANISM OF THE BEHAVIORAL RESPONSES OF ANIMALS FED DISPROPORTIONATE QUANTITIES OF AMINO ACIDS?

As mentioned earlier, if an animal is allowed a choice between an imbalanced or deficient diet and a protein-free diet, the animal will quickly avoid the imbalanced diet and choose the protein-free diet, even though the animal can not survive on the protein-free diet (Leung *et al.,* 1968b). The imbalanced diet, although causing an initial food intake depression, will support a slow rate of growth after the animal has been adapted to it for a few days. If, after the animals have learned to avoid the imbalanced diet and established their preference for the protein-free diet, the corrected (complete) diet is substituted for the imbalanced diet, the animals still avoid the corrected diet (Fig. 24). This appears to be a result of the close resemblance of olfactory and oral pharyngeal properties and textures of the imbalanced and corrected diets. This response appears to be a form of learned aversion, because the animal learns to associate the adverse metabolic response of the imbalanced diet with its particular smell and taste as well as its physical properties. The animal apparently sampled the "corrected diet" but would not take in enough quantity to recognize (metabolically) the difference between it and the imbalanced diet, which the animal has learned to avoid. However, if the protein-free diet is withdrawn for 1 day, so that the animal is forced by its drive to eat for energy

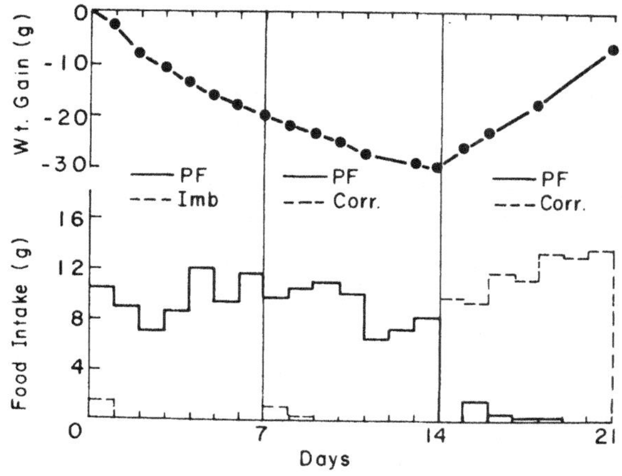

Fig. 24. Changes in body weight and food preference of rats offered a protein-free (PF) diet and an imbalanced (Imb) diet or a protein-free (PF) diet and a corrected (Corr.) diet. [Taken with permission from Leung *et al.* (1968b).]

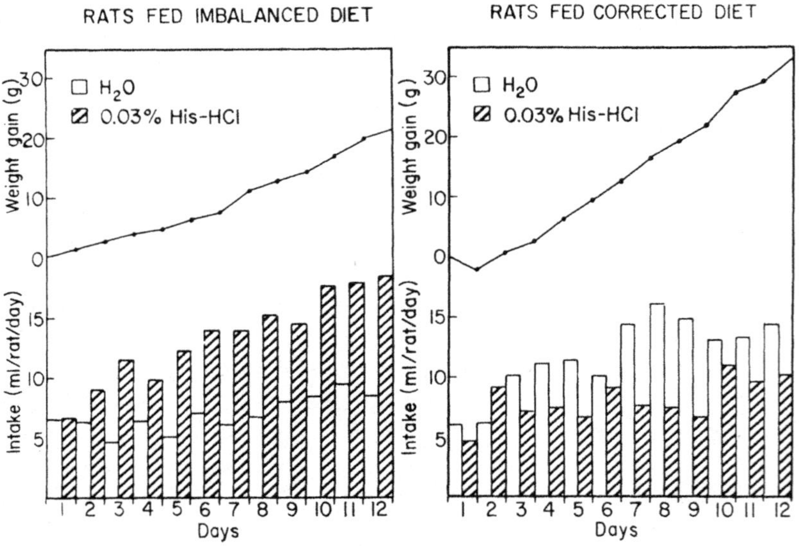

Fig. 25. Left: weight gain and fluid intake of rats fed the imbalanced diet and offered a choice between water and 0.03% His–HCl solution. Right: weight gain and fluid intake of rats fed the corrected diet and offered a choice between water and 0.03% His–HCl solution. [Taken with permission from Rogers and Harper (1970).]

to consume the "corrected" diet, the animal then detects the positive metabolic response from the corrected diet and will subsequently reject the protein-free diet when it is again presented the next day together with the "corrected diet" (Fig. 24). Thus, it appears that the major mechanism whereby an animal learns to select the proper diet, at least with regard to amino acid balance, is via "learned aversions."

It is not clear how the "learned aversion" involving the ingestion of disproportionate quantities of dietary amino acids relates to the classic learned taste aversions, which are often determined with the use of lithium chloride. The common link may be via identification with some adversive gastrointestinal consequence (nausea, vomiting, etc.) (Garcia and Koelling, 1966; Rose, 1938).

Another mechanism involved in the selection of the "corrected" diet with recovery from deficiency could be a "learned preference" (Rogers and Harper, 1970; Rozin and Kalat, 1971). Figure 25 illustrates an example of "learned preference," in which the animal fed a histidine-imbalanced diet limiting in histidine will select histidine–hydrochloride in solution over plain water, whereas the animal ingesting the "corrected" diet supplemented with adequate histidine to correct the deficiency prefers the water to the histidine solution. It would appear that "learned preference" is used less frequently by animals in their dietary selections for protein and amino acids, since animals in the wild tend to eat foods that are relatively complete nutritionally. It may be that a "learned preference" is involved when animals select a higher level of protein in a two-choice situation in which both diets are completely adequate. It is doubtful, therefore, that the mechanism involved in amino acid-specific hungers is via a specific innate hunger as occurs for that of sodium in sodium-deficient rats (Rozin and Kalat, 1971; Richter, 1936).

WHAT SENSES ARE IMPORTANT AND/OR ESSENTIAL IN THE RESPONSE OF ANIMALS TO AMINO ACIDS?

Although our primary emphasis has not been to study in depth the involvement of the use of senses—taste and smell—in animals ingesting disproportionate amounts of amino acids in the control of food intake, we have studied the food intake responses and dietary choice of anosmic animals fed an amino acid-imbalanced or -deficient diet or a diet containing an excess amount of protein. Figure 26 shows that the intact rats discriminated against a low-protein diet containing lemon or cherry flavor when they were allowed a choice between the same diet with or without the flavorings. After olfactory bulbectomy, however, the anosmic rats con-

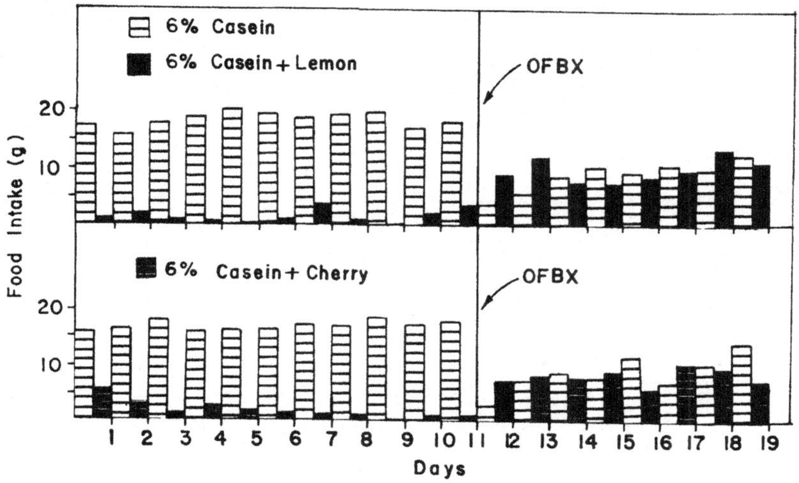

Fig. 26. Discriminated food selection by rats simultaneously fed flavored and unflavored 6% casein diets before olfactory bulbectomy (OFBX), and loss of the discrimination after the operation. [Taken with permission from Leung *et al.* (1972).]

sumed about equal quantities of both the flavored and the nonflavored diet without showing any preference (Fig. 26). Other flavorings we have used, such as vanilla, banana, anise, or maple, in conjunction with a low-protein (6% casein) diet, did not affect the dietary preference of the animals before or after olfactory bulbectomy (Fig. 27). Although olfactory cues are apparently used in the behavioral response of animals to disproportionate quantities of dietary amino acids, and the effective locations of the prepyriform lesions mentioned earlier are immediately adjacent to the primary olfactory projection areas, which in turn have projections directly to the hypothalamus as well as major components to the limbic system (Raisman, 1966), olfactory-bublectomized rats still show a food intake depression similar to that of intact animals when they were fed the imbalanced or the deficient diet (Fig. 28). Also, anosmia, although rendering the animals incapable of discriminating the flavors they usually avoided, still did not prevent the animals from effectively rejecting the imbalanced or deficient diet.

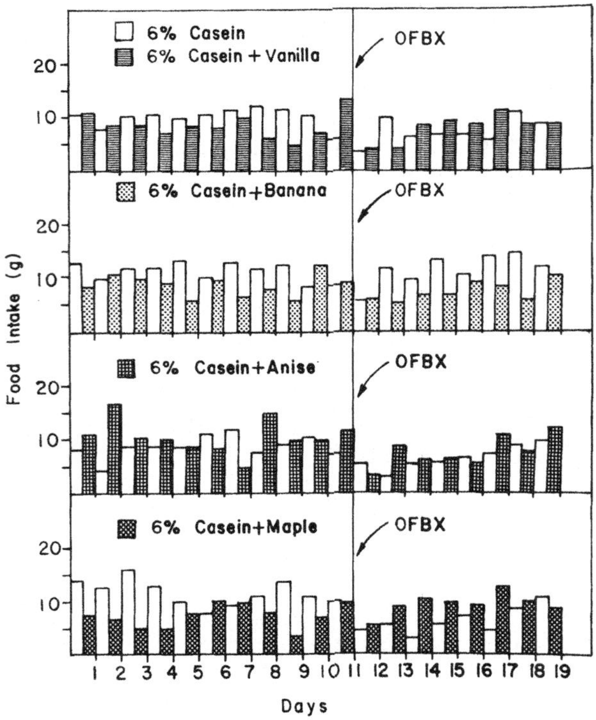

Fig. 27. Indiscriminate food selection by rats simultaneously fed flavored and unflavored 6% casein diets before and after olfactory bulbectomy (OFBX). [Taken with permission from Leung *et al.* (1972).]

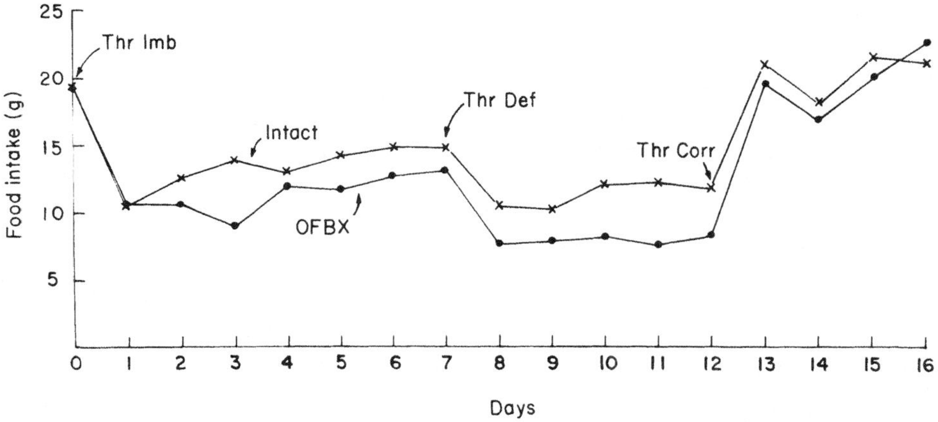

Fig. 28. Food intake of intact and olfactory-bulbectomized (OFBX) rats fed a threonine-imbalanced (Thr Imb), -deficient (Thr Def), or -corrected diet (Thr Corr). [Taken with permission from Leung *et al.* (1972).]

The fact that the animals selected the "corrected" diet over the imbalanced or deficient diet indicates that metabolic responses, originating from the ingestion of the various deficient diets, are probably the dominant factors influencing their choice of the various diets. Also, the anosmic rats did not readily continue to avoid the "corrected" diet when it was substituted in place of the deficient diet in the choice study (Fig. 29). This would indicate that the anosmic animals, when denied the ready use of the olfac-

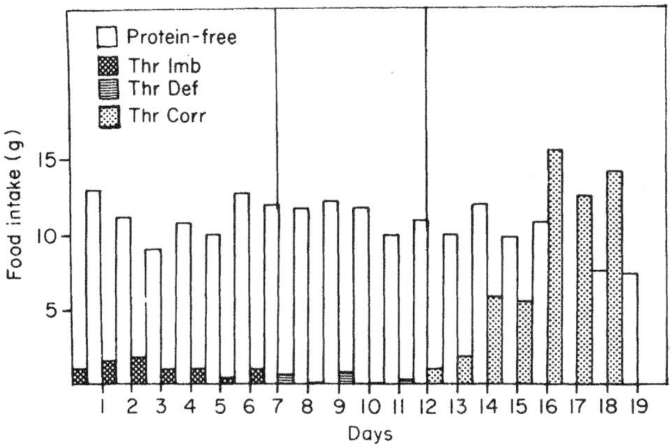

Fig. 29. Dietary choice by olfactory-bulbectomized rats of protein-free versus a threonine-imbalanced (Thr Imb), threonine-deficient (Thr Def), or threonine-corrected (Thr Corr) diet. [Taken with permission from Leung, *et al.* (1972).]

tory cue in their dietary choice, apparently are forced to sample the "corrected" diet in larger quantities, and in so doing, would soon detect the difference in metabolic characteristics between the deficient and the "corrected" diet, and hence facilitate the selection of the latter diet. It therefore appears that rats deprived of olfaction may be more dependent upon metabolic (internal) cues. It would be difficult to conceive, under these circumstances, the use of the taste as the sole cue by which the animals distinguish between the deficient diet and the "corrected" diet. The "corrected" diet contains only a small difference of one amino acid among the vast amounts of other amino acids. The olfactory-bulbectomized rats also showed a similar depression in food intake as the intact rats when fed a diet containing an excess quantity of protein (75% casein) (Leung *et al.*, 1972). All these results suggest that olfaction may be used by animals to facilitate their discrimination against the imbalanced or deficient diet, but olfaction is not essential in the control of food intake and dietary choice of rats fed diets containing disproportionate amounts of amino acids.

The role of taste in the dietary selection of diets involving disproportionate amounts of amino acids has been more difficult to study. In experiments where we employed quinine or saccharin as taste cues in the dietary choice of animals involving amino acids, we have observed that when quinine, a taste cue which the animals normally avoid, was incorporated into the "corrected" diet or into the protein-free diet, the animals would choose the quinine-containing diets over the imbalanced or deficient diet which contained saccharin, a taste cue that the animals normally prefer (Larson *et al.*, 1971). Thus, animals will avoid the adverse metabolic consequence of the imbalanced or deficient diet even to the extent of selecting a bitter taste (quinine). These choices are in direct contrast to their normal taste preferences. Repeated trials in which saccharin is associated with the aversive diet (diet containing disproportionate amounts of amino acids) and quinine is associated with the more metabolically beneficial diets result in a loss of the rat's normal preference and/or aversion to saccharin and quinine. It would appear then that when smell and taste can no longer be utilized by the animals to facilitate the identity of a particular diet in their dietary choice involving amino acids, other cues (visual?) or markings may be employed to enable them to properly select the most balanced diet.

Figure 30 illustrates the results of an experiment in which animals were given a choice between a "corrected" diet (food cup placed in a no. 2 size, tall can with a hole in the side for accessible purpose) and an imbalanced diet limiting in threonine (food cup placed in a short, open can). Both diets were flavored with maple. Although the rats did not make a clear choice on the first day, they clearly began to make a selection for the

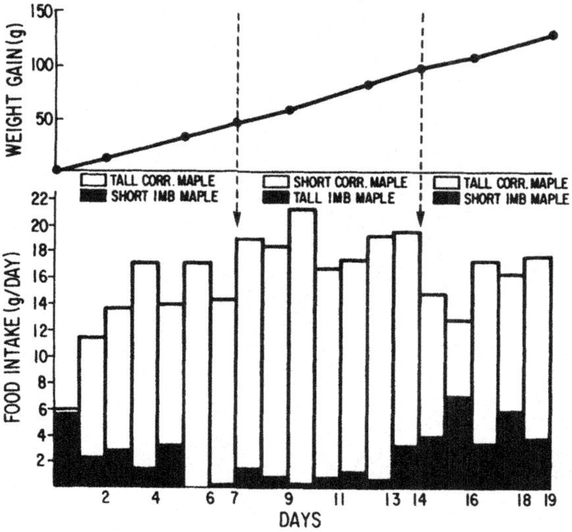

Fig. 30. Changes in body weight and food preference of rats (six rats/group) offered a choice between the threonine-imbalanced (Imb) and the threonine-corrected (Corr.) diets. Both diets contained maple flavoring. Food cups containing the maple-flavored corrected and imbalanced diets were placed in a tall container can (Tall) and a short can (Short), respectively. Food cups were switched among tall and short container cans in the course of the experiment. The times for the switches are indicated by arrows. Imbalanced diet (low in threonine): 8.38% dispensable amino acid mixture + 8.26% indispensable amino acid mixture + 0.22% alanine + 0.2% threonine. Corrected diet (high in threonine): 8.38% dispensable amino acid mixture + 8.26% indispensable amino acid mixture + 0.5% threonine. The dispensable amino acid mixture used in all the amino acid diets consists of (in % of diet): L-arginine·HCl, 1.0; L-asparagine, 1.0; L-serine, 0.5; L-proline, 1.0; glycine, 1.0; L-glutamic acid, 3.0; L-alanine, 0.5; and sodium acetate, 0.38. The indispensable amino acid mixture used in all the amino acid diets consists of (in % of diet): L-histidine·HCl, 0.6; L-isoleucine, 0.9; L-leucine, 1.2; L-lysine·HCl, 1.4; L-methionine, 0.55; L-cystine, 0.4; L-phenylalanine, 0.75; L-tyrosine, 0.5; L-tryptophan, 0.2; L-valine, 0.9; and sodium acetate, 0.85. The composition of the diets apart from the amino acid mixtures mentioned above has been reported (Leung and Rogers, 1969).

"corrected" diet on the second day. After the animals had established their choice for the "corrected" diet and completely rejected the imbalanced cup (after 7 days), the food cup containing the corrected diet was switched to the short can and vice versa for the cup with the imbalanced diet. After the switch, the animals continued to maintain their choice for the corrected diet. Although each of the six rats sampled the imbalanced diet on the eighth day, it is not clear whether it was sufficient to enable the rat to identify the diet via the metabolic response. We wonder whether the rat had in some way placed an identifying mark on the cup containing the imbalanced diet. Again, after 7 days on this second choice regimen, the cups containing the corrected and the imbalanced diets were switched back to their original containers. The animals again maintained their preference for the corrected diet, but not before sampling considerably greater quantities of the imbalanced diet (Fig. 30). Another dietary choice experiment with another group of rats in which the animals were allowed to make a similar choice, as described previously, except that the corrected diet was flavored with maple while the imbalanced diet was flavored with anise, is shown in Fig. 31. The animals again maintained their choice for the corrected diet through the entire experimental period, even though the food cups and container cans were switched back and forth as mentioned earlier. The differences in flavors between the corrected and imbalanced diet enabled the animals to maintain a choice for the corrected diet without necessarily even sampling the imbalanced diet in each switch. Figure 32 illustrates a dietary choice experiment in which the food cup containing maple-flavored corrected diet was placed in the tall can, while the food cup containing the anise-flavored imbalanced diet was placed in the short

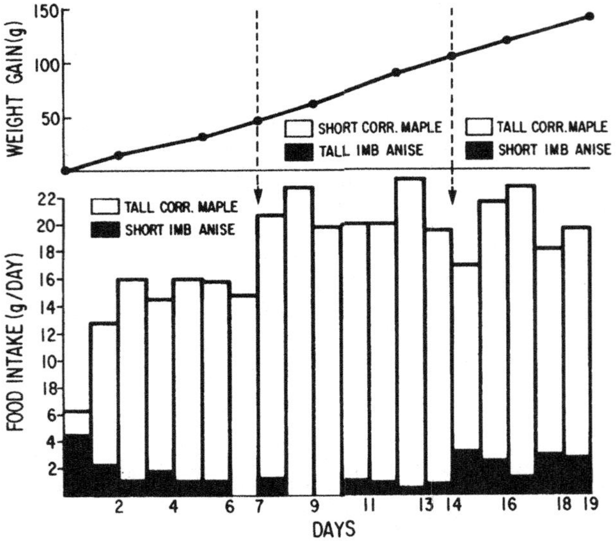

Fig. 31. Changes in body weight and food preference of rats (six rats/group) offered a choice between the threonine-imbalanced (Imb) (low in threonine) and the threonine-corrected (Corr.) (high in threonine) diets. Food cups containing the maple-flavored corrected diet and the anise-flavored imbalanced diet were placed in a tall and short can, respectively. Food cups were switched among tall and short cans during the course of the experiment. The times for the switches are indicated by arrows. The complete compositions of the amino acid mixtures used in the diets are described in Fig. 30.

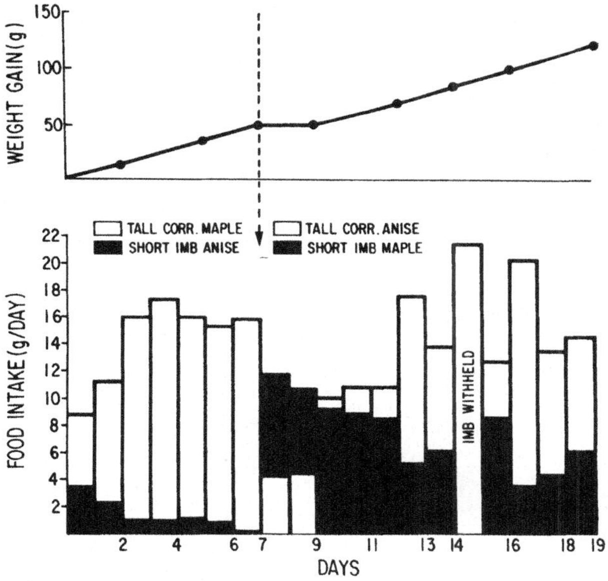

Fig. 32. Changes in body weight and food preference of rats (six rats/group) offered a choice between the threonine-imbalanced (Imb) (low in threonine) and the threonine-corrected (Corr.) (high in threonine) diets. Food cups containing the maple-flavored corrected diet and the anise-flavored imbalanced diet were placed in a tall and short can, respectively. The flavor components of the respective corrected and imbalanced diets were switched during the course of the experiment. The time for the switch is indicated by an arrow. The complete compositions of the amino acid mixtures used in the diets are described in Fig. 30.

one, and the animals were allowed to make the usual choice. The animals, consistent with previous observations, made their choice for the corrected diet after the first day, and maintained a distinct preference by rejecting the imbalanced diet almost completely. When, however, the flavor components of the corrected and the imbalanced diet were switched (i.e., the maple and anise flavorings were switched) leaving the container cans arrangement the same, the rats followed the flavor rather than avoiding the "imbalanced diet." Although the animals eventually selected more of the corrected diet 5 days after the flavor switch, it is apparent that the flavor was the cue being used and thus was a complicating factor for the rat. Apparently the animals learned to associate the maple flavor with the corrected diet, and were seriously misguided during the flavor switch. Seven days after the flavor switch, when most of the animals had resumed their preference for the corrected diet, the maple-flavored imbalanced diet was withdrawn for 1 day, so that the animals could consume only the anise-flavored corrected diet. After the imbalanced diet flavored with maple was returned to the cage, some of the animals returned to consume some of the imbalanced diet.

Thus it appears that although the animals make use of the olfactory and/or oralpharyngeal senses to enable them to select a proper diet with regard to amino acid balance in a choice situation, these senses, although helpful, do not appear indispensable in the control of food intake of animals ingesting disproportionate amounts of dietary amino acids. It does appear that animals use flavor cues most effectively in establishing learned aversions involving disproportionate quantities of amino acids; however, if the animal cannot distinguish between two diets by flavor, it is not clear how he makes the distinction, only that he does.

REFERENCES

Anderson, H. L., Benevenga, N. J., and Harper, A. E. (1968). Associations among food and protein intake, serine dehydratase, and plasma amino acids. *Am. J. Physiol.* **214,** 1008.

Garcia, J., and Koelling, R. A. (1966). Relation of cue to consequence in avoidance learning. *Psychon. Sci.* **4,** 123.

Harper, A. E. (1967). Effect of dietary protein content and amino acid pattern on food intake and preference. *Handb. Physiol. Sect. 6: Aliment. Canal* **1,** 399–410.

Harper, A. E., and Rogers, Q. R. (1966). Effect of amino acid imbalance on rats maintained in a cold environment. *Am. J. Physiol.* **210,** 1234.

Harper, A. E., Benevenga, N. J., and Wohlhueter, R. M. (1970). Effects of ingestion of disproportionate amounts of amino acids. *Physiol. Rev.* **50,** 428.

Larson, D. M., Leung, P. M. B., and Rogers, Q. R. (1971). The role of taste in dietary choice with amino acid imbalanced or deficient diet. *Fed. Proc., Fed. Am. Soc. Exp. Biol.* **30,** 233.

Leung, P. M. B., and Rogers, Q. R. (1969). Food intake: Regulation by plasma amino acid pattern. *Life Sci.* **8,** 1.

Leung, P. M. B., and Rogers, Q. R. (1970). Effect of amino acid imbalance and deficiency on food intake of rats with hypothalamic lesions. *Nutr. Rep. Int.* **1,** 1.

Leung, P. M. B., and Rogers, Q. R. (1971). Importance of prepyriform cortex in food intake response of rats to amino acids. *Am. J. Physiol.* **221,** 929.

Leung, P. M. B., and Rogers, Q. R. (1973). Effect of amygdaloid lesions on dietary intake of disproportionate amounts of amino acids. *Physiol. Behav.* **11,** 221.

Leung, P. M. B., and Rogers, Q. R. (1974). Effect of infusing the growth limiting amino acid or ammonia into portal vs jugular vein on food intake of rats fed amino acid imbalanced or low protein diet. *Fed. Proc., Fed. Am. Soc. Exp. Biol.* **33,** 651.

Leung, P. M. B., Rogers, Q. R., and Harper, A. E. (1968a). Effect of amino acid imbalance in rats fed *ad libitum,* interval-fed, or force-fed. *J. Nutr.* **95,** 474.

Leung, P. M. B., Rogers, Q. R., and Harper, A. E. (1968b). Effect of amino acid imbalance on dietary choice in the rat. *J. Nutr.* **95,** 483.

Leung, P. M. B., Rogers, Q. R., and Harper, A. E. (1968c). Effect of cortisol on growth, food intake, dietary preferences and plasma amino acid pattern of rats fed amino acid imbalanced diets. *J. Nutr.* **96,** 139.

Leung, P. M. B., Rogers, Q. R., and Harper, A. E. (1968d). Effect of amino acid imbalance on plasma and tissue free amino acids in the rat. *J. Nutr.* **96,** 303.

Leung, P. M. B., Larson, D. M., and Rogers, Q. R. (1972). Food intake and preference of olfactory bulbectomized rats fed amino acid imbalanced or deficient diets. *Physiol. Behav.* **9,** 553.

Maramatsu, K., Odagiri, H., Morishita, S., and Takeuchi, H. (1971). Effects of excess levels of individual amino acids on growth of rats fed casein diets. *J. Nutr.* **101,** 1117.

Noda, K. (1975). Possible effect of blood ammonia on food intake of rats fed amino acid imbalanced diets. *J. Nutr.* **105,** 508.

Osborne, T. B., and Mendel, L. B. (1916). A quantitative comparison of casein, lactalbumin and edestin for growth or maintenance. *J. Biol. Chem.* **26,** 1.

Panksepp, J., and Booth, D. A. (1971). Decreased feeding after injections of amino acids into the hypothalamus. *Nature (London)* **233,** 341.

Pant, K. C., Rogers, Q. R., and Harper, A. E. (1974). Plasma and tissue free amino acid concentration in rats fed tryptophan-imbalanced diet with or without niacin. *J. Nutr.* **104,** 1584.

Peng, Y., Tews, J. K., and Harper, A. E. (1972). Amino acid imbalance, protein intake, and changes in rat brain and plasma amino acids. *Am. J. Physiol.* **222,** 314.

Peng, Y., Gubin, J., Harper, A. E., Vavich, M. G., and Kemmerer, A. R. (1973). Food intake regulation: Amino acid toxicity and changes in rat brain and plasma amino acids. *J. Nutr.* **103,** 608.

Raisman, G. (1966). Neural connections of the hypothalamus. *Br. Med. Bull* **22,** 197.

Richter, C. P. (1936). Increased salt appetite in adrenalectomized rats. *Am. J. Physiol.* **115,** 155.

Rogers, Q. R., and Harper, A. E. (1968). Significance of tissue pools in the interpretation of changes in plasma amino acid concentrations. *In* "Protein Nutrition and Free Amino Acid Patterns" (J. H. Leatham, ed.), pp. 107–126. Rutgers Univ. Press, New Brunswick, New Jersey.

Rogers, Q. R., and Harper, A. E. (1970). Selection of a solution containing histidine by rats fed a histidine-imbalanced diet. *J. Comp. Physiol. Psych.* **72,** 66.

Rogers, Q. R., and Leung, P. M. B. (1973). Influence of amino acids on the neuroregulation of food intake. *Fed. Proc., Fed. Am. Soc. Exp. Biol.* **32,** 1709.

Rogers, Q. R., Tannous, R. I., and Harper, A. E. (1967). Effects of excess leucine on growth and food selection. *J. Nutr.* **91,** 561.

Rose, W. C. (1931). Feeding experiments with mixtures of highly purified amino acids. I. The inadequacy of diets containing nineteen amino acids. *J. Biol. Chem.* **94,** 155.

Rose, W. C. (1938). The nutritional significance of the amino acids. *Physiol. Rev.* **18,** 109.

Rozin, P., and Kalat, J. W. (1971). Specific hungers and poison avoidance as adaptive specializations of learning. *Psych. Rev.* **78,** 459.

Russek, M. (1971). Hepatic receptors and the neurophysiological mechanisms controlling feeding behavior. *Neurosci. Res.* **4,** 213.

Sauberlich, H. E. (1961). Studies on the toxicity and antagonism of amino acids for weanling rats. *J. Nutr.* **75,** 61.

Sharma, K. N., and Nasset, E. S. (1962). Electrical activity in mesenteric nerves after perfusion of gut lumen. *Am. J. Physiol.* **202,** 725.

Spolter, P. O., and Harper, A. E. (1961). Leucine–isoleucine antagonism in the rat. *Am. J. Physiol.* **200,** 513.

Striker, E. M., and Zigmond, M. J. (1976). Brain catecholamine and the lateral hypothalamic syndrome. *In* "Hunger: Basic Mechanisms and Clinical Implications" (D. Novin, W. Wyrwicka, and G. Bray, eds.), pp. 19–32, Raven Press, New York.

Wurtman, R. J., and Fernstrom, J. D. (1975). Control of brain monoamine synthesis by diet and plasma amino acids. *Am. J. Clin. Nutr.* **28,** 638.

DISCUSSION

DeWys: In the United States you looked at insufficiencies of essential amino acids. Will you comment on levels in blood of nonessential amino acids such as in the situation of energy deprivation in which alanine is converted to glucose and then maybe some lowering of the glutamine levels occurs.

Rogers: Alanine and glutamic acid can be left out of the diet completely without depressing normal food intake and normal growth.

DeWys: What if the blood level is depressed?

Rogers: I am not familiar with any experiments where a positive growth response has been found by adding either of these amino acids to the diet. However, if nitrogen is limiting per se, and you add nitrogen as an amino acid or even as ammonia (ammonium citrate), you will get a positive response.

DeWys: We have to assume that all essential amino acids are active in some way?

Rogers: Using free amino acid diets, an amino acid imbalance can be produced with any one of the essential amino acids being limiting.

Kare: Your report on the selection of the 25% casein as opposed to the 50% casein raises a special problem. We did some work on the acceptability of various proteins by rats. One thing became clear. Casein is uniquely offensive among all the proteins to rats.

Rogers: I would have to accept that an animal might select a higher level of meat than casein. There are also other factors involved. The level an animal will select may well depend upon pretreatment. If the animals are prefed a high-protein diet, they may well tend to select a higher level of protein than if they are prefed a low-protein diet.

Sharma: How does the prepyriform cortex affect your area of olfaction?

Rogers: So far as I know, the prepyriform cortex does not have anything to do with olfaction. We impinged upon the olfactory tract so it became important that we eliminate olfaction as the essential key component mediating the responses that we had observed. It appears that the anterior prepyriform cortex is important only for amino acid imbalances and deficiencies. We have found that lesions in the anterior prepyriform cortex do not abolish the food intake depressing effect of a high-protein diet, a high-leucine diet, or a thiamine-deficient diet. The animals lesioned in the anterior prepyriform cortex also still show a learned aversion.

Unknown: I think you do get an attenuation of learned aversion with any anterior cortical lesion or actually an olfactory bulb lesion.

Rogers: Olfactory bulbectomy produced animals which were actually somewhat better in establishing a taste avoidance.

Fisher: You mentioned at the outset that all indispensable amino acids deficiencies give the same response curve, but I am under the impression that lysine is peculiar in that it does not give rise to as severe a depression in food intake. Have you had experience with that?

Rogers: Lysine has a different response curve in terms of the slope than the other amino acids, and that is the reason it is reutilized more efficiently. I would therefore expect less severe imbalances, but we have not studied lysine imbalance ourselves. There are examples of lysine imbalances in the literature.

Nicolaidis: What about the case in which there is a missing element?

Rogers: It is very subtle, but I think it is there. If you look at the response curves of amino acids, there are a lot of marginal deficiencies which will cause a small increase of food intake. If you plot food intake versus percentage of indispensable amino acids, you will find a low level of food intake without any amino acids in the diet. As you add more protein, weight gain increases and often there is a hump just prior to or at the requirement. At this point, body composition has a higher lipid content than that found with a higher or lower level of protein, reflecting the disposal of the excess calories as fat as the animal attempts to eat for more amino acids.

PERCEPTION AND EXPERIENCE

Chapter 12

THE ONTOGENY, EVOLUTION, AND STIMULUS CONTROL OF FEEDING IN HUMANS AND REPTILES

GORDON M. BURGHARDT

INTRODUCTION

Diversity in food choice parallels the diversity of organic evolution. Animals vary from vegetarians to omnivores to carnivores; they may specialize on only one or a few food items or they may be extremely opportunistic in food choice. Food selection may also vary from active to passive, from rigid to easily modifiable, and from an individual to a social affair. Further, all these permutations may occur in the same species, or even within an individual over its life span. Generalizations are thus seemingly impossible and extrapolations across various taxa, especially to humans, a venture ill-fated for success. But one can identify processes and mechanisms at work at the behavioral and perceptual level that may, through parametric testing, eventually allow a model of some hueristic value. To review all the approaches, possibilities, and relevant literature is beyond the scope of this report. Here I will discuss some of the factors involved mainly through our research on lizards and snakes, chemoreception being important to them in obtaining and selecting food.

Nutrition and the chemical senses are necessarily considered complementary topics. Food that provides good nutrition is only good if it is eaten, and it is usually ingested only if it is palatable. Thus, a concern that food be tasty is common to the prepared foods industry, plant breeders searching for new strains of vegetables and fruits, and livestock producers and fisheries about to introduce new types of meat or seafood. Unfortunately, it seems that in recent years only a minimum palatability is often strived for, as many new foods are developed more for the producer's and seller's convenience in terms of harvesting, shipping, storing, and profit margins rather than out of an honest concern with gastronomy or nutrition. Sometimes one aspect is accentuated to the exclusion of others; sweetness or crunchiness becomes a supernormal sign stimulus.

Nonetheless, while it is important to ask how tasty a target population regards a food, we must remember that even nutritious and savory food may be rejected. Beyond the question of nutrition, smell, taste, appearance, and texture there are metabolic and cultural differences that need to be considered. An example of a metabolic factor would be the problem some adult human populations apparently have in digesting milk products (Johnson *et al.*, 1974). As a cultural factor the name of a food and its associations may loom important. How many Americans would find domestic dogs or snakes acceptable fare, or even be willing to ingest them, regardless of how well they tasted and regardless of nutrition, once they were told what they were eating. Yet in many parts of Asia dogs and snakes are gourmet delicacies.

THE INFLUENCE OF COLOR ON ODOR RECOGNITION

Another problem, and one that we have been interested in, is the relationship between various cues associated with food. Little is known about the interaction of odors and other senses. Harper *et al.* (1968, p. 93) pointed out that even with a familiar substance, removal of auxiliary cues may lead to failure in identifying even the "simplest and most common odors." Certainly, in the real world far removed from the strict controls of the laboratory, visual cues are present and useful. Does not the color as well as the odor of certain common foods indicate to an individual whether they are fit for consumption or spoiled? Of course these properties need not be intrinsic to the ingested substance. How many people could really discriminate between all the various bland American beers without the aid of the bottle or can?

In unpublished work by Bob Armbruster (1973) from our laboratory various fruit odors with different associated colors were studied. Five natural fruit extracts were used as odor stimuli: lemon, orange, strawberry, grape, and blueberry. Food coloring was used to color the resulting solutions which were kept in clear glass vials. Forty-eight male and female undergraduates were tested individually with the odors under three different visual conditions: control, blindfold, and conflict (Table I). In the control condition all of the stimuli were colored as they would be expected to appear. The subjects were also tested on the same five stimuli blindfolded, thus eliminating all visual cues. In two conflict conditions the coloring of the solutions was interchanged so that the color and odor were incongruous.

Each solution consisted of approximately 1 ml of the concentrated flavoring added to 6 ml of distilled water. All were well above the human threshold for an olfactory stimulus and all subjects smelled something from each vial, erroneous though their choices may have been. Each of the 24 male and 24 female subjects were administered five presentations of all five odors in the control condition, the blindfold condition, and one of the conflict variations.

For half the subjects (12 male and 12 female), all stimuli, regardless of condition, were randomized with the restriction that no odor could be presented twice in a row. For the other subjects a blocked presentation was used where each condition was administered in its entirety before the next condition was begun. All possible orderings of the three conditions were used in counterbalanced fashion. Within each condition all stimuli were presented in a random fashion, again with the restriction that no odor could be presented consecutively.

For both presentations the interstimulus interval was 10 sec with a 5-min break between conditions or between blocks of 25 stimuli. All subjects received the same instructions and no further information was provided.

TABLE I

Coloring of Olfactory Stimuli by Experimental Condition[a]

	Color cue			
Stimulus	Control	Blindfold	Conflict I	Conflict II
Lemon	Yellow	—	Red	Orange
Orange	Orange	—	Blue	Yellow
Strawberry	Red	—	Yellow	Purple
Grape	Purple	—	Orange	Blue
Blueberry	Blue	—	Purple	Red

[a] From Armbruster (1973).

TABLE II

Percentage of Correct Responses to Fruit Odors by Condition, Presentation, and Subject Group[a]

Presentation and subject group	Condition		
	Control	Blindfold	Conflict
Blocked males	69.6	48.3	25.6
Blocked females	70.6	44.3	20.3
Random males	72.6	47.0	27.6
Random females	73.6	53.3	14.6
All subjects	71.6	48.2	22.0

[a] From Armbruster (1973).

The verbal identification to the stimulus as well as the latency of the response were recorded.

The percentage of correct responses was highest in the control condition and lowest in the conflict condition (Table II). For all four groups of subjects, blocked males, blocked females, random males, and random females, the differences in the percentage correct between any condition with any other were significant ($p < 0.005$, Wilcoxon Matched-Pairs Signed-Ranks Test).

Inspection of Table II indicates that the random females were superior to the random males in the blindfold condition, but inferior in the conflict conditions. This tantalizing sex difference did not appear in the blocked subjects and was not statistically significant.

No significant differences between groups were found in the percentage correct on each odor in any condition, although the strawberry stimulus in the blindfold condition came close to being significant due to the increased recognition by blocked females. But in each condition, for all groups, significant differences were found between all possible pairings of stimuli except between the lemon and orange stimuli (Figs. 1–3).

Differences in percentage of correct responses within conditions, taking into account the order of presentation of the conditions in the 24 blocked subjects, demonstrated only one significant difference, that being between the percentage of correct scores when the conflict condition was presented first (12.8% correct) and when the conflict condition was presented second following the control condition (34.5% correct) ($p < 0.02$, Mann–Whitney U test).

No significant differences for individual stimuli in the two conflict condi-

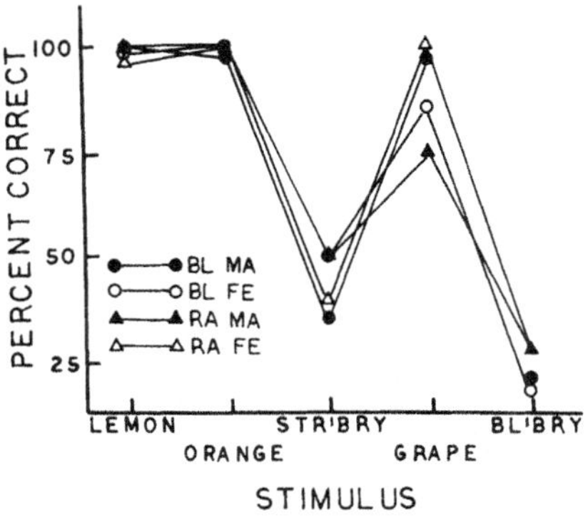

Fig. 1. Correct identifications of fruit odors by four groups of subjects: control condition (BL MA, blocked males; BL FE, blocked females; RA MA, random males; RA FE, random females). (From Armbruster, 1973.)

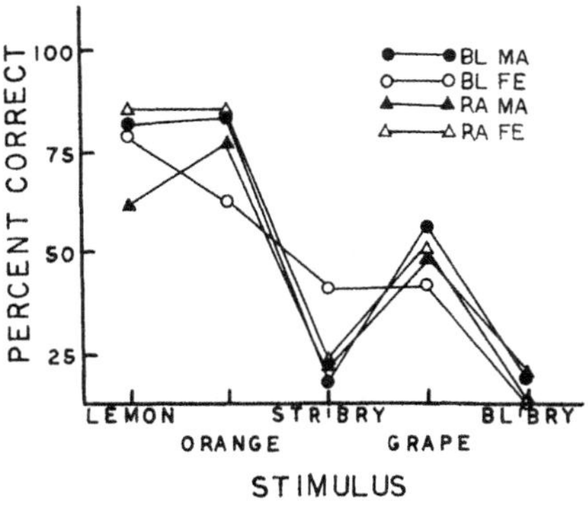

Fig. 2. Correct identifications of fruit odors by four groups of subjects: blindfold condition. (See Fig. 1.)

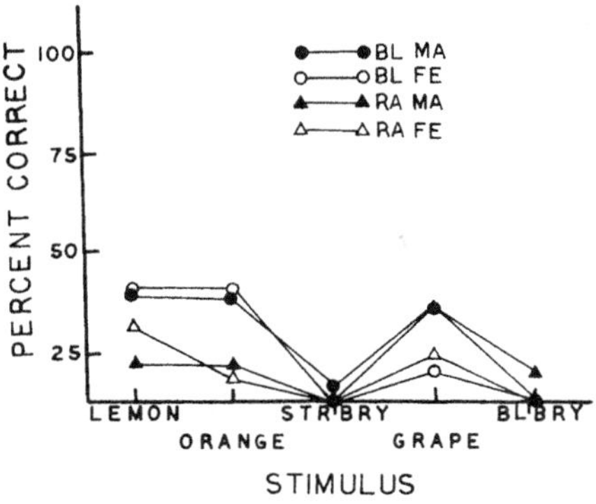

Fig. 3. Correct identifications of fruit odors by four groups of subjects: conflict condition. (See Fig. 1.)

tions were found, although grape was misidentified much more in conflict condition one than in conflict condition two (Fig. 4).

It was reasoned at the outset of the study that if subjects found certain stimuli more difficult to identify, they would take a longer time to respond. This reasonable hypothesis was not confirmed, although the trend was in its favor. The subjects apparently did not realize when they were responding incorrectly, as most seemed to respond quickly on the basis of the color of the solution. Subjects appeared extremely conscious of the coloring of the stimuli and of the time it was taking them to respond, even though the instructions specifically stated that the task was to "identify certain liquids according to their odors."

This experiment shows how, in humans at least, visual cues can function to greatly enhance or suppress the accurate discrimination of odors. Such results are similar to those by Pangborn (1960), Pangborn and Hansen (1963), Pangborn *et al.* (1963), and Maga (1974) which dealt with changes in taste thresholds due to the color of test liquids. More such behavioral studies are needed to assess cross-modal effects on chemosensory responding.

Nevertheless, these results are typical of those which turn scientists to the use of animals in order to minimize the role of nontarget cues. This is especially true when studying the chemical senses which are less critical in human behavior than in the behavior of many other species. Also, when we work with relatively nonsocial "lower" vertebrates it is feasible to assume that cultural factors are of much less importance than in humans. But whether this means that the senses do not interact in food choice is not at all obvious. Less complexity in information processing and the greater reliance on single "marker" cues may be expected in some "simple"

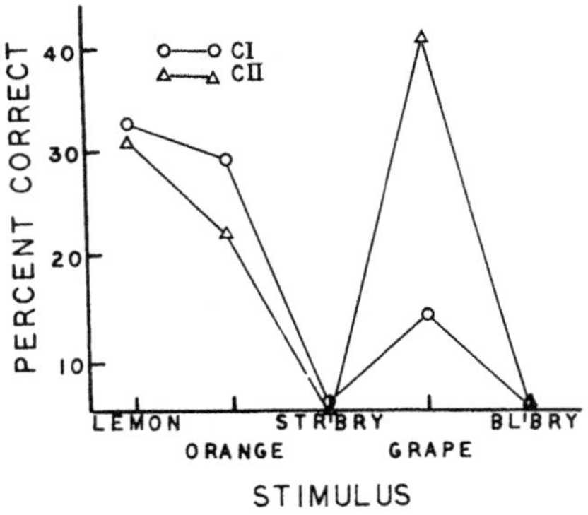

Fig. 4. Correct identification of fruit odors by all subjects for the two conflict conditions. CI, conflict condition I; CII, conflict condition II. (From Armbruster, 1973.)

organisms. Dethier's work (Chapter 24) certainly shows this in blow flies. Extrapolation even to other insects, such as honey bees whose foraging involves sensory abilities so outrageous that they intrude into philosophical discussions of what "Man" is (Griffin, 1976), is unwarranted.

SNAKES AND DIGESTION

In spite of the above I will now change our focus from people to serpents. Snakes, as far as we know, do not have a highly developed cultural, social, linguistic, or cognitive life. But even the Bible claims they have much wisdom, and, presumably, much to teach us. All snakes are carnivores and we do know that many species rely a great deal on the chemical senses (Burghardt, 1970a). We also know that different species of snakes have quite widely varying and nonoverlapping food habits, even when residing in the same habitat. Thus, while many predators are viewed as opportunistic, snakes are more or less selective. That is, they choose only a small number of the total possible species within their preferred size class available to them. Why is this? One reason may be that the digestive system of many snakes has become specialized for processing certain types of prey more readily than others. Little experimental work has been done to demonstrate the mechanism or extent of these digestive limitations, but they doubtless occur. Some snakes of the family Colubridae eat insects; most do not. Earthworm-loving eastern garter snakes *(Thamnophis sirtalis)* that through chemical deception were induced to swallow normally uneaten prey such as soft-bodies grubs either regurgitated them or passed them through the alimentary system undigested—this in spite of the common notion that snakes can digest anything, which has no doubt stemmed from the fact that mammal-eating snakes digest the bones of their prey. In addition, some potential prey items such as insects, toads, frogs, and salamanders often possess distasteful or poisonous properties which more actively lead to aversions or metabolic revolts, but which some snakes have physiologically overcome. For example, the eastern hognose snake *(Heterodon platyrhinos)* eats virtually nothing but toads, whereas most snakes shun toads or eat them rarely. But the hognose is different from other snakes in that it has enlarged adrenal glands and apparently requires more food to maintain its weight and growth (Farlow, 1975). Both differences may be related to a toxic toad diet (Spaur and Smith, 1971).

But what about digestive incompatibility with food classes based on larger taxon-typical properties such as fur in mammals or chitin in insects? Primates eat insects, fruits, vegetables, and even red meat with generally no ill effects. The black bear *(Ursus americanus),* a species that we have been working with the last few years (Pelton and Burghardt, 1976), is perhaps the ultimate omnivore yet is descended from a carnivore with a carnivore digestive system. It eats a vast range of fruits, berries, nuts, grasses, fish, insects, and almost anything found in a picnic or lunch basket, but is digestively remarkably inefficient—fruits they eat come out the other end looking like jam or preserves of the same type, at least in the cubs we reared at home (Burghardt and Burghardt, 1972). Most of a bear's food intake is passed through in such a way that it is easily identifiable.

Perhaps the trade-off is that a snake is a more efficient food processor simply because it ingests fewer different items than the more omnivorous endotherm. This attractive idea is rendered less likely by the many mammals and birds which eat the same items on as specialized a basis. For instance, rat snakes *(Elaphe)* and boas digest almost an entire mouse, rat, chicken, or rabbit, including bones, excreting a highly concentrated mass of feces (see Pope, 1961). Owls, also highly specialized carnivores on warm-blooded prey, must periodically regurgitate balls of fur and other undigested prey parts. Carnivorous mammals pick and choose parts of the carcass. Cows are specialized grazers, what with their rumens and time-consuming cud-chewing, yet even they extract little more than half of the available nutrients from grass.

This is really a surprising situation. Here we have mammals and birds with high metabolic rates that one would expect must extract the last molecule of protein, vitamin, and mineral from their food, but which seem much less efficient processors than is "the reptile," the prototype to many of my friends of the lethargic, lazy, dull, and behaviorally uninteresting animal. The specific ecological and evolutionary factors involved in this difference remain to be studied, and our knowledge of reptile metabolism and digestion is in a very rudimentary state. But I do want to establish that nutrition is not only a matter of the caloric, vitamin, protein, or mineral content of the food ingested, but is also influenced by how much the animal can derive from the food regardless of its nutritional value as determined in the chemical laboratory. That is, the same food item may have a vastly varying nutritional value depending on the species ingesting it. It could be that being capable of less sustained activity than endotherms, reptiles have found it more advantageous to expend their limited energy extracting nutrition from what is in their gut than have many endotherms. Possibly, individual as well as phylogenetic factors are involved since it has been demonstrated that *Xenopus* tadpoles utilize suspended food differentially, being more efficient when particles are less dense (Wasserzug, 1972).

PREY RECOGNITION IN NEONATE SNAKES

Before specialized digestive systems evolved, behavioral preferences must have developed. That is, specializations of the internal economy necessarily followed the reliance of certain snakes on one type of prey more than another. Thus feeding behavior in a predator depends not only on sensory mechanisms, but upon a host of metabolic, structural, sensory, and behavioral adaptations. Hopefully our sensory and behavioral research on snakes may eventually be integrated with nutrition, growth, and metabolism studies. The latter are unfortunately rare (e.g., Myer and Kowell, 1971).

As we frequently use newborn snakes, the role of maternal guidance of food choice and uncontrolled individual prior experience can be effectively discounted. Of course, many birds and mammals can be hand-reared without the mother, but the time-consuming nature of such rearing makes long-term studies not only tedious and masochistic but rather inefficient, complicated, and open to charges of social and sexual improperties.

Our animal research, then, involves the study of prey preferences: sensory control, ontogeny, modification, inter- and intraspecific differences, and the relationship to ecology and phylogeny. A great deal of information has accumulated since I began this work in 1962, some of which was reviewed several years ago (Burghardt, 1970a). That paper also reviews our knowledge of chemical perception in food choice of reptiles. While not repeating that information here, a brief overview of our early work will help make the greater details of our recent findings on behavior patterns associated with chemoreception that shift in evolution and with individual experience more digestable. Unreferenced studies are documented in the review chapter.

We have found that newborn snakes of a number of species respond with a characteristic prey-attack response to water-based extracts of prey species. For example, newborn eastern garter snakes will respond to extracts prepared from the surface substances of earthworms, fish, amphibians, and leeches. They will largely ignore extracts prepared from baby mice, insects, slugs, and other food items not normally eaten. How do we show this? Very simply, we take the prey item and drop it into warm distilled water, usually at 60°C. The normal prey weight-to-water ratio is 3 g of prey item for each 10 ml of distilled water. The animal remains in the water for 2 min, after which it is removed. The resulting liquid is centrifuged or filtered and the colorless, odorless supernatant frozen or refrigerated in small vials until use.

The typical method by which the stimuli are presented to the animal is to dip a cotton swab into an extract, rub or flick off the excess, and slowly introduce the swab into the isolated animal's enclosure. The swab is moved slowly in toward the animal and maintained in front of or to one side of it at about a centimeter distance. Tongue flicks toward and away from the swab are noted as well as the occurrence of actual attack behavior, which is timed to the nearest tenth of a second. Snakes can be repeatedly tested on such extracts. We use tongue flicks as a measure of interest in prey extracts as compared to distilled water, a measure appropriate to hungry snakes in which chemoreception is important in the initial locating and striking of prey but not in species where visual or thermal cues are critical, as in vipers or rattlesnakes (Burghardt, 1970a; Chizar *et al.,* 1977b; Kubie and Halpern, 1975).

The attack response of newborn garter snakes is triggered primarily by the tongue–Jacobson's (vomeronasal) organ system, with nasal olfaction and vision playing minor roles (Wilde, 1938; Burghardt and Pruitt, 1975). The tongue picks up volatile and nonvolatile chemicals and transfers them to the openings of Jacobson's organ in the anterior roof of the mouth. Gustation is unimportant to snakes until prey is already in the jaws. Habituation does occur but its effects are normally short-lived. Studies of tongueless animals and psychophysical data obtained with varying extract concentrations support these conclusions (e.g., Fig. 5). Tongue-flicking is a general arousal and exploratory behavior, and thus its chemosensory control is not exclusively food related as compared to prey attack (Chizar and Carter, 1975; Chizar *et al.,* 1977a). However, our observations of feeding show that prey attack in newborn garter snakes is always preceded by tongue flicks, and that one of the flicks makes contact with the prey. This demonstrates the relative importance of nonvolatiles in triggering the prey attack.

But what about different snakes with different and more or less specialized feeding habits? We have found that newborn snakes of various species differ in the prey they will attack. While eastern garter snakes *(Thamnophis sirtalis)* and plains garter snakes *(T. radix)* respond to a wide variety of prey extracts, but not to insects, newly hatched green snakes *(Opheodrys vernalis blanchardi)* respond to crickets and other insects but not to any of the prey extracts attacked by *Thamnophis,* a difference completely congruent with the species-characteristic feeding habits. Some neonate water snakes *(Natrix sipedon)* respond to extracts of fish and frogs, but not crayfish. Other water snakes (e.g., *Regina grahami)* respond to crayfish but not fish and frogs. Newly hatched rodent-eating rat snakes *(Elaphe)* respond to mouse extracts (Burghardt and Abeshaheen, 1971).

The extract swab technique is quick, cheap, and gross; it is also artificial and genuine arguments can be raised against its ecological and behavioral validity when the prestrike appetitive behavior of snakes is considered.

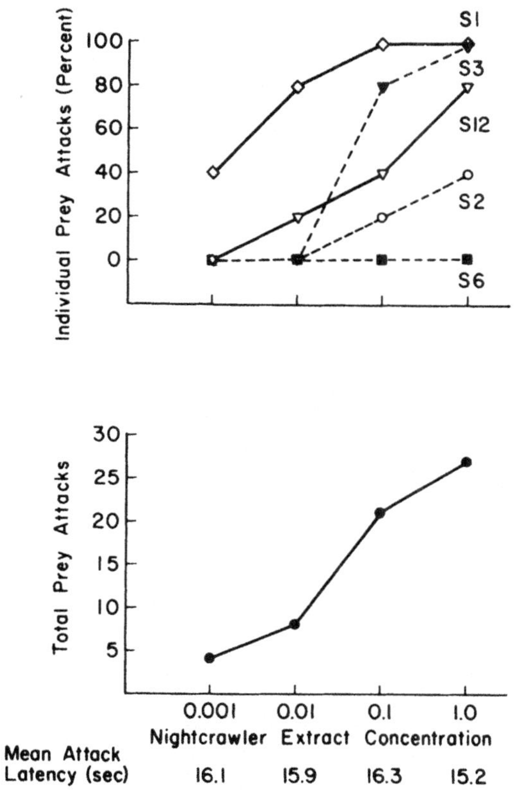

Fig. 5. Combined and selected individual prey attack profiles to various concentrations of nightcrawler extract in a litter of 13 unfed newborn *Thamnophis butleri*. Each snake was tested on the four worm extracts plus distilled water, in balanced "random" orders, for 5 consecutive days. Upper graph gives the individual curves for five selected subjects.

John Abeshaheen and I thus developed a method in which prey odors are introduced into an animal's enclosure by passing air over prey in flasks at a low rate (500 ml/min) (Fig. 6). Using activity and tongue flicking as indicators of response to odor laden air, we showed that naive garter snakes responded to earthworm odor more than to baby mouse (Table III), while the reverse held for tongue flicking, at least, in newly hatched rodent-eating corn snakes. These data also indicate the relative importance of an intact tongue–Jacobson's organ system and the relative unimportance of the external nares when they were covered.

Animals deprived of extract or feeding experience for up to six months still show the basic innate template evidenced in newborn litter mates. Furthermore, maternal experience has no demonstrated enhancing effect

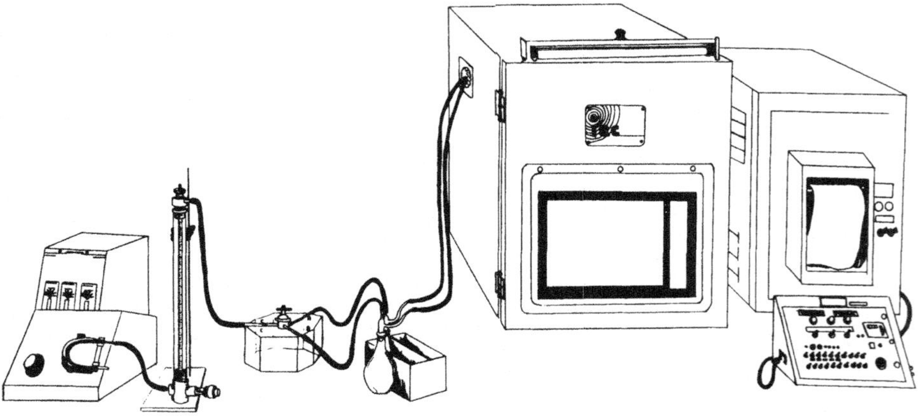

Fig. 6. Olfactometer and associated equipment to observe responses of snakes to airborne prey odors. Items from left to right include the pump, flowmeter, Teflon Y stopcock, flasks for prey, isolation chamber with viewing port, and event recorder with keyboard.

in the viviparous *T. sirtalis* (Burghardt, 1971). More recent data with oviparous corn snakes, *Elaphe guttata,* born to parents hatched in the laboratory and reared entirely on lab mice, indicate the same thing, although tongue flicking is more of a marker than is attack behavior (Table IV).

Newborn snakes from female *T. sirtalis* caught in different geographical areas of the species' range also show significant differences; not so much in the type of prey attacked as in the relative preference for it (Burghardt,

TABLE III

Mean Activity and Tongue Flick Scores of Garter Snakes *(T. radix)* to Airborne Odors of Earthworms and Newborn Mice[a]

Litter (no. of subjects)	Mean activity score (sec)		Mean tongue flick score	
	Worm	Mouse	Worm	Mouse
1. Naive subjects				
Normal (9)	214.6	153.9	46.6	27.4
Anosmic (9)	219.1	171.0	30.3	12.9
2. Naive subjects				
Normal (4)	311.8	67.3		
Tongueless (4)	120.3	103.3		

[a] 10-min tests.

TABLE IV

Mean Tongue Flick Responses to Prey Extracts by Newly Hatched Corn Snakes Born to Parents Raised on Lab Mice[a]

	Stimulus							
	H_2O	Cricket	Domestic chick	Quail chick	Lab mouse	Peromyscus	Microtus	Gerbil pup
Score	0	1.9	9.6	13.5	4.8	10.9	2.0	7.9

[a] Eight subjects tested once each.

1970b), as S. Arnold (personal communication) has also subsequently demonstrated in other *Thamnophis*. More recently I have shown that preference differences appear to exist within naive littermates of the same species (Burghardt, 1975). That is, while most members of a litter may prefer worm extract to fish extract, some individuals will show the reverse preference and significantly prefer fish to worm. Such individual variability is also found in responses to varying concentrations of prey extracts in *T. sirtalis* (Burghardt, 1975) and T. butleri (Fig. 5). These appear to be biases and tendencies found in the ingestively naive animal. Other aspects of feeding behavior could also be studied profitably. For instance, adult garter snakes can be quite readily trained to accurately follow trails of prey, such as those made from earthworm extracts (Kubie and Halpern, 1975). Whether newborn snakes have this ability has not yet been shown.

ONTOGENETIC CHANGES IN PREY PREFERENCES

In spite of the specific and presumably largely genetically based chemical prey preferences of newborn snakes, modification of feeding behavior is rather easy, within limits, if one goes about it correctly. For instance, after being presented with a prey extract that elicits an attack, garter snakes tongue flick at a subsequently presented distilled water swab much more than usual (Burghardt, 1969). Feeding a garter snake fish or earthworms will enhance responses to extracts prepared from the prey experienced (Fuchs and Burghardt, 1971). On the other hand, making the animal ill after ingesting prey, as with lithium chloride, leads to an aversion to the specific prey type eaten both in adult (Burghardt *et al.*, 1973; Czaplicki *et al.*, 1975) and newborn (Burghardt and Czaplicki, unpublished observations) snakes. Unpublished data have also shown that experience with specific prey enhances the activity and tongue-flicking behavior of

TABLE V

Mean Activity and Tongue Flick Scores to Airborne Fish and Worm Odors by Garter Snakes Reared for 2 Years Exclusively on Fish or Worm (Less "No Odor" Control Scores)[a]

	Mean activity score (sec)		Mean tongue flick score	
Subject no.	Worm	Fish	Worm	Fish
9 (fish reared)	−35	72	0	34
10 (fish reared)	−79	53	7	17
19 (worms reared)	123	−17	25	−39
20 (worms reared)	78	−40	13	−4

[a] Five 10-min tests with each prey odor.

hungry snakes in the olfactometer (Table V). Some of these processes also appear to take place in wild populations. We have found that banded water snake populations in different habitats but less than 80 km apart have quite different prey extract preferences that seem at least partially due to postnatal experience (Gove and Burghardt, 1975).

Thus we have established that in many species of snakes, the chemical senses are crucially important in prey identification prior to any experience with the food object itself. Furthermore, the information that the animals receive from their prey is quite specific. Intraspecific and interspecific variability are what one would expect with any trait under at least partial genetic control and hence subject to the operation of natural selection. Such behavior, however, is easily modified at least to the extent of altering relative preference for some prey in those species with a diverse feeding repertoire such as garter snakes. Plasticity in diet selection in other species is far less known and could in fact be rather limited or restricted to certain classes of prey. Field and laboratory studies of closely related species groups containing both highly specialized and less specialized members may give us ideas about ways to go about introducing new foods to other animals, such as endangered species, as well as to different cultures in our own species that differ in the current diversity of diet.

VISUAL CUES AND PREY SELECTION IN SNAKES

Chemoreception is not everything, and I want to briefly mention some recent work in our laboratory which goes back to the question of visual influences addressed in our fruit odor study with people. In the species we have studied, movement is a potent cue in attracting snakes' attention and increasing tongue flicking, but not in triggering a predatory strike in the absence of appropriate chemical cues. In some lizards (e.g., *Ameiva)*, nonmoving prey are eaten only after considerable tongue activity while moving prey are attacked without any tongue flicks or at most one or two (e.g., D. Gove, personal communication). Of course, such studies do not rule out the role of chemoreception but do show the facilitating effect of prey movement. In toads and frogs, by contrast, movement of small artificial objects or even two-dimensional figures through slits in a screen are sufficient to trigger a strike, usually a rapid, directed flipping out of the tongue (Ingle, 1971, 1973).

Work in our laboratory on this topic with Dallas Denny and Ruthie Saunders utilized precisely controlled stimuli rotating at speeds of 0.5 to 4000 rpm. Earthworm-loving *Thamnophis* were used in these experiments. A 1.5-cm section of worm was placed on a hook attached to a motor below the animal's home cage firmly enough to stay attached during rotation but loose enough for the snake to remove when it attacked. Attacks were made at all speeds tested up to 2048 rpm, but the optimal speed was 16 rpm. At this speed covering the stimuli with glasses to eliminate differential "wind"-generated odor gradients at different rotation speeds still led to choosing the moving over the nonmoving worm. Realistic plastic worms did not lead to actual attack in the uncovered condition except occasionally when earthworm odor permeated the environment. These experiments show that while in garter snakes prey movement is not essential for locating and attacking prey, it can facilitate matters.

Vision can aid in other ways, too. Arboreal snakes have particularly well-developed visual systems. We have observed in the laboratory the feeding strategy of lizard-eating colubrid snakes. A comparison of two of them will show the different strategies employed. *Oxybelis aeneus* is a brown, slender, widespread neotropical snake that remains in low shrubs (Henderson, 1974). It has a slightly bifid tongue which can be extended or retracted painfully slowly as well as in typical serpent fashion (Keiser, 1975). It slowly approaches its normal prey, anoline lizards, using its extraordinary cryptic coloration and body movements reminiscent of a branch swaying in the breeze. The unusual tongue movements may be related to picking up and concentrating normally subthreshold vomeronasal stimuli as hypothesized by Keiser (1975), but I am more inclined to view the slow tongue movements as a necessary cryptic adaptation to visually alert prey. When the snake gets close to the lizard it swiftly strikes, but if it misses it does not pursue. *Leptophis ahaetulla,* a beautiful green snake, often living in the same environment as *Oxybelis,* is by contrast a quick attacker and pursuer of lizards. If its first strike misses (as usually happens since it "telegraphs its punch") it is relentless in the chase. Here we have two different visually controlled strategies for dealing with the same prey, yet chemoreception is apparently necessary in both. In short, I think snakes and lizards will prove particularly useful in the study of interactions between visual cues and odors.

THE EVOLUTION OF FEEDING BEHAVIOR

Constriction

The topography of the behavior mediated by chemoreception is certainly important but has been neglected by most workers. Even in our published snake work, for instance, it may appear as if all the snakes attacked and ingested their prey the same way. But this is not necessarily true; studies in progress by Harry Greene and others in our laboratory illustrate how careful descriptive studies can help us formulate ideas on the evolution of feeding behavior.

Perhaps the most dramatic feeding behavior of snakes is found in the constrictors, those snakes that throw one or more coils around their prey to subdue it. Yet constricting behavior has apparently evolved independently in two major families of snakes, the Boidae and the Colubridae. Boas and pythons are the archetypal constrictor—very ancient snakes still having vestigal hind limbs. The colubrids, on the other hand, are the progressive vanguard of serpent evolution. All boids constrict but probably less than 10% of the colubrids do so.

TABLE VI

Variables in Constricting Behavior

I. Type of movement	IV. Coil orientation
Winding	vertical
Wrapping	horizontal
II. Initial neck twist	angle
yes	V. Part of body used
no	anterior
III. Surface used	middle
right lateral	posterior
left lateral	tail
ventral	combination
combination	

TABLE VII

Constricting Behavior in Snakes[a]

Mode	Occurrence
Anterior coil, wound with initial twist, horizontal or angle	All boids, some *Lampropeltis*
Anterior coil, wound with initial twist, vertical	*Arizona*, some *Lampropeltis*
Anterior coil, wound without initial twist, vertical	*Boaedon, Chrysopelea, Elaphe, Gonyosoma, Pituophis, Trimorphodon*, some *Lampropeltis*
	Some *Spilotes*
Anterior coil, wrapped with initial twist, horizontal or vertical	Some *Lampropeltis*, some *Spilotes*
Posterior coil, wrapped with body and/or tail, with or without initial twist, angle or vertical	Some *Boiga*, some *Lampropeltis*, some *Spilotes*

[a] H. W. Greene, unpublished observations.

In our studies we record a variety of variables (Table VI). These lead to the following results in those species observed up to now (Table VII). Note that all boids use the first method whereas none of the colubrids use it regularly. Also note the wide variability in the same colubrid taxa; this is found even within individuals. Boids are thus much more stereotyped both as a group and as individuals. Our ontogeny results go along with this finding. Newborn boids are almost as machinelike as the adults, whereas young colubrids are often much more sloppy, variable, and inefficient, many months elapsing before they reach adult proficiency.

Such behavioral divergence is not due to ecological constraints in terms of size, diet, or habitat. Indeed, the boids seem to have a wider range of habitat type, size, and diet than do the constricting colubrids. Yet all boids in all environments and of all sizes apparently constrict all prey the same general way; they are an extreme example of evolutionary inertia. Also constricting like a boid is the Java Wart snake *(Acrochordus javanicus)*, a member of a miniscule family and probably one of the most primitive living snakes, even more so than boids (S. B. McDowell, manuscript). Yet today the colubrids are geographically much more successful outside the tropics. Clearly, such data pose interesting questions and we hope to follow up on many of them.

Tongue Movements

Snakes are universally considered to have evolved from lizards—the specific lizard stock being more controversial. Using the extract techniques we developed we have also shown that newly hatched lizards can identify species-characteristic prey on the basis of chemical cues. This is true in skinks *(Eumecis fasciatus)* (Burghardt *et al.*, 1973) and even more so in alligator lizards (H. Greene, unpublished data), young of which initially attack insect extracts but not those from earthworms or mice (Table VIII). These findings are of interest because the families of which these species are members are close to the lineage from which snakes are considered to have evolved from lizards. However, the rates of tongue flicking are much lower than in most snakes tested, and the response to the stimuli in the alligator lizards habituated by the second day of testing.

Feeding adaptations are considered critical in the shift from lizard to snake and involve many alterations in head morphology (Gans, 1961). The tongue is important as a snake can move it in many ways. All snakes observed can extend the tongue far out of the mouth, all use it extensively in feeding and exploratory behavior, all but one species, apparently, can do multiple oscillations (up–down movements during one extension), orient the entire tongue to one side without moving the head, and control the movements of the two forked segments (D. Gove, unpublished observations).

In our laboratory Gove has also extensively observed tongue flicking in lizards. She finds that here the tongue is often used in drinking, manipulating food in the mouth, and licking and cleaning particles from around the jaws and eyes, etc. There also is a great range of structure of the tongue, from thick and fleshy to long, thin, and snakelike (McDowell, 1972). Even from a strictly anatomic approach plausible ideas on the relationship of various lizards to snakes can be formulated. Gove has compared many groups of lizards. The progression through various groups to the more snakelike condition anatomically is paralleled by both the maximum extension of the tongue (relative to snout size) and the area it is possible for the lizard to sample with the tongue (Table IX). But even the most snakelike lizards do not perform the tongue movements characteristic of

TABLE VIII

Prey Attacks and Tongue Flick Rates in Two Naive Alligator Lizards *(Gerrhonotus liocephalus)*

Stimulus	Attacks	Tongue flicks/min
Cricket	2	16.0
Mealworm	1	11.5
Newborn mouse	0	9.0
Earthworm	0	7.0
Distilled water	0	4.5

TABLE IX

Area Sampled by Tongue Extensions of Various Lizards by Family[a]

Family	Species observed	Mean relative area sampled	Range	Species
Gekkonidae	1	81		*Eublepharis macularis*
Lacertidae	1	450		*Lacerta pityusensis*
Scincidae	4	619	300–825	*Eumeces laticeps, Tiliqua rugosa, T. gerrardi, T. scincoides*
Teiidae	1	2025		*Tupinambis tequixin*
Anguidae	6	694	300–920	*Abronia deppii, Gerrhonotus gadovi, G. kingi, G. liocephalis, G. multicarinatus, Ophisaurus ventralis*
Helodermatidae	2	1075	950–1200	*Heloderma horridum, H. suspectum*
Varanidae	4	2221	1630–2825	*Varanus exanthematicus, V. prasinus, V. griseus, V. varius*

[a] D. Gove, unpublished observations; at least 10 samples for each species.

all snakes, such as multiple up–down oscillations of the tongue (the latter being found so far only in *Ophisaurus* and one skink). The only exception to multiple oscillations found in snakes to date is the aquatic constricting Java wart snake. In snakes the tongue seems completely devoted to sensory roles (vomeronasal organ sampling and perhaps picking up tactile or vibration cues), while in all lizards the tongue has functions other than sensory ones, yet we are beginning to see some behavioral overlap between lizards and snakes that makes plylogenetic and, we hope, ecological sense.

CONCLUSION

Taken together these studies typify the problems inherent in trying to study a topic as broad as the determinants of food choice in a group of animals, human or nonhuman. They do, I believe, show that truly satisfactory answers will be gained only after the study of many species with diverse food repertoires, morphology, and ecology. Snakes are important to the topics addressed in this volume because (a) the chemical senses are important (but not all important), (b) marker behaviors such as tongue flicking are easily observed and quantified, (c) experience can be easily controlled in the laboratory, and (d) the chemical bases of their preferences should be within reach methodologically. Yet reptiles are only one of the many groups of animals that can provide important information for those primarily interested in human nutrition and food discrimination and preferences. A truly comparative approach, regardless of whether it concentrates on rodents, primates, snakes, or flies, will, I believe, be useful in generating ideas to solve problems faced by the diverse food habits and habitats characteristic of different populations of our own species.

ACKNOWLEDGMENTS

Much of the research described in this report was supported by research grants from the National Institute of Mental Health and the National Science Foundation. Some of the writing took place while I was on a John Simon Guggenheim Fellowship. The following students allowed use of their unpublished data or participated in some of the studies: John Abeshabeen, Robert Armbruster, Dallas Denny, Doris Gove, Harry Greene, Steve Jones, Ruthie Saunders, and Robert Tanner. C. Abeshabeen sketched the apparatus in Fig. 6.

REFERENCES

Armbruster, R. C. (1973). The role of vision in the recognition of olfactory cues. M.A. Thesis, University of Tennessee, Knoxville (unpublished).

Burghardt, G. M. (1969). Comparative prey-attack studies in newborn snake of the genus *Thamnophis*. *Behaviour* **33,** 77–114.

Burghardt, G. M. (1970a). Chemical perception in reptiles. In "Communication by Chemical Signals" (J. W. Johnston, Jr., D. G. Moulton, and A. Turk, eds.), pp. 241–308. Appleton, New York.

Burghardt, G. M. (1970b). Intraspecific geographical variation in chemical food cue preferences of newborn garter snakes *(Thamnophis sirtalis)*. *Behaviour* **36,** 246–257.

Burghardt, G. M. (1971). Chemical-cue preferences of newborn snakes: Influence of prenatal maternal experience. *Science* **171,** 921–923.

Burghardt, G. M. (1973). Chemical release of prey attack: extension to naive newly hatched lizards, *Eumeces fasciatus*. *Copeia* **1973,** 178–181.

Burghardt, G. M. (1975). Chemical prey preference polymorphism in newborn garter snakes, *Thamnophis sirtalis*. *Behaviour* **52,** 202–225.

Burghardt, G. M., and Abeshaheen, J. P. (1971). Responses to prey chemical stimuli in newly hatched snakes of the genus *Elaphe*. *Anim. Behav.* **19,** 486–489.

Burghardt, G. M., and Burghardt, L. S. (1972). Notes on the behavioral development of two female black bear cubs: The first eight months. *In* "Bears—Their Biology and Management" (S. Herrero, ed.), No. 23, pp. 255–273. Int. Union Conserv. Nat. Nat. Resour. (I.U.C.N.), Morges, Switzerland.

Burghardt, G. M., and Pruitt, C. H. (1975). The role of the tongue and senses in feeding of newborn garter snakes. *Physiol. Behav.* **14,** 185–194.

Burghardt, G. M., Wilcoxon, H. C., and Czaplicki, J. A. (1973). Conditioning in garter snakes: Aversion to palatable prey induced by delayed illness. *Anim. Learn. Behav.* **1,** 317–320.

Chizar, D., and Carter, T. (1975). Reliability of individual differences between garter snakes *(Thamnophis radix)* during repeated exposures to an open field. *Bull. Psychon. Soc.* **5,** 507–509.

Chizar, D., Carter, T., Knight, L., Simonsen, L., and Taylor, S. (1977a). Investigatory behavior in the plains garter snake (Thamnophis radix) and several additional species. *Anim. Learn. Behav.* (in press).

Chizar, D., Scudder, K., and Knight, L. (1977b). Rate of tongue flicking by garter snakes *(Thamnophis radix haydeni)* and rattlesnakes *(Crotalus v. viridis, Sistrurus catenatus tergeminus* and *S. c. edwardsi)* during prolonged exposure to food odors. *Behav. Biol.* (in press).

Czaplicki, J. A., Porter, R. H., and Wilcoxon, H. C. (1975). Olfactory mimicry involving garter snakes and artificial models and mimics. *Behaviour* **54,** 60–71.

Farlow, J. O. (1975). Observations of a captive Tuatara *(Sphenodon punctatum). J. Herpetol.* **9,** 353–355.

Fuchs, J., and Burghardt, G. M. (1971). Effects of early feeding experience on the responses of garter snakes to food chemicals. *Learn. Motiv.* **2,** 271–279.

Gans, C. (1961). The feeding mechanism of snakes and its possible evolution. *Am. Zool.* **1,** 217–227.

Gove, D., and Burghardt, G. M. (1975). Responses of ecologically dissimilar populations of the water snake, *Natrix s. sipedon,* to chemical cues from prey. *J. Chem. Ecol.* **1,** 25–40.

Griffin, D. R. (1976). "The Question of Animal Awareness: Evolutionary Continuity of Mental Experience." Rockefeller Univ. Press, New York.

Harper, R., Bate Smith, E. C., and Land, D. G. (1968). "Odor Description and Odor Classification." Elsevier, New York.

Henderson, R. W. (1974). Aspects of the ecology of the neotropical vine snake, *Oxybelis aeneus* (Wagler). *Herpetologica* **30,** 19–24.

Ingle, D. (1971). Prey-catching behavior of anurans toward moving and stationary objects. *Vision Res., Suppl.* **3,** 447–456.

Ingle, D. (1973). Size preferences for prey-catching in frogs: Relationship to motivational state. *Behav. Biol.* **9,** 485–491.

Johnson, J. D., Kretchmer, N., and Simoons, F. J. (1974). Lactose malabsorption: Its biology and history. *Adv. Pediatr.* **21,** 197–237.

Keiser, E. D., Jr. (1975). Observations on tongue extension of vine snakes (Genus *Oxybelis*) with suggested behavioral hypotheses. *Herpetologica* **31,** 131–133.

Kubie, J., and Halpern, M. (1975). Laboratory observations of trailing behavior in garter snakes. *J. Comp. Physiol. Psychol.* **89,** 667–674.

McDowell, S. B. (1972). The evolution of the tongue of snakes and its bearing on snake origins. *Evol. Biol.* **6,** 191–273.

Maga, J. (1974). Influence of color on taste thresholds. *Chem. Senses Flavor* **1,** 115–119.

Myer, J. S., and Kowell, A. P. (1971). Eating patterns and body weight change of snakes when eating and when food deprived. *Physiol. Behav.* **6,** 71–74.

Pangborn, R. M. (1960). Influence of color on the discrimination of sweetness. *Am. J. Psychol.* **73,** 229–238.

Pangborn, R. M., and Hansen, B. (1963). The influence of color on discrimination of sweetness and sourness in pear nectar. *Am. J. Psychol.* **76,** 315–317.

Pangborn, R. M., Berg, H. W., and Hansen, B. (1963). The influence of color on discrimination of sweetness in dry table wine. *Am. J. Psychol.* **76,** 492–495.

Pelton, M. R., and Burghardt, G. M. (1976). Black bears of the Smokies. *Nat. Hist., N.Y.* **85**(1), 54–63.

Pope, C. H. (1961). "The Giant Snakes." Knopf, New York.

Spaur, R. C., and Smith, H. M. (1971). Adrenal enlargement in the hognosed snake *Heterodon platyrhinos. J. Herpetol.* **5**, 197–199.

Wilde, W. S. (1938). The role of Jacobson's organ in the feeding reaction of the common garter snake, *Thamnophis sirtalis sirtalis* (Linn.). *J. Exp. Zool.* **77**, 445–465.

Wasserzug, R. (1972). The mechanism of ultraplantonic entrapment in anuran larvae. *J. Morphol.* **137**, 279–288.

DISCUSSION

Bosma: What is the functional difference between the vomeronasal organ and the true nose?

Burghardt: There is an old study by Wilde where the nerve from the vomeronasal organ to the accessory olfactory bulb was severed and it seems that in garter snakes, at least, it is the organ critical for prey attack. If the snake has a functional nose but the vomeronasal system is messed up, it will not attack unless you perhaps use some conditioning procedures. Nothing at all will happen in the ingestively naive newborn snake. It will starve.

Bosma: The lizard uses its true nose, you might say, for evaluation of food?

Burghardt: Well, some may. But the vomeronasal system is used in those lizards that seem to be more in the line of snake evolution and which often possess a bifurcated tongue that is long, thin, and specialized for vomeronasal functioning. There are some lizards, such as the chameleon, that have atrophied olfactory and vomeronasal systems; the tongue is specialized for snapping up prey.

Mugford: Is there any gathering of food by the mother for the young?

Burghardt: There is no maternal care in these snakes. They either lay eggs or give birth to up to 50 or 60 young. That is how we get so many subjects. But the young snakes may stay together in the same area for a while after birth.

Kare: Have you separated out by gas chromatography what is in the earthworm's surface that is uniquely attractive? Is it a simple compound?

Burghardt: We have not used gas chromatography as our interest was primarily in the nonvolatiles. But we have determined that it is not a simple compound. It is heterogeneous, about 5000 molecular weight.

Pfaffmann: Would you say development of the vomeronasal system in the reptile reaches its peak development?

Burghardt: In snakes, yes.

Pfaffmann: From there on, it is on the decline?

Burghardt: Yes. I am convinced that if you really want to find out what the vomeronasal organ does, it is best to start out with an animal that has the organ well developed and is highly dependent upon it. Work coming out from some other laboratories on sexual recognition and trailing show that the organ is also important here, especially for nonvolatile compounds.

Pfaffmann: Continuing that point, the adaptability that you did describe is only for certain species.

Burghardt: Individual variability has only been demonstrated in the eastern garter snake and in the banded water snake.

Pfaffmann: Majority of the species are not that adaptable?

Burghardt: They have not been looked at.

Cagan: Is a species that eats just a particular type of prey, such as crickets, and only responds to the cricket extract responding positively only to cricket and simply ignoring the others? What happens, for example, with mixtures of cricket extract plus extracts of some of the other potential prey?

Burghardt: They would probably perceive the other odors but tune them out. In all the species we have tested, with all the extracts, only once have we found an unlearned avoidance response. That was testing a litter of newborn racers *(Coluber constrictor mormon)* to scorpion extract. Some snakes withdrew from the stimulus and did a little dance. Mixed prey extracts presented to *Thamnophis* were about intermediate in effectiveness.

Cagan: With some of the species which are highly selective, have you ever attempted to restrict the availability of food to something they are normally averse to, in order to see if they would attack and eat it?

Burghardt: In laboratories and zoos over the world the most prevalent problem with most snakes is getting them to eat anything. So far as I know, the answer is the snake will starve. However, offering diverse foods to exotic snakes should be a more general practice and I think some surprises will result, not because the animals become more willing to take unpalatable prey, but because of our limited knowledge as to the feeding habits of the vast majority of snakes. For instance, we discovered that the vine snake *(Oxybelis aeneus)*, the arboreal lizard specialist, with a long, thin, narrow snout actually caught and ate live fish placed in its water dish.

Chapter 13

Conditioned Responses to Food Odor and Taste in Rats and Wild Predators

JOHN GARCIA AND LINDA P. BRETT

TASTE AND ODOR CUES IN FOOD REGULATION

An intuitive analysis of the roles of smell and taste in food regulation might suggest that these two chemical senses serve similar functions. Personal experience tells us that the aroma emanating from the contents of our dinner plate can greatly enhance or seriously detract from the enjoyment of that meal. In fact, many people hold the opinion that odor is a larger component of food flavor than is taste. Olfaction might therefore be expected to be at least as important as taste in adjusting eating behavior commensurate with visceral consequences.

However, an ethological analysis reveals that olfaction and taste actually operate within different spheres of influence. While an odor may lead us to food and prompt us to put that food into the mouth, it is taste which ultimately determines whether the food is then swallowed and additional mouthfuls taken. Functionally, olfaction is in many ways more similar to vision and audition than to taste. It is specifically suited to identifying and locating external objects and events. For example, a rat might follow his nose to a food bin, to a sex partner, or away from a predator. He will also display a healthy wariness or neophobia when he encounters nonidentifiable or unfamiliar odors. A common feature of all of these plastic "coping" behaviors, which can be broadly characterized as approach/appetitive and avoid/defensive responses, is that they are linked closely in time and space with their instigating sensory events and ultimate effects. When food is consumed, however, the internal homeostatic monitors may not register the nutrient or toxic effects until several minutes, or even several hours, later. To effectively regulate the internal environment, the animal must therefore possess a highly reliable and adjustable mechanism for "evaluating" food prior to its ingestion. For mammals, this behavioral role is primarily a function of the gustatory system.

An animal that experiences illness after ingesting a distinctive tasting, nontoxic fluid will develop a conditioned aversion for the flavor of that fluid (Garcia *et al.*, 1955, 1974). This conditioned aversion is often acquired with a single flavor–illness pairing even when the experimentally induced illness is delayed for several hours following ingestion (Smith and Roll, 1967). Moreover, the animal's rejection of the flavor is independent of the particular circumstances in which it is tested; its aversion is just as strong when it drinks in the home cage as it is when it drinks in the place where it became sick (Garcia *et al.*, 1970). A palatable flavor has become unpalatable.

Shifts in palatability are also found when ingestion of a flavor is followed by positive visceral effects. For example, thiamin-deficient rats will show an increased preference for a flavor that has been experimentally paired with thiamin injections (Garcia *et al.*, 1967). In fact, rats that acquire an aversion for an arbitrary flavor they consumed just prior to a bout of apomorphine-induced nausea will also show an increased preference for another flavor they consumed just prior to recovery from the nausea (Green and Garcia, 1971).

At a purely descriptive level, we can say that the rat treats a flavor consumed prior to illness as if it were poison and a flavor consumed prior to recuperation as if it were medicine. However, it is erroneous to assume that the tastes are functioning as signals for the negative or positive visceral events in the same way that a buzzer signals an electric shock to the feet. Rats will form aversions for a flavor they have already consumed even though they are not conscious of the illness, i.e., when they are unconscious, due to either anesthesia (Roll and Smith, 1972) or chemical depression of the cortex (Burešová and Bures, 1973), during the entire course of the illness. Moreover, humans acquire aversions for flavors ingested just prior to an unrelated bout of illness, although they are well aware that the flavored food was not the source of their discomfort. Conditioned aversions seem curiously independent of cognitive, conscious processes, as palatability shifts can occur with or without awareness of the conditioning contingencies.

Anatomic evidence also suggests a unique relationship between the gustatory system and the internal homeostatic monitors. Taste afferents converge with fibers from the viscera and from the internal monitors of the area postrema (which includes the chemoreceptive trigger zone for emesis in vomiting animals) at the nucleus solitarius in the brainstem. Gustatory fibers are then relayed to the pons, where one branch projects to the feeding areas of the hypothalamus and the other to the cortex. In contrast, primary olfactory fibers do not directly converge with visceral afferents (White, 1965), but project to the limbic system. These projections suggest that affective responses to external events, such as the abovementioned fear of novel odors, are an important part of the behavioral role of olfaction.

The majority of the attempts to establish illness-based aversions to odors support this analysis. For example, while rats readily form aversions for a very mildly odorous saccharine solution in a single trial, even if the illness is delayed for several hours, they do not become averted to a strong odor with repeated trials and a relatively short illness delay of 30 min (Hankins *et al.*, 1973). Even when illness immediately follows exposure to the odor, aversions are generally weak compared with those typically seen with taste–illness conditioning (Garcia and Koelling, 1967; Lorden *et al.*, 1970; Supak *et al.*, 1971; Hankins *et al.*, 1973, 1976; Simson and Booth, 1973). Taukulis (1974) has reported aversions over long odor–illness delays in a single trial when a stream of amyl acetate vapor was directed at the rat's muzzle as it licked at a water spout. Thus, if a powerful odorant is employed, perhaps a potent odor aversion can be conditioned with illness. It is conceivable, however, that molecules of the amyl acetate vapor went into solution on the animal's moist tongue and created an effective taste stimulus. Anosmic control groups are therefore necessary to sort out the real contributions of odor and taste to so-called odor aversions.

Two experiments that employed anosmic controls (Hankins *et al.*, 1973) provide the most convincing evidence so far that olfaction has, at best, a minor role in the acquisition of food aversions. Peripheral anosmia in rats was produced by perfusion of zinc sulfate through the nasal cavity. The ability of these rats to form an illness-based aversion for a solution with distinct taste and odor properties (apple juice, experiment I) or with distinct taste but weak odor properties (0.1% saccharin, experiment II) was compared to that of intact rats. The performances of the anosmic and intact groups are graphically presented in Fig. 1.

If odor aversions are an important component of food aversions, we would expect that the rats that could smell the solutions would show the greater amount of conditioned suppression, especially in the apple juice experiment. However, the anosmic animals actually show slightly stronger

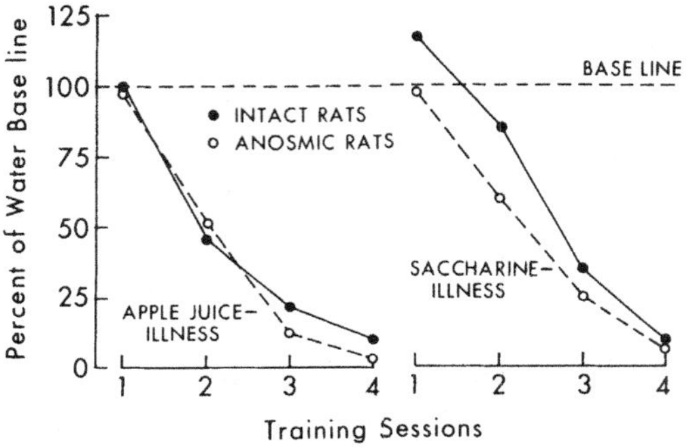

Fig. 1. Apple juice–illness and saccharin–illness conditioning. Mean fluid intake (percent of pretreatment water base line) for intact and anosmic groups. Replotted from Hankins *et al.*, 1973.

aversions than the intact animals, though this difference is not statistically significant. Also, the intact rats in the apple juice–illness experiment became increasingly hesitant over trials in taking their first lick at the spout. The lick latencies of the anosmic rats did not change. This indicates that the intact animals began to recognize the odor, disrupting their approach to the "poison" fluid. The anosmic animals could not use this defensive strategy. None of the groups in the saccharin–illness experiment showed increased lick latencies, suggesting that the odor was too weak to serve as a discriminative cue.

Other studies of the food-related behavior of anosmic rats further support the argument that the gustatory system alone is sufficient to adaptively guide consummatory feeding responses. For example, Richter (1942–1943, 1953) found that inability to smell did not interfere with the rat's selection of an appropriate diet. Similary, Barnett and his associates (1975) report that wild rats that learned to reject food containing a strongly odorous poison continued to select identical but nonadulterated food over poisoned food after being made peripherally anosmic. Thus, we can conclude that odor is (1) not a necessary cue for food–illness conditioning, and (2) a sufficient cue only under the most optimal training conditions.

TASTE AND ODOR CUES IN PAIN AVOIDANCE

We have argued that the unique behavioral role of the olfactory channel is the direction of approach and withdrawal responses to external stimuli. This role overlaps to a large degree with the adaptive roles of vision and audition, but each of these external channels may be more or less specialized for dealing with different types of events. For example, there is some evidence that olfactory cues may be especially important in conditioning that involves sexual behavior (Carr *et al.,* 1970; Marr and Gardner, 1965). Sights and sounds, which can be very rapidly located in time and space, may be the favored cues for conditioning that involves avoidance of painful assaults to the body surface.

With repeated trials and brief conditioned stimulus–unconditioned stimulus (CS–US) intervals, taste also can serve as a signal for pain. However, the animal's response to the taste is qualitatively different from that found in flavor–illness conditioning. Rats shocked while drinking saccharin actually increase their home cage saccharine intake, even though they avoid the same fluid in the shock apparatus (Dietz and Capretta, 1967; Garcia *et al.,* 1966). Rather than forming an aversion, the shocked rats respond to the taste as one component of a complex discriminative stimulus for defensive behavior. When the flavored fluid is made available outside of the experimental context, the animals respond to its positive hedonic properties and eagerly drink.

In attempting to compare the cue effectiveness of odor and taste in pain avoidance learning, we encounter the same confound prevalent in flavor–illness experiments: Tasty solutions may have distinctive odors, and vice versa. For example, Krane and Wagner (1975) recently reported decreased drinking in rats following a single pairing of a strong (0.7%) saccharine solution and an intense (4.0 mA for 1.5 sec) shock delayed for several seconds. These investigators argue that the decrease represents a taste aversion, and on this basis dismiss the notion of associative biases and neural specificity. However, this interpretation underestimates the complexity of their conditioned stimulus.

Again, we can sort out the contributions of odor and taste by comparing the performance of smelling and nonsmelling animals. Hankins *et al.* (1976) tested anosmic and intact rats in a shuttlebox shock avoidance task with a variety of olfactory and gustatory cues. Some of the experimental groups, which are analogous to the illness-reinforced groups shown in Fig. 1, were shocked when they chose the incorrect spout and drank apple juice or 0.1% saccharin for longer than 2 sec. The performances of the anosmic and intact animals are compared in Fig. 2. The intact animals in both the apple juice and saccharine conditions show a rapid improvement in avoiding shocks, indicated by a reduction in shock frequency over trials. In contrast, the animals that cannot utilize odor cues show much less improvement and are still getting a large percentage of shocks during the final training session. These results indicate that taste alone is a relatively poor cue for shock avoidance. This conclusion is further supported by the performance of groups that had a very weak (0.025%) saccharine solution

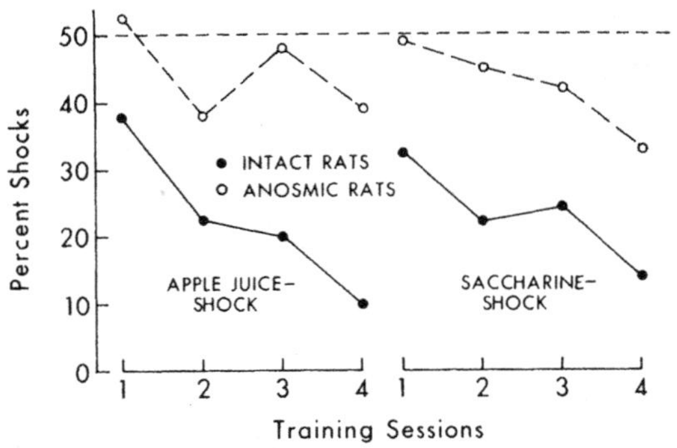

Fig. 2. Apple juice–shock and saccharin–shock conditioning. Mean percent of trials on which shock was delivered to intact and anosmic groups in shock-shuttle experiments (50% = chance performance). Replotted from Hankins *et al.*, 1976.

paired with shock. Neither intact animals nor anosmics could use this cue to avoid the shock, but both acquired aversions for this taste when it was paired with illness in another experiment in the same study.

Conditioning was finally obtained in anosmic rats tested with a strong (0.7%) saccharine solution and intense (0.8 mA for 0.2 sec) shock, which illustrates this well-known axiom: When conditioning stimuli are intense and parametric arrangements are simple, even severely handicapped animals can be conditioned, with less adequate sensory channels and neural mechanisms playing a compensatory role.

TASTE AND ODOR CUES IN PREDATORY BEHAVIOR

The evidence for dissociation of odor and taste raised a fascinating research question: If a wild carnivore becomes ill after consuming its prey (or a prey–meat bait), will the resultant aversion inhibit killing of that prey, or will it kill and then reject it after tasting? This issue has practical as well as theoretical significance, since food aversion conditioning could offer an effective, inexpensive, and humane method of controlling predation on livestock.

Gustavson and his associates (1974, 1976) have found that feral coyotes and wolves readily acquire illness-based aversions for the flesh of natural prey, such as rabbits and sheep. First, consider the sheep-averted animals. Three coyotes (Gustavson *et al.*, 1974) and two wolves (Gustavson *et al.*, 1976), all sheep killers prior to treatment, were fed sheep flesh baits on one or two occasions. These baits, which consisted of chopped mutton wrapped in a fresh woolly sheep hide, were either laced with or followed by an emetic dose of lithium chloride. This treatment produced vomiting in all of the animals. A few days later the canines were tested with live sheep. One coyote, who had only had a single lithium treatment, ran up to its lamb, stopped and sniffed at it, and then retreated and retched. The two other coyotes each received two lithium treatments. When posttested, they also broke off attack after sniffing the lamb. Moreover, both of these animals were observed eating grass during the test periods, a response which may reflect the hungry coyotes' strong simultaneous impulses to approach and to avoid the former prey.

A different type of conflict response was seen when the wolves were tested together after their single lithium treatment. The animals charged the sheep as they had before treatment and made oral contact several times. However, instead of carrying out a prolonged biting attack and finally killing the sheep, the wolves would bite and then immediately back off. Within several minutes, the sheep became the dominant animal in the pen. When the wolves approached, the sheep charged, driving them further away and eliciting puppylike submissive gestures (see Fig. 3). This "social" response to the sheep suggests that when feeding is blocked, the animal may respond to other stimulus properties of the prey.

Predatory attacks were also inhibited in eight coyotes that had consumption of a rabbit carcass or rabbit bait paired with lithium illness once or twice. For some of these animals, the rabbit was the only prey available in the test situation, whereas others had the option of preying on a chicken. One additional coyote, a timid wild creature that was initially tested without prey options in its home cage, killed several rabbits after treatment but ate none. This suggested that the coyote might be responding to the rabbit as an intruder rather than as food. With persistence, this animal was finally tested in an outdoor arena. A tethered rabbit and a tethered chicken were simultaneously available. After thoroughly sniffing the rabbit, she rejected it and proceeded to kill and eat the chicken.

The availability of alternative prey in the test situation also appeared to increase resistance to extinction. The six coyotes tested in the choice conditions failed to kill rabbits in the six to nine tests conducted over several weeks. Coyotes that had only the aversive prey available during the test, whether averted to rabbits or to sheep, showed much greater variability. Of these latter animals, the three rabbit-averted coyotes showed the greatest recovery in the consumption of the prey. The pattern of this recovery, which was identical for the three, illustrates palatability adjustment in the positive direction. After killing a rabbit, the coyote did not eat it all at once. It first ate off the ears and then waited about half an hour before returning to the carcass. Next, it ate the rabbit's head and retired for another half-hour or so. Finally, the coyote ate what remained.

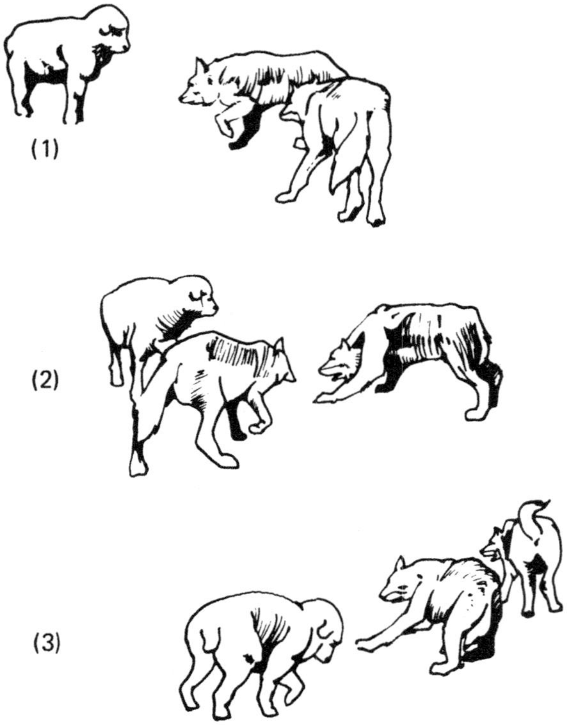

Fig. 3. Two wolves encounter a live sheep following a single mutton bait–lithium pairing. The wolves direct puppylike submissive gestures toward the sheep (1 and 2) and retreat when it charges (3), indicating that they are responding to nonfood-related stimulus features of the former prey. Drawn by J. E. Garcia.

The parts first sampled by the hungry animals are significant, since they contain relatively small amounts of the fatty acids that give meat its characteristic taste and odor. Since these milder tasting portions are consumed without ill effects, the palatability of the prey is increased and the animal proceeds to the fattier portions.

We have elsewhere proposed (Garcia *et al.*, 1973) a two-phase process for mammalian prey–illness conditioning that fits the data presented above and describes the special roles of odor and taste. When the animal becomes ill after eating flesh with distinctive odor and taste properties, the palatability of the prey abruptly drops (phase one). When he next encounters the live prey, its odor and appearance may still elicit attack, but the animal will not eat its kill. In the second phase of conditioning, the animal associates the external properties of the prey, including its smell, with the now aversive taste. Since the prey is unfit to eat, attack would be a wasted effort, and would also necessitate oral contact. Approach and attack are therefore inhibited.

The fact that several of the animals inhibited killing after a single prey–lithium treatment is not inconsistent with the two-phase hypothesis. The odor of the meat is paired with the taste while the animal is eating and also when he is regurgitating, providing ample opportunity for association. It should be noted that distal cues such as odor may also be directly associated with the illness to some extent, as is evidenced by the conditioned retching seen in one coyote. However, these effects are not seen in all animals and extinguish rapidly compared to avoidance reactions to the taste.

Additional support for the two-phase notion is provided by studies of aversive conditioning in predatory birds. In contrast to the mammalian predators we have discussed, the olfactory and gustatory abilities of birds are generally less developed. Only a few avian species, such as the kiwi (Wenzel, 1971) and turkey vulture (Stager, 1964), have been shown to use odor cues in food location, and comparative study reveals that birds have fewer taste receptors than mammals (Kare, 1965). A large part of the behavioral role served by olfaction in mammals is served by the very keen visual system of birds. There is also some evidence that avian vision serves part of the role of mammalian taste in certain species. Wilcoxin *et al.* (1971) found that bobwhite quail made sick after drinking blue, sour water subsequently avoided blue water more than sour water. A bias toward associating visual cues with visceral effects would seem to be an efficient survival strategy for seed-eating birds, since their food is typically covered by tough, bland seed coats and may be swallowed whole. However, some omnivorous and carnivorous birds, whose diet may be richer in taste information, appear to have evolved the same strategy as the rat, wolf, and coyote.

Brower (1969; Brower and Glazier, 1975) reports that blue jays that become ill after eating toxic monarch butterflies *(Danaus plexiapus)* may subsequently seize toxic monarchs with their beaks, but will then release them. This release is probably a response to the distinctively bitter flavor of the cardiac glycoside poison sequestered by these butterflies. After a few experiences with bad-tasting monarchs, the blue jays may reject both toxic and nontoxic monarchs on sight. Thus, taste is the critical cue in blocking consumption in these omnivorous birds, with visual cues guiding their attack behavior.

While blue jays and the mammalian predators must taste their prey when attacking, carnivorous hawks kill with their taloned feet. This difference in style made us wonder whether visual cues might be especially salient in establishing prey aversions in these birds. To examine this question, we tested several red-tailed hawks and one American rough-legged hawk with three different types of "poison" prey (Brett *et al.*, 1976). One group was given an emetic dose of lithium chloride after consuming a dead black mouse treated with a mild quinine solution (taste + color condition). After a single lithium treatment, both of the birds in this group refused to seize and eat the dead "poison" mouse (see Fig. 4), although they would still eat "safe" unflavored white mice. A series of generalization tests showed, however, that the taste of the mouse was a stronger cue for avoidance than was color. When offered live black bitter mice, the birds most often killed and then rejected the prey after tasting the bitter fur, although one bird showed considerable vacillation in his attack behavior. When tested with the separate components of the taste–color cue, both birds rejected white mice with a bitter flavor, but rapidly consumed non-bitter black mice.

The performance of hawks that were made sick after eating mice that differed from safe white mice only in taste or only in color further supports the primacy of gustatory cues in establishing aversions to prey. Birds in the "taste only" condition ($N = 2$) rejected bitter white mice after a single lithium treatment, but continued to consume unflavored white mice. In contrast, "color only" hawks ($N = 3$) required two or three lithium

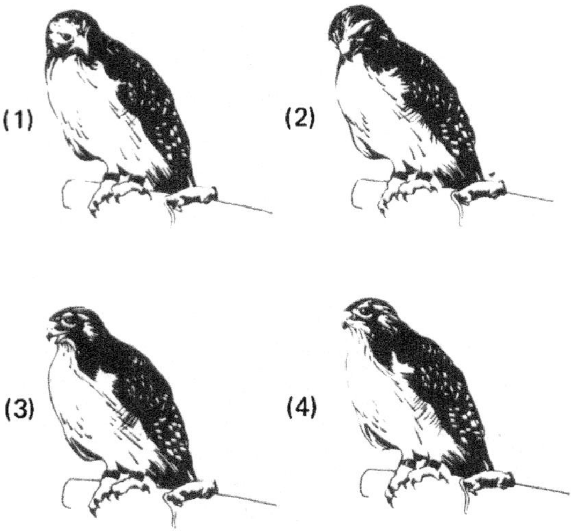

Fig. 4. This red-tailed hawk received a single bitter black mouse–lithium pairing. Several days later, she is again offered a dead bitter black mouse, and shows signs of conditioned digust. The hawk looks at the mouse (1) and then averts her gaze (2). She "smacks" her beak repeatedly (3), a behavior seen just after vomiting during the treatment trial. Finally, she stares away from the mouse, glassy-eyed (4). Drawn by J. E. Garcia.

treatments before consumption of black mice was completely inhibited. Moreover, all of the birds in this group generalized their aversion to some degree to white mice, suggesting the acquisition of an aversion for the taste of the prey.

It appears that the potency of the color cue may have been enhanced when it was paired with a distinctive taste cue. Following a single prey–illness pairing, our "taste + color" hawks avoided bitter black mice purely on the basis of appearance, whereas the "color only" birds seized and ate their "poison" black mice. Rusiniak (1976) has recently found a similar interaction in rats, using taste and odor cues. Rats that were made sick after drinking a fluid with a distinctive almond odor subsequently showed only a weak aversion. Other rats that were made ill after drinking a fluid with the same almond odor, but with a distinctively sweet taste, later showed strong avoidance on the basis of either cue when taste and odor were presented separately. These results may necessitate a revision of the two-phase hypothesis to include a parallel conditioning process, where an effective taste cue potentiates normally inadequate or less effective cues.

In summary, the food-related behavior of mammals and some birds is guided by a simple feedback system, with taste and the external receptors participating at different points in the loop. The animal locates food with the external receptors and is rewarded for its efforts by the taste of food in the mouth. If it swallows the food and it is nutritious, the palatability of that food will automatically increase, as will the probability that the animal will seek out and consume that food again. If consumption is followed by toxic effects, the animal will form an aversion for the taste of the food and approach tendencies will be diminished. Food that has distinctive odor or visual properties may be recognized when next encountered, in which case the wary animal will avoid making oral contact with the unpleasant taste. However, if alternative food sources are scarce and the animal is very hungry, it may sample the aversive food and the adjustment process will be repeated.

Some animals, such as the seed-eating quail, may have evolved variations on this basic scheme which are peculiar to the demands of their dietary niche. However, these instances should not be regarded as exceptions to a rule. Rather, they reflect a selective process which has endowed each species with the propensity to learn biologically important relationships.

ACKNOWLEDGMENT

This research was supported by USPHS Grant 1 RO1 NS 11618 and Program Project Grant HD 05958.

REFERENCES

Barnett, S. A., Cowan, P. E., Radford, G. G., and Prakash, I. (1975). Peripheral anosmia and the discrimination of poisoned food by *Rattus rattus L. Behav. Biol.* **13,** 183–190.

Brett, L. P., Hankins, W. G., and Garcia, J. (1976). Prey-lithium aversions. III. Buteo hawks. *Behav. Biol.* **17,** 87–98.

Brower, L. P. (1969). Ecological chemistry. *Sci. Am.* **220,** 22–29.

Brower, L. P., and Glazier, S. C. (1975). Localization of heart poisons in the monarch butterfly. *Science* **188,** 19–25.

Burešová, O., and Bureš, J. (1973). Cortical and subcortical components of the conditioned saccharin aversion. *Physiol. Behav.* **11,** 435–439.

Carr, W. J., Krames, L., and Costanzo, D. J. (1970). Previous sexual experience and olfactory preference for novel *vs* original sex partners in rats. *J. Comp. Physiol. Psychol.* **71,** 216–222.

Dietz, M. N., and Capretta, P. J. (1967). Modification of sugar-saccharin preferences in rats as a function of shock to the mouth. *Proc. Am. Psychol. Assoc.* **1,** 161–162.

Garcia, J., and Koelling, R. A. (1967). A comparison of aversions induced by X-rays, toxins and drugs in the rat. *Radiat. Res., Suppl.* **7,** 439–450.

Garcia, J., Kimeldorf, D., and Koelling, R. A. (1955). A conditioned aversion towards saccharin resulting from exposure to gamma radiation. *Science* **122,** 157–159.

Garcia, J., Ervin, F. R., and Koelling, R. A. (1966). Learning with prolonged delay of reinforcement. *Psychon. Sci.* **5,** 121–122.

Garcia, J., Ervin, F. R., Yorke, C. H., and Koelling, R. A. (1967). Conditioning with delayed vitamin injections. *Science* **155,** 716–718.

Garcia, J., Kovner, R., and Green, K. F. (1970). Cue properties versus palatability of flavors in avoidance learning. *Psychon. Sci.* **20,** 313–314.

Garcia, J., Clarke, J. C., and Hankins, W. G. (1973). Natural responses to scheduled rewards. *In* "Perspectives in Ethology" (P. P. G. Bateson and P. H. Klopfer, eds.), pp. 1–41. Plenum, New York.

Garcia, J., Hankins, W. G., and Rusiniak, K. W. (1974). Behavioral regulation of the milieu interne in man and rat. *Science* **185,** 824–831.

Green, K. F., and Garcia, J. (1971). Recuperation from illness: Flavor enhancement for rats. *Science* **173,** 749–751.

Gustavson, C. R., Garcia, J., Hankins, W. G., and Rusiniak, K. W. (1974). Coyote predation control by aversive conditioning. *Science* **184,** 581–583.

Gustavson, C. R., Kelly, D. J., Sweeney, M., and Garcia, J. (1976). Prey-lithium aversions. I. Coyotes and wolves. *Behav. Biol.* **17,** 61–72.

Hankins, W. G., Garcia, J., and Rusiniak, K. W. (1973). Dissociation of odor and taste in baitshyness. *Behav. Biol.* **8,** 407–419.

Hankins, W. G., Rusiniak, K. W., and Garcia, J. (1976). Dissociation of odor and taste in shock-avoidance learning. *Behav. Biol.* **18,** 345–358.

Kare, M. R. (1965). The special senses. *In* "Avian Physiology" (P. Sturkie, ed.), 2nd ed., pp. 406–446. Cornell Univ. Press, Ithaca, New York.

Krane, R. V., and Wagner, A. R. (1975). Taste aversion learning with a delayed shock US: Implications for the "generality of the laws of learning." *J. Comp. Physiol. Psychol.* **88,** 882–889.

Lorden, J., Kenfield, M., and Braun, J. J. (1970). Response suppression to odors paired with toxicosis. *Learn. Motiv.* **1,** 391–400.

Marr, J., and Gardner, L. (1965). Early olfactory experience and later social behavior in the rat: Preference, sexual responsiveness, and care of young. *J. Genet. Psychol.* **107,** 167–174.

Richter, C. P. (1942–1943). Total self regulatory functions in animals and human beings. *Harvey Lect.* **38,** 63–103.

Richter, C. P. (1953). Experimentally produced behavior reactions to food poisoning in wild and domesticated rats. *Ann. N.Y. Acad. Sci.* **56,** 225–239.

Roll, D. L., and Smith, J. C. (1972). Conditioned taste aversion in anesthetized rats. *In* "Biological Boundaries of Learning" (M. E. P. Seligman and J. Hager, eds.), pp. 98–102. Appleton, New York.

Rusiniak, K. W. (1976). Roles of olfaction and taste in appetitive and consummatory behavior during illness aversion conditioning. Doctoral Dissertation (unpublished).

Simson, P., and Booth, D. (1973). Effect of CS-US interval on the conditioning of odor preferences by amino acid loads. *Physiol. Behav.* **11,** 801–808.

Smith, J. C., and Roll, D. L. (1967). Trace conditioning with X-rays as the aversive stimulus. *Psychon. Sci.* **9,** 11–12.

Stager, K. E. (1964). The role of olfaction in food location by the turkey vulture *(Cathartes aura). Los Angeles County Mus. Contrib. Sci.* **63,** No. 81, 1–63.

Supak, T., Macrides, F., and Chorover, S. (1971). The baitshyness effect extended to olfactory discrimination. *Commun. Behav. Biol.* **5,** 321–324.

Taukulis, H. (1974). Odor aversions produced over long CS-US delays. *Behav. Biol.* **10,** 505–510.

Wenzel, B. M. (1971). Olfactory sensation in the kiwi and other birds. *Ann. N.Y. Acad. Sci.* **188,** 183–193.

Wilcoxon, H. C., Dragoin, W., and Kral, P. A. (1971). Illness induced aversions in rat and quail: Relative salience of visual and gustatory cues. *Science* **171,** 826–828.

White, L. (1965). Olfactory bulb projections of the rat. *Anat. Rec.* **152,** 465–480.

DISCUSSION

Moskowitz: Do individuals who are externally punished for overeating by social cues, such as "You're bad, you're bad," change their eating behavior?

Jordan: In general, it does not change the amounts of food eaten. Individuals who are criticized for eating may resort to eating while alone.

Garcia: If you use an external stimulus, what do animals do? Just as people try to figure out how to get around the external stimulus, so do animals. You give fresh hamburger with lithium to a coyote and it makes him sick. Next time you give it to him he acts as if he ate something rotten.

Moskowitz: There should be a way of doing that with people, too.

Jordan: Dr. Stunkard studied the formation of food aversions in humans. Most aversions are initiated in childhood. If they get sick following a meal it is always the novel food in the situation that they identify as the one that made them sick. Never the food they are familiar with, which means that if someone really has a highly preferred food, it will be very difficult to get that person to become averted to it.

Maller: You probably know about using antabuse to treat alcoholics. Anxiety following outcome studies really have not been very successful.

Jordan: As long as you can get the person to take the pill they will stay off the alcohol, but that is because they know that when they drink the alcohol they will get sick.

Garcia: Another problem is that when people become averted to the booze it is very specific. You get all kinds of people who will tell you that sweet drinks make them sick.

Pfaffmann: I am still not sure of the special difference between the alcohol and the illness associated with drinking alcohol. Is there a contradiction?

Garcia: It has been working out very well in research on rats or coyotes. The animals aren't shy of the poison; they are shy of the bait. Drinkers do not become shy of alcohol, but of the mixture in which it is found. People get sick on a martini and then they switch to another cocktail.

Schrire: I am more interested in what the sheep did than what the wolves do. There is a sheep led into a pen; presumably the sheep that you are using are aware that they are in a predator/prey situation. This is not the first time that sheep ever saw or smelled a wolf.

Garcia: I don't know. It might have been.

Schrire: I would have thought that if you had used a kind of prey population like the caribou/wolf relationship, if you had then introduced a caribou to a couple of wolves that had been primed on lithium, one would assume the caribou would surely not, within minutes, become the dominant animal. What one has got in mind is two populations that coexist and are ecologically suited to each other. The caribou recognized the danger of the wolf, and its response would surely be to get out of the situation as quickly as possible.

Garcia: Well, they quickly learn that they cannot get away in this situation. I think they learn these things very rapidly. That sheep might have known a lot about wolves. Studies show that this kind of learning is social learning. Habituation toward a fear object even in primates is shown to be a function of the group. You can get the same curve when you interchange animals in the group. Everybody apparently checks with everybody else to see if it is all right to go that way. I think animals do that. They read the intention movements of the other one, so just by looking at fright responses of other sheep, this animal could learn a lot about wolves.

Mugford: Lithium sulfate has a long history in human psychiatry. We are not talking about humans but do you know of any animal studies that have used lithium sulfate?

Garcia: I think Nachman looked at that. He says a therapeutic dose will produce an aversion in a rat.

Pfaffmann: I mentioned the use of hedonic responses by Harvey Grill; he demonstrated that if you viewed a videotape frame-by-frame analysis in the rat, you would note a very clear difference between the sugar responses. With lithium poisoning, presumably, sugar produces in the rat the quinine-aversive response. This is essentially another demonstration of what you were saying.

DeWys: Do pregnant women who suffer with morning sickness develop aversions?

Garcia: Yes, that is where you get some of your best data. The other interesting thing is in anorexia nervosa, which may also be an aversion thing, because you see it in adolescent girls who maybe have something like morning sickness, and especially if that interacts with eating, they become averted and then start elaborating a big defense of their behavior. I suspect that in many cases they have a generalized food aversion.

Gershoff: There are conflicting studies about morning sickness. It increases one's vitamin B_6 requirement and you can reverse morning sickness in lots of women by giving them additional vitamin B_6.

Chapter 14

The Development of Flavor Preferences in Humans: A Review

GARY K. BEAUCHAMP AND OWEN MALLER

It is commonly held that childhood experiences and associations with food influence adult food preferences and aversions. Yet, while this generalization has merit, remarkably little experimental data exist which bears directly on the role of experience in establishing or altering flavor preferences. It is the purpose of this chapter to provide primarily an overview of some of the data in the literature relevant to this topic. By highlighting several of the important research questions as well as possible directions for future research, it is hoped that further research will be stimulated to answer some of the intriguing questions which have been raised.

Although temperature, texture and visual characteristics of food are important variables, we will be using the term "flavor" only in reference to the chemical senses, taste and olfaction. "Experience" is broadly conceived of as any preceding environmental input. Major consideration will be directed to the question of whether and to what extent experience with a food or flavor effects a persistent flavor preference. Consideration will be given to (1) a summary of evidence concerning the role of experience in food preferences, (2) the role of experience and taste preferences, (3) the role of experience and olfactory preferences, and (4) the problems of and possible directions for future research.

EXPERIENCE AND FOOD PREFERENCES

Nonhuman animal research documents the influence of experience on subsequent food preferences. Burghardt (Chapter 12) has provided data on nonmammalian forms.

Striking evidence of this phenomenon in humans derives from the considerable differences in food habits among groups of people (cf. Rozin, 1976a,b, for an excellent discussion). Food habits are acquired dispositions to select and eat certain food items on a regular basis for a long period of time. Although genetic differences could account for some of these differences in food habits (e.g., Greene, 1974), it is widely assumed that aspects of previous experience are prepotent in controlling food preferences.

Food habits are extremely conservative and one of the last characteristics of a cultural group to change (Rozin, 1976a,b). Yet they do change and the investigation of these changes as well as their initial formation provide hypotheses concerning experiential effects on food (and flavor) preferences.

Two dramatic changes of food habits occur at birth, when the newborn consumes only milk, and at weaning, when he changes from a solitary food item to a diet containing a variety of flavors. The first change is abrupt, the second more gradual. In evaluating the role postweaning experiences may have on food preference, we need data concerning the initial preferences. The best such data available on humans come from the classic experiments of Davis (1928) on self-selection of food among three newly weaned children. These three children were given a variety of high-quality foods from which to choose during each of three daily meals. No effort was made to control what and how much of any food was ingested. It was concluded that the newly weaned child is able to select a nutritious and balanced diet over time. Processed foods, however, were not included in the choices; a study planned to test whether these foods would result in a disruption of accurate self-selection was apparently preempted by the Depression of the 1930's (Davis, 1939). While each of these children was able to select a diet on which growth was adequate, there were interestingly large individual differences among the children in what foods were preferred (Fig. 1).

Are such individual differences the result of some inherent individual variation in sensory and/or postingestional responses to the different food items? While it would seem that previous intake experience could play no role, as all children were taken at weaning, there were differential feeding experiences among the three children prior to the initiation of the experiment: One child had no previous experience with fruit juice, the second

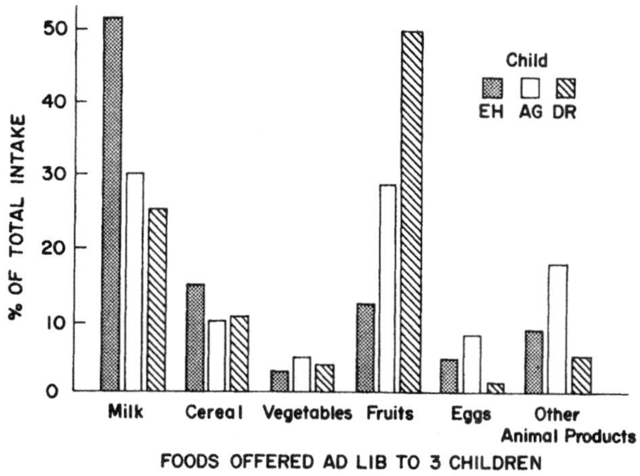

Fig. 1. Percentage of total intake accounted for by various classes of foods consumed by three children. Figure drawn from data in Davis (1928).

had been getting juice for a few weeks prior to beginning the experiment, while the third had had juice for several months before the experiment began. While admittedly only three infants are involved, the correlation between the length of experience with fruit juice and the level of fruit preference is perfect. Whether differential early experience with fruit juices alters subsequent preferences for fruit could be examined experimentally.

One widely held hypothesis is that early experience (occurring before the organism is mature) is somehow more important than later experience. "More important" in this context usually means more influential and more resistant to change. Often a corollary to this hypothesis is that there is a critical or sensitive period during ontogeny (usually *early* in ontogeny) when some form of experience will have its most profound effects (e.g., Hess, 1973; Scott, 1962). The belief in this hypothesis is reflected in the special concern for establishing the proper food habits in the young child. For example, Leininger (1970) reports the following words during the birth ceremony of the Gadsup people (from New Guinea) as the father's brother places food on the infant's tongue: "We give you these foods from our gardens so that you will want these foods and work hard to grow them."

Considerable difficulties exist in proving that early experiences are really different from later experiences. There is also a methodological problem in separating the effect of early exposure from the effects of frequency. Davis' (1928) observations indicate that the young, newly weaned child is willing to attempt to eat almost anything. Apparently, it is not until somewhat later that the child develops its well-known finickiness. Perhaps it is during this early explorative phase of eating that the familiar and the unfamiliar become defined and, once this period has come to an end, the unfamiliar food (and flavor) is rejected. It would be at this point that the human would exhibit a neophobia toward new foods similar to that described in the rat (Rozin, 1976b). Perhaps even early experience with a large variety of foods may predispose an individual to be more willing to try new foods, as has been demonstrated for the rat (Capretta *et al.*, 1975). One study found that American children, having more experience than Japanese children with more spices, liked a greater variety of spices (Muto *et al.*, 1969).

Evidence bearing directly on these points is scanty. However, Japanese primate studies (e.g., Itani and Nishimura, 1973) suggest that the youngest animals are more likely than adults to acquire and retain new food habits. For example, in one experiment where a troupe was provisioned with candy for the first time, a much larger percentage of young animals began eating candy during the experimental period than did other animals (Fig. 2). Parallel to this, Bavly's (1966) work on introduction of new recipes into families of different national origin now living in Israel (Fig. 2) showed that the children were most responsible for the introduction of new recipes and

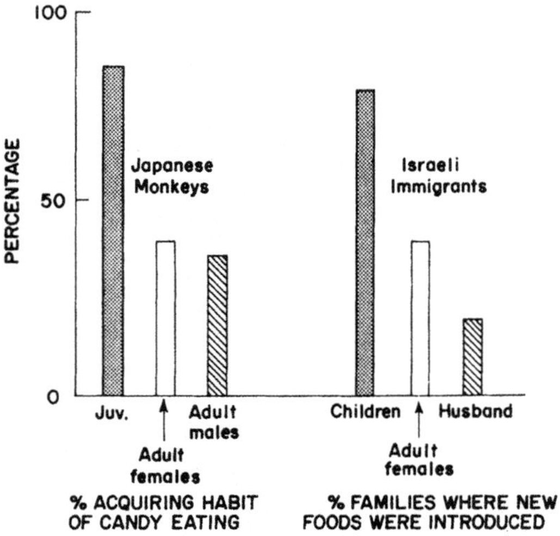

Fig. 2. Comparison of influences of individuals of different ages in the development of new food habits. The data for Japanese monkeys (taken from Itani and Nishimura, 1973) show the percentage of animals of three different classes which acquired the new habit of eating candy during an experimental period. The data for the Israeli immigrants (taken from Bavly, 1966) show the percentage of families where new recipes were introduced by each of three classes of people.

hence the acquisition of new food habits (cf. also Wenkam and Wolff, 1970). While substantial differences separate these two types of data, they are suggestive of the relative receptiveness of the young members of a species to accepting new foods.

Other important factors in the establishment and alteration of food preferences in humans are social and cognitive factors. Early studies by Duncker (1938) and Marinho (1942) demonstrated that food preferences of children over 3 years of age could be altered by observing an important person choosing the initially preferred food. These investigations found that younger children were more ready to imitate older ones, that stories of heroes eating a less preferred food were effective in altering food preferences, and that such effects could be extremely long-lasting (up to 1 year). Duncker concluded that those manipulations succeeded entirely through a social effect, there being no actual alteration in the flavor preferences of the children. That is, the child still liked best the flavor he initially preferred, but he chose a less pleasant flavor due to social influences. Duncker concluded by saying, "It would be interesting to find out whether, under conditions such as ours, and after a long enough time, this sort of secondary quality would finally have 'converted itself' into a primary sensory quality, so that the object would still have been preferred after the social experience had long since been forgotten" (Duncker, 1938, p. 504). Almost 40 years later, this question is still valid and unanswered. While cognitive or social experiences almost certainly influence food choices, whether they alter flavor preferences and to what extent is not known.

While there is some evidence that young individuals may be more open to new experience and that such experiences may be especially important, there is no solid evidence that early experiences result in permanent irreversible attachments to a food or flavor. Such strong "imprinting" effects, in fact, have not been demonstrated in mammals. In omnivores, who typically must depend upon a changing array of available foods, such addictions would appear to be unlikely and maladaptive.

From this brief overview of the role of experience in food preferences, several hypotheses concerning experience and flavor preferences can be suggested. Due to the difficulties in examining the ontogeny of human food habits, these hypotheses may, perhaps, be more readily tested using simpler systems, namely, the taste and olfactory systems. Data derived in turn from such experimental work could then be used to predict how experience regulates food habits. The hypotheses are as follows:

1. Preferences for certain items or classes of foods extend throughout an individual's lifetime.

2. Preferences are related to previous intake experience.
3. Experiences early in life (i.e., during the early years) are especially important in determining later food preferences.
4. Preferences are modified by cognitive factors.

THE ROLE OF EXPERIENCE IN TASTE PREFERENCES

In this and the next section, the four hypotheses developed in the previous section will be considered as they relate to taste and olfaction. In the fourth section the interrelations between the taste and olfactory systems will be briefly explored. For convenience, the taste system will be considered as being made up of four major qualities, namely, sweet, sour, salty, and bitter, even though other qualities may also exist (Meiselman, 1975).

Do Individual Taste Preferences Extend Throughout an Individual's Lifetime?

While longitudinal studies are lacking, there are some cross-sectional data which bear on this question. Desor *et al.* (1975) have shown that preferences determined by four-cup preference tests for sucrose, lactose, and salt were higher among 9–15 year olds than among adults (Fig. 3). These data suggest that over the long term, absolute preferences do not remain stable but, for these substances, actually decrease. However, the stability of relative levels of preference has not yet been determined. That is, individuals demonstrating a high level of preference for a substance when 9–15 years old compared to other 9–15 year olds may (or may not) also tend to exhibit a high level of preference when they become adults, compared to other adults.

Considerable individual differences in taste preferences have been reported (e.g., Pfaffmann, 1961; Pangborn, 1970; Pangborn *et al.,* 1970). Preferences for both salt and sucrose exhibit a bimodal distribution. Data for the same test carried out several days apart (Pfaffmann, 1961) or with two different methods (hedonic scale ratings and paired comparison method; Pangborn, 1970) indicate that these differences are reliable. However, the ontogeny of such individual differences or even whether such differences are found in younger people has not been reported. Further work, especially of a longitudinal nature, is required to see if taste preference differences among individuals exhibit similar consistency to those evident in food preferences.

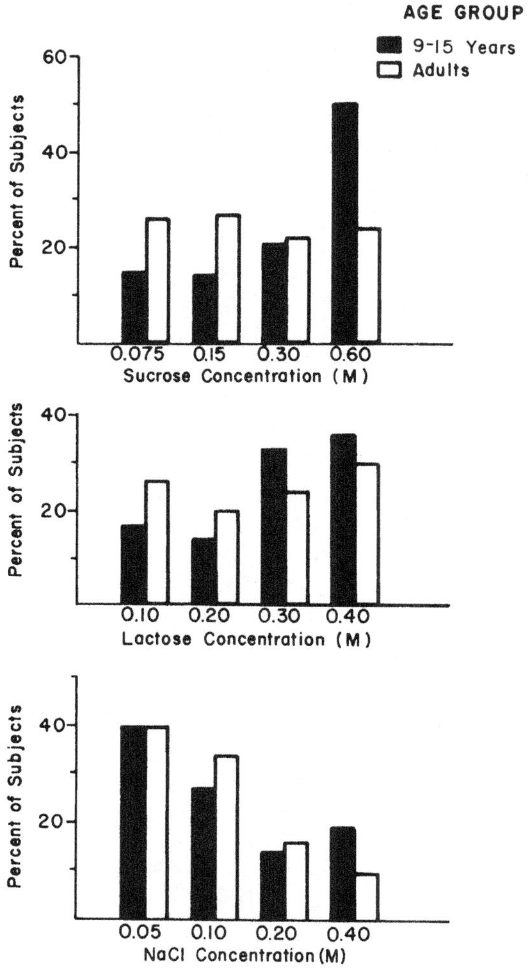

Fig. 3. Percentages of 9- to 15-year olds ($N = 618$) and adult ($N = 140$) subjects who selected different concentrations of sucrose, lactose, and sodium chloride as their most preferred concentration (from Desor *et al.*, 1975, with permission).

Are Taste Preferences Related to Previous Intake Experience?

Moskowitz *et al.* (1975) have reported data which suggest that previous food intake experiences may play an important role in determining taste preferences. These investigators found that Indian laborers, who reportedly ingest large quantities of very sour foods, find citric acid to be increasingly pleasant with increasing concentration and quinine sulfate very pleasant at relatively low concentrations. These findings contrast with the generally unpleasant responses to citric acid and quinine by Indian medical students (Moskowitz *et al.*, 1976) and Western populations who do not have a diet as abundant in sour foods. While the pleasantness of citric acid may be based upon previous experience with sour foods and foods flavored with sour condiments, the anomalous response to bitter is more problematical. Other explanations for these population differences in taste preferences, such as genetic differences (due to inbreeding) and/or general debilitation of the laborers, need to be examined. However, this very interesting study will hopefully stimulate a great deal of productive cross-cultural research.

Aside from this study, there is no direct experimental work linking previous intake to taste preferences. Desor *et al.* (1975), however, have shown racial differences in response to salt and sweet (sucrose and lactose), especially among 9–15 year olds. The young blacks prefer more concentrated solutions than the whites (Fig. 4). These data might be explained on the basis of differences in previous intake. Greene *et al.* (1975) found that heritability measures for taste preferences were low for *both* monozygous and dizygous twin pairs. These puzzling findings were interpreted as suggesting that the expression of genetic differences in taste preferences is greatly modified by environmental (presumably intake) factors during the course of growth and development. Further research separating differences in taste preferences for salty and sweet solutions and differences in intake of salty and sweet foods is required to examine this hypothesis for explaining racial differences.

Are Experiences Early in Life Especially Important in Influencing Taste Preferences?

There is a widespread popular belief that large, early intake of such substances as sucrose or salt predisposes children to become addicted to these substances. For example, Dahl (1972), a medical investigator interested in the etiology of essential hypertension, has suggested that (1) *excessive* salt appetite is acquired, and (2) that the critical factor in the link between excessive salt intake and hypertension is *excessive early* intake. This strong view has been challenged by others (e.g., Lauer *et al.*, 1976), but no human data exist with which the first hypothesis can be tested. Similar hypotheses exist to account for excessive sucrose intake.

Investigations of developmental changes in preferences may suggest areas where experience could be playing a role. Desor *et al.* (1973) have demonstrated that human newborns prefer to ingest sugar solutions compared to water. In addition, the infant's preference for various sugars

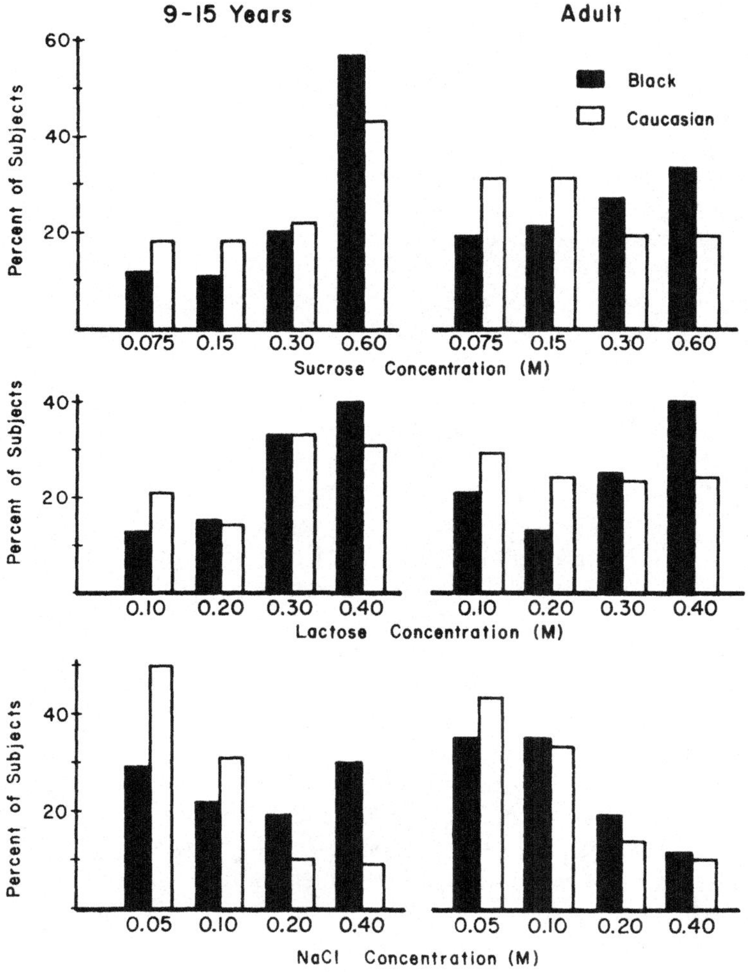

Fig. 4. Percentages of 9- to 15-year-old blacks ($N = 310$) and whites ($N = 308$) and adult blacks ($N = 52$) and whites ($N = 88$) who selected different concentrations of sucrose, lactose, and sodium chloride as their most preferred concentration (from Desor *et al.*, 1975, with permission).

corresponds to adult judgment of sweetness (Maller and Desor, 1974). These data indicate that the preference for sweet is inherent in the species rather than acquired.

When older (approximately 2 months and 6 months) children were tested with 0.2 *M* fructose, there was no evidence of change during these 6 months in response to this sugar (Desor and Maller, 1977). However, since neither longitudinal data on individual children nor detailed intake data were obtained in this study, one cannot be sure that intake experiences may not have had some influence on preference levels; that is, the level of sweetness one prefers (e.g., high, low, or moderate), rather than whether or not one likes sweets.

The data on salt (NaCl) preferences are more suggestive of changes during ontogeny. Maller and Desor (1974) found that newborn infants were indifferent to salt solutions at concentrations between 3 m*M* and 0.2 *M*. Desor (1975) subsequently reported that by 6 months of age, infants exhibited a preference for salt at some concentrations.

Salt preferences were tested (G. K. Beauchamp, unpublished) with children of two age groups, 6 months (20–28 weeks) and 1½ to 3 years (18–36 months), using methodology similar to that employed by Maller and Desor (1974). In this research the 1½–3-year-old children were offered, for 30 sec each, four cups containing 75 ml of a taste substance. Three of the cups contained salt water (0.05, 0.10, 0.20 *M*) and the fourth contained only deionized water. Presentation order was balanced. The same design and procedures were used for 6-month-old children with three exceptions. The solutions were presented in specially designed baby bottles (Desor *et al.*, 1973), each stimulus was presented for 60 sec, and there was a 30-sec pause between stimulus presentations. With the older children, the experimenter handed each cup to the child and attempted to coax him or her to sample the solution it contained. No effort was made to force the child to finish a sample. If a child refused to taste any one of the four samples, the data for this subject were not used. Similar procedures were employed with the 6-month-old group (see Desor *et al.*, 1973, for further descriptions of the methods of testing newborn children which were identical to those employed here unless otherwise stated). The study was conducted double-blind.

The results (Fig. 5) indicate that, using these methods, 6-month-old children consumed the same amount of these NaCl solutions as they did of water. However, subjects in the 1½–3-year-old group rejected salt water compared to deionized water ($F = 7.79; p < 0.01$). There were no concentration effects. The older children ingested larger quantities than the younger ones (means of 67 ml versus 24 ml), so the data were converted to proportions. Analysis of the proportions indicated a significant difference ($F = 13.03, p < 0.01$) between the two age groups (Fig. 5).

These data indicate that there is a shift in preference between 6 months (when infants are indifferent to or perhaps prefer NaCl solutions compared to water) and 1½–3 years (when these solutions are rejected). Several explanations for this age-related change are possible. The change in re-

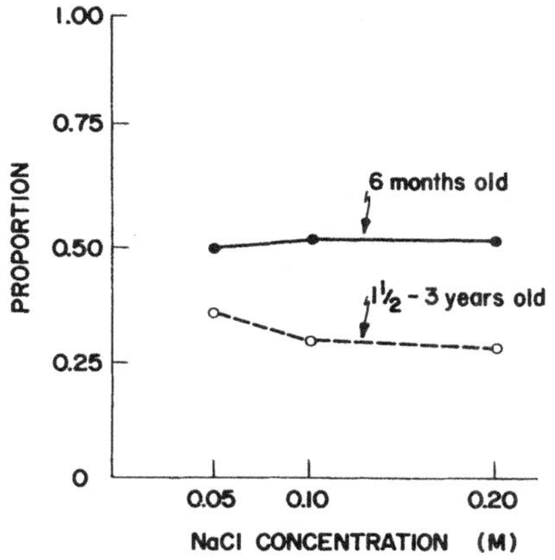

Fig. 5. Mean proportion of salt (NaCl) water consumed by children 6 months of age (20–28 weeks) and children 1½–3 years of age. Each proportion was calculated by dividing the sum of the intake of deionized water plus the intake of that concentration of salt solution into the intake of that concentration of salt solution. If salt solution was neither preferred nor rejected compared to deionized water, the expected proportion would be 0.50.

sponse may be to water and not to salt; that is, water has become more palatable to the older children. More likely, this change reflects an alteration in response to salt in a water solution. For example, by 1½–3 years, the context in which salt is presented may have an important influence on its palatability. Salt in soup may be highly preferred by children in this age group while salt in water is clearly not preferred. This differential response may, in turn, be conditioned by previous intake experiences. A third possible explanation for these differences is that children 6 months old cannot taste salt at these concentrations. Arguing against this interpretation are data reported by Desor (1975) indicating that some concentrations within the same range are actually preferred by infants this age. Further research is required to determine what role, if any, intake experiences play in age-related changes in salt preferences.

Are Taste Preferences Modified by Cognitive Effects?

Duncker (1938) concluded that the alteration of food preferences by social variables did not influence flavor preferences as discussed above. For taste preferences, we are aware of no data directly relevant to this question. The crucial function of gating that the taste system apparently performs may result in it being relatively insensitive to such effects.

THE ROLE OF EXPERIENCE IN OLFACTORY PREFERENCES

While most researchers tentatively accept somewhere between four and six primary qualities for the taste system, no such agreement exists for olfaction. One of the basic parameters of an olfactory stimulus, however, is its hedonic quality (Schiffman, 1974). Further, while the major, if not the only, role of taste involves nutrition, the olfactory system is involved in other physiological functions (i.e., in reproduction and other forms of social interaction). The precise role that odors have in controlling food intake remains unclear. In light of these problems and ambiguities, the hypotheses generated above are even less clearly tested by existing data on olfaction than in the taste system.

Do Olfactory Preferences Extend throughout an Individual's Lifetime?

There are several reports that relative preferences for samples of odorants remain quite stable over short intervals (Moncrieff, 1966) and, for one observer at least (Kenneth, 1928), over a period of 5 years. The stability of odor preferences as a function of age of respondant has been investigated several times, but this is an area littered with methodological complications (Engen, 1974). Most work in this area has involved comparisons in pleasantness between odors of different quality rather than, as has been the case with taste, between different concentrations of the same substances. For example, Moncrieff (1966) has shown that the relative preference of ten odorants seems to vary as a function of age. In general, among children (7–12 years old) flower odors are less preferred compared to the other odors, whereas fruit smells were more preferred. In adults, this was reversed. Engen (1974), using more sophisticated psychophysical techniques, has explored this area most extensively. He suggested that, on an absolute scale, the psychological range between the most and least pleasant odor is much smaller for children than for adults. However, it would appear, based on Moncrieff's (1966) work, that at least as a class relative odor preferences remain relatively stable throughout adulthood, this being especially the case for those substances found either very pleasant or very unpleasant.

Are Odor Preferences Related to Previous (Intake) Experience?

It is widely presumed that the pleasantness of an odor is directly related to previous experience with that odor. Psychoanalytic and learning theorists alike point to the importance of previous experience in determining hedonics of odors. One generalization that has been made is that the more familiar an odor, the more pleasant it is. For example, Engen (1974) showed that if an odor is unfamiliar to a person it is likely to be rated as unpleasant. Conversely, familiar odors are generally described as pleasant. To explain this finding, Engen theorized that as a result of classic conditioning, an odor acquires the hedonic value of the stimulus (e.g., good or spoiled food) with which it is associated. Evidence that odor memory is especially long-lasting (Lawless and Cain, 1975; Engen and Ross, 1973) provides ample scope for such associative effects. However, data dealing directly with the role of experience in determining food odor preferences are lacking.

Are Experiences Early in Life Especially Important in Determining Later Odor Preferences?

Engen *et al.* (1963) have shown that human infants are sensitive to olfactory stimuli. Recently, Russell (1976) has reported that by 6 weeks of age, infants discriminate between odors left on breast pads by their own mothers and odors left on breast pads by unfamiliar mothers. All infants were being breast-fed. While this was a preliminary report, it does suggest that experience in the very young child is involved in the attachment of meaning (and probably hedonic value) to odorous stimuli. Whether the source of the chemicals on which the individual discriminations were made was the milk, other secretions or excretions of the mother, or even the kind of soap the mothers used is apparently not yet known.

In a widely cited study, Stein *et al.* (1958) reported that whereas 3–4-year-old children found amyl acetate, synthetic feces odor, and synthetic sweat odor equally pleasant, by the age of 6–8 years, the feces and sweat odors were distinctly unpleasant. This change to unpleasantness may, according to the authors, "be a temporary regression to pregenitality, with the development of reaction formations as defenses against pregenital impulses" (Stein *et al.*, 1958, p. 265). While Engen (1974) has shown this study to be confounded by response bias (young children are likely to say "I like it" in response to any stimulus), and while the above "explanation" may appear at best speculative, it is still quite possible that for many unpleasant (and pleasant) odors, early learning experiences serve to establish the level of pleasantness.

Are Odor Preferences Modified by Cognitive Factors?

There is considerable research on the role of expectations on flavor identification. For example, the effects of colors on flavor labeling have been investigated (cf. Burghardt, Chapter 12; Stone and Pangborn, 1968).

However, studies of how input from other senses as well as internal state influence odor preferences are not plentiful. G. K. Beauchamp and J. A. Desor (unpublished data) examined the pleasantness and edibility of objects based on the objects' odors only, odors plus visual characteristics, visual characteristics only, or on a verbal description of the objects. One of the goals of this research was to see how contextual, cognitive factors influence the perceived pleasantness of odors.

Four groups of seven subjects each (naive humans ages 19–33 years of both sexes) were tested. In group 1, subjects were presented with 24 opaque jars each containing an odorous object. For each, the subject was instructed to sample the odor emanating from the jar and then rate (a) how likely they were to eat an object with the odor, and (b) how pleasant the odor of the object was. Ratings were placed anywhere along a scale between end points labeled very likely to eat/very pleasant–very unlikely to eat/very unpleasant. These subjects could not see the objects and were not asked to identify them. Group 2 subjects were tested in exactly the same manner except they could see the objects but could not smell them. For group 3, the objects could be both seen and smelled while group 4 subjects made the same ratings based on a written description of the objects.

The 24 stimulus objects consisted of eight fresh foods (cantelope, lemon, green pepper, honey, bread, tuna fish, fried liver, ham—all fresh), eight "rotten" foods (orange, tomato, apple, green pepper, egg, milk, raw beef, liverwurst—all held at room temperature for 6 days before being used), and eight inedible objects (lavender, cigar, motor oil, crayons, rubber, manure, pine needles, ivy leaves). The 24 objects were presented in a random order. In pilot work it was found that subjects were able to identify about 50% of the fresh and inedible objects on the basis of odor alone; only 16% of the rotten objects could be correctly identified. However, when they were allowed to see as well as smell the objects, approximately 90% from all three classes could be correctly identified.

Major results of this experiment as they relate to pleasantness of odors are illustrated in Fig. 6. For fresh foods, there were no differences in pleasantness based on whether olfaction alone, vision alone, olfaction plus vision, or the questionnaire provided the stimuli to be rated. However, for rotten food and nonfood objects, there were significant differences based on the modality of the stimuli. Of greatest interest in the present context are the responses to rotten foods. People tended to rate these objects as more pleasant when they could obtain olfactory information compared with the condition in which no such information (vision only, questionnaire) was available. The availability of *both* olfactory and visual cues resulted in a response not significantly different from olfaction

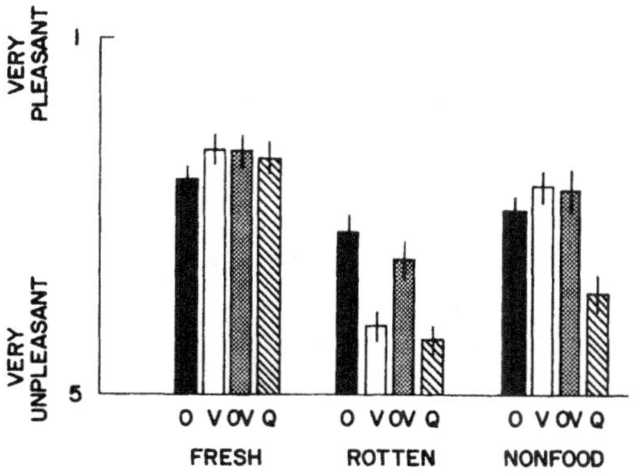

Fig. 6. Mean (± SEM) pleasantness (1 = very pleasant; 5 = very unpleasant) judged for three classes (fresh foods, rotten foods, nonfoods) of objects (N = 8 objects/class). Four groups of subjects were used: Group O used olfaction only, group V used vision only, group OV used both vision and olfaction, and group Q used a verbal description to judge the pleasantness of the objects.

alone ($p > 0.10$), while both resulted in more pleasant judgment ($p < 0.05$) than vision alone.

We interpret these results as indicating that olfactory characteristics have a very powerful influence on the perceived pleasantness of an object, and for the rotten food items used in this study, visual (i.e., cognitive, contextual) factors play a much smaller role than might have been expected. That is, when a relatively unpleasant visual input (the substances could be identified as, for example, rotten tomatoes 90% of the time) was placed in conflict with a relatively more pleasant olfactory input, the hedonic qualities of the olfactory component were dominant. These results lend further support to the generalization that hedonic quality is among the most basic components of an odor (Schiffman, 1974). Considerably more research using a wide variety of odors is required to determine the generality of these findings.

CONCLUSIONS

Previous intake experience may play a crucial role in determining later flavor preferences. However, good experimental studies are not available from which one could determine the important parameters of these effects. Moreover, the mechanisms mediating the effects remain speculative at this time. For example, the formation of conditioned aversions (see Garcia, Chapter 13) could be involved in determining human flavor preferences. However, based on the one survey on the occurrence of human taste aversions (Garb and Stunkard, 1974), it would appear unlikely that formation of conditioned aversions could explain the vast majority of human food choices (and aversions). Only 38% of the sample questioned reported having even one taste aversion. Furthermore, these aversions were very specific. While it would be possible to argue that many minor, unrecognized aversions directed toward a large number of foods experienced over one's life play a role in individual patterns of food dislikes, this seems a fairly unlikely hypothesis.

Learned flavor preferences based on "curative" powers of substances presented to animals when recovering from a deficiency have been much more difficult to demonstrate than conditioned aversions. Several recent papers (e.g., Revusky, 1967, 1968; Christensen *et al.*, 1974) have provided evidence that such effects do take place. Thus, preferences could develop for foods (with particular flavors) which are eaten during need states. Since the first and most extensive satisfaction of need would be by early food intake, such experiences would have special importance. An alternate or complimentary mechanism for the role of experience is based not on reduction of physiological need but on mere familiarity (Zajonc, 1968). According to this hypothesis, a food or flavor becomes preferred due only to repeated exposure to that food or flavor.

A crucial but not yet fully answered question is what role flavor plays in regulating food preferences. More specifically, what are the respective roles of taste and olfaction in food choice? Taste is undeniably important. Most people will quickly reject very bitter, very sour, and very salty foods although even here expectations play an extremely crucial role (consider the preference many individuals have for quinine water, pickles, and caviar). Perhaps even more dramatic is the powerful role played by sweet taste in stimulating ingestion (Pfaffmann, 1961). Yet to many, these seem rather crude factors. Most cuisines or food habits of individuals are not characterized, basically, by differences in sweetness, sourness, saltiness, or bitterness. Furthermore, Mozell *et al.* (1969) have demonstrated that foods cannot be identified on the basis of taste; olfaction appears crucial in making distinctions based on flavor.

Thus, one might argue that as long as the taste of a food is within an acceptable range, olfaction is the crucial sense in food choice and flavor preferences. However, this generalization has several limitations. While people cannot identify foods on the basis of taste, even using odor cues results in a surprisingly low level of identification (Desor and Beauchamp, 1974). Thus, olfactory identification of foods may not be the precise tool, for most of the human population, that it is often assumed to be. One study tested directly the relative importance of taste and "flavoring" (presumably olfactory stimuli) in increasing the intake of a food of low palatability. Grewal *et al.* (1973) evaluated the ability of increased sweetness and/or added flavor to increase intake of a high-calorie protein food supplement (CSM) among children 4–7 years old. When the most preferred flavor among the seven examined (almond) was added to CSM, there was no increase in intake. However, sugar (14.85% sucrose) significantly increased intake. The determination of the relative roles and mode of interaction of taste and odor stimuli in food choices require considerable further work.

A number of methods now exist for determining hedonic qualities of taste and olfactory stimuli. While reliability may be high for these methods, validity studies are badly needed. When considering the role of taste and olfaction in food intake, the major question is to what extent these factors control the amount of substances ingested. However, whether measurement of hedonic parameters of model systems of taste and/or smell predicts ingestive responses to real foods has not often been evaluated. Moskowitz *et al.* (1974) have made one effort to partially bridge this gap by showing that degree of sweetness and pleasantness of sweetness both in model taste systems and in foods conform to topographically similar systems. Further work investigating whether taste and olfactory preference differences, as determined by a number of methods using model substances and real foods, actually predict differences in the quantity or volume of food and fluids consumed would be extremely valuable.

More generally, cognitive factors (such as expectations as to whether or not an object should be good food) may often override flavor characteristics in controlling ingestion (cf. Wenkam and Wolff, 1970). Experiential factors may determine how a substance is labeled or perceived to fit into the food category more than how highly preferred the flavor is. If this were found to be the case, then model animal studies on flavor and ingestion would be of limited value in understanding human food choices.

Clearly, there is considerable need for future research which examines the role of long-term intake experiences on human flavor preferences. Longitudinal, developmental studies would be very helpful and informative. One could study the possible relationship between changes in taste preferences in human children, and their intake of sweet, salty, sour, and bitter substances could be evaluated. Further, cross-cultural studies could contribute greatly to our understanding of the etiology of food and flavor preferences. The work by Moskowitz *et al.* (1975) on Indian laborers is only a beginning. Further progress on the analysis of the importance of early compared with later experience could also be investigated in cross-cultural studies. Finally, while the emphasis here has been on the role of experience in the establishment and modification of preferences, cross-cultural studies could also contribute to our understanding of the possible role of innate differences in flavor preferences among groups.

REFERENCES

Bavly, S. (1966). Changes in food habits in Israel. *J. Am. Diet. Assoc.* **48,** 488–495.

Capretta, P. J., Petersik, J. T., and Steward, D. J. (1975). Acceptance of novel flavors is increased after early experience of diverse tastes. *Nature (London)* **254,** 689–691.

Christensen, C. M., Caldwell, D. F., and Oberleas, D. (1974). Establishment of a learned preference for a zinc-containing solution by zinc-deficient rats. *J. Comp. Physiol. Psychol.* **87,** 415–421.

Dahl, L. K. (1972). Salt and hypertension. *Am. J. Clin. Nutr.* **25,** 231–244.

Davis, C. M. (1928). Self-selection of diets by newly-weaned infants. *Am. J. Dis. Child.* **36,** 651–679.

Davis, C. M. (1939). Results of the self-selection of diets by young children. *Can. Med. Assoc. J.* **41,** 257–261.

Desor, J. A. (1975). Taste preferences in children during the first 6 months. Presented at the annual meeting of the Eastern Psychological Association, New York.

Desor, J. A., and Beauchamp, G. K. (1974). The human capacity to transmit olfactory information. *Percept. Psychophys.* **16,** 551–556.

Desor, J. A., and Maller, O. (1977). In preparation.

Desor, J. A., Maller, O., and Turner, R. E. (1973). Taste in acceptance of sugars by human infants. *J. Comp. Physiol. Psychol.* **84,** 496–501.

Desor, J. A., Greene, L. S., and Maller, O. (1975). Preferences for sweet and salty in 9 to 15-year-old and adult humans. *Science* **190,** 686–687.

Duncker, K. (1938). Experimental modification of children's food preferences through social suggestion. *J. Abnorm. Soc. Psychol.* **33,** 489–507.

Engen, T. (1974). Method and theory in the study of odor preferences. *In* "Human Responses to Environmental Odors" (A. Turk, J. W. Johnston, and D. G. Moulton, eds.), pp. 121–141. Academic Press, New York.

Engen, T., and Ross, B. M. (1973). Long-term memory of odors with and without verbal descriptions. *J. Exp. Psychol.* **100,** 221–227.

Engen, T., Lipsitt, L. P., and Kayne, H. (1963). Olfactory responses and adaptation in the human neonate. *J. Comp. Physiol. Psychol.* **56,** 73–77.

Garb, J. L., and Stunkard, A. J. (1974). Taste aversions in man. *Am. J. Psychiatry* **131,** 1204–1207.

Greene, L. S. (1974). Physical growth and development, neurological maturation and behavioral functioning in two Ecuadorian Andean communities in which goiter is endemic. II. PTC taste sensitivity and neurological maturation. *Am. J. Phys. Anthropol.* **41,** 139–151.

Greene, L. S., Desor, J. A., and Maller, O. (1975). Heredity and experience: Their relative importance in the development of taste preference in man. *J. Comp. Physiol. Psychol.* **80,** 279–284.

Grewal, T., Gopaldas, T., Hartenberger, P., Ramakrishnan, I., and Ramachandran, G. (1973). Influence of sugar and flavour on the acceptability of instant CSM: Trials in young children from an urban orphanage. *J. Food Sci.* **10,** 149–152.

Hess, E. H. (1975). "Imprinting." Van Nostrand-Reinhold, New York.

Itani, J., and Nishimura, A. (1973). The study of infrahuman culture in Japan. *Proc. Int. Congr. Primatol. 4th, 1972* Vol 1, pp. 26–50.

Kenneth, J. H. (1928). A few odor preferences and their constancy. *J. Exp. Psychol.* **11,** 56–61.

Lauer, R. M., Filer, L. J., Reiter, M. A., and Clark, W. R. (1976). Blood pressure, salt preference, salt threshold, and relative weight. *Am. J. Dis. Child.* **130,** 493–497.

Lawless, H. T., and Cain, W. S. (1975). Recognition memory for odors. *Chem. Senses Flavor* **1,** 331–337.

Leininger, M. (1970). Some cross-cultural universal and non-universal functions, beliefs and practices of food. *In* "Dimensions of Nutrition" (J. duPont, ed.), pp. 153–179. Colorado Assoc. Univ. Press, Denver.

Maller, O., and Desor, J. A. (1974). Effect of taste on ingestion by human infants. *In* "Oral Sensation and Perception: Development in the Fetus and Infant" (J. Bosma, ed.), pp. 279–291. US Govt. Printing Office, Washington, D.C.

Marinho, H. (1942). Social influence in the formation of enduring preferences. *J. Abnorm. Soc. Psychol.* **37,** 448–468.

Meiselman, H. (1975). Human taste perception. *In* "Handbook of Flavor Ingredients" (T. E. Furia and N. Bellanca, eds.), 2nd ed., Vol. 1, pp. 2–43. CRC Press, Cleveland, Ohio.

Moncrieff, R. W. (1966). "Odour Preferences." Leonard Hill, London.

Moskowitz, H. R., Kluter, R. A., Westerling, J., and Jacobs, H. L. (1974). Sugar sweetness and pleasantness: Evidence for different psychological laws. *Science* **184,** 583–585.

Moskowitz, H. R., Kumraiah, V., Sharma, K. N., Jacobs, H. L., and Sharma, S. D. (1975). Cross-cultural differences in simple taste preferences. *Science* **190,** 1217–1218.

Moskowitz, H. R., Kumraiah, V., Sharma, K. N., Jacobs, H. L., and Sharma, S. D. (1976). Effects of hunger, satiety and glucose load upon taste intensity and taste hedonics. *Physiol. Behav.* **16,** 471–475.

Mozell, M. M., Smith, B. P., Smith P. E., Sullivan, R. L., and Swender, P. (1969). Nasal chemoreception in flavor identification. *Arch. Otolaryngol.* **90,** 367–373.

Muto, S., Muzuno, K., and Kobayashi, Y. (1969). Dietary patterns of Japanese and American preschool children in Tokyo. *J. Am. Diet. Assoc.* **55,** 252–256.

Pangborn, R. M. (1970). Individual variation in affective responses to taste stimuli. *Psychon. Sci.* **21,** 125–126.

Pangborn, R. M., Crisp, R. B., and Bertolero, L. L. (1970). Gustatory, salivary and oral thermal responses to solutions of sodium chloride at four temperatures. *Percept. Psychophys.* **8,** 69–75.

Pfaffmann, C. (1961). The sensory and motivational properties of taste. *In* "Nebraska Symposium on Motivation" (M. Jones, ed.), pp. 71–108. Univ. of Nebraska Press, Lincoln.

Revusky, S. H. (1967). Hunger level during food consumption: Effects on subsequent preference. *Psychon. Sci.* **7,** 109–110.

Revusky, S. H. (1968). Effects of thirst level during consumption of flavored water on subsequent preference. *J. Comp. Physiol. Psychol.* **66,** 777–779.

Rozin, P. (1976a). Psycho-biological and cultural determinants of food choice. *In* "Dahleon Workshop on Appetite and Food Intake" (T. Silverstone, ed.), Dahleon Konferenzen, Berlin, in press.

Rozin, P. (1976b). The selection of foods by rats, humans and other animals. *Adv. Study Behav.* **6,** 21–76.

Russell, M. J. (1976). Human olfactory communication. *Nature (London)* **260,** 520–522.

Schiffman, S. S. (1974). Physiochemical correlates of olfactory quality. *Science* **185,** 112–117.

Scott, J. P. (1962). Critical periods in behavioral development. *Science* **138,** 949–958.

Stein, M., Ottenberg, P., and Roulet, N. (1958). A study of the development of olfactory preferences. *Arch. Neurol. Psychiatry* **80,** 264–266.

Stone, H., and Pangborn, R. M. (1968). Intercorrelation of the senses. *In* "Basic Principles of Sensory Evaluation," pp. 30–46. Special Technical Publication No. 433, American Society for Testing and Materials.

Wenkam, N. S., and Wolff, R. J. (1970). A half century of changing food habits among Japanese in Hawaii. *J. Am. Diet. Assoc.* **57,** 29–32.

Zajonc, R. B. (1968). Attitudinal effects of mere exposure. *J. Person. Soc. Psychol.* **9,** 1–27.

DISCUSSION

Mugford: I cannot see the evidence for the conservatism of cuisine among Western people. For instance, for the last 2 years in England there has been a shortage of potatoes, and 5 or 10 years ago it would have been inconceivable that there would have been such a rapid uptake of pastas and rice products to substitute for potatoes, but this had in fact happened. The development of the whole convenience food industry has produced thousands and thousands of products, hundreds of new ones each year—all of which are novel.

Beauchamp: The argument contrary to that is that where you find the new food being introduced, it is disguised as an old food. Rozin especially speaks of the special role spices play in maintaining a cuisine. You can introduce a new food to American society as long as you have a lot of ketchup on it.

Cagan: Does that not simply mean that if you are looking for similarities in such a situation that you can always find them? How does it really address the question?

Moskowitz: I would like to answer Cagan's statement. If you examine the current wave of new food items, you find that in general the ones that have the most lasting impact are those that are the blandest and that provide the minimum of cultural shock for the individual. You might have a new food that is very spiced, a snack food, but the probability of it surviving in the market is incredibly low as compared with another kind of potato chip which has the same kind of flavor as the one that you are used to. I think our food habits do resist cultural changes. It is inevitable that we have changes but we are very resistant if not extraordinarily resistant to them.

I think even with children there is some early evidence by Dean Forster, who published in 1950 the results of some ratings for a number of different odors for a variety of children of different ages as well as adults, and he found something very anomalous. I think it was oil of wintergreen or methyl salicylate which was rated to be extremely unpleasant by a group of British children in 1950. He found this to be anomalous when he was looking at data from other countries, and he says that it is the result of having had oil of wintergreen paired with cough medicine, and this specific pairing was an absolute turn-off signal, that it was totally unexpected.

Gershoff: I think on one of the illustrations (Desor *et al.,* 1975) I saw something that was rather interesting. You had a high lactose concentration and it seemed to me that adult blacks preferred it as compared to adult Caucasians. In view of all the propaganda that we have had to listen to in recent years about lactose deficiency, it seems to me that that is a rather interesting observation.

Beauchamp: I would wonder whether the lactose, although it is the milk sugar, is identified with having something to do with milk. It is also true, I think, that the preference for milk among black male adults (Meiselman) was not found to be any different than it was among whites.

Senseman: You mentioned Rozin's idea that spices may play a special role in disguising a novel food as a familiar, "safe" food. Characteristically, spices confer not only a distinctive flavor but also an intense flavor to the food items to which it is added. What I am wondering is this. There is some experimental work and much anecdotal observations to suggest that some bland foods become less palatable after frequent presentation. In general, animals adapt faster to weak stimuli than to strong stimuli. Might not spices serve a dual role to increase the taste intensity of a relatively bland food and thereby decrease the rate at which a person becomes tired of its taste?

Beauchamp: That is a possibility.

Kinnard: I want to talk about the subject that Gershoff referred to earlier. There were three substances: lactose, sodium chloride, and sucrose, that blacks showed a greater preference for than whites—at least at a higher concentration. Were other factors carefully controlled, like the socioeconomic status and level of educational achievement?

Beauchamp: Greene *et al.* (1975) did census tracking and balanced the socioeconomic level of blacks and whites. Geographically, they were all from the Philadelphia area. Obviously, there were some factors that were not controlled, but socioeconomic scale was controlled.

Pager: This is just an idea, but perhaps there is a dichotomy between the social odors that you are talking about and odors involved in foods. Perhaps social ones have these attachments, but many food odors may not. I think that because there are many effects of experience with odors does not prove that experiences will affect all.

Pangborn: I would like to reiterate what has been said in your presentation and by some of the discussants as well as what was said yesterday about hedonic responses. It is important to distinguish between the hedonic responses to the concept or attitude versus the hedonic response to the actual substance. Next, one should distinguish between whether the substance is presented on a single occasion or during prolonged exposure. As an example, I think you could take orange flavor. Ask someone if he likes or dislikes orange flavor and you will get a response in a sort of hypothetical setting. Then ask if he likes or dislikes this specific orange flavoring. Then, does he like it just to rinse his mouth with in solution or to consume it?

FLAVOR RECOGNITION

Chapter 15

Food Preference Ratings of Congenitally Anosmic Humans

RICHARD L. DOTY

People who have lost their sense of smell commonly report, as a result of this loss, that both the flavor sensations and the enjoyment derived from eating are diminished (cf. Kneeland, 1851; Mozell *et al.*, 1969; Ogle, 1870; Patrick, 1899). As noted by Douek (1974), it is generally believed that the

> Perception of flavour requires perception of the odour of foods. In the absence of the sense of smell it is only possible to experience the quality of a substance in terms of its simplest modalities of saltiness, sweetness, sourness, and bitterness as well as its texture and temperature. The whole range of subtle qualities which are referred to as "flavour" is lost when the sense of smell is absent.

Given the important role of olfaction in effecting the flavorfulness of foods, it is of interest to establish whether the congenital absence or presence of this modality influences man's long-term food preferences and habits. Unfortunately, no studies have addressed this problem directly. Individuals with *acquired* anosmia are able to recognize a broad range of foods on the basis of taste and textural cues alone (Clark and Dodge, 1955a,b; Crosland *et al.*, 1926, 1928), although their ability to recognize certain foods is reduced or eliminated when the food items are relatively homogeneous in texture (Mozell *et al.*, 1969). Whether this holds true for congenital anosmics is not known, but appears likely.

During the last 2 years we have been testing anosmic patients (most of whom were available for testing only during a short period as out-of-town inpatients at the Hospital of the University of Pennsylvania) to establish the molecular properties of volatile chemicals which produce intranasal trigeminal stimulation under normal sniffing conditions (e.g., Doty, 1975; Doty *et al.*, 1977). As a preliminary approach to establishing the possible influences of olfaction upon food preferences I decided to poll, by a simple survey procedure, the preferences of a selected group of such anosmics. Although a food preference survey has numerous limitations, it is a logical first step in providing data on this issue, given that food preference ratings of anosmics located in various parts of the country, as well as abroad, can be reliably and economically established by telephone.

The present surveys had two major goals. The first was to determine whether *congenitally anosmic* adults report lower general food preferences than normal ones for various foodstuffs, particularly those relatively homogeneous in texture (e.g., beverages, sherberts, puddings). The second was to establish whether the relative rank ordering of reported food preferences within various food classes differed markedly between the anosmic and normal individuals.

SUBJECTS

Fifteen anosmics and 15 controls were used in this study. Eleven of the anosmics were males (mean age = 22.91 years; SD = 6.93) diagnosed at the Department of Medicine, Endocrine Section, University of Pennsylvania, as having Kallman's syndrome, a congenital disorder characterized by hypogonadism and anosmia due to aplasia of the olfactory bulbs and/or stalks (cf. Schneider, 1974). An additional three anosmic males (24, 25, and 38 years) and an anosmic female (23 years) who reported never having been able to smell were also tested. Presumably, the loss of smell in these individuals was congenital. All of the anosmics were unable to detect a large number of volatile compounds via cranial nerve I, although intranasal detection of a number of vapors was possible via cranial nerve V. Details of the psychophysical tests used to verify the general anosmia are presented elsewhere (Doty, 1975; Doty *et al.*, 1977). The Kallman's syndrome patients, who were typically not residents of the Philadelphia area, were psychophysically tested for the verification of anosmia during a brief stay for endocrine studies at the Hospital of the University of Pennsylvania. The remaining anosmics were referred for psychophysical testing by physicians in the tristate area of the Delaware Valley.

The 15 control subjects were matched to the anosmics on the basis of age, sex, and ethnic background. The majority were students, although several were employees of the Monell Center.

FOOD PREFERENCE SURVEYS

In the initial survey 94 food items were rated on the nine-point hedonic scale described by Peryam and Pilgrim (1957) and used in numerous armed forces food preference surveys (e.g., Meiselman *et al.*, 1974; Moskowitz *et al.*, 1972; Waterman *et al.*, 1974). The food items were

TABLE I

Items of Questionnaires Administered to Congenitally Anosmic and Normal Individuals

Food items[a]

- Beverages: Beer,* Coca-Cola,* Grape Soda,* Ice Coffee,* Ice Tea,* Hot Coffee,* Hot Tea,* Hot Chocolate, Orange Soda,* Root Beer,* 7-Up,* Wine*
- Dairy Products: American Cheese, Buttermilk, Chocolate Milk, Cheddar Cheese, Cottage Cheese, Evaporated Milk, Light Cream, Limburger Cheese, Margarine, Philadelphia Cream Cheese, Swiss Cheese
- Fruit Juices: Apple,* Cranberry,* Grape,* Lemonade,* Orange,* Pineapple,* Prune,* Tomato*
- General Food Classes: American,* Chinese,* Italian,* Mexican*
- General Seasonings: Dill Pickles,* Ketchup,* Mayonnaise,* Green Olives,* Salt, Sweet Pickles,* Worcestershire Sauce
- Hot Seasonings: Cinnamon,* Garlic Salt,* Horse Radish,* Mustard,* Onion Salt,* Pepper, Red Peppers, Tabasco Sauce*
- Meats & Fish: Anchovies,* Bacon,* Flounder,* Hamburgers, Hot Dogs, Steak,* Trout,* Tuna,* Uncooked Frankfurters*
- Miscellaneous Foods: Apples, Cherries, Rye Bread,* Corn Bread,* Melba Toast*
- Potatoes: Baked,* Boiled,* Chips, French Fries,* Hash Browns,* Mashed,* Scalloped,* Sweet
- Puddings: Butterscotch,* Chocolate,* Lemon,* Rice, Tapioca
- Salads: Cole Slaw,* Jello,* Kidney Bean,* Macaroni,* Potato Salad,* Tossed Lettuce*
- Sherberts: Lemon,* Lime,* Orange,* Raspberry,* Strawberry*
- Snacks: Peanut Brittle,* Fritos, Graham Crackers,* Ginger Snaps,* Pizza, Pretzels, Popcorn, Potato Chips
- Soups: Bouillion,* Chicken Noddle,* Cream of Mushroom,* Tomato,* Vegetable*
- Sweets: Cotton Candy,* Fudge,* Gum Drops,* Hershey Bar, Honey,* Marshmallows,* Peanut Brittle,*† Red Licorice, Rock Candy,* Taffee,* Tootsie Rolls*
- Vegetables: Asparagus,* Broccoli,* Carrots, Corn,* Cucumbers, Peas,* Spinach,* Tomatoes

Nonfood items

- Activities: Cooking, Doing Dishes, Doing Schoolwork, Going to Motion Pictures, Listening to Music, Mowing the Lawn, Playing Tennis, Reading, Sailing, Watching Television, Writing
- Automobiles: Cadillac, Ford, Rolls Royce, Toyota, Volkswagen
- Colors: Blue, Green, Purple, Red, Yellow

[a] Asterisk indicates items used on both first and second surveys; dagger indicates item used in more than one food class.

chosen to cover a variety of general food classes, including soups, salads, beverages, dairy products, deserts, and assorted snacks. In addition, preferences for the general food groups of American, Chinese, Italian, and Mexican foods were also polled. In the second survey (the major purpose of which was to provide test–retest reliability information), 34 food items were added to the 94 items of the initial survey, along with 21 items not related to food preferences per se (e.g., colors, activities, automobiles). These control items were added to establish whether a slight, statistically insignificant tendency for normals to assign larger ratings than anosmics to some of the food classes on the first survey simply reflected a general response bias in the use of the hedonic scale. Fifty of the food items on the second survey had been used previously by Meiselman *et al.* (1974), whereas the remainder were taken from a food item list developed for this study. The items included in the two surveys are presented in Table I.

SURVEY ADMINISTRATION PROCEDURES

The food preference surveys were administered primarily by telephone, although in a few cases they were administered in our laboratory. The subjects were asked to write the numbers 1 through 9 down the front of a plain piece of paper, and were then asked to insert the words "dislike extremely," "dislike very much," "dislike moderately," "dislike slightly," "neither like nor dislike," "like slightly," "like moderately," "like very much" and "like extremely," after the respective numbers. The task of the subjects was to rate their preferences for each food item on this hedonic scale. The food items were read to each subject in irregular order, with the stipulation that no more than two foods from the same general food category be read in succession. The administration of the food surveys occurred at various times of the day, depending upon when the anosmics could be contacted, with equivalent numbers of normals and anosmics being tested at each period to counterbalance possible administration time effects across the two groups. The time between the administration of the two surveys ranged from 7 to 11 days.

RELIABILITY OF THE FOOD PREFERENCE SURVEYS

Both the individual and group test–retest reliabilities of the survey were assessed by Pearson product–moment correlations between the scale value scores for each item common to the two surveys (94 items). The anosmic and normal group reliability coefficients were 0.94 and 0.96, respectively. The anosmic individual coefficients ranged from 0.72 to 0.95

(median = 0.82; mean = 0.82; SD = 0.06), whereas those of the normals ranged from 0.75 to 0.96 (median = 0.87; mean = 0.87; SD = 0.06) (all p's < 0.001).

CORRELATION OF FOOD PREFERENCE DATA WITH RESULTS FROM SIMILAR PREFERENCE INVENTORIES

As a verification of the validity of our survey administration procedures, the responses of the subjects tested in our survey were correlated with those previously obtained from 3800 United States Armed Forces personnel (Meiselman *et al.*, 1974). Pearson correlations computed between the scores of the items common to our survey and this Armed Forces survey (n = 50 items) were 0.85 and 0.84 for the anosmic and normal subjects, respectively (p's < 0.001).

MAJOR FINDINGS OF PRESENT SURVEYS

To determine whether the responses of the anosmics differed from those of the normals within each of the general food categories listed in Table I (e.g., meats, beverages), separate two-way analyses of variance (anosmic/normal versus food items, repeated measures on the latter factor) were performed on the data within each general food class. On the first administration of the questionnaire, the normals gave larger preference ratings than the anosmics to the fruit juices ($F = 4.56$, $df = 1/28$, $p = 0.042$, % variance accounted for = 2.13) and puddings ($F = 7.00$, $df = 1/28$, $p = 0.014$, % var. = 14.31). They also reported higher preferences for the four general food groups (i.e., American, Chinese, Mexican, Italian) than did the anosmics ($F = 4.58$, $df = 1/28$, $p = 0.042$, % var. = 3.56). On the second survey, the normals again gave larger preference ratings than the anosmics to the four general food groups and to the puddings (respective F's = 6.66, 7.88, $df = 1/28$, p's < 0.05, % var. = 5.60 and 11.80, respectively), but not to the fruit juices ($F = 2.97$, $df = 1/28$, $p = 0.096$, % var. = 1.53). In addition, a significant difference appeared in the sherbert category ($F = 4.43$, $df = 1/28$, $p < 0.05$, % var. = 9.39).* The means and standard deviations of the ratings given to the items within these food classes are presented in Table II. No other

* Nonparametric analyses of medians (e.g., Kruskal-Wallis one-way analysis of variance, normals versus anosmics) were in general support of these findings. The scale values rarely differed significantly from normal distributions on either skewness or kurtosis (see McNemar, 1955, for a description of the measures of skewness and kurtosis used in our computer program).

TABLE II

Means and Standard Deviations of Pleasantness Ratings of Normals and Anosmics to Questionnaire Items[a,b]

	First administration				Second administration			
	Normals		Anosmics		Normals		Anosmics	
	Mean	SD	Mean	SD	Mean	SD	Mean	SD
Fruit juices								
Apple	6.5	(2.0)	6.7	(1.6)	6.9	(1.8)	7.1	(1.4)
Cranberry	5.9	(1.6)	4.5	(2.2)	5.7	(1.4)	4.5	(2.2)
Grape	7.6	(1.0)	6.5	(1.5)	7.7	(1.0)	6.3	(2.0)
Lemonade	7.7	(1.2)	7.3	(1.2)	7.5	(1.2)	7.3	(1.2)
Orange	7.9	(0.9)	7.3	(1.1)	8.1	(0.9)	7.4	(1.0)
Pineapple	6.2	(1.9)	5.7	(1.7)	6.4	(1.9)	5.7	(2.2)
Prune	4.1	(2.4)	2.7	(1.4)	3.5	(2.6)	2.7	(1.3)
Tomato	6.1	(2.5)	6.1	(2.5)	5.9	(2.1)	6.3	(1.8)
General food categories								
American	7.9	(0.4)	7.7	(0.9)	7.6	(1.6)	7.1	(0.8)
Chinese	7.7	(0.9)	6.7	(1.9)	7.5	(1.4)	6.8	(2.0)
Italian	7.9	(1.3)	7.5	(1.3)	8.1	(0.9)	7.1	(1.1)
Mexican	6.1	(2.1)	5.1	(1.7)	6.0	(2.0)	4.9	(1.8)
Puddings								
Butterscotch	5.9	(2.2)	4.4	(2.1)	5.5	(2.4)	4.1	(2.3)
Chocolate	7.0	(1.4)	4.5	(2.1)	6.9	(1.3)	4.9	(2.4)
Lemon	5.3	(1.9)	4.1	(2.1)	5.1	(1.8)	3.8	(1.8)
Rice	—	(—)	—	(—)	5.3	(2.1)	4.3	(2.3)
Tapioca	—	(—)	—	(—)	5.4	(1.5)	3.5	(1.7)
Sherberts								
Lemon	6.1	(1.9)	5.9	(2.0)	5.9	(1.8)	4.8	(2.1)
Lime	5.7	(2.1)	5.7	(2.2)	6.0	(1.7)	5.4	(2.2)
Orange	6.7	(1.9)	6.0	(2.1)	6.7	(1.5)	5.7	(2.5)
Raspberry	6.8	(1.2)	5.4	(1.9)	6.7	(1.1)	4.7	(1.9)
Strawberry	6.5	(1.1)	5.2	(1.6)	6.3	(1.3)	5.0	(2.3)

[a] Only food classes for which significant ANOVA's were obtained are presented.

[b] Boxes indicate normal and anosmic group means differ significantly from one another on the basis of analyses of variance.

significant differences were present between the anosmic and normal food preference ratings for any of the remaining food classes, or for the nonfood control items (e.g., colors, activities). It should be noted that even though statistically significant differences appeared between the responses of the normals and those of the anosmics in some food classes, the proportion of total variance accounted for by the anosmic/normal factor was quite small in most of the cases.

In all of the food categories except soup (first survey), sherberts (second survey), and fruit juices (second survey), the food item main effect was significant (all p's < 0.01). None of the interactions between the subject type (anosmic/normal) and the food items within a food class were significant (p's > 0.20), suggesting a relative similarity in the ordering of responses of the anosmics and normals to the various food items. This

TABLE III

Pearson Product–Moment Correlation Coefficients between Mean Preference Ratings of Anosmic and Normal Individuals within Food and Nonfood Categories of Preference Inventory[a]

	First administration	Second administration
Activities	—	0.97[d] (11 items)
Autos	—	0.82ns (5 items)
Beverages	0.52ns (11 items)	0.82[c] (12 items)
Colors	—	0.81ns (5 items)
Dairy products	—	0.93[d] (13 items)
General foods	0.97[b] (4 items)	0.98[b] (4 items)
General seasonings	0.13ns (5 items)	0.20ns (7 items)
Hot seasonings	0.90[b] (6 items)	0.81[b] (8 items)
Juices	0.93[d] (8 items)	0.92[d] (8 items)
Meats	0.94[c] (7 items)	0.92[d] (9 items)
Miscellaneous foods	—	0.66ns (5 items)
Potatoes	0.82[b] (6 items)	0.90[b] (8 items)
Puddings	0.93ns (3 items)	0.79ns (5 items)
Salads	0.83[b] (6 items)	0.90[b] (6 items)
Sherberts	0.27ns (5 items)	0.11ns (5 items)
Snacks	—	0.72[b] (9 items)
Soups	0.64ns (5 items)	0.29ns (5 items)
Sweets and condiments	0.87[d] (12 items)	0.82[d] (15 items)
Vegetables	0.99[d] (5 items)	0.78[b] (8 items)

[a] ns = not significant with $p < 0.05$.
[b] $p < 0.05$.
[c] $p < 0.01$.
[d] $p < 0.001$.

similarity is clearly exemplified by the positive and relatively strong Pearson correlation coefficients between the preference ratings of the normals and anosmics within the majority of the food categories (Table III). The correlations were also positive in the nonfood item classes, although significant only for the activity category, which contained the largest number of items.

DISCUSSION OF SURVEY RESULTS

Two general conclusions appear warranted from the results of these surveys. First, at least for some foods, normals report slightly larger, statistically significant preference ratings than do anosmics. However, this difference was not found for most of the food classes examined. Interestingly, the food classes exhibiting this difference on one or both of the surveys (i.e., puddings, fruit juices, sherberts) appear relatively homogenous in texture compared to the other food classes. Second, normals and anosmics give relatively similar ordering of food preferences across a wide variety of foodstuffs, as indicated by the Pearson correlations between their preference ratings within most of the food classes (Table III).

Although considerable information exists concerning food preferences of adult humans (cf. Meiselman *et al.,* 1974; Rozin, 1975; Waterman *et al.,* 1974), little is known about the relative importance of sensory and cultural factors (or their interactions) in influencing such preferences. It is likely that both genetic and experiential factors play a role (Amerine *et al.,* 1965), as illustrated by marked inter- and intracultural differences in the ability to taste and metabolize specific components of foodstuffs. This is classically exemplified by the distribution of "tasters" and "nontasters" of phenylthiocarbamide (PTC) in various human populations (cf. Cohen and Ogdon, 1949), as well as the differences in the abilities of human groups to metabolize lactose (e.g., Flatz *et al.,* 1969). The data of the present study suggest that congenital lack of smell ability does not markedly affect the relative food preference reports of humans across a number of general food items, although slight differences in overall preference for specific food items may exist between anosmic and normal individuals.

Several explanations for the relative similarity of reported food preferences of the normals and anosmics can be hypothesized. First, it is conceivable that taste, texture, temperature, auditory, and visual cues provide enough information upon which stable estimates of the hedonicity of food can be established. If this is true, olfactory cues may be redundant, adding only "overtones" to the overall taste experience. On the other hand, this similarity may mean that the sensory cues provided by smell and perhaps the other senses are overridden to a large extent by culturally determined food preference influences, such as early conditioning in the home or by peer group experiences (cf. McCarthy, 1935). Heritability estimates of preferences for various concentrations of lactose, fructose, and sodium chloride are uniformly low in American Black and Caucasian populations (Greene *et al.*, 1975), possibly adding support to this notion. In most cases it is likely that both cultural and more rigidly defined sensory processes work together in establishing food preferences.

It is important to note that, while not statistically significant, the preference ratings of the normals had a tendency in some food classes such as sweets to be larger than those of the anosmics. This was not true, however, of all of the food categories, nor was it true for the nonfood items. Perhaps the use of larger samples and/or more refined psychometric measuring instruments than the nine-point category rating scale would reveal significant differences between the preferences of anosmics and normals in these cases. Since the choice of items within a food class is somewhat arbitrary, differently composed food classes might yield somewhat different results. Similarly, more refined food item categories (e.g., rare T-bone steak, medium rare filet mignon smothered in red wine sauce) might produce clear differences in the anosmic and normal preference ratings.

Although the present study is a first step in categorizing the food preferences of anosmics, it cannot be viewed as definitive. Future studies should test the preferences of anosmics and normals directly in the laboratory using food items which vary systematically on taste, temperature, and texture dimensions. Furthermore, such studies should utilize a larger number of congenital anosmics who do not have associated disorders (e.g., hypogonadism). Despite the limitations of the present survey, however, its findings suggest that the general food preferences of anosmics and normals are quite similar across a wide range of foodstuffs, although subtle differences between these groups are likely present for the preferences of certain food items. Overall, these data indicate that congenital absence of olfaction does not result in markedly aberrant food preferences in human beings.

REFERENCES

Amerine, M. A., Pangborn, R. M., and Roessler, E. B. (1965). "Principles of Sensory Evaluation of Food." Academic Press, New York.

Clark, E. C., and Dodge, H. W. (1955a). Effect of anosmia on the appreciation of flavor. *Neurology* **5**, 671–674.

Clark, E. C., and Dodge, H. W. (1955b). Extraolfactory components of odor. *J. Am. Med. Assoc.* **159**, 174–1723.

Cohen, J., and Ogdon, D. P. (1949). Taste blindness to phenyl-thio-carbamide and related compounds. *Psychol. Bull.* **46**, 490–498.

Crosland, H. R., Goodman, M., and Hockett, A. (1926). Anosmia and its effects upon taste perceptions. *J. Exp. Psychol.* **9,** 398–408.

Crosland, H. R., Miller, R. C., and Broadway, W. E. (1928). Oral perceptions in relation to anosmia. *J. Exp. Psychol.* **11,** 161–166.

Doty, R. L. (1975). Intranasal trigeminal detection of chemical vapors by humans. *Physiol. Behav.* **14,** 855–859.

Doty, R. L., Brugger, W. E., Jurs, P. C., Orndorff, M. M., Synder, P. J., and Lowry, L. D. (1977). A comparison of psychometric responses from normal and anosmic humans to intranasal trigeminal stimulation from inhaled chemicals, with an analysis of the structure–activity relationships involved. (Submitted for publication.)

Douek, E. (1974). "The Sense of Smell and its Abnormalities." p. 229. Livingstone, Edinburgh.

Flatz, G., Saegudom, C., and Sanguanbhokhai, T. (1969). Lactose intolerance in Thailand. *Nature (London)* **221,** 758–759.

Gilland, A. R. (1921). The taste sensitivity of an anosmic subject. *J. Exp. Psychol.* **4,** 318–326.

Greene, L. A., Desor, J. A., and Maller, O. (1975). Heredity and experience: Their relative importance in the development of taste preference in man. *J. Comp. Physiol. Psychol.* **89,** 279–284.

Kneeland, (1851). Case of an affection of the olfactory nerve, with total loss of smell and taste; with probable identity of these two senses. *Am. J. Med. Sci.* **21,** 41–46.

McCarthy, D. (1935). Children's feeding problems in relation to the food aversions in the family. *Child Dev.* **6,** 277–284.

McNemar, Q. (1955). "Psychological Statistics," p. 28. Wiley, New York.

Meiselman, H. L., Waterman, D., and Symington, L. E. (1974). Armed forces food preferences. *U. S. Army Natick Lab., Tech. Rep.* **75-68-FSL.**

Moskowitz, H. R., Nichols, T. L., Meiselman, H. L., and Sidel, J. L. (1972). Food preferences of military men, 1967. *U. S. Army Natick Lab., Tech. Rep.* **72-70-PR.**

Mozell, N. M., Smith, B. P. Smith, P. E., Sullivan, R. L., and Swender, P. (1969). Nasal chemo-reception in flavor identification. *Arch. Otolaryngol.* **90,** 131–137.

Ogle, W. (1870). Anosmia; or, cases illustrating the physiology and pathology of the sense of smell. *Med.-Chir. Trans.* **53,** 263–290.

Patrick, G. T. W. (1899). On the confusion of tastes and odors. *Psychol. Rev.* **6,** 160–162.

Peryam, D. R., and Pilgrim, F. J. (1957). Hedonic scale method of measuring food preferences. *Food Technol.* **11,** 9–14.

Rozin, P. (1976). The selection of foods by rats, humans, and other animals. *Adv. Study Behav.* **6,** 21–76.

Schneider, R. A. (1974). Newer insights into the role and modifications of olfaction in man through clinical studies. *Ann. N. Y. Acad. Sci.* **237,** 217–223.

Waterman, D., Meiselman, H. L., Branch, L. G., and Taylor, M. S. (1974). The 1972 Westover Air Force Base food preference survey and reliability study. *U. S. Army Natick Lab., Tech. Rep.* **75-25-FSL.**

DISCUSSION

Senseman: Do the anosmics have any problems with taste perception per se?

Doty: I really do not know for sure in all cases. We have taste-tested some using the procedure disclosed by Desor and Maller, and the sensitivity fell within the normal range.

Gilland (1921) tested one anosmic for detection taste thresholds to sweet, sour, salty, and bitter substances and found them to be in the normal range.

Senseman: I am more interested in their ability to discriminate subtle differences in taste quality. We have the notion that the gustatory sense is rather limited in its ability to discriminate in contrast to the capacity of the olfactory sense. Yet I am wondering if a person deprived of olfaction might display a marvelous ability to discriminate between food items using taste cues alone?

Doty: That is certainly possible. I have seen people saying things both ways.

Naim: Basically, humans do not respond to odors as well as do rats. Consequently, in humans, the type of food diet selection due to odors could be really different. Just yesterday, Garcia showed that rats can detect the odors of saccharin if the concentration is high. We have found that anosmic rats can not distinguish between diets containing 1% saccharin versus diets containing 0.5% saccharin, whereas normal rats can.

Garcia: I think if you take a took at all the rat data we have been through, the impressive thing is that the nose learns from the mouth all it knows about food. The mouth could do just as well without the nose in selecting the diet. There is just no doubt about it. It is a higher order of conditioning. Everything you bring to your mouth is putting an olfactory gradient up in the perfect classic conditioning setup. You reinforce it with the taste within 450 msec, which is the optimum time that this learns.

Moskowitz: I would suggest that there is a different way of approaching this study. Anosmics and normals have different sensory systems. Both of them have taste (anosmics and normals) and texture perception. Would it not be possible in an experiment like this to study to what degree satisfaction is due to the kinesthetic or the texture perception and to what degree to the taste perception? Even if there are no significant differences in many of the food classes, perhaps normals and anosmics are weighting the input from different sensory systems differently.

Doty: I think that is an excellent point and certainly identifies an area requiring more research. One thing I should mention is that most of these patients were referred from various clinics to me for tests to see if they can smell or not. I could only get them for a couple of hours at the most. Many of them come from all over the country, as well as abroad. Thus, it is not easy to get a large local group that can be brought back for repeated testing. In another year, I hope to have a large enough local population of anosmics. Then I can start getting answers to the questions you pose.

Chapter 16

Sweet Taste Sensitivity in Japanese Macaques

MASAYASU SATO, YASUTAKE HIJI, HIROSUMI ITO, AND TOSHIAKI IMOTO

Species differences in taste sensitivity, especially to sweet-tasting substances among various mammals, has been demonstrated by both behavioral and electrophysiological experiments. For example, rats prefer saccharin but do not prefer dulcin to water, while squirrel monkeys prefer sucrose and dulcin but show an aversion to saccharin (Carpenter, 1965; Fisher *et al.,* 1965). Rhesus monkeys, on the other hand, show a preference for saccharin at low concentrations (Weiskrantz, 1960).

Previously, we have examined taste sensitivities in crab-eating macaques and rhesus monkeys to various chemicals by recording responses in the whole chorda tympani. Our results, which have been reported, indicate that these monkeys respond well not only to stimuli representing the four primary tastes such as sucrose, NaCl, HCl and quinine, but also to saccharin, dulcin, and lead acetate (Ogawa *et al.,* 1972). Further analysis of impulse discharges in single chorda tympani nerve fibers of crab-eating macaques has revealed two facts: first, although a majority of single taste nerve fibers respond to two or more kinds of the four basic stimuli, relatively high specific sensitivity is present in fibers responsive to sucrose or quinine; and second, sweet-tasting compounds such as saccharin, dulcin, and lead acetate yield across-fiber correlation profiles similar to that produced by sucrose (Sato, 1975; Sato *et al.,* 1975). Thus, our results on macaque monkeys point out the presence of a specifically labeled coding system for sucrose and quinine taste and a broad sensitivity to a variety of chemicals that taste sweet to humans and are of little nutritional value. These characteristics of monkey taste receptors are considered to be a consequence of the phylogenetic evolution of mammals. In keeping with the results on crab-eating macaques mentioned above, Brouwer *et al*. (1973) have recently reported that the sweet-tasting proteins, Monellin and Thaumatin, elicited a significant response in the chorda tympani of the green monkey, while little or no response could be obtained in the guinea pig and rat.

Based on the above-mentioned facts we further examined responses in Japanese macaques *(Macaca fuscata)* to various sweeteners, including aspartame and stevioside, using electrophysiological and behavioral methods. Aspartame is a dipeptide about 100–200 times sweeter than sucrose in aqueous solutions (Cloninger and Baldwin, 1970; Dastoli and Price, 1966), while stevioside is a glycoside extracted from the leaves of the plant *Stevia rebaudiana* and is about 100–300 times sweeter than sucrose (Ishima and Katayama, 1974). We then proceeded to examine responses of the extract from the tongue epithelium of the monkey occurring on addition of the sweeteners.

CHORDA TYMPANI NERVE RESPONSES IN JAPANESE MACAQUES TO VARIOUS SWEETENERS INCLUDING ASPARTAME AND STEVIOSIDE

Good responses to the four basic stimuli and sweeteners such as saccharin, dulcin, and sodium cyclamate were obtained from the chorda tympani of Japanese macaques. The threshold concentrations for the latter stimuli were about 0.1 m*M*, 0.3 m*M*, and 1 m*M*, respectively, and saccharin at any concentrations elicited a response larger than that produced by NaCl. Responses to and thresholds for these sweeteners are essentially similar to those in crab-eating macaques and rhesus monkeys (Ogawa *et al.*, 1972).

As expected from our previous experiments, we could record good responses to aspartame and stevioside from the chorda tympani of a majority of Japanese macaques. An example of some records of integrated responses to 500 m*M* sucrose, 30 m*M* aspartame, 3.8 m*M* (3% w/v) stevioside, and 50 m*M* NaCl in monkeys is shown in Fig. 1. The top record, which shows the most frequently encountered response characteristics, indicates that 30 m*M* aspartame and 3.8 m*M* stevioside produced responses which are much larger than the responses to 500 m*M* sucrose. On the other hand, when we examined responses to aspartame and

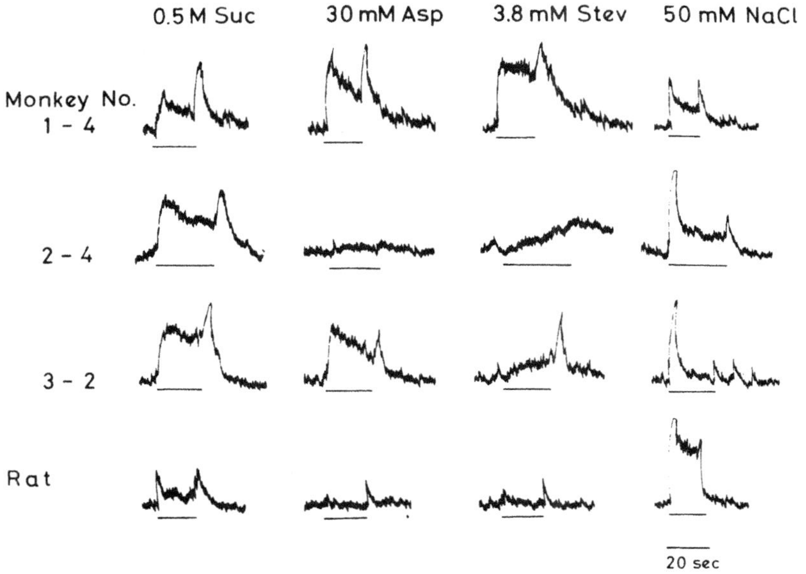

Fig. 1. Integrated responses to 500 m*M* sucrose, 30 m*M* aspartame, 3.8 m*M* stevioside, and 50 m*M* NaCl in the chorda tympani nerve of three Japanese macaques and an albino rat. The horizontal bar underneath each record indicates the approximate period of stimulation.

stevioside in the rat chorda tympani, little response was recorded as is shown in the bottom trace in Fig. 1.

We examined responses in the chorda tympani to sucrose, aspartame, and stevioside in ten macaques consisting of nine Japanese macaques and one crab-eating macaque. As described above, a majority of the macaques responded well to the three sweeteners. However, responses to aspartame, stevioside, or both are very poor in a few macaques, although they elicited good responses to sucrose. Records of responses to the three sweeteners in such a monkey are shown in the middle of Fig. 1, where little response to 30 m*M* aspartame was recorded from monkey no. 2-4. Monkeys no. 2-4 and 3-2 did show responses to stevioside which increased in magnitude very slowly, and the response levels attained during stimulation were much smaller than that obtained by 0.5 *M* sucrose. Magnitudes of responses to 30 m*M* aspartame and 3.0 or 3.8 m*M* stevioside expressed relative to that for 500 m*M* sucrose in the ten macaques examined are presented in Fig. 2. Although there is a fairly large variation in the relative value among individual macaques, the magnitude of response to aspartame is especially small in monkey no. 2-4, while those responses to

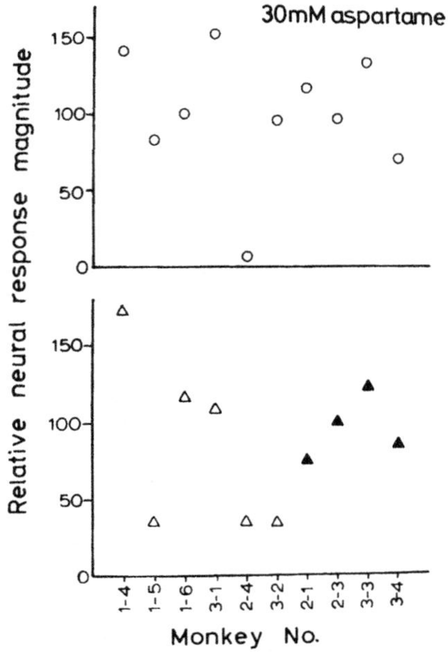

Fig. 2. Magnitudes of integrated chorda tympani nerve responses to 30 m*M* aspartame (circles) and 3 m*M* (filled triangles) and 3.8 m*M* (open triangles) stevioside, expressed relative to that for 500 m*M* sucrose, in ten macaques consisting of nine Japanese macaques and a crab-eating macaque (1-6).

stevioside are small in monkeys no. 1-5, 2-4, and 3-2. Thus, we found three monkeys among ten which possess normal taste sensitivity to sucrose but abnormally low sensitivity to aspartame and stevioside, or stevioside only.

Fig. 3 presents an example of concentration–response relations for stevioside, aspartame, and sucrose obtained from a Japanese macaque

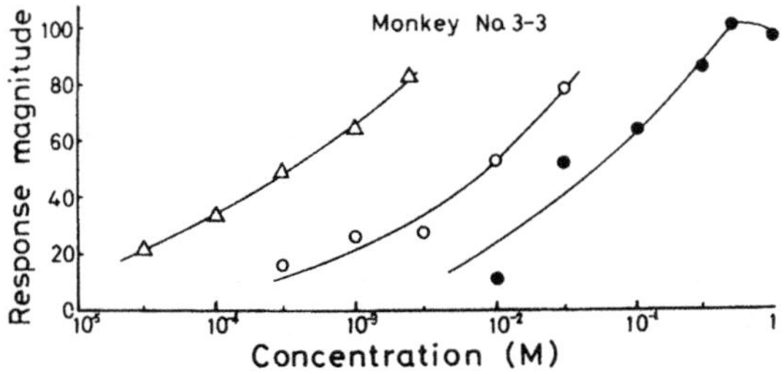

Fig. 3. Relationship between concentrations of stevioside (triangles), aspartame (open circles), and sucrose (filled circles), and the magnitudes of chorda tympani responses, obtained from a Japanese macaque.

which possesses normal sensitivities to the three chemicals. In this figure concentrations of aspartame and stevioside that produced a response nearly equal in magnitude to that elicited by 0.1 and 0.3 *M* sucrose are about 18 and 40 m*M* for aspartame and 1 and 4 m*M* for stevioside, respectively, indicating that stevioside is approximately 100 times more effective in eliciting gustatory impulses than sucrose, while aspartame is only 6–7 times stronger in effect than sucrose. Although there is a variation in the relative effectiveness of aspartame and stevioside among individual macaques, it can generally be stated that in macaque monkeys, as in humans, aspartame and stevioside are much more effective in producing a taste response, but that the effectiveness of aspartame on monkey taste receptors relative to that of sucrose is somewhat less than that demonstrated in human sensory tests.

During the experiments we examined responses to various sugars in the chorda tympani of six macaques. All the macaques responded well to nine kinds of sugars. The relative magnitudes of responses to sugars in individual macaques are shown in Table I. In the table there is a large variation in the relative response magnitude among individual monkeys. For example, responses to fructose and mannose are very small in monkey no. 2-3 compared with those in other macaques, while responses to galactose are

TABLE I

Relative Taste Effectiveness of Sugars, as Measured from the Neural Responses in Japanese Macaques[a]

Sugar	Monkeys 2–4	3–2	2–1	2–3	3–3	3–4	Mean ± SD	Order
Sucrose	100	100	100	100	100	100	100	2(2)
Maltose	130	100	103	130	96	94	108.9 ± 16.7	1(1)
Lactose	60	41	60	60	63	100	64.0 ± 19.4	8(8)
Fructose	100	75	80	26	82	106	78.0 ± 28.3	4(5)
Galactose	150	78	93	140	67	100	104.6 ± 33.4	3(4)
Glucose	80	52	107	125	63	93	86.7 ± 27.3	5(3)
Mannose	50	79	73	19	63	111	65.8 ± 30.7	6(9)
Sorbose	80	73	80	45	63	93	72.4 ± 16.6	7(6)
Xylose	60	60	40	60	63	98	63.6 ± 18.9	8(7)

[a] All the sugars were 0.5 *M* and values presented are initial maximum magnitudes of integrated responses, expressed relative to the mangitude of response to sucrose. The order of effectiveness for monkeys was determined by using the Kendall coefficient of concordance. Numerals inside parentheses indicate the order determined from the magnitude of the steady-state response.

very larger in monkeys no. 2-4 and 2-3. However, monkey no. 2-4, which did not respond well to aspartame and stevioside, and no. 3-2, in which response to stevioside was small, yielded good responses to all the sugars.

An interesting fact in our results on Japanese macaques, shown in Table I, is that the relative effectiveness for maltose and galactose is large while that for fructose is small compared with the effectiveness of these sugars in other mammals. The order of effectiveness is sucrose > fructose > sorbose > mannose > glucose > maltose ≈ xylose > rhamnose > galactose in the rat (Noma *et al.*, 1971); sucrose > fructose > mannose > glucose > sorbose > rhamnose > maltose > galactose > xylose in the hamster (Noma *et al.*, 1974); fructose > sucrose > lactose > maltose > galactose ≈ glucose in the squirrel monkey (Snell, 1965); and fructose > sucrose > glucose > mannose ≈ galactose > maltose > lactose in humans (Schutz and Pilgrim, 1957). Whether such a difference is a species difference or bears any physiological significance cannot be answered here.

INGESTIVE RESPONSES IN JAPANESE MACAQUES TO VARIOUS SWEETENERS

In order to assess how macaques taste various sweeteners that produce neural responses in the chorda tympani, ingestive responses to the sweeteners in Japanese macaques were measured in 30-min, single stimuli tests before electrophysiological experiments were carried out. The experimental procedure was similar to that reported by Maller (1973). Animals, transferred from the Takazaki Natural Zoo, were housed in individual cages and fed monkey pellets. Prior to testing, the animals were trained to sample a solution immediately after a bottle was placed in the cage. Each test solution was presented on two consecutive days for a 30-min period each day, and intake of the fluid was measured. Before the two testing days ingestion of water was determined by presenting water in a bottle for 30 min for 1 or 2 days. Water was not given to the monkey except for this testing. The taste testing was done in the afternoon approximately between 3:30 and 4:00 PM.

First, several kinds of sweeteners of a given concentration were given to the monkeys successively for 2 days. Figure 4 demonstrates intake volumes of 30 m*M* sucrose, 6 m*M* saccharin, 10 m*M* aspartame, 0.5 m*M* stevioside, and water ingested by four monkeys. The upper four figures show intake volumes of individuals while the lower one indicates their mean values. Monkeys, in general, distinctly preferred sucrose, aspartame, and stevioside to water, but they did not show significant preference

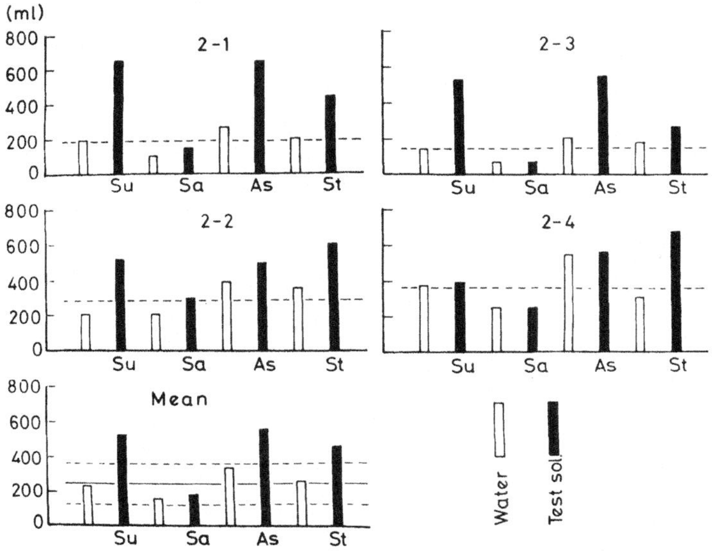

Fig. 4. Upper figures: Intake volumes of 300 m*M* sucrose (Su), 6 m*M* saccharin (Sa), 10 m*M* aspartame (As), and 0.5 m*M* stevioside (St) measured during 30-min periods in four Japanese macaques. Each value represents the mean of measurements on 2 consecutive days. In each figure, empty blocks represent intake volumes of water, measured before successive 2 testing days, and the horizontal line indicates the mean of water intakes in each animal. Bottom figure: Mean intake volumes and water intakes in four macaques. The horizontal lines represent the mean of all the water intakes (solid line) and ±SD (dashed line).

for 6 m*M* saccharin. For example, the ingestive responses of monkey 3-3 to six kinds of sweeteners is shown in Fig. 5. In this test also the animal showed preferences for 300 m*M* sucrose, 10 m*M* aspartame, and 0.5 m*M* stevioside. Preferences for 3 m*M* dulcin and 10 m*M* cyclamate were not significant, and he showed aversion to 3 m*M* saccharin. Another monkey, 3-4, with which similar measurements were performed, showed strong preferences not only for sucrose, aspartame, and stevioside, but also for saccharin and dulcin; he did, however, show aversion to cyclamate.

Second, ingestions of varying concentrations of a sweetener were measured by presenting to the monkey test solutions in an ascending concentration series. Ingestion of water was measured immediately before and after the series of tests for a given sweetener. Intake curves for saccharin and sucrose of varying concentrations, obtained from two monkeys, are shown in Fig. 6. Since in these tests intake of water was measured only before and after the tests for sweeteners, assessment of statistical significance for the difference of the intake value of each solution from

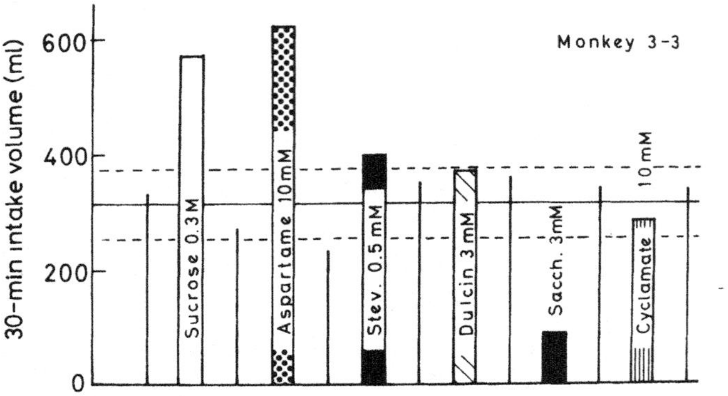

Fig. 5. Intake volumes of 300 m*M* sucrose, 10 m*M* aspartame, 0.5 m*M* stevioside, 3 m*M* dulcin, 3 m*M* saccharin, and 10 m*M* cyclamate measured during 30-min periods on 2 consecutive days in a Japanese macaque. Each point is a mean of 2 days' measurements. Vertical bars represent intake volumes of water measured between successive 2 testing days. The horizontal bars represent the mean of water intakes (solid line) and ±SD (dashed line).

water intake cannot be made, but sucrose appears to be preferred to water at 10 m*M*, and the intake volume increased with ascending concentration, reaching at 300 and 500 m*M* a value about four times greater than the water intake. Saccharin below 1 m*M* was preferred to water but was rejected at 30 m*M*. In two other monkeys, saccharin was also preferred below 1 m*M*, but one of them rejected 10 m*M* saccharin and the other showed neither preference nor aversion from 3–30 m*M* saccharin. In one animal intakes of cyclamate solutions of varying concentrations were measured. The animal preferred 0.3 and 1 m*M* cyclamate, but rejected 10 m*M*.

Weiskrantz (1960) reported that rhesus monkeys showed a moderate preference for 0.7 m*M* saccharin, a clear and strong preference for 2–10 m*M*, but an aversion to 40 m*M* saccharin. Maller and Hamilton (1968), using two-bottle 24-hr preference tests, demonstrated in rhesus monkeys a strong preference for 0.83–12.4 m*M* but an aversion to 25 m*M* saccharin. In our tests Japanese macaques showed a moderate preference for 1–3 m*M* saccharin to water but showed an aversion to 3-10 m*M* saccharin. Therefore, our results are qualitatively the same as those for rhesus monkeys, but Japanese macaques appear to show a less pronounced preference for saccharin than do rhesus monkeys.

Preliminary experiments to determine the intake volumes of aspartame and stevioside of varying concentrations were carried out. In one macaque aspartame was preferred to water approximately above 1 m*M*, but the intake volume for 10 m*M* aspartame, compared with water intake, was not as large as that shown in Figs. 4 and 5. In another macaque, where water intake was measured between tests for each concentration, the monkey showed an intake of 0.1 m*M* aspartame larger than the mean water intake, and an intake of 10 m*M* aspartame about twice that of the mean water intake. In one macaque stevioside above 0.3 m*M* was preferred to water, and the intake volume increased with increasing concentration. Similar preference for stevioside was demonstrated in two other macaques.

The results of our measurements on the ingestive responses of Japanese macaques to various sweeteners demonstrate that they show high preference for sucrose, aspartame, and stevioside. Although thresholds for the responses to these chemicals could not be determined accurately, the macaques responded behaviorally to 10 m*M* sucrose, 1 m*M* aspartame, and 0.01 m*M* stevioside, and the intake volumes increased gradually with

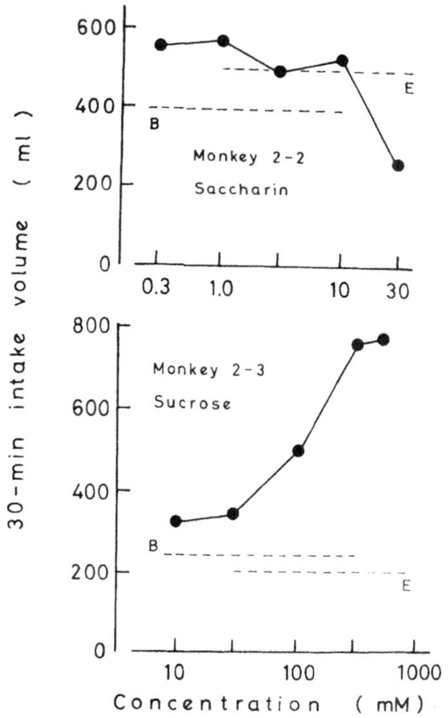

Fig. 6. Intake volumes of saccharin (upper figure) and sucrose (lower figure) of varying concentrations. Each point is a mean of measurements during 30-min periods on 2 consecutive days in two Japanese macaques. Water intake was measured for 2 days before and after the tests, the mean value of respective measurements being represented by horizontal dotted lines (B and E).

ascending concentrations. These results are in qualitative agreement with the results of neurophysiological experiments, where neural responses were recorded with sucrose above 10 m*M*, aspartame at 0.3 m*M*, and stevioside at 0.03 m*M*. However, the ingestion magnitudes for the three chemicals do not parallel those of the neural response magnitudes. These sweeteners are effective in the order of 10 m*M* aspartame > 0.3 *M* sucrose > 0.5 m*M* stevioside when the intake volumes were compared, but the effectiveness for the neural response is 0.3 *M* sucrose > 0.5 m*M* stevioside > 10 m*M* aspartame. The discrepancy suggests a slight difference in taste quality information mediated by the chorda tympani among these three chemicals.

INTERACTION OF PROTEIN EXTRACTED FROM MONKEY TONGUE EPITHELIUM WITH SWEETENERS

Since the work by Dastoli and Price (1966) a number of studies have recently been reported which indicate the presence in the tongue epithelia of several mammals of a protein that complexes with sugars, saccharin, and some amino acids (Dastoli, 1974; Dastoli and Price, 1966; Hiji *et al.*, 1971; Hiji and Sato, 1973). The complex formation of the protein extracted from the epithelium of the denervated rat tongue with labeled fructose decreases to about 1/20 of the value found in the normal rat tongue (Hiji and Sato, 1973). The relative magnitudes of the increase in UV absorbancy which occur at about 280 mμ on addition of various sugars to the 60% ammonium sulfate fraction of the extract of rat tongue epithelium parallels the relative magnitude of chorda tympani nerve responses to varying sugars in the rat (Hiji *et al.*, 1971). These findings and others reported by several investigators indicate that the binding protein in the 60% ammonium sulfate fraction extracted from the tongue epithelium is the taste receptor protein that reacts with sweet-tasting substances at the receptor membrane (Dastoli, 1974).

Based on the results of such biochemical experiments on the extracted proteins from bovine and rat tongues, and on those of the behavioral and neurophysiological experiments indicating that taste receptors in the rat respond only to sugars and saccharin while those in the macaque monkey respond to various sweeteners, we reasoned that the protein extracted from the rat tongue would interact with neither aspartame nor stevioside while that from the monkey tongue should respond to both sweeteners. In order to see a correlation between neurophysiological and biochemical experiments, we sacrificed the monkey immediately after the electrophysiological experiment, obtained the 60% ammonium sulfate fraction from the homogenate of the epithelium of the monkey tongue, and measured the difference spectral absorbancy at 278 mμ (Δ OD_{278}) on addition of sucrose, aspartame, and stevioside to the fraction. For comparison, similar experiments were carried out on the protein extracted from the rat tongue epithelia.

Figure 7 presents difference spectral absorbancies (Δ OD_{278}) obtained on addition of 500 m*M* sucrose, 30 m*M* aspartame, and 3.8 m*M* stevioside to the 60% ammonium sulfate fractions from the tongue epithelia of monkeys no. 1-4 and 3-2 as well as of 70 rats. In the rat there is an increase in Δ OD_{278} on addition of sucrose but none upon addition of aspartame or stevioside. On the other hand, not only sucrose but also aspartame and stevioside produced a marked increase in Δ OD_{278} in the extract from monkey no. 1-4, and a similar increase can be seen on addition of sucrose and aspartame to that from monkey no. 3-2, although in this case the increase produced by stevioside is much smaller compared with those by sucrose and aspartame. These results on the extracted protein are in accordance with those obtained by neurophysiological experiments on the two monkeys. As shown in Fig. 1, monkey no. 1-4 presented good responses to both aspartame and stevioside, while in monkey no. 3-2 the response to stevioside is very small in magnitude compared with those to sucrose and aspartame.

In order to examine in more detail correlations between the results of neurophysiological experiments and those of the biochemical experiments, Δ OD_{278} values on addition of sucrose, aspartame, and stevioside

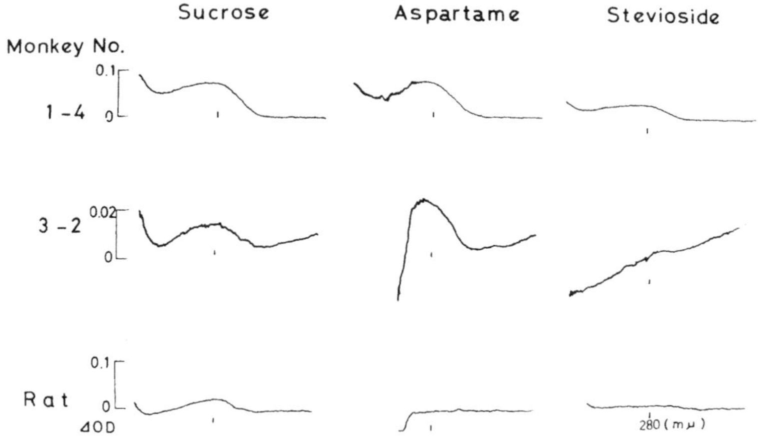

Fig. 7. UV difference spectral absorbancy obtained on addition of 0.5 *M* sucrose, 30 m*M* aspartame, and 3.8 m*M* stevioside to 60% ammonium sulfate fraction of the extract from the tongue epithelia of two Japanese macaques (no. 1-4 and 3-2) and an albino rat.

of varying concentrations to the protein extracted from the tongue epithelia of monkeys were measured. Figure 8 is an example of the results of such experiments, in which relationships between the concentration of the sweeteners and the relative Δ OD_{278} values were obtained from monkey 3-4. In this figure sucrose produced an increase in Δ OD_{278} above 1 m*M*, aspartame above 0.3 m*M*, and stevioside above 0.03 m*M*. The three relationships in Fig. 7 are rather similar in threshold and general features to those between the neural response magnitude and the concentrations presented in Fig. 3.

Interrelationships between neural responses and the results of biochemical experiments were further examined by comparing the concentration–response curves for sucrose, aspartame, and stevioside with those relating the relative Δ OD_{278} value to the concentration of the three chemicals, both sets of curves having been constructed from the average values obtained from a number of experiments on monkeys which showed good sensitivity to the sweeteners. The results are illustrated in Fig. 9, which again indicates a reasonably good correlation between the two sets of experimental results. The relative Δ OD_{278} values for aspartame and stevioside are almost linearly related to the neural response magnitudes; the values for sucrose, however, are slightly greater at the middle range of the concentrations used. The agreement between neural response magnitudes for sweeteners and the magnitude of response of the protein extracted from the tongue epithelium to these chemicals provides further evidence indicating that the protein is really a receptor protein molecule at the taste receptor membrane, and, in addition, confirms the hypothesis that the interaction between molecules of sweeteners with the receptor protein is the initial step in taste receptor stimulation, which somehow leads to initiation of impulses at the afferent nerve terminals.

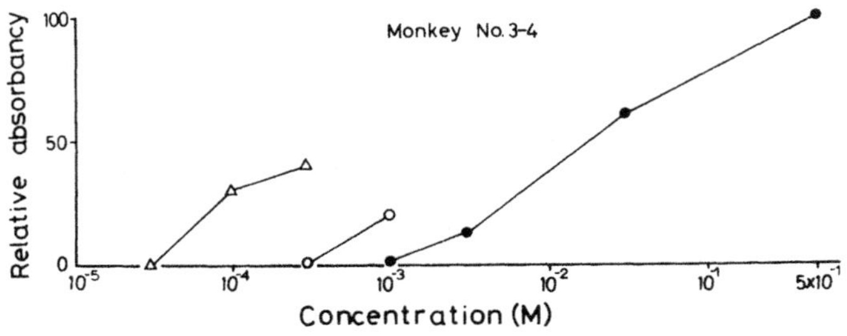

Fig. 8. Relationships between the concentration of stevioside (triangles), aspartame (open circles), and sucrose (filled circles) and the relative Δ OD_{278} value obtained on addition of the chemicals to 60% ammonium sulfate fraction of the extract from the tongue epithelium of a Japanese macaque. Each Δ OD_{278} value is expressed relative to that for 500 m*M* sucrose.

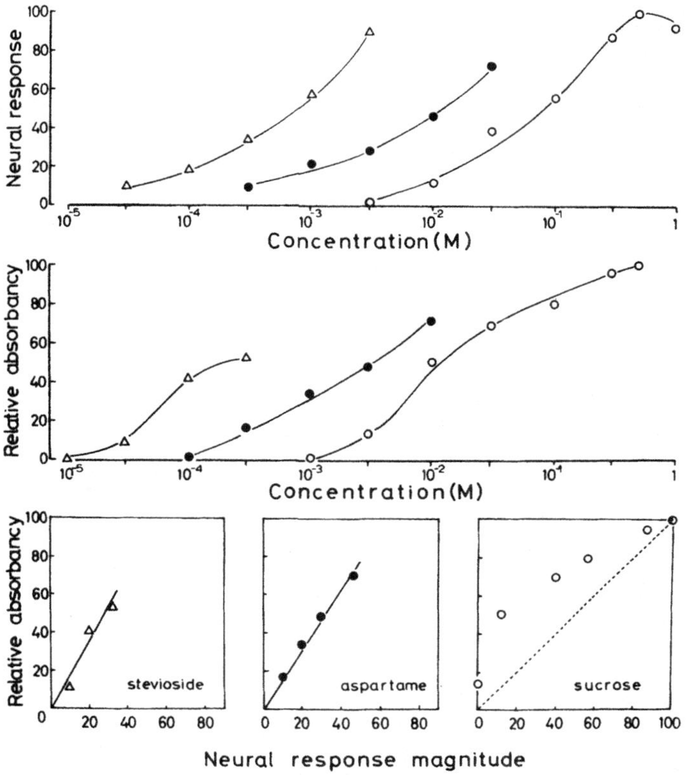

Fig. 9. Upper figure: Relationships between the concentration of stevioside (triangles), aspartame (filled circles), and sucrose (open circles) and the relative magnitude of chorda tympani nerve responses in Japanese macaques. Each point is a mean of 2–3 measurements for sucrose, 3 measurements for aspartame, and 4–5 measurements for stevioside. Middle figure: Relationships between the concentration of stevioside, aspartame, and sucrose and the relative Δ OD_{278} values obtained from the extract of tongue epithelium in Japanese macaques. Each point is a mean of 2–4 measurements for sucrose, of 2–5 for aspartame, and 2 for stevioside. Lower figure: Relationships between mean response magnitudes in the chorda tympani and the relative magnitude of absorbancy on addition of stevioside, aspartame, and sucrose to extracts from monkey tongue epithelia. Each point in this figure was replotted from those in the upper and middle figures.

CONCLUSIONS

I have attempted to answer the question of how the taste of food is recognized by the nervous system by describing the results of our experiments on macaque monkeys using electrophysiological, behavioral, and biochemical methods. Good responses to sugars, saccharin, aspartame, and stevioside were demonstrated in the chorda tympani of a majority of Japanese macaques, and such neural responses were well correlated with their ingestive responses.

Taste receptor proteins, extracted from the tongue epithelium of Japanese macaques, showed an increase in UV difference spectral absorbancy upon addition of sucrose, aspartame, and stevioside. A reasonably good correlation was demonstrated between the neural response magnitude and the increase in UV difference spectral absorbancy. This supports the concept that the complex formation of the receptor protein with sweeteners is the initial step in stimulating the sweet taste receptor cells, which ultimately leads to initiation of impulses at the afferent nerve terminals.

A few monkeys exhibited abnormally low sensitivity to aspartame and stevioside, or stevioside only, although they elicited good neural responses to sugars. Relatively low sensitivity to stevioside was also demonstrated in the protein extracted from the tongue epithelium of a monkey, which had yielded a small neural response to stevioside compared with that of sucrose. Such a result suggests that the receptor sites in the sweet receptor protein for sucrose, aspartame, and stevioside may be different from one another.

ACKNOWLEDGMENTS

The authors thank Miss Chieko Saku for her technical assistance. They also express their gratitude to Ajinomoto Co. for kindly supplying them with aspartame, and to Tama Biochemical Institute for supplying them with stevioside. This work was supported in part by a grant (No. 010607 and 048235) from the Ministry of Education in Japan.

REFERENCES

Brouwer, J. N., Hellekant, G., Kasahara, K., Van Del Wel, H., and Zotterman, Y. (1973). Electrophysiological study of the gustatory effects of the sweet proteins Monellin and Thaumatin in monkey, guinea pig and rat. *Acta Physiol. Scand.* **89,** 550.

Carpenter, J. A. (1965). Species differences in taste preferences. *J. Comp. Physiol. Psychol.* **49,** 139.

Cloninger, M. R., and Baldwin, R. E. (1970). Aspartylphenylalanine methyl ester: A low-calorie sweetener. *Science* **170,** 81.

Dastoli, F. R. (1974). Taste receptor proteins. *Life Sci.* **14,** 1417.

Dastoli, F. R., and Price, S. (1966). Sweet-sensitive protein from bovine taste buds; isolation and assay. *Science* **154,** 905.

Doetsch, G. S., and Erikson, R. P. (1970). Synaptic processing of taste-quality information in nucleus tractus solitarius of the rat. *J. Neurophysiol.* **33,** 490.

Fisher, G. L., Pfaffmann, C. and Brown, E. (1965). Dulcin and saccharin taste in squirrel monkeys, rats and men. *Science* **150,** 506.

Hiji, Y., and Sato, M. (1973). Isolation of the sugar-binding protein from rat taste buds. *Nature (London), New Biol.* **244,** 91.

Hiji, Y., Kobayashi, N., and Sato, M. (1971). Sweet-sensitive protein from the rat tongue: Its interaction with various sugars. *Comp. Biochem. Physiol. B* **39,** 367.

Ishima, T., and Katayama, O. (1974). Taste of a sweetener, stevioside. *Abstr. Symp. Sensory Tests, 4th,* 113 p. (in Japanese).

Maller, O. (1973). Quality and intensity of sugar solutions as determinants of ingestion by the rhesus monkey *(Macaca mulatta). Folia Primatol.* **20,** 72.

Maller, O., and Hamilton, C. L. (1968). Sucrose and caloric intake by normal and diabetic monkeys. *J. Comp. Physiol. Psychol.* **66,** 444.

Nofre, C., and Sabadi, J. (1972). A propos de la proteine linguale dite "sensible aux socres." *C. R. Acad. Sci., Ser. D* **274,** 2913.

Noma, A., Goto, J., and Sato, M. (1971). The relative taste effectiveness of sugars and sugar alcohols for the rat. *Kumamoto Med. J.* **24,** 1.

Noma, A., Sato, M., and Tsuzuki, Y. (1974). Taste effectiveness of anomers of sugars and glycosides as revealed from hamster taste responses. *Comp. Biochem. Physiol. A* **48,** 249.

Ogawa, H., Yamashita, S., Noma, A., and Sato, M. (1972). Taste responses in the macaque monkey chorda tympani. *Physiol. Behav.* **9,** 325.

Ostretsova, I. B., Safarian, E. K., and Etingof, R. N. (1975). Localization of proteins on the tongue which bind glucose. *Proc. Acad. Sci. U.S.S.R.* **223,** 1484.

Sato, M. (1975). Response characteristics of taste nerve fibers in macaque monkeys: Comparison with those in rats and hamsters. *Olfaction Taste, Proc. Int. Symp., 5th, 1974* p. 23.

Sato, M., Ogawa, H., and Yamashita, S. (1975). Response properties of macaque monkey chorda tympani fibers. *J. Gen. Physiol.* **66,** 781.

Schutz, H. G., and Pilgrim, F. J. (1957). Sweetness of various compounds and its measurement. *Food Res.* **22,** 206.

Snell, T. C. (1965). The responses of the squirrel monkey chorda tympani to a variety of taste stimuli. Thesis, Brown University, Providence, Rhode Island.

Weiskrantz, L. (1960). Effects of medial temporal lesions on taste preference in the monkey. *Nature (London)* **187,** 879.

DISCUSSION

Pfaffmann: The experiments pinpointed pretty well the thing that has turned up sporadically in other experiments, i.e., the role of the synthetic sweeteners or nonsugar stimuli in these experiments with a sensitivity closest to man and closest to man in its evolutionary relationship. It poses some interesting questions about the nature of the sugar receptors as to just what is the nature of the difference in such lower forms as, for example, hamsters or any other rodent.

DeWys: In the experiment in which you looked at water consumption of the sweetened solution, there was a striking difference between saccharin and the other three. Could, or did, you analyze saccharin *vis-a-vis* the protein in looking at absorbants, and correlating that with the neurophysiological reports. It seems to me that on the water consumption three were the same and one was different, and it would be of great interest to know how that different one came out in some of the other kinds of experiments you did.

Sato: In neurophysiological experiments, we got good responses to saccharin above 0.3 m*M*, but ingestion of saccharin of 3 m*M* or 6 m*M* is below the water intake. Saccharin of less than 1 or 3 m*M* was preferred by the monkey, and this comes from the fact that saccharin taste is not quite the same as sucrose or aspartame. I mean that it may be slightly bitter when the concentration of saccharin is very low. As you know, it is difficult for us to measure the interaction between saccharin and protein because of a large absorbancy due to saccharin, so we did not make any measurements on saccharine effect in biochemical experiments.

Cagan: There is a critical technical aspect which I want to stress. Using the difference spectral technique for assessing a specific interaction of sugars and other sweeteners with these extracts is actually a very nonspecific technique. Some of us who work in the biochemistry of taste have tried to use that method and found it to be nonspecific. For example, you can obtain difference spectra in the ultraviolet region with many types of compounds added to many types of proteins, whether or not taste stimuli and taste receptors are involved. Data refuting the original studies of this type, based on refractive index changes, were published by Nofre and Sabadi (1972). Recently a paper by Ostretsova *et al.* (1975) refuted the findings of sugar interaction with taste tissue using the difference spectral technique. However, using binding of radioactive sugars, they were able to demonstrate taste-related binding. In summary, the technique employing spectral differences is indirect, nonspecific, and has not been substantiated in taste studies by other laboratories. A more direct approach is to use radioactively labeled stimulus molecules to measure these interactions.

Chapter 17

A Framework for the Mechanisms of Action of Special Taste Substances: The Example of Monosodium Glutamate

ROBERT H. CAGAN

Providing adequate nutrition to people involves a wide range of considerations. The flavor* of food is one area to which increased emphasis will have to be given, as highly nutritious but unpalatable materials are often not utilized to their fullest extent. The possible increased use of substances that can enhance flavor could serve, for example, to accentuate desirable flavor notes that may be already present in the foodstuffs or perhaps to enhance the flavor of another material that can be added in only a small amount, such as one that is limited by solubility.

The flavor enhancer monosodium glutamate† (MSG) is widely used in cookery, especially in Oriental cuisine. Despite its long history of usage, little if anything is understood of its mechanism of action. As a basis upon which to study the mechanism of action of MSG, the existing knowledge of interactions of other chemical stimuli with sensory receptors may provide a useful starting point. This framework could possibly also be

* Flavor is a term used to denote the complex information deriving from taste, olfaction, touch, temperature, and mild pain.

† The free acid of this compound is L-glutamic acid.

useful in studying other types of special taste substances, e.g., taste modifiers such as gymnemic acid and miraculin (Kurihara, 1971). It is at present not clear whether MSG acts in some manner as a taste modifier in addition to its effect as a taste stimulus. The details of the interaction between a taste stimulus molecule and a receptor site (Cagan, 1974) are only beginning to be understood. Recent biochemical findings (Krueger and Cagan, 1976) point to the initial event in taste sensation being a weak reversible binding interaction at the surface of the receptor cell. This has long been postulated based upon electrophysiological recordings (Beidler, 1962). The transductive events subsequent to the binding interaction, which intervene between it and generation of the nerve impulses, are even less well known. For example, it is not known if the processes which lead to activation of the primary afferent nerve fiber occur strictly through membrane events or whether intracellular metabolic events also occur between the initial membrane stimulation and subsequent membrane depolarization. The first synaptic transfer occurs at the level of the taste receptor cell and the primary afferent nerve. Later events, including processing in the central nervous system, are also not yet well understood (Beidler, 1971).

Details of the receptor and neural processes need further clarification to provide a more complete picture of the cellular and molecular events upon which studies of mechanisms of special taste substances could build. The approach in this paper is to examine the existing body of information about MSG and to discuss the various levels in the taste system in terms of where MSG might have an effect: (1) receptor site interactions; (2) transduction processes within receptor cells; (3) peripheral synaptic transmission; (4) neural transmission and central processing.

TASTE EFFECTS OF MONOSODIUM GLUTAMATE

L-Glutamic acid is widely consumed, being a constituent of proteins. It also occurs as the free amino acid (0.01–0.2%) in a variety of fresh and cooked fresh vegetables and in meats and seafood (Hac *et al.*, 1949; Maeda *et al.*, 1958, 1961). Milk from lactating human females contains an appreciable amount of free L-glutamate (78–208 μmoles/100 ml) (Stegink *et al.*, 1972), in addition to the amount present as a constituent of the proteins.

MSG was discovered to be present in a material used in Japanese cuisine which had long been appreciated for its special flavor characteristics. This was the seaweed, sea tangle (*Laminaria* sp.). In addition, soya sauce was found to contain glutamate (ammonium salt). Soon after its isolation and identification, MSG entered commercial production in the early 1900's. The history of use and study of MSG was reviewed by several authors (Marshall, 1948; Crocker and Sjöström, 1948; Amerine *et al.*, 1965; Kuninaka, 1967; Oeda, 1963), including discussions of the important early contributions of Ikeda (1908, 1909, 1912).

The sensory effects of MSG are especially interesting because it has been described both as a distinctive taste stimulus and as a flavor enhancer; its precise sensory effects, however, have still not been rigorously defined. The designations used for the taste of MSG itself generally place it into a separate taste category. This special flavor of MSG is reflected by the terminology in English-language publications, where it is referred to as "distinctive" (Mosel and Kantrowitz, 1952), "distinct taste" (Amerine *et al.*, 1965), or "unique" (e.g., Solms, 1969; Solms *et al.*, 1965; Mosel and Kantrowitz, 1954), and in Japanese studies where it is assigned to a special category of *umami* (Yoshida and Saito, 1969; Yamaguchi *et al.*, 1971), which means "delicious taste" or "savory taste" (Komata, 1976). The taste of certain other amino acids has been referred to as including an "MSG-like" taste (Kirimura *et al.*, 1969). The taste of MSG was also evaluated using a "like–dislike" rating; substantial individual variation was reported in that study (Pangborn, 1970).

Among the amino acids, some D-enantiomers taste sweet while their L-enantiomers are tasteless or bitter. For example, of the common amino acids, four of these have D-isomers which taste sweet and L-isomers which taste bitter (leucine, phenylalanine, tryptophan, and tyrosine) (Kaneko, 1938, 1939; Solms *et al.*, 1965; Solms, 1969). Stereospecificity appears to apply also to the activity of MSG. It seems generally accepted that the L-isomer of glutamic acid is the effective form as a flavor enhancer (e.g., Manning and Buchanan, 1948). The data of Kaneko (1963) provide an experimental basis for this. Reference has also been made (Marshall, 1948) to the unsuitability of manufacturing processes that resulted in racemization. Other evidence (Fagerson, 1954) suggests that only the form of glutamate with both carboxyl groups ionized is active as a taste substance. This suggestion was based on a comparison of the titration behavior of L-glutamate with the taste effects of MSG; the most pronounced taste effects occur between pH 5.5 and 8.0.

MSG is itself a taste stimulus. Its taste threshold for humans is in the millimolar concentration range (Table I). Electrophysiological measurements in rats show that MSG is effective in this range of concentration for stimulating the chorda tympani nerve (Sato and Akaike, 1965).

The tastes of different amino acids were evaluated in multidimensional scaling studies. In these studies (Ninomiya *et al.*, 1965, 1966, as cited by Yoshida and Saito, 1969), the tastes of various amino acids were described

TABLE I

Taste Threshold for Monosodium Glutamate (MSG)

Investigators	No. of subjects	Taste threshold (m*M*)
Knowles and Johnson (1941)	33	1.2[a]
Lockhart and Gainer (1950)	8	1.5
Mosel and Kantrowitz	4	0.95
(1954)	38	0.75

[a] Glutamic acid rather than MSG was used. Yoshida *et al.* (1966) also determined taste thresholds for a number of amino acids.

by the subjects in terms of sweet, salty, sour, bitter, and *umami* or "delicious taste." MSG showed by far the highest rating of *umami* (71.4% of the taste), and some saltiness (13.3%) and sweetness (9.8%) as well. Nearest in *umami* was glutamic acid (24.8%). Yoshida and Saito (1969) reanalyzed the published data of Ninomiya *et al.* (1965, 1966) and extended the results by evaluating subjective similarities among amino acids determined directly. Graphic representation of the earlier results revealed that MSG fell into the "space" separately from the other groupings but near the sour category. The results from the similarity ratings appeared more difficult to interpret; in this case MSG showed no clear-cut "space," leading the authors to suggest that it has a "unique dimension of taste" (Yoshida and Saito, 1969).

MSG AS A FLAVOR ENHANCER

Considerable effort has been directed toward studies of the effects of MSG on the flavors of various foods (e.g., Girardot and Peryam, 1954; Hanson *et al.*, 1960; Sjöström and Crocker, 1948, Cairncross, 1948; Nair, 1948). These reports describe a favorable effect of MSG on the flavors of many foods, including meats, poultry, seafoods, and some soups and vegetables but not, for example, on fruits and fruit juices or on cereals. MSG was claimed (Cairncross, 1948) to enhance sweetness in foods that are less than optimally sweet, but other results (Girardot and Peryam, 1954) showed no improvement of the sweet foods included in the study. In general, predicting which specific foods will be improved by MSG appears not to have a firm basis. A simple interpretation, such as an effect due to the slight saltiness contributed by the Na^+ in MSG (Yoshida and Saito, 1969; Bartoshuk *et al.*, 1974), does not appear from the various reports able to account for the changes in sensory quality.

Studies with foods have also shown that MSG and certain 5′-ribonucleotides together have a synergistic effect on enhancing food flavors (Sakaguchi, 1961; Kuninaka *et al.*, 1964; Kuninaka, 1967; Yamaguchi, 1967; Yamaguchi *et al.*, 1971; Komata, 1976). It appears likely that the sensory effect of added MSG is actually due, at least in part, to this synergism with ribonucleotides that are already present in the food. To understand the mechanism of action of MSG we may actually need to understand the synergistic action of MSG and ribonucleotides, rather than the effect of MSG alone.

Few animal studies with MSG in the diet have been reported. Scott and Quint (1946) found that 1% MSG in the diet did not influence the food preference of rats. This lack of effect was recently confirmed, but it was demonstrated that rats preferred solutions of MSG to water (Naim *et al.*, 1977). Preferences by rats were greater for solutions of L-MSG than for D-MSG paired with water (Hiji and Sato, 1967). Waldern and Van Dyk (1971) found that weanling calves ate more of a 0.2% MSG-containing diet beginning at 3 weeks of age and continuing through the sixth week. A recent study (Mehren and Church, 1976) with pigmy goats suggested that MSG reduced their taste activity for the two compounds tested, sucrose and quinine. Preliminary reports described effects of MSG added to diets fed to pigs (Henson *et al.*, 1962; Klay, 1964).

Foods generally have complex flavors which are difficult to examine in a manner suited to define specific taste-enhancing effects of MSG. The few studies in which MSG has been used with pure individual taste stimuli are reviewed below (Table II). In studies by Lockhart and Gainer (1950), the taste threshold of MSG was determined to be around 1.5 m*M* (see Table I). The effect of subliminal concentrations of MSG (0.5 and 1 m*M*) did not affect the intensity level of sweetness of sucrose (36–52 m*M*) or saltiness of NaCl (23 m*M*) (Table II). Mosel and Kantrowitz (1952) reported that MSG (4 m*M*) had no effect on threshold sensitivity to salty (NaCl) or sweet (sucrose), but it did increase sensitivity (i.e., lowered the value of the taste threshold) to sour (tartaric acid), and it greatly increased sensitivity to bitter (quinine). However, the results are difficult to interpret because of the experimental design in which pretreatment with MSG was studied for its subsequent effect on taste. First, they determined the taste thresholds of the subjects for the salty, sweet, bitter, and sour stimuli. The subjects then rinsed with MSG (4 m*M*) for 7 sec and rinsed again with water. Their taste thresholds to the four stimuli were reevaluated 5 min later. Unless it can be shown that MSG persists at the site of action (which

TABLE II

MSG as a Taste Modifier with Solutions of Individual Taste Stimuli

Investigators	Procedure	Findings
Lockhart and Gainer (1950)	Mixtures	Subliminal concentrations of MSG (0.5 and 1 m*M*) did not affect intensity of sweetness (sucrose 36–52 m*M*) or saltiness (NaCl, 23 m*M*)
Mosel and Kantrowitz (1952)	Pretreatment	MSG (4 m*M*) had no effect on threshold sensitivity to salty (NaCl) or sweet (sucrose), and increased sensitivity to sour (tartaric acid) and bitter (quinine)
Pilgrim *et al.* (1955)	Mixtures	MSG (12 m*M*) increased intensity of saltiness (NaCl) and bitterness (quinine), but did not affect intensity of sweetness (sucrose) or sourness (HCl)
	Pretreatment	MSG (4 m*M*) decreased threshold sensitivity to sweet (sucrose) and sour (HCl)
		MSG (10 m*M*) had no effect on differential sensitivity to salty (NaCl) or bitter (caffeine)

appears to be commonly assumed), they may not have actually studied the effect of MSG as a possible taste modifier in a manner comparable to the situation with MSG present in foods. An effect in this case is also difficult to distinguish from that due to adaptation, unrelated to the possibility of a specific taste-modifier action. Pilgrim *et al.* (1955) found that MSG (12 m*M*) added to solutions of taste stimuli increased the subjective intensity of saltiness (NaCl) and bitterness (quinine), but did not affect sweetness (sucrose) or sourness (HCl). In order to determine the effect of MSG on threshold sensitivity to taste stimuli, they used a prior rinse with 4 m*M* MSG as had Mosel and Kantrowitz (1952). Threshold sensitivity to sweet and sour decreased. Prior rinsing with 10 m*M* MSG had no effect on differential thresholds for salty (NaCl) and bitter (caffeine). Their results generally did not confirm, and in one case (sour) were opposite to, those of Mosel and Kantrowitz (1952). The results of these investigations (Table II) therefore do not support a role of MSG as a general taste modifier which affects sensory responses to all other taste stimuli.

That MSG affects the flavors of many foods (Girardot and Peryam, 1954; Hanson *et al.*, 1960; Sjöström and Crocker, 1948; Cairncross, 1948; Nair, 1948) seems undisputed, and it has led to large commercial usage of this substance. It appears that a clear description of the effects of MSG on the sense of taste for other specific stimuli is still not well established. This area could need reinvestigation with more rigorous approaches of psychological testing ("psychophysics"), which would be helpful in providing a basis for studies of the mechanism of action of MSG as a flavor enhancer.

Electrophysiological experiments with rats tend to confirm some of the observations in humans. MSG is a taste stimulus; in addition, a synergistic effect occurs when a mixture of MSG and ribonucleotide is used as the stimulus (Sato and Akaike, 1965; Adachi *et al.*, 1967b). This synergistic effect has also been demonstrated electrophysiologically in cats (Kawamura *et al.*, 1964; Kawamura and Adachi, 1965; Adachi *et al.*, 1967a). Sato and Akaike (1965) investigated the effects of various ribonucleotides added to MSG, and found that 5′-GMP, 5′-IMP, 5′-UMP, and 5′-CMP each exerted a synergistic effect with MSG on the rat chorda tympani nerve response. The pyrimidine nucleotides, 5′-UMP and 5′-CMP, however, are apparently not effective flavor enhancers in humans (Kuninaka *et al.*, 1964), a difference in results which is currently not explained. Electrophysiological studies in cats (Kawamura *et al.*, 1964; Adachi, 1964; Adachi *et al.*, 1967a) showed some effects of MSG in mixtures with standard taste stimuli. The electrophysiological studies were extended to recordings from single fibers in the rat chorda tympani nerve (Sato *et al.*, 1970; Yamashita *et al.*, 1973). In common with response patterns to other stimuli, fibers were not found that were *exclusively* responsive to MSG. An interesting observation was that mixtures of MSG and 5′-ribonucleotides tended to be most stimulatory when recording from fibers that were most responsive to sucrose, but not from those most responsive to NaCl. A single fiber study in cats, however, reported the existence of a particular fiber type specific for MSG or IMP (Adachi, 1962), and additional single fiber studies in cats measured the responses of salt fibers to MSG and ribonucleotides (Kawamura and Adachi, 1965).

POSSIBLE SITES OF ACTION OF MONOSODIUM GLUTAMATE

Without a clear description of the specific effects of MSG on the tastes of individual taste stimuli, it is at best perilous to formulate detailed hypotheses for possible mechanisms of action of MSG. It would be possible with existing psychological testing procedures to determine with certainty whether MSG affects either the sensitivity (taste threshold) to particular stimuli or affects the slope of the stimulus–response relationship. The well-established synergism of MSG and the 5′-ribonucleotides, however, may offer an important clue to a unique action on the taste system. It is suggested that this synergistic effect be exploited as an important criterion in studying the mechanistic basis of the effect of MSG.

As a starting point, it is perhaps useful to suggest some of the possible *sites* of action which could be focal points for studies of MSG mechanisms.

Receptor Site Interactions

Binding Interaction of a Stimulus Molecule with Its Receptor Site

The effect of MSG on the *affinities* of particular taste stimulus molecules for their receptor sites is not known. Taste thresholds deduced from psychophysical studies would provide a guide for such an approach. Taste thresholds to each taste stimulus should be determined both in the absence of MSG and in the presence of a low level (e.g., 0.5 m*M*) of MSG. The results of the earlier data of effects of MSG on threshold sensitivities (see Table II) do not support the hypothesis of a general effect of MSG on taste threshold sensitivity. Although the available data are not conclusive, a consistent change in affinity of taste receptor sites appears unlikely to explain the taste effects of MSG. It would be of interest, however, to determine taste thresholds for various taste stimuli in the presence of combinations of MSG plus 5′-ribonucleotides to determine if the combination is effective as a taste-modifying agent.

The limited psychological information currently available suggests that the effects of MSG may not be on affinity constants of receptor sites for other types of taste stimuli.

In addition to that possibility, the affinity of the binding interaction of MSG itself as a taste stimulus molecule is an important point to establish. Enhancement of this affinity by 5′-ribonucleotides would thus offer a mechanistic basis for the synergistic taste effect observed with mixtures of MSG and 5′-ribonucleotides.

Accessibility of Receptor Sites

To distinguish accessibility from affinity (above), attention is called to the possibility that MSG (or MSG and ribonucleotide) might affect the *number* of receptor sites available for interacting with particular taste stimuli, rather than affecting the affinity of a particular site for its stimulus. This could suggest that some receptor sites exist which are normally "buried" and are made "accessible" by the presence of MSG, or that MSG and ribonucleotides facilitate accessibility of each other to the receptors. This concept may be related to the hypothesis of "spare receptors'' in pharmacological research (Van Rossum and Ariëns, 1962). In addition, such a mechanism would likely involve a cooperative interaction. The isolated taste cell preparation described recently (Brand and Cagan, 1976) might be useful for studies of the affinity and accessibility of receptor sites.

Transduction Processes within Receptor Cells

The transduction processes in taste sensation are virtually unknown. For example, the possibility that cyclic AMP could serve as a "second messenger" (Robison *et al.*, 1971) has not received experimental support in the taste system (Cagan, 1976). Other possible intermediates, such as cyclic GMP, have not been studied.

The centrality of glutamate in intermediary metabolism might be considered in terms of a possible involvement of MSG action. Glutamate is involved in many key biochemical steps, such as the conversion of glutamate to α-ketoglutarate by glutamic dehydrogenase (EC 1.4.1.3), the reaction of glutamate with pyruvate to form alanine and α-ketoglutarate, which is catalyzed by glutamic-alanine transaminase (EC 2.6.1.2), the reaction of glutamate with oxaloacetate catalyzed by glutamic-aspartic transaminase (EC 2.6.1.1), and the decarboxylation of glutamate to γ-aminobutyrate (EC 4.1.1.15). Because of the lack of information on the enzymology and control of metabolic pathways in taste receptor cells, it is not possible at present to assess these as possible sites of action for a taste-modifier effect of MSG.

It is also possible that membrane events alone are responsible for converting the initial stimulus–receptor interaction to depolarization of the membrane; any metabolic perturbations in this case would be simply epiphenomena. There are currently few guides, with none of the biochemical processes involved in transduction in taste receptors known.

Peripheral Synaptic Transmission

The initial synapse in the taste system is at the peripheral receptor cell in the taste bud. This is unlike the olfactory system, in which the receptor cell is a bipolar neuron, with the first synapse occurring in the olfactory bulb of the brain. In the mammalian taste system, although morphology suggestive of synapses is observed, the existence of classic synaptic structures between taste cells and nerve fibers seems questionable (e.g., Farbman, 1965; Gray and Watkins, 1965; Murray, 1973); the term "neurogustatory junction" is therefore proposed here for this connecting point.

Electrophysiological studies (Taglietti *et al.*, 1969; Miller, 1971) showed that several neighboring taste papillae are interconnected and that neural activity originating by stimulating one papilla can be influenced when the nearby interconnected papillae are also stimulated. Thus, the initial taste information is being altered at the periphery of the taste system.

In addition to the connections among neighboring papillae, there is anatomic evidence showing that extensive branching of nerve fibers occurs both below and within individual papillae (Beidler, 1969). Of the nerve fibers which branch within the papilla, some branches terminate without innervating taste buds. The rat fungiform papilla contains but a single taste bud, and approximately 25% of the nerve fibers in the papilla enter this taste bud. Individual cells within the taste bud generally make contact with many nerve fibers. From measurements made on the content of individual fibers in the chorda tympani nerve, it was concluded that the individual nerve fibers also branch extensively prior to entering the papilla. Thus an effect of MSG at the level of the initial neurogustatory junction could have a profound effect on processing of taste information. An effect at the level of the neurogustatory junction would offer the possibility for subtle and finely-tuned sensory effects.

Most of the evidence pertaining to neurotransmitters in the taste bud is histochemical. The results of Ellis (1959) demonstrated that acetylcholinesterase is associated with the nerve fibers which innervate the lingual papillae and is in the taste buds. Five primate species and several other mammalian species were examined in the study, which might suggest but did not directly demonstrate a role for acetylcholine within the periphery of the taste system. This enzyme activity was also reported (El-Rakhawy and Bourne, 1961) in human taste buds. Vij *et al.* (1972) showed that transecting the taste nerve resulted in loss of this enzymatic activity in the taste papilla. Recently, Paran and Mattern (1975) and others used electron microscopy to localize the cholinesterase activity. They confirmed in the rat the localization of acetylcholinesterase activity associated with nerves both beneath and within the taste bud, and also reported activity in the taste pore region between the microvilli. The results of Duncan (1964), however, cast doubt on the involvement of acetylcholine in transmission in the afferent nerves. Perfusion of the frog tongue with inhibitors of the esterase failed to inhibit activity in the nerve upon stimulation of the tongue with taste stimuli. He concluded that cholinergic transmission was unlikely, in contrast to an earlier suggestion that acetylcholine was involved in taste receptor transmission (Landgren *et al.*, 1954).

Gabella (1969) used histochemical methods to show the presence of adrenergic fibers underlying the taste buds of the rabbit, although he did not find them entering the taste buds. This was also observed in the mouse (Geerdink and Drukker, 1973). Histochemical studies in the frog claimed to show catecholamines within taste cells (DeHan and Graziadei, 1973; Savushkina *et al.*, 1974), with support from electrophysiological studies in the frog recently appearing (Morimoto and Sato, 1975). In mammals, however, catecholamine fluorescence was not observed in the taste buds (Gabella, 1969; Geerdink and Drukker, 1973), although uptake into the taste buds of the mouse of administered L-dopa was reported (Geerdink and Drukker, 1973). Recently, evidence from fluorescence histochemistry led Nada and Hirata (1975) to suggest that certain taste bud cells in the rabbit contain 5-hydroxytryptamine (serotonin).

Detailed understanding of neurotransmitters in the taste system is unfortunately lacking, and new data in this area could have an important bearing upon experimental studies to understand the action of MSG. The intriguing possibility of a relationship between amino acids as neurotransmitters and the activity of MSG as a possible taste modifier also arises in this connection, and should be considered in formulating an experimental approach.

Neural Transmission and Central Processing

Information generated at the periphery of the taste system is transmitted via nerve fibers to the brain. Specific patterns of action potentials ("coding") that may be involved in transmitting a message to the brain are still not clearly established. Two major hypotheses under consideration are "across-fiber patterning" (Erickson, 1963) and "labeled lines," although the two hypotheses may not be mutually exclusive (Pfaffmann, 1974). Recordings from single nerve fibers in response to stimulating the tongue with various stimuli are necessary for this analysis, and a number of investigators have reported such data. The effect of MSG on the responses of the nerves to various taste stimuli has not been systematically examined (see section on MSG as a Flavor Enhancer). Sato *et al.* (1970) studied single-fiber responses in the rat chorda tympani nerve to MSG and ribonucleotides, and found that the response to MSG alone was correlated with that to NaCl, while the response to MSG plus ribonucleotides correlated with that to sucrose.

The possibility must be considered that neural information due to MSG is transmitted to the brain, where it then affects neural processing of other specific taste information, presumably via synaptic facilitation or inhibition. Whether this would require a "labeled line" for MSG cannot be definitely stated. The single fibers of the rat chorda tympani nerve responding to mixtures of MSG and ribonucleotides correlated with those responding to sucrose; the highest correlations observed were for these stimuli. The potentiation of the response to MSG caused by ribonucleotides was observed in recordings from both the whole chorda tympani nerve (Sato and Akaike, 1965) and single fibers (Sato *et al.,* 1970), evidence which suggests that peripheral processing occurs, although it does not rule out effects on central processing in addition.

Effect on Volatile Stimulus Concentration

Still another possible site of action does not involve the taste pathway. If MSG affects the stimuli in the volatile fraction of a food, then the olfactory properties (and hence the flavor) of the food could thereby be influenced. This possibility should be further investigated, particularly in view of a report by Maga and Lorenz (1972) that 3 m*M* MSG added to a beef broth increased the amount of volatiles in the vapor phase measured by gas chromatography, although it should be considered as a secondary type of MSG action.

SUMMARY

Monosodium glutamate (MSG), which occurs widely in foods, is also added to many foods to enhance their flavor. MSG is itself a taste stimulus, being effective at millimolar concentration levels, and it evokes a taste sensation described as "unique," "distinct," and in Japanese *umami,* which means "delicious taste" or "savory taste." Its effect is synergistic when present in mixtures with certain 5′-ribonucleotides, such as 5′-GMP and 5′-IMP.

Despite the long history of usage of MSG, virtually nothing is known of the mechanisms underlying its taste activity. Several loci within the taste system are considered in relation to the possible site of action of MSG. These are the following: (1) receptor site interactions; (2) transduction processes within receptor cells; (3) peripheral synaptic transmission; (4) neural transmission and central processing. It is suggested that receptor site interactions and peripheral synaptic transmission are two areas that appear promising as potential sites of action of MSG, and are ones which are also experimentally amenable to investigation.

ACKNOWLEDGMENTS

Dr. Carol Christensen of the Monell Center provided very helpful constructive criticism of this manuscript, for which I am most appreciative. I thank Dr. Yoshi-hisa Sugita and other members of the staff of Ajinomoto, Inc. for kindly providing a number of references in the Japanese literature and English translations of several articles.

REFERENCES

Adachi, A. (1962). Study of the taste mechanisms of sodium glutamate and sodium inosinate. *J. Physiol. Soc. Jpn.* **24,** 607–613 (in Japanese with English summary).

Adachi, A. (1964). Neurophysiological study on taste effectiveness of seasoning. *J. Physiol. Soc. Jpn.* **26,** 347–355 (in Japanese with English summary).

Adachi, A., Funakoshi, M., and Kawamura, Y. (1967a). Neurophysiological studies on taste effectiveness of chemical taste enhancers. *Wenner-Gren Cent. Int. Symp. Ser.* **8,** 411–413.

Adachi, A., Okamoto, J., Hamada, T., and Kawamura, Y. (1967b). Taste effectiveness of mixtures of sodium 5′-inosinate and various amino acids. *J. Physiol. Soc. Jpn.* **29,** 65–71 (in Japanese with English summary).

Amerine, M. A., Pangborn, R. M., and Roessler, E. B. (1965). "Principles of Sensory Evaluation of Food," pp. 115–120. Academic Press, New York.

Bartoshuk, L. M., Cain, W. S., Cleveland, C. T., Grossman, L. S., Marks, L. E., Stevens, J. C., and Stolwijk, J. A. J. (1974). Saltiness of monosodium glutamate and sodium intake. *J. Am. Med. Assoc.* **230,** 670.

Beidler, L. M. (1962). Taste receptor stimulation. *Prog. Biophys. Biophys. Chem.* **12,** 107–151.

Beidler, L. M. (1969). Innervation of rat fungiform papilla. *Olfaction Taste, Proc. Int. Symp., 3rd,* 1968 pp. 352–369.

Beidler, L. M., ed. (1971). "Handbook of Sensory Physiology," Vol. 4, Part 2. Springer-Verlag, Berlin and New York.

Brand, J. G., and Cagan, R. H. (1976). Biochemical studies of taste sensation. III. Preparation of a suspension of bovine taste bud cells and their labeling with a fluorescent probe. *J. Neurobiol.* **7,** 205–220.

Cagan, R. H. (1974). Biochemistry of sweet sensation. *In* "Sugars in Nutrition" (H. L. Sipple and K. W. McNutt, eds.), pp. 19–36. Academic Press, New York.

Cagan, R. H. (1976). Biochemical studies of taste sensation. II. Labeling of cyclic AMP of bovine taste papillae in response to sweet and bitter stimuli, *J. Neurosci. Res.* **2** (in press).

Cairncross, S. E. (1948). Effect of monosodium glutamate on food flavor. *In* "Flavor and Acceptability of Monosodium Glutamate," pp. 32–38. Quartermaster Food and Container Institute for the Armed Forces, and Associates, Food and Container Institute, Chicago, Illinois.

Crocker, E. C., and Sjöström, L. B. (1948). Observational study of monosodium glutamate. *Food Res.* **13,** 450–455.

DeHan, R. S., and Graziadei, P. (1973). The innervation of frog's taste organ. "A histochemical study." *Life Sci.* **13,** 1435–1449.

Duncan, C. J. (1964). Synaptic transmission at taste buds. *Nature (London)* **203,** 875–876.

Ellis, R. A. (1959). Cholinesterases in the mammalian tongue. *J. Histochem. Cytochem.* **7,** 156–163.

El-Rakhawy, M. T., and Bourne, G. H. (1961). Cholinesterases in the human tongue. *Bibl. Anat.* **2,** 243–255.

Erickson, R. P. (1963). Sensory neural patterns and gustation. *Wenner-Gren Cent. Int. Symp. Ser.* **1,** 205–213.

Fagerson, I. S. (1954). Possible relationship between the ionic species of glutamate and flavor. *J. Agric. Food Chem.* **2,** 474–476.

Farbman, A. I. (1965). Fine structure of the taste bud. *J. Ultrastruct. Res.* **12,** 328–350.

Gabella, G. (1969). Taste buds and adrenergic fibers. *J. Neurol. Sci.* **9,** 237–242.

Geerdink, H. G., and Drukker, J. (1973). Uptake of L-dopa by cells in the taste buds of the vallate papilla of the mouse. *Histochemie* **36,** 219–223.

Girardot, N. F., and Peryam, D. R. (1954). MSG's power to perk up foods. *Food Eng.* **26,** 71–72 and 182, 185.

Gray, E. G., and Watkins, K. C. (1965). Electron microscopy of taste buds of the rat. *Z. Zellforsch. Mikrosk. Anat.* **66,** 583–595.

Hac, L. R., Long, M. L., and Blish, M. J. (1949). The occurrence of free L-glutamic acid in various foods. *Food Technol.* **3,** 351–354.

Hanson, H. L., Brushway, M. J., and Lineweaver, H. (1960). Monosodium glutamate studies. I. Factors affecting detection of and preference for added glutamate in foods. *Food Technol.* **14,** 320–327.

Henson, J. N., Bogdonoff, P. D., and Thrasher, G. W. (1962). Levels of monosodium glutamate in pig starter preference. *J. Anim. Sci.* **21,** 999–1000 (abstract).

Hiji, Y., and Sato, M. (1967). Preference-aversion function for sodium monoaminodicarboxylates in rats. *J. Physiol. Soc. Jpn.* **29,** 168–169.

Ikeda, K. (1908). The method for the preparation of seasoner which has glutamate as a main component. Japanese Patent 14,805 (in Japanese) (as cited by Kuninaka *et al.,* 1964).

Ikeda, K. (1909). On a new seasoning. *J. Tokyo Chem. Soc.* **30,** 820–836 (as cited by Kuninaka, 1967, and by Kuninaka *et al.,* 1964) (in Japanese).

Ikeda, K. (1912). On the taste of the salt of glutamic acid. *Orig. Comm., Int. Congr. Appl. Chem., 8th, 1912* Vol. 18, p. 147 (abstract).

Kaneko, T. (1938). The relationship between taste and structure of α-amino acids. I. Tyrosine and its derivatives. *J. Chem. Soc. Jpn.* **59,** 433–439 (in Japanese).

Kaneko, T. (1939). The relationship between taste and structure of α-amino acids. II. Stereochemistry of α-amino acids. *J. Chem. Soc. Jpn.* **60,** 531–538 (in Japanese).

Kaneko, T. (1963). Stereochemical configuration and its relationship to physiological properties. *Kagaku (Kyoto)* **18,** 947–956 (in Japanese).

Kawamura, Y., and Adachi, A. (1965). Single taste nerve responses to the chemical taste enhancers. *J. Physiol. Soc. Jpn.* **27,** 279–284 (in Japanese with English summary).

Kawamura, Y., Adachi, A., Ohara, M., and Ikeda, S. (1964). Neurophysiological studies on taste effectiveness of flavor enhancing substances. *Symp. Amino Acid Nucleic Acid* **10,** 168–178 (in Japanese) (as cited by Kuninaka, 1967).

Kirimura, J., Shimizu, A., Kimizuka, A., Ninomiya, T., and Katsuya, N. (1969). The contribution of peptides and amino acids to the taste of foodstuffs. *J. Agric. Food Chem.* **17,** 689–695.

Klay, R. F. (1964). Monosodium glutamate in pig creep rations. *J. Anim. Sci.* **23,** 598 (abstract).

Knowles, D., and Johnson, P. E. (1941). A study of the sensitiveness of prospective food judges to the primary tastes. *Food Res.* **6,** 207–216.

Komata, Y. (1976). Utilization of nucleic acid-related substances—utilization in foods. *In* "Microbial Production of Nucleic Acid-Related Substances" (K. Ogata, S. Kinoshita, T. Tsunoda, and K. Aida, eds.), pp. 299–319. Wiley, New York.

Krueger, J. M., and Cagan, R. H. (1976). Biochemical studies of taste sensation. 4. Binding of L-[^{3}H]alanine to a sedimentable fraction from catfish barbel epithelium. *J. Biol. Chem.* **251,** 88–97.

Kuninaka, A. (1967). Flavor potentiators. *In* "Symposium on Foods: The Chemistry and Physiology of Flavors" (H. W. Schultz, E. A. Day, and L. M. Libbey, eds.), pp. 515–535. Avi Publ., Westport, Connecticut.

Kuninaka, A., Kibi, M., and Sakaguchi, K. (1964). History and development of flavor nucleotides. *Food Technol.* **18,** 287–293.

Kurihara, K. (1971). Taste modifiers. *Handb. Sensory Physiol.* **4,** Part 2, 363–378.

Landgren, S., Liljestrand, G., and Zotterman, Y. (1954). Chemical transmission in taste fibre endings. *Acta Physiol. Scand.* **30,** 105–114.

Lockhart, E. E., and Gainer, J. M. (1950). Effect of monosodium glutamate on taste of pure sucrose and sodium chloride. *Food Res.* **15,** 459–464.

Maeda, S., Eguchi, S., and Sasaki, H. (1958). The content of free L-glutamic acid in various foods. *J. Home Econ. (Tokyo)* **9,** 163–167 (in Japanese).

Maeda, S., Eguchi, S., and Sasaki, H. (1961). The free L-glutamic acid content of foods. *J. Home Econ. (Tokyo)* **12,** 105–106 (in Japanese).

Maga, J. A., and Lorenz, K. (1972). The effect of flavor enhancers on direct headspace gas-liquid chromatography profiles of beef broth. *J. Food Sci.* **37,** 963–964.

Manning, P. D. V., and Buchanan, B. F. (1948). Quality production of glutamate. *In* "Flavor and Acceptability of Monosodium Glutamate," pp. 15–25. Quartermaster Food and Container Institute for the Armed Forces, and Associates, Food and Container Institute, Chicago, Illinois.

Marshall, A. E. (1948). History of glutamate manufacture. *In* "Flavor and Acceptability of Monosodium Glutamate," pp. 4–14. Quartermaster Food and Container Institute for the Armed Forces, and Associates, Food and Container Institute, Chicago, Illinois.

Mehren, M. J., and Church, D. C. (1976). Influence of taste-modifiers on taste responses of pygmy goats. *Anim. Prod.* **22,** 255–260.

Miller, I. J., Jr. (1971). Peripheral interactions among single papilla inputs to gustatory nerve fibers. *J. Gen. Physiol.* **57,** 1–25.

Morimoto, K., and Sato, M. (1975). Noradrenaline as a chemical transmitter from taste cells to sensory nerve terminals in frog. *Proc. Jpn. Acad.* **51,** 347–352.

Mosel, J. N., and Kantrowitz, G. (1952). The effect of monosodium glutamate on acuity to the primary tastes. *Am. J. Psychol.* **65,** 573–579.

Mosel, J. N., and Kantrowitz, G. (1954). Absolute sensitivity to the glutamic taste. *J. Gen. Psychol.* **51,** 11–18.

Murray, R. G. (1973). The ultrastructure of taste buds. *In* "The Ultrastructure of Sensory Organs" (I. Friedmann, ed.), pp. 1–81. Am. Elsevier, New York.

Nada, O., and Hirata, K. (1975). The occurrence of the cell type containing a specific monoamine in the taste bud of the rabbit's foliate papilla. *Histochemistry* **43,** 237–240.

Naim, M., Ohara, I., and Kare, M. R. (1977). Acceptability of monosodium glutamate by the rat. *Fed. Proc., Fed. Am. Soc. Exp. Biol.* **36,** 1132 (abstract).

Nair, J. H. (1948). Evaluation of glutamate in food specialties. *In* "Flavor and Acceptability of Monosodium Glutamate," pp. 48–53. Quartermaster Food and Container Institute for the Armed Forces, and Associates, Food and Container Institute, Chicago, Illinois.

Ninomiya, T., Ikeda, S., Yamaguchi, S., and Yoshikawa, T. (1965). Studies on the taste of amino acids. *Rep. Sens. Eval. Symp., Jpn. Union Sci. Eng., 6th, 1965* pp. 16–37 (as cited by Yoshida and Saito, 1969) (in Japanese).

Ninomiya, T., Ikeda, S., Yamaguchi, S., and Yoshikawa, T. (1966). Studies on the taste of various amino acids. *Rep. Sens. Eval. Symp., Jpn. Union Sci. Eng., 7th, 1966* pp. 109–123 (in Japanese) (as cited by Yoshida and Saito, 1969).

Oeda, H. (1963). Monosodium glutamate. *Kirk-Othmer Encycl. Chem. Technol., 2nd Ed.* Vol. 2, pp. 198–212.

Pangborn, R. M. (1970). Individual variation in affective responses to taste stimuli. *Psychon. Sci.* **21,** 125–126.

Paran, N., and Mattern, C. F. T. (1975). The distribution of acetylcholinesterase in buds of the rat vallate papilla as determined by electron microscopic histochemistry. *J. Comp. Neurol.* **159,** 29–43.

Pfaffmann, C. (1974). Specificity of the sweet receptors of the squirrel monkey. *Chem. Senses Flavor* **1,** 61–67.

Pilgrim, F. J., Schutz, H. G., and Peryam, D. R. (1955). Influence of monosodium glutamate on taste perception. *Food Res.* **20,** 310–314.

Robison, G. A., Butcher, R. W., and Sutherland, E. W. (1971). "Cyclic AMP." Academic Press, New York.

Sakaguchi, K. (1961). "Outline and Characteristics of Japanese Fermentation Industries." Rikaguku Kenkyusho, Tokyo (as cited by Amerine *et al.,* 1965).

Sato, M., and Akaike, N. (1965). 5′-Ribonucleotides as gustatory stimuli in rats. Electrophysiological studies. *Jpn. J. Physiol.* **15,** 53–70.

Sato, M., Yamashita, S., and Ogawa, H. (1970). Potentiation of gustatory response to monosodium glutamate in rat chorda tympani fibers by addition of 5′-ribonucleotides. *Jpn. J. Physiol.* **20,** 444–464.

Savushkina, M. A., Krokhina, E. M., and Esakov, A. I. (1974). Investigation of the taste buds of the frog tongue by luminescence microscopy. *Byull. Eksp. Biol. Med.* **78,** 118–121 (translation).

Scott, E. M., and Quint, E. (1946). Self selection of diet. II. The effect of flavor. *J. Nutr.* **32,** 113–119.

Sjöström, L. B., and Crocker, E. C. (1948). The role of monosodium glutamate in the seasoning of certain vegetables. *Food Technol.* **2,** 317–321.

Solms, J. (1969). The taste of amino acids, peptides, and proteins. *J. Agric. Food Chem.* **17,** 686–688.

Solms, J., Vuataz, L., and Egli, R. H. (1965). The taste of L- and D-amino acids. *Experientia* **21,** 692–694.

Stegink, L. D., Filer, L. J., Jr., and Baker, G. L. (1972). Monosodium glutamate: Effect on plasma and breast milk amino acid levels in lactating women. *Proc. Soc. Exp. Biol. Med.* **140,** 836–841.

Taglietti, V., Casella, C., and Ferrari, E. (1969). Interactions between taste receptors in the frog tongue. *Pfluegers Arch.* **312,** 139–148.

Van Rossum, J. M., and Ariëns, E. J. (1962). Receptor-reserve and threshold-phenomena. II. Theories on drug-action and a quantitative approach to spare receptors and threshold values. *Arch. Int. Pharmacodyn. Ther.* **136,** 385–413.

Vij, S., Kanagasuntheram, R., and Krishnamurti, A. (1972). Enzymic changes in taste buds of monkey following transection of glossopharyngeal nerve. *J. Anat.* **113,** 425–432.

Waldern, D. E., and Van Dyk, R. D. (1971). Effect of monosodium glutamate in starter rations on feed consumption and performance of early weaned calves. *J. Dairy Sci.* **54,** 262–265.

Yamaguchi, S. (1967). The synergistic taste effect of monosodium glutamate and disodium 5′-inosinate. *J. Food Sci.* **32,** 473–478.

Yamaguchi, S., Yoshikawa, T., Ikeda, S., and Ninomiya, T. (1971). Measurement of the relative taste intensity of some L-α-amino acids and 5′-nucleotides. *J. Food Sci.* **36,** 846–849.

Yamashita, S., Ogawa, H., and Sato, M. (1973). The enhancing action of 5′-ribonucleotide on rat gustatory nerve fiber response to monosodium glutamate. *Jpn. J. Physiol.* **23,** 59–68.

Yoshida, M., and Saito, S. (1969). Multidimensional scaling of the taste of amino acids. *Jpn. Psychol. Res.* **11,** 149–166.

Yoshida, M., Ninomiya, T., Ikeda, S., Yamaguchi, S., Yoshikawa, T., and Ohara, M. (1966). Studies on the taste of amino acids. Part I. Determination of threshold values of various amino acids. *J. Agric. Soc. Jpn.* **40,** 295–299 (in Japanese with English summary).

DISCUSSION

Pangborn: Relative to the article where you cited an increase in gas–liquid chromatography volatile activity in a solution of broth containing monosodium glutamate, I presume this was head space analysis. Did they also run a sodium chloride control?

Cagan: They did not include a NaCl control.

Pangborn: Because there could be simply a matter of a salting out or a gassity effect which is pushing the molecules from the aprons into the vapor stage, and it may not be an increase in volatiles.

Cagan: In general it is of course important to determine whether any effect is characteristic of the addition of glutamate.

Moss: You identified the correlation between your direct binding to the receptor site of various substances. I am very much interested in the blood sugar levels for your sugar application. What is the correlation between blood sugar level and taste receptor response? Is there a capability of taste being a barometer of hyperglycemia or hypoglycemia?

Cagan: The presentation by Nicolaidis addressed part of your question. In addition, there are studies showing that intravenous substances, including glucose, can elicit responses of the taste nerve. The characteristics of the binding interaction determine the initial information entering the taste system, but as we have seen, superimposed on that are feedback loops.

Pfaffmann: Hagstrom and I reduced the blood sugar in a recording preparation in peripheral nerve of taste sensitivity in the hamster. We were recording from the taste system because gymnemic acid blocked about 50% of the taste response to sugar, which indicated we were dealing with sweet responses. Then we dropped the blood sugar to 50% by insulin and there were no changes in the electrophysiological message coming up from the sensory nerve. That was using total nerve response, not individual fibers. At least at that gross a recording level there was no indication of a modulation of receptor.

Cantor: With reference to MSG, do I understand you correctly that you think there is an effect of concentration of MSG, and your conclusion comes largely from the food literature, but that you have not been able to find a corresponding effect?

Cagan: MSG is itself a taste stimulus. To evaluate the possibility that it also acts as a taste *modifier,* in addition to its effect as a stimulus, I recommend that we study mixtures in which MSG is present at a concentration too low to taste the MSG itself.

Cantor: In the early days of the introduction of MSG in quantity into our food system, which is about 35 or 40 years ago, the amounts that were recommended for use were relatively small. There followed a large amount of experimental work on MSG effects on vegetables and other materials. The objective was to determine what the market might be if major foodstuffs were able to be taste-stimulated by MSG. Then someone took a look at per capita consumption in Japan—huge relative to the United States—and recipes started calling for a half teaspoon of MSG, then a tablespoon. Currently, I think the amount of MSG per capita used in this country is still under Japanese consumption, but it is substantially greater than it was.

Cagan: In Japan a seaweed (sea tangle, *Laminaria* sp.) was used as a seasoning, which was later discovered to contain a large quantity of glutamate.

Maller: There is a literature on diabetics, half of which shows that diabetics who have a very high circulating level of blood sugar tend to have a decreased sensitivity to sugars, and I think it is limited to sugar as opposed to salty, bitter, or sour taste. It is messy because the other half of the literature shows that there is an increased sensitivity.

Naim: You mentioned the catfish as a good experimental model for studying mechanisms of taste. Is glutamic acid by itself a taste stimulus in the catfish?

Cagan: We have not used glutamic acid in our studies on catfish up to now. A number of amino acids are able to stimulate the catfish taste system, as judged by nerve responses, and we have measured binding of several amino acids. In the catfish, glutamate is a less effective stimulus than alanine, measured electrophysiologically. The nerve recording technique, however, does not tell us how the intact animal responds to it. We have no information about whether glutamate serves as a taste stimulus, as a taste modifier, or both in this species.

DYSFUNCTIONS

Chapter 18

Disordered Oral Sensation and Appetite

VERNON J. BRIGHTMAN

INTRODUCTION

For the last few years, we have been investigating a group of patients who have in common some abnormality of oral sensation. While our studies were not directed primarily at the dietary habits of these patients, or their capacity for, or feelings about, food,* we were stimulated to look into this aspect of their problems when asked to contribute to this work. What initial data we have suggest that this group of individuals can provide an opportunity for further examination of some of the oropharyngeal factors associated with human ingestive behavior that have been explored by others with animals, in which taste, smell and other oropharyngeal sensations were experimentally abolished. Oral sensation probably plays only a minor role in the *regulation of food intake,* and patients with disordered oral sensation, like animals who eat without benefit of oral sensation, might be expected to show little evidence of nutritional disturbance. On the other hand, in experimental animals, oral sensa-

* Appetite is here defined as "capacity for, or feelings about, food." Among the various definitions for appetite provided by several dictionaries, this, which is one of three given in the *Concise Oxford English Dictionary,* is useful in the context of studies of human ingestive behavior, since "capacity for" and "feelings about food" are both experimentally measurable variables.

tions have a strong effect on the *selection of food* and the *motivation to eat and drink,* and further study of these variables in patients with disordered oral sensation is planned. Since taste serves not only as a discriminatory stimulus but also as an affective stimulus with a definite hedonic value, studies of appetite in patients with disordered oral sensation offer certain advantages over studies in experimental animals, in which hedonic response is usually judged by preference–aversion behavior. The highly developed mimetic (facial) musculature in man, which along with taste sensation in the anterior part of the mouth is supplied by branches of the facial nerve, also provides (by means of the "gustofacial reflex") another opportunity for measuring taste response in man. This paper reports data concerning appetite, nutritional status, and oral sensory complaints in the patient sample, and also describes some initial experiments concerning the gustofacial reflex and hedonic response to tastants in a sample of young healthy adults.

LITERATURE REVIEW

The natural response of a patient confronted with the possibility of disease or mutilating surgery that affects oral function is intense concern about two processes, eating and speaking (Huxley, 1968). As far as the former is concerned, there is considerable evidence from patients who have oropharyngeal cancer and who receive therapeutic radiation that loss of oropharyngeal general and special sensory function (through section of branches of cranial nerves V, VII, and IX, or as the result of the destruction of sensory and gustatory receptors through xerostomia and radiation) leads to considerable difficulty with eating. In fact, oral disability, including difficulty with eating, is one of the major complications of the treatment of head and neck cancer. While decreased taste sensitivity and xerostomia following radiation have been documented in humans (Conger, 1973; Kalmus and Farnsworth, 1959; Brown *et al.,* 1976), there are very few detailed descriptions of the oral disability these patients experience, or what effect it has on their ingestive behavior. One of the few descriptions that I am aware of, and certainly one of the most moving, was published in the *Lancet* as a posthumous report from a physician who had been treated on several occasions with radiotherapy for cancer of the pharynx (MacCarthy-Leventhal, 1959). From her own personal experience, this woman described "mouth blindness" (loss of taste for food) and "hallucinatory tastes of copper and bicarbonate of soda" when she drank liquids. All food was likened to "cinders" or "charcoal biscuits." Her appetite seemed controlled by her stomach, yet in the absence of oral stimuli and saliva, she felt her esophagus unable to relax to admit the "cindery bolus," which could be ingested only with copious fluid. She described absence of hunger sensation, and exhaustion and despair concerning eating. In the absence of oral sensory information, "panic and nausea" were experienced at the thought of anything objectionable in the food. Increasing the strength of alcohol and condiments failed to remove the hallucinatory tastes and simply added a sensation of burning. Only the savory aroma of hot and steaming food soothed the stomach and stimulated the appetite.

While subject to the bias that is present in any anecdote, this paper does indicate that in the human (a) familiar oral and olfactory stimuli are important for appetite; (b) oral perception is an important guardian of what is ingested as food and what is rejected as nonfood, decreased oral perception engendering severe anxiety; and (c) difficulty with eating decreases hunger and leads to depression. Also, it is apparent that the author considered her inability to enjoy food a "bitter disappointment," and an abnormality that "alienated" her from others.

Unfortunately, radiation to the oral cavity causes widespread mucosal damage as well as xerostomia, and it is difficult to identify relationships between individual oral sensory dysfunctions and ingestive behavior in patients with head and neck cancer. A variety of other considerations also make them poor candidates for extended investigation, and we have turned to a group of patients with less extensive oral sensory abnormalities for our studies. Before describing our study population, however, it is important to review the literature dealing with ingestive behavior in animals, including those which have been experimentally deprived of oral sensation, since the individual contributions of oral sensory stimuli and salivary secretion to ingestion have been documented in such experiments.

Three components of ingestive behavior have been identified: (1) regulation of the amount and nutrient composition of solid and liquid food necessary to maintain the normal nutrient and water composition of the body; (2) selection (i.e., identification and choice) of particular foods, and (3) motivation for feeding. In general, oral stimuli play little part in (1), but their contribution to (2) and (3) is considerable. The evidence is as follows (Epstein, 1967).

The composition of food can be radically altered as regards its taste, smell, texture, concentration, and nutrient content without significant change in bodily composition. Both rats and dogs adjust the volume that they ingest to compensate for variations in dietary composition, this mechanism failing only when adulterants in the diet either reach a toxic concentration or strain the capacity of the animal to obtain adequate nutrients. Rats whose oral sensations are abolished by topical oral application of cocaine, or by section of oral sensory and gustatory nerves, continue to regulate their food intake, and dogs in which the IXth nerve and chorda tympani, the Vth nerve, or the olfactory bulbs have been sectioned also drink normally. Oral factors do, however, limit the success of "cafeteria experiments," the animals' selection of food from the variety offered being "nutritionally wise" only when palatable (and multiple) sources of protein are offered.

The extensive series of experiments that Epstein and Teitelbaum carried out with rats, in which the contribution of oral sensory stimuli was eliminated by use of intragastric self-injection, also demonstrated that "taste, smell and other oropharyngeal sensations are not essential for the arousal of hunger and thirst, for satiation of both, for the control of the frequency and size of meals, and for the control of the total amount of food or water ingested daily for prolonged periods of time."

On the other hand, there is no doubt that animals identify and distinguish among foods largely by taste and smell. Both innate and acquired food preferences are linked to the oral and olfactory cues characterizing the food, and food habits associated with familiar tastes and smells are often maintained even in the face of nutritional catastrophe. Oral factors are also important cues for learned aversions, and preference–aversion curves based on palatability (the animal equivalent of hedonic evaluation) are due to oral factors only. In contrast to the regulation of food intake, preference–aversion behavior is greatly modified by both oral sensory denervation and intragastric feeding. Specific hungers for salt and sugar are also dependent on the animal's ability to detect these substances in the diet, i.e., on gustatory sensitivity.

Taste and smell are also powerful motivators of feeding, drinking, and learning, both taste and smell, for example, being capable of motivating ingestive behavior in the absence of natural or artificially induced hunger or thirst. Taste and smell are thus not simply discriminatory stimuli, but like pain and contact sensations they are also affective stimuli: Animals, like humans, eat and drink certain substances because they like them and not solely to meet their nutritional needs.

Elimination of saliva (by surgical extirpation of the salivary glands, ligation of the ducts with section of the auriculotemporal nerve, or by administration of atropine-like drugs) leads to a phenomenon known as postprandial drinking, when dry food is offered to the animal. Aqueous and oily mash diets eliminate the phenomenon, which is generated by the dryness of the mouth and is presumably dependent on at least some oropharyngeal sensory information. Surgically induced (Lawson *et al.*, 1974), radiation-induced (Tanikawa, 1963), and drug-induced (Galili and

Maller, 1976) xerostomia also lead to changes in preferences for the taste of liquid and solid food, probably as a result of damage to taste receptors.

Analogies between these phenomena in animals and those reported by patients with oropharyngeal cancer are easily made. However, clear identification of the exact abnormality underlying these patients' complaints (which is necessary before rational treatment of their problem is possible) can be achieved only when the various components of feeding behavior identified in animal studies (i.e., regulation, selection, and motivation) are examined separately in humans with a variety of oral sensory abnormalities. The following data represent our initial studies with such patients.

STUDY POPULATION

The patients included in this study were all referred to the Oral Medicine Research Clinic at The Philadelphia General Hospital for investigation of some chronic oral problem. The majority of the studies carried out on these patients were completed during a 1-week period of admission to the hospital's Clinical Research Center for the purpose of establishing a diagnosis for the patient's oral problem. A reasonable explanation for their symptoms was found in about half of the patients following fairly intensive clinical study; equally intensive efforts failed to provide an explanation for the oral symptoms of the remainder. Three major categories of patients were distinguished:

1. *Patients with oral pain syndromes,* comprised of those with recognized *neuropathy* (i.e., evidence of trauma or viral-induced damage to an oral sensory nerve); those with evidence of TMJ myofascial dysfunction syndrome without intrinsic joint disease (referred to as *temporomandibular joint patients*); and those with symptoms of oral pain and burning often of unusual anatomic distribution and character and without evidence of adequate stimuli to account for the symptom (referred to as *atypical facial pain* and *glossalgia,* respectively).
2. Patients with *chronic oral mucosal disease* such as erosive lichen planus, recurrent aphthous ulcers, and fusospirochetal (Vincent's) gingivostomatitis.
3. Patients with *taste syndromes,* i.e., patients who complained primarily of some altered taste sensation that was often without adequate explanation *(idiopathic dysgeusia),* or who complained of taste problems in association with *salivary gland disease*.

The clinical management of these patients is complicated by the fact that most of them have multiple oral sensory complaints which succeed each other as the primary problem over time. Thus, a patient may be treated for a complaint of taste abnormality at one time, as a patient with atypical facial pain at another, and yet again for a complaint of dryness. Moreover, symptoms of burning, pain, dryness, or taste abnormality are not confined specifically to any one diagnostic grouping and may be found in patients in all three groups. In an attempt at standardizing the reporting of oral complaints by these patients, we developed an oral sensation questionnaire which enquired about symptoms of pain, burning, numbness, taste, smell, and salivary abnormality. The patient was also asked to rank the intensity of his current symptoms on a scale of 0–5 (0, none at all; 2.5, mild; 5.0, very bad), which followed the question about each symptom.

Table I summarizes the responses of 39 patients to this questionnaire at the time of their admission to the Clinical Research Center. It can be seen, for example, that complaints of taste abnormality were not restricted to those who listed this as their primary complaint (i.e., the ten in the idiopathic dysgeusia and sialadenitis category), but such complaints were also listed by about two-thirds of the patients in the oral pain syndromes and chronic mucosal disease categories, respectively. Since the questionnaire also contained two questions relating to taste abnormality, viz., (a) Have you noticed any unusual tastes in your mouth, or any difficulty tasting? and (b) Does food taste different from what it used to?, a further subdivision of patients concerning the nature of their taste abnormality was possible (Table II).

Two separate usages of the word "taste" may be distinguished in reference to physiological (versus psychological, e.g., "a man of good taste") processes. On the one hand, taste is used by the sensory physiologist to refer to the sensations of taste that are transmitted via the gustatory nerves in response to the application of drops of sapid solutions on the

TABLE I

Responses of 39 Patients with Disordered Oral Sensation to an Oral Sensation Questionnaire

		% with symptoms of					
Diagnostic category	No. of patients	Pain	Burning	Numbness	Taste	Smell	Salivation
Oral pain syndromes	22	100	64	66	66	5	50
Chronic mucosal disease	7	100	56	14	70	28	14
Idiopathic dysgeusia and sialadenitis	10	40	50	20	100	50	70

TABLE II

Nature of Complaint for 27 Patients Admitting to Taste Aberrations

Diagnostic category		No. of patients	Nature of complaint	
			Sensory	Perceptual
Pain syndromes	neuropathy	2	0	2
	glossalgia	2	2	0
	temporomandibular joint	3	3	2
	atypical facial pain	5	5	3
Chronic mucosal disease		5	4	3
All pain–mucosal syndromes		17	14 (82%)	10 (59%)
All taste syndromes		10	10 (100%)	6 (60%)

tongue; on the other, in everyday speech, "taste," particularly when applied to food, refers to the entire spectrum of oral sensory stimuli produced by introducing food into the mouth. Unless otherwise qualified, it is likely that question (b), by specifying food, will evoke responses concerned with this latter usage, which I have referred to as the *perception of taste*. Question (a), by comparison, is more likely to be interpreted in terms of altered or diminished taste sensations, when a familiar-tasting substance is taken into the mouth or when the mouth is empty; this usage I will refer to as the *sensation of taste*. While identical interpretations of these questions cannot be expected from all patients, these two questions do serve to classify the nature of the taste abnormality that these patients complain of. The data suggest that certain types of complaints may be more common among patients in some diagnostic categories, but it is also apparent that as a group the patients with pain syndromes and chronic mucosal disease ("all pain–mucosal syndromes"), for whom taste abnormality is only a subsidiary complaint, report similar types of taste complaint as those in the taste syndrome category, for whom taste abnormality is their primary concern. For this reason, we have combined all patients who admit to taste abnormality on the questionnaire as "dysgeusics" and distinguish them from the "nondysgeusics," i.e., those who *deny* any taste abnormality.

We have previously reported on the results of a psychophysical test of taste sensitivity carried out on "dysgeusics" and "nondysgeusics" (Brightman *et al.*, 1973), which showed a statistically significant difference in the ability to distinguish threshold concentrations of salt and hydrochloric acid between the two groups. "Dysgeusics" could not taste these solutions as well as "nondysgeusics," although no significant differences were noted in this regard for sugar and quinine. This taste test (Desor and Maller, 1975) also included an evaluation of the effect of biasing instructions on taste sensitivity. Following determination of the threshold concentration of sucrose, each patient was presented with two further series of cups containing either water, or water with sucrose at just below the threshold concentration, and was requested to repeat the test following advice that "Some people with your problem think all of these taste like plain water"; and again for the second series, with the advice "Some people with your problem think all of these taste like sugar water." Since all patients showed little biasing of their test results with these instructions, and since a difference did exist regarding taste threshold between "dysgeusics" and "nondysgeusics," it would seem likely that the "dysgeusic" patients in our series are experiencing some sensory loss regarding taste rather than an imaginary symptom.

This was an important distinction, since we had previously noted a very high frequency of depression among our patients (Gessel and Brightman, 1971—see Table III), and the chronic and somewhat bizarre oral sensory complaints that they reported were interpreted by some as simply evidence of increased help-seeking behavior characteristic of depression.

It is not the purpose of this paper to describe this study population in great detail, but it is important to point out the high frequency of blood sugar and blood lipid abnormalities in this group of patients (see Tables IV and V). A family history of diabetes mellitus exists in many of them, and based on the results of a 3-hr glucose tolerance test with prior carbohydrate loading, 42% of the patients were found to have blood sugar abnormalities or diabetes. Additional patients with blood lipid abnormalities

TABLE III

Psychiatric Characteristics of 57 Patients Referred to the Oral Medicine Clinic at The Philadelphia General Hospital[a]

Primary complaint	No.	Depression	Anxiety	Schizophrenia	Stress reaction	Character defect	None
Disordered oral sensation	22	19 (86%)	0	0	1	0	2
Recurrent aphthous ulcers	19	6 (35%)	2	0	2	2	7
Miscellaneous diagnoses	16	1 (7%)	3	3	0	2	7

[a] Gessel and Brightman, 1971.

TABLE IV

Blood Glucose and Serum Lipid Abnormalities in 38 Patients with Chronic Orofacial Sensory Abnormalities[a]

	Group I[b]		Group II[c]	
	Pain	Dysgeusia	Pain	Dysgeusia
Sex (M/F)	1/6	1/4	1/4	2/3
Average age	42	53	54	46
Family history diabetes	22%	40%	37%	0
Preexisting diabetes	0	22%	7%	0
Hyperglycemia	28%	20%	7%	0
Diabetes[d]	7%	40%	35%	40%
Average cholesterol (mg%)	225	221	244	207
Average triglycerides (mg%)	175	124	120	135
Lipoprotein abnormality	25%	0	38%	33%

[a] Brightman *et al.*, 1976.
[b] Reasonable explanation for symptoms.
[c] No explanation for symptoms.
[d] Diabetic by three or more of the criteria used.

brought the total frequency of blood sugar and lipid abnormalities in both groups of patients (those whose symptoms were explained and those with unexplained symptoms) to 63 and 53%, respectively. While these data and those of other workers (Brody *et al.*, 1971) are suggestive, there are, however, no convincing data at this time to account for these patients' oral sensory abnormalities on the basis of diabetic peripheral neuropathy.

TABLE V

Blood Glucose and Serum Lipid Abnormalities in 38 Patients with Chronic Orofacial Sensory Problems[a]

	Group I[b]	Group II[c]
Blood glucose abnormalities	42%	42%
Blood lipid abnormalities	21%	10%[d]
Total number of patients with either abnormality	12/19	10/19
% Patients with either abnormality	63%	53%

[a] Brightman *et al.*, 1976.
[b] Reasonable explanation for symptoms.
[c] No explanation for symptoms.
[d] Two patients with blood glucose abnormalities also.

INGESTIVE BEHAVIOR OF PATIENTS WITH DISORDERED ORAL SENSATION

For the purpose of this presentation we asked the following questions about 39 patients with disordered oral sensation. Ten of these 39 had a primary complaint of taste abnormality (referred to as patients with taste syndromes) and the other 29 constituted two control groups: 22 with a primary complaint of chronic orofacial pain rather than a taste problem (referred to as patients with oral pain syndromes) and seven with chronic oral mucosal disease (recurrent aphthous ulcers or erosive lichen planus).

Question 1: Do Oral Sensory Abnormalities Interfere with Eating?

Conceivably, it might be possible to have an oral sensory abnormality (e.g., a persistent bad taste) intense enough to be bothersome when the mouth contained saliva alone, but not of sufficient intensity to persist when strongly flavored or attractive food was masticated. Symptoms of this intensity would presumably not interfere with eating. Patients with such mild symptoms were the minority in our study population, however, since 60% of the "dysgeusics" complained that "food tasted different from what it used to" (Table II). We do not have any detailed dietary information from these patients at the moment, but it would certainly be of interest to know if these patients varied their diets as a result of the food tasting different. The dietician who worked with these patients during their hospital admission described them as "finicky eaters" with "bizarre ideas about food and its relation to their symptoms," but under the conditions specified for their hospital admission (high carbohydrate diet emphasizing soft-textured foods to minimize their oral discomfort), specific data on food selection or intake could not be obtained.

The two patients with diagnosed (trigeminal) neuropathy who complained of taste abnormality are of interest (see Table II), since they denied sensory taste aberration but described "food tasting different." Neither recognized any olfactory deficit and presumably their reference to "taste" is not to a specific sapid quality of food.

Question 2: Do Patients with Disordered Oral Sensation Show Evidence of Nutritional Inadequacy?

While no thorough evaluation of the nutritional status of these patients has been carried out, the few observations we have made incidental to other studies on them suggest that a general nutritional inadequacy is unlikely. Occasional patients in both the "dysgeusic" and "nondysgeusic" groups were thin almost to the point of emaciation, but both groups of patients, on the average, were judged "well nourished" clinically, this impression being borne out by an average ponderal index* of 12.5 for both groups, and by the number of individuals in each group who had an index of < 12.5 and were thus defined as "obese." A complete blood count, SMA-12 and SMA-6, were performed on all patients, and these tests also revealed no abnormalities except for those individuals who were shown to have specific disease states, such as diabetes, neutropenia, and multiple myeloma.

Question 3: Do Oral Sensory Abnormalities Affect Appetite?

While we have suggested a workable definition for "appetite" in the introduction to this paper, we did not attempt to impose this definition on our patients, and simply recorded their responses to two questions about appetite that are contained in the Cornell Medical Index (item 45) and the Minnesota Multiphasic Personality Inventory (item 2), respectively:

CMI No. 45: "Is your appetite always poor?" (Yes, No)

MMPI No. 2: "I have a good appetite" (Yes, No)

While some patients may undoubtedly interpret "appetite" differently from others, this approach has the advantages that, first, the question is asked in two different ways, and, second, the questions regarding appetite are placed among other health-related questions, no single one of which is directly identified with the patient's oral problem when the questionnaire is presented to him.

Only 16% of the "dysgeusic" patients recognized appetite change, and only 25% of the "nondysgeusics" (i.e., those with oral symptoms that did not include taste abnormality) did so (Table VI). As might be expected from the knowledge of the great variety of cues available for appetite stimulation in animals, aberration in those from the oropharyngeal region seems well compensated for by the available olfactory and visual stimuli.

There are no great discrepancies between these data and those that might be predicted from the results of animal experimentation. The critical experiment that is suggested by the results of animal experiments, however, has not been carried out, and it still remains important to determine whether food intake, selection, and preference are affected in these patients.

* Height in inches divided by $[\text{weight (lb)}]^{1/3}$.

TABLE VI

Comparison of 24 Dysgeusic and 24 Nondysgeusic Patients

	Dysgeusic	Nondysgeusic
Female/male	$^3/_1$	$^5/_1$
Average age	53 (19–75)	48 (16–76)
Frequency of depression	>75%	>75%
"Food tastes different"	63%	10%
Loss of appetite[a]	16%	25%

[a] Cornell Medical Index No. 45, M.M.P.I. No. 2.

If such experiments confirm that capacity for food and nutritional state are not severely altered by these oral sensory problems, a further question must be answered, viz.,

Question 4: What Is the Nature of the Complaint That Causes These Patients to Seek Repeated Consultations about Their Oral Problem?

On the average the patients with disordered oral sensation who are referred to us have had 4.2 consultations in the preceding 3 years, which is the average duration of their symptoms. By contrast, patients with recurrent aphthous ulcers, another vaguely defined and generally poorly managed oral problem, average only 3.8 consultations over a much longer period of 10 years for their average symptom duration (Gessel and Brightman, 1971). What is their disability?

To some extent, pain may be their motivating factor. With use of the McGill pain questionnaire (Melzack, 1975) and the intensity rating for each symptom on the oral sensation questionnaire we have been able to show: (a) that the pain experienced by the patients with the oral pain syndromes in our population equals and even exceeds the intensity of pain elsewhere in the body as far as some pain qualities are concerned (Table VII), and (b) that in general, our patients rank their symptoms of pain higher on an intensity scale than their symptoms of taste aberration (Table VIII). In fact, some patients in the idiopathic dysgeusia and sialadenitis category, whose primary complaint and help-seeking concern taste, appear to consider their mouth problem as "painful."

It is doubtful, however, that pain is the entire explanation, and the failure of topically and systemically administered analgesics to control these patients' oral problems would also argue for some other type of

TABLE VII

Comparison of Average Pain Intensity For Various Categories of Oral and Nonoral Pain

Diagnostic category		No. of patients	Pain intensity[a]		
			PPI	NWC	PRI (R)
Oral	neuropathy	4	2.2	10.0	26.0
	glossalgia	4	2.5	7.3	15.7
	temporomandibular joint	5	3.6	8.0	21.8
	atypical facial pain	11	2.1	6.7	15.8
	dental	15	2.3	8.3	19.5
Nonoral[b]	menstrual	25	2.4	6.7	17.5
	arthritis	19	1.9	8.1	18.8
	backache	14	2.6	10.9	26.3
	phantom limb	6	3.0	8.3	25.0
	postherpetic	5	3.0	10.4	22.6
All oral categories		39	2.5	8.1	19.8
All nonoral categories		69	2.6	8.9	20.0

[a] PPI, present pain intensity; NWC, number of words chosen; PRI (R), pain rating index. See Melzack, 1975.
[b] Melzack, 1975.

TABLE VIII

Comparative Intensity of Oral Symptoms for 39 Patients with Disordered Oral Sensation as Scored from Responses To Oral Sensation Questionnaire

Diagnostic category	No. of patients	Intensity of symptoms[a]					
		Pain	Burning	Numbness	Taste	Smell	Salivation
Oral pain syndromes	22	3.43 (22)	3.46 (3)	2.99 (12)	3.14 (12)	—	2.83 (8)
Chronic mucosal disease	7	3.03 (7)	1.67 (4)	—	1.63 (5)	3.30 (2)	—
Idiopathic Dysgeusia and sialadenitis	10	3.13 (4)	3.94 (5)	2.50 (2)	2.34 (10)	3.00 (5)	2.60 (7)

[a] 0, "none at all"; 2.5, "mild"; 5.0, "very bad." Numbers in parentheses refer to number of patients in each category who responded affirmatively to each symptom.

disability. Since oral sensations, and notably taste, are recognized as strong hedonic stimuli, we suggest that this is another avenue that should be studied with these patients. To this end, we have adapted the techniques used by Steiner in defining the gustofacial reflex in infants (Steiner, 1973) to a study of the same phenomena in adult human subjects at the same time as the subjects complete a written hedonic evaluation of the test food item. Currently, our data collection has been restricted to healthy young adults, but it is clear that the technique provides two simultaneous evaluations of taste response: that provided consciously by the patient (intensity or hedonic ratings), and that expressed in changes in his facial musculature which are largely unconscious and, once recorded on videotape, are available for scoring by independent observers (Brightman *et al.,* 1975). The gustofacial reflex is present in adult humans; it is probably communicatory in nature, and like other behaviors will be more fully expressed under certain experimental conditions (group eating is a particularly useful model). We have used the consistency with which observers score a videotaped record of the subject's facial expressions as a measure of the intensity of the response to the tastant, and with this method divergence of the subjects' hedonic response curve from that for observer's rating of facial expression was noted when different concentrations of "sugar sandwiches" were ingested. This approach is suggested as a useful method for further study of "feelings about food," another important component of appetite.

CONCLUSIONS

Abnormalities of taste are a problem not only for those patients who specifically seek help for such symptoms, but also for about two-thirds of patients with various chronic oral pain syndromes or ulcerative diseases of the oral mucosa. Taste abnormalities interfere with eating, at least to the extent that about 60% of patients with these symptoms complain that food tastes different. (This altered taste perception while eating is not due to pain, since among patients with chronic oral pain syndromes who deny taste problems, complaints that food tastes different are rare.) Despite their awareness of altered taste perception, these patients have similar height–weight ratios to those of patients with chronic oral pain syndromes without taste abnormality, and only a small percentage (16 and 25%, respectively, for the two groups) recognize appetite change. These data tend to confirm the results obtained from studies on experimental animals to the effect that motivation for eating is linked to a variety of cues, among which taste sensation probably plays a minor role.

The nature of the disability experienced by patients with taste abnormalities that causes them to seek continued medical care is not clear. Some of these patients experience oral pain that they rank of intensity approximating that reported for pain elsewhere in the body, and initial evidence suggests that their pain is more of a concern to them than taste abnormality. Depression, which is very common among these patients, may also account for their increased help-seeking behavior. However, the effect of altered taste perception on such factors as selection of food, hedonic response, and facial expression has not been explored and may provide further clues to these patients' problems.

ACKNOWLEDGMENTS

The author wishes to acknowledge the assistance provided by Dr. Owen Maller, Dr. Jeannette Desor, and Dr. Jacob Steiner in the planning and evaluation of these studies, as well as the contributions of the following staff of the Department of Oral Medicine at The Philadelphia General Hospital who assisted in the management of these patients: Dr. Arnold Gessel, Dr. Chris Nichols, Dr. Michael Silvert, Dr. Robert Arm, Dr. Bruce Blasberg, Dr. Levente Bodak, Dr. Adrian Garfunkel, Dr. Douglas Peterson, and Dr. Roger McWilliams; Mrs. Suzanne Fisher, Mrs. Janice Garber, and Mrs. Anne Moskovitz; and Dr. Arthur Segal and Miss Patty Werther (gustofacial reflex studies).

Supported by USPHS DE03764 to The Philadelphia General Hospital.

REFERENCES

Brightman, V., Desor, J., and Maller, O. (1973). Taste sensitivity in dysgeusics and non-dysgeusics. *J. Dent. Res.* **52,** 270 (Abstr. No. 845).

Brightman, V., Segal, A., Werther, P., and Steiner, J. (1975). Ethologic study of facial expression in response to taste stimuli. *J. Dent. Res.* **54,** L141 (abstr. No. L564).

Brody, H. A., Prendergast, J. J., and Silverman, S. Jr. (1971). The relationship between oral symptoms, insulin release and glucose intolerance. *Oral Surg., Oral Med. Oral Pathol.* **31,** 777–782.

Brown, L. R., Dreizen, S., Rider, L. J., and Johnston, D. A. (1976). The effect of radiation-induced xerostomia on saliva and serum lysozyme and immunoglobulin levels. *Oral Surg., Oral Med. Oral Pathol.* **41,** 83–92.

Conger, A. D. (1973). Loss and recovery of taste acuity in patients irradiated to the oral cavity. *Radiat. Res.* **53,** 338–347.

Desor, J. A., and Maller, O. (1975). Taste correlates of disease states: cystic fibrosis. *J. Pediatr.* **87,** 93–96.

Dethier, V. (1976). "The Hungry Fly." Academic Press, New York.

Epstein, A. N. (1967). Oropharyngeal factors in feeding and drinking. *Handb. Physiol. Sect. 6: Aliment. Canal* **1,** 197–218.

Galili, D., and Maller, O. (1976). Xerostomia and taste preferences. *J. Dent. Res.* **55,** B324 (Abstr. No. 1049).

Gessel, A., and Brightman, V. J. (1971). Depression in patients with disturbed oral sensation. *J. Dent. Res.* **49,** A268 (abstr. No. 864).
Halpern, B. P., and Nichols, T. L. (1975). Human drinking: Orbicularis oris EMG patterns and liquid contact durations. *Physiologist* **18,** No. 3.
Huxley, L. A. (1968). "This Timeless Moment—A Personal View of Aldous Huxley," pp. 86–89. Farrar & Rinehart, New York.
Kalmus, H., and Farnsworth, D. (1959). Impairment and recovery of taste following irradiation of the oropharynx. *J. Laryngol. Otol.* **73,** 180–182.
Lawson, W. B., Hagstrom, E. C., and Walter, G. F. (1974). Salt preference in desalivate rats. *Physiol. Behav.* **12,** 733–739.
MacCarthy-Leventhal, E. M. (1959). Post-radiation mouth blindness. *Lancet* **2,** 1138–1139.
Melzack, R. (1975). The McGill pain questionnaire: Major properties and scoring methods. *Pain* **1,** 277–299.
Moncrieff, R. W. (1951). "The Chemical Senses." Leonard Hill, London.
Steiner, J. (1973). The human gustofacial response. In *Fourth Sympos.* "Oral Sensation and Perception" (J. F. Bosma, ed.), U.S. Dept. of Health, Education, and Welfare, Bethesda, Maryland.
Tanikawa, K. (1963). Clinical and experimental studies on radiation injury to taste. *Nippon Acta Radiol.* **23,** 704–712.

DISCUSSION

Moskowitz: Are you familiar with Bruce Halpern's studies of facial expression while drinking? He has shown that the lip musculature exhibits very specific activities with both pleasant and unpleasant stimuli (Halpern and Nichols, 1975).

Brightman: Yes, I have had an opportunity to see some of Halpern's films of both rat and human subjects this year while I was at Cornell. He uses a camera that focuses mainly on the subject's lips, and, of course, obtains much more detail of that area than we can secure from videotape photographs. To achieve this, one has to keep the subject's face more or less rigidly fixed so that the mouth stays in focus, and this tends to make the testing situation even more artificial. Under these conditions, adult human subjects usually require several sessions before they will relax and freely exhibit a gustofacial response. About 18 months ago, when we had an opportunity to discuss our experiments, we had both just realized the importance of placing the subject in a situation where he could express the particular behavior (facial expression) we wished to observe—having the subjects eat together rather than alone solved the problem for both of us.

Bosma: On this matter of facial expression there is a center of intensive study of that at the University of California at San Francisco. Our particular contact is Helen Oster, who is doing quite sophisticated studies in infant oral gestures.

Garcia: I was not surprised at the manner in which your subjects used a paper napkin to express aversion to a salt sandwich (a photograph of a subject wiping his lips vigorously with a napkin after eating a salt sandwich was shown as part of Dr. Brightman's oral presentation). Throughout the animal kingdom, when you have any animal that has an appendage to use, you get an equivalent response even though a napkin may not be involved.

Dethier: These responses are not restricted to vertebrates. In fact, it can be demonstrated that if you give a fly an aversive taste stimulus, it will wipe its mouth with its foot (Dethier, 1976).

Pfaffmann: This same response, as well as gaping of the lower jaw, is also part of the behavior pattern of rats in response to aversive taste stimuli. It is followed by a wiping movement in which the rat's chin is dragged along the ground; finally it shucks its feet, completing a whole series of actions related to aversive response. This is the way you read the rat's mind.

Jordan: Since many of these people are depressed, have they been given antidepressants, and do they respond to them? There is a certain population you see in psychiatry with chronic facial pain that responds very beautifully to antidepressants. The pain goes away as you alleviate the depression. I am curious whether your patients have been evaluated psychiatrically and treated?

Brightman: All of the patients in our series have received a structured psychiatrically oriented interview from either Dr. Arnold Gessel or Dr. Aaron Katcher, the two psychiatrists on the faculty of the University of Pennsylvania School of Dental Medicine. We also have each patient complete the Cornell Medical Index and the MMPI. As I showed in Table III, 86% of the patients with disordered oral sensation were felt to be depressed. We have not carried out a controlled trial to test the efficacy of antidepressant medications for these problems, although we make frequent use of antidepressant therapies in managing them. Except for those patients in whom an underlying systemic process (such as diabetes) is discovered, we have very few methods of treatment available to us. Since the majority of them are depressed, treatment often focuses on this aspect of their problems. Our clinical impression is that some of them respond very, very well. Others return with the same oral symptoms despite what appear to be adequate trials of one or more tricyclic antidepressants or MAO-inhibitors.

Jordan: The other thing that I was curious about was whether there were different subpopulations in this group. For instance, there is appetite suppression in 25% roughly. Very often, when somebody's depressed that is one of the cardinal signs of depression. I am wondering if there is any correlation between those that do have loss of appetite and the responses to antidepressant therapy.

Brightman: Unfortunately, our emphasis to date has been on accumulating enough patients for the study. As you can see, I had to scratch a little bit to get respectable numbers in some of these groups. As more data accumulate, we can begin to ask those questions.

Bosma: The interlinkage of oral sensory variations, anorexia, and depression, I think, has to be emphasized. Is it still true, as was reported a few years ago, that anorexia is the most consistent prologue of juvenile suicide?

Jordan: I really do not know.

Senseman: I was interested in the fact that you said some of your patients complained about a metallic taste. In your opinion is that a frequent complaint?

Brightman: I cannot tell you how many of them referred to a metallic taste in describing their oral symptoms. Metallic taste is, of course, referred to in the literature on taste physiology as possibly a fifth taste quality (Moncrieff, 1951). Whether such sensations are experienced in the absence of metallic salt solutions or are a normal physiological response, or only a pathological phenomenon, is not clear.

Bosma: There may be one particular phenomenon shared by many of these patients that separates them from normal people for whom the mouth summates to silence when we are not giving attention to it. In some of these patients there is the phenomenon of persistent oral sensibility. They will have difficulty describing it to you, but there is always some feeling in their mouth; they will struggle to name it. Their mouth is not silent, and they never experience relief from these oral sensations. Some of them, for instance, chew gum continually and some report difficulty in getting to sleep. This is a most peculiar item. I think it deserves particular mention when we come to this matter of oral sensory aberrations.

Brightman: I think this is a very common problem and not only for the population of patients I have described. In any dental clinic there are a large number of people who are there not because they think they might have a cavity, but because their mouth is not quiet and they are hoping the dentist will find some way to help. It is a very profitable area to study.

Bosma: Profitable?

Brightman: Certainly not from the point of view of the practicing dentist. But it should be a profitable field for the scientist studying oral sensation.

Moskowitz: The literature rarely describes any experiments or observations of what happens when these people taste or eat something. They either describe what Jim is talking about, e.g., a continued harsh or bitter taste, or they complain that their food tastes like dishwater. Are these true tastes, that is to say, are they ongoing tastes when there is nothing in your mouth, and do they summate with what you are eating to produce a different impression; or in fact, do they go away while the patient is eating?

Brightman: Based on anecdotal reports from the patients. Many of these bad tastes do apparently continue, i.e., the hallucinatory taste is there, not only when they are not eating, but it bothers them during eating as well. However, they are heterogeneous as regards their symptoms, and I do not quite know what emphasis to place upon symptoms at this time. We tend to lump them all together as patients having oral sensory abnormality related to ingestive behavior. This is more the definition I am thinking of when I now speak of dysgeusia. I am not focusing just on taste, but on all of the ingestive abnormalities where the mouth just does not feel right.

Chapter 19

Changes in Taste Sensation in Cancer Patients: Correlation with Caloric Intake

WILLIAM D. DEWYS

Anorexia (reduced appetite) may be a very distressing problem for the patient with advanced cancer. Anorexia has been described qualitatively but its pathophysiology is poorly understood (DeWys, 1970). We have recently presented observations on changes in taste sensation in patients with malignancy and have correlated these observations with patient's symptoms and the extent of the patient's disease. The present report will review these observations and will also present observations on caloric intake in this study population.

A series of 50 patients with a spectrum of malignancies of variable extent was studied (DeWys, 1974; DeWys and Walters, 1975). Patient symptoms were evaluated with a semistructured interview, and taste sense was measured using the technique of Henkin (Henkin *et al.*, 1963, 1971). Control values were obtained from 17 normal volunteers and six patients with nonneoplastic diseases who were similarly taste tested. Data were subjected to significance testing using the rank sum test or chi-square with Yates' correction (Hill, 1961).

A general reduction in the pleasurable aspect of taste was noted by 25 patients. Sixteen patients reported an aversion for meat; these included 13 patients who had also indicated a general reduction in taste sensation. Detection thresholds for NaCl and HCl were essentially similar in the

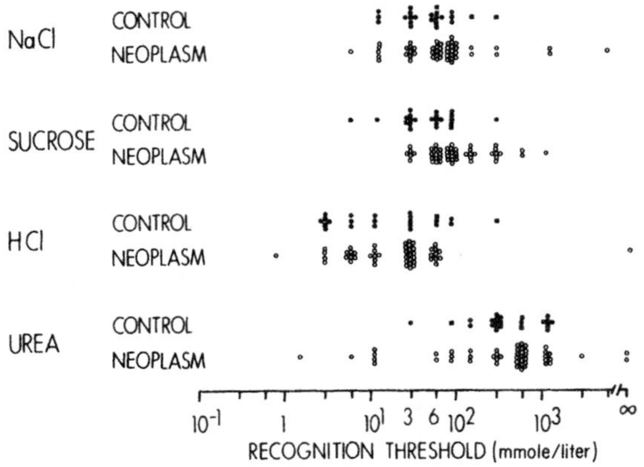

Fig. 1. Recognition thresholds for four tastants in control and neoplasm groups. In the control columns, the symbols are normal volunteers (circles) and patients with nonneoplastic diseases (squares).

control group and the tumor group (DeWys and Walters, 1975). The detection threshold for sucrose showed an upward skewing in the tumor patient group compared to the control population, and there was a suggestion of downward skewing in the patient group for detection threshold for urea.

Similar patterns were seen for the recognition thresholds and are shown in Fig. 1. Recognition thresholds for NaCl and HCl for the patient group were not significantly different from controls. The median recognition threshold for sucrose was higher in tumor patients than in controls, and 17 tumor patients had sucrose threshold values higher than all but one of the control group ($p < 0.001$). The results of testing with urea demonstrated a downward skewing in the tumor patients; eight tumor patients had values lower than all but one of the controls ($p < 0.005$).

The sucrose recognition results for the cancer patients could be correlated with the presence of the symptom of decreased taste sensation (Fig. 2). Patients who did not exhibit this symptom had sucrose recognition results nearly similar to controls, whereas in the patients with this symp-

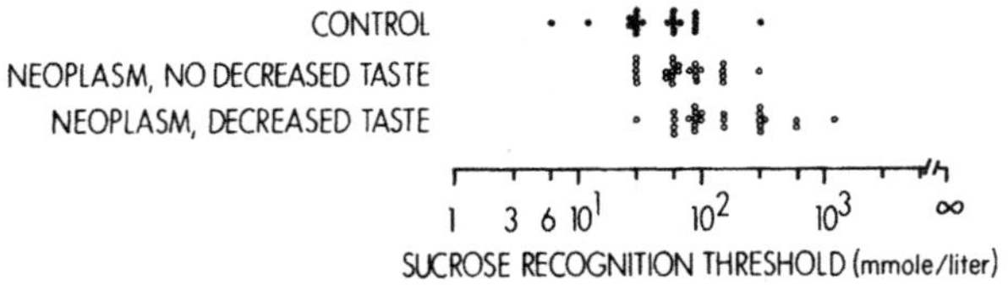

Fig. 2. Sucrose recognition threshold data divided as to the presence or absence of the symptom of decreased taste. Legend same as in Fig. 1.

tom there was a trend toward higher thresholds; 24 out of 25 symptomatic patients had thresholds above the median of normal. Also, 12 out of 25 symptomatic patients had thresholds above all but one of the patients in the control group ($p < 0.01$).

The urea recognition results for the cancer patients could be correlated with the presence of the symptom of meat aversion (Fig. 3). In the patients who reported meat aversion, 12 out of 16 had urea recognition thresholds below 300 mmole/liter, which is significantly different from the two out of 34 in the patients who denied the symptom ($p < 0.001$) or the four out of 23 in the control group ($p < 0.002$).

The interrelations between the symptoms of loss of taste or meat aversion and the recognition thresholds for urea and sucrose were explored (DeWys and Walters, 1975). It was possible to identify four subgroups of patients as to the presence or absence of these two symptoms and to correlate the symptomatic groupings with taste testing results. A group of three patients reported only an aversion for meat; their taste test results disclosed a low-urea threshold and normal sucrose threshold. Both symptoms were reported by 13 patients; their taste test results revealed low-urea and high-sucrose thresholds, with some overlap of the normal range. In three of the five in this symptomatic group whose taste testing results were in the normal range, gastric carcinoma with partial obstruction or colon carcinoma with large bowel obstruction was present, suggesting the possibility of a conditioned aversion in their symptom development.

A group of 12 patients reported only a general taste loss. These patients had a sucrose threshold that was above the 95% expected interval for normal (six patients) or above the normal median (the other six), while their urea threshold was normal or elevated. The group denying these two symptoms (22 patients) had thresholds for urea and sucrose which were, with one exception, within the 95% expected interval for normal.

The probability of having an abnormality of taste could be correlated with the patient's body burden of tumor (Fig. 4). Tumor involvement by organ system was scored on a 0–3 scale, and the sum of the scores was

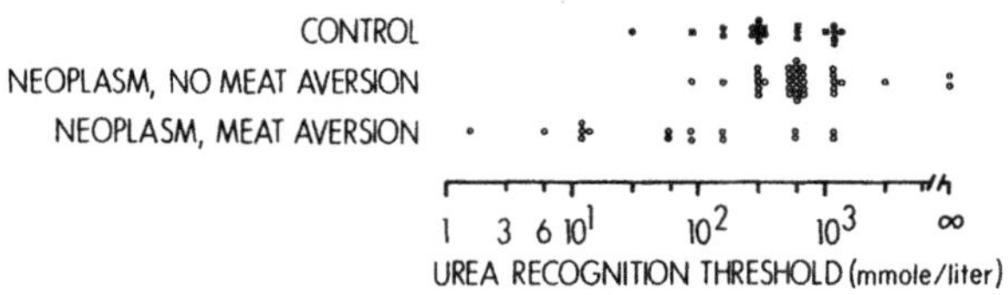

Fig. 3. Urea recognition threshold data divided as to the presence or absence of the symptom of meat aversion. Legend same as in Fig. 1.

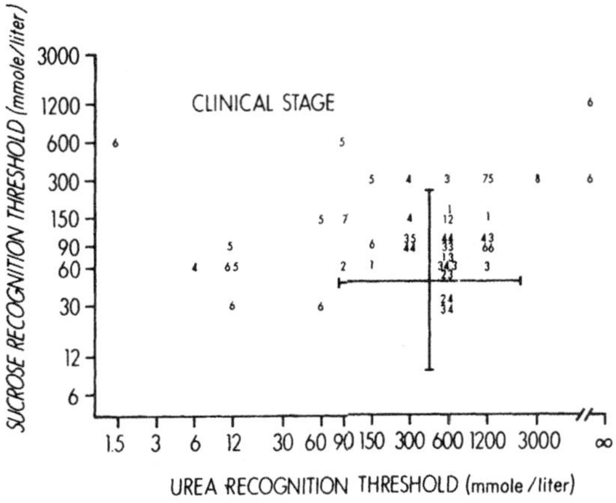

Fig. 4. Correlation of tumor extent score (DeWys and Walters, 1975) and recognition threshold for urea and sucrose. Prediction interval of 95% for normal is shown as the intersecting lines. Patients with limited disease (score 1, 2) had normal taste, whereas the majority of patients with more extensive disease (score 5 or more) had abnormal taste.

called the tumor extent score (DeWys and Walters, 1975). All nine patients with limited disease (tumor extent score 1 or 2) had normal recognition thresholds for both sucrose and urea (Fig. 4). Of 21 patients with slightly more advanced disease (tumor extent score 3 or 4), three had an abnormality of recognition for either sucrose or urea. Patients with advanced disease (tumor extent score 5 or more) had a very high probability of having an abnormality of taste; 15 out of 20 had an abnormality of recognition for sucrose or urea or both. This correlation of taste abnormality with advanced disease is highly significant (tumor extent score 1–4 versus 5 or more; $\chi^2 = 19.3$; $p < 0.001$).

In a series of patients, taste sensation was tested before and after a course of antineoplastic treatment, and the serial results were divided into three categories: clinically improved, clinically status quo, or clinically worse. The categorization of response to treatment was made without knowledge of the serial taste testing results. In nine patients who responded to therapy with reduction in their tumor size, there was a trend toward normalization of taste abnormalities; four patients changed from low urea threshold toward normal values and five patients shifted from high normal or elevated values for sucrose toward lower values (DeWys, 1974). In six patients whose tumor was essentially unaltered by treatment, there was no significant change in their taste testing results on serial determinations (DeWys, 1974). A group of 11 patients had progression of their tumor while receiving treatment. The results are less clear-cut, but it is noteworthy that two patients had increases of their glucose threshold into the abnormal range, and that two patients had decreases of their urea threshold into the abnormally low range.

In 40 patients of the original series of 50 we were able to estimate their daily caloric intake and could correlate caloric intake with their taste testing results. Patients recorded a 5-day log of food intake, and this information on food intake was converted to caloric intake by reference to standard tables of food composition (U.S. Department of Agriculture, 1963). To allow for differences in body size caloric intake was expressed as calories per kilogram body weight per day (Fig. 5). Patients whose taste test results fell within the 95% prediction range of normal were divided into two groups as to the presence or absence of other causes of decreased caloric intake. Patients having symptoms of pain, nausea, bowel obstruction, or brain metastases had significant (column 2 versus 1; $p < 0.02$) reduction in caloric intake compared with patients not having these symptoms (hereafter called the asymptomatic group). Patients with an abnormally low urea threshold had caloric intakes which were significantly lower than the asymptomatic group (column 3 versus 1; $p < 0.02$) and also lower than the total group with normal taste (column 3 versus 1 + 2; $p < 0.03$). In the group with an elevated sucrose threshold several patients had low caloric intake, but the caloric intake of the whole group compared with the asymptomatic group did not reach statistical significance. One patient who had an abnormally low-urea threshold and an abnormally high-sucrose had a fairly good caloric intake. This patient seemed highly motivated and was apparently able to maintain caloric intake in spite of altered taste sensation. When all the patients with altered taste sensation are lumped together their caloric intake was significantly lower than the asymptomatic group (column 3, 4, +5 versus 1; $p < 0.02$), while the comparison with all patients with normal taste sensation approached statistical significance (column 1 + 2 versus 3, 4, +5; $p \simeq 0.07$).

The finding of an abnormality on taste testing could be correlated with the likelihood of patient weight loss before and/or after testing. Of 17 patients having a high-sucrose and/or a low-urea threshold, 16 had significant weight loss (2.3 kg in a 2-month period before and/or after testing), while 22 of 33 patients with normal thresholds had similar degrees of weight loss. Looked at another way, 12 patients did not have weight loss and 11 of these had normal taste thresholds. This distribution approaches statistical significance ($\chi^2 = 3.25; p \simeq 0.07$).

The anorexia of malignancy has been described qualitatively previously, but the physiological determinants of this syndrome have not been delineated. The present studies suggest that abnormalities of taste sensation may be an important physiological determinant of the anorexia of

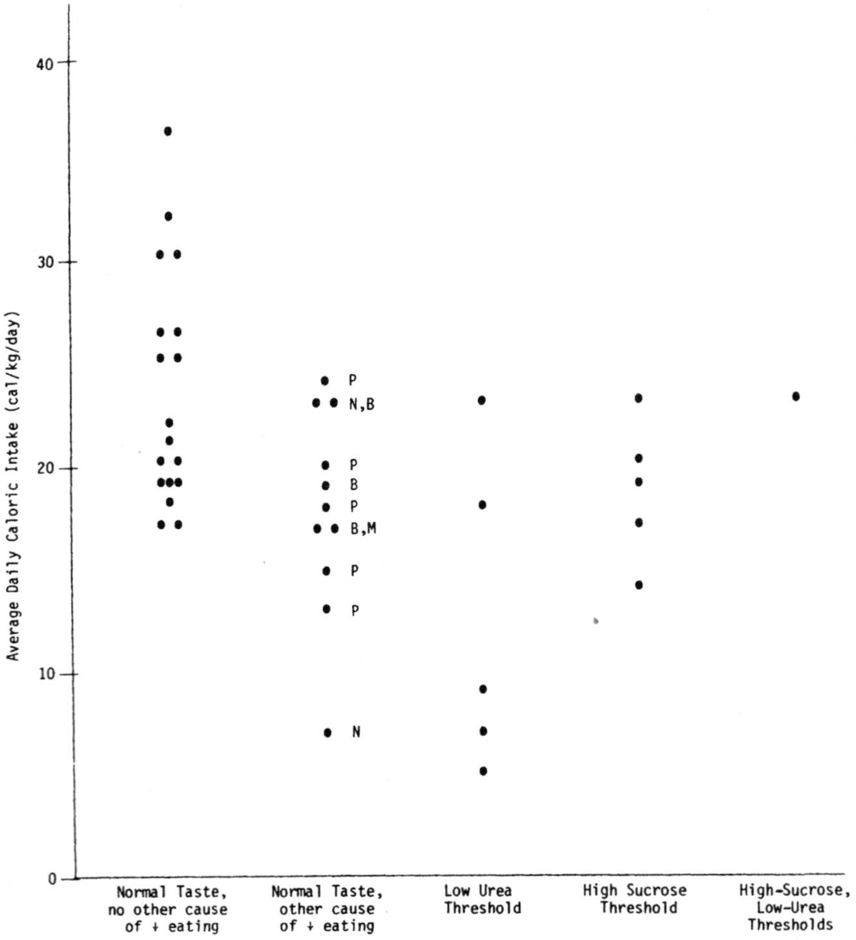

Fig. 5. Correlation of caloric intake and recognition thresholds for urea or sucrose. Each point represents data from one patient. For patients in the second column other causes of decreased eating included pain (P), nausea (N), bowel obstruction (B), and brain metastases (M). ↓, decreased.

malignancy. It is well known that the hedonic value of taste input influences food intake behavior.

A positive taste stimulus triggers multiple physiological reflexes, many of which are contributory to the intake and digestion of food. The nature of the taste sensation modifies the volume and character of saliva (Kare, 1969). Appealing foods result in copious gastric secretions (Kare, 1969; Pavlov, 1910). Positive orogastric stimulation also causes a variety of physiological changes conducive to food intake behavior, which include alterations in respiratory quotient and blood glucose (Nicolaidis, 1969). Patients with elevated taste thresholds may not receive a sufficiently positive taste stimulus and thus may not satisfactorily trigger these physiological reflexes, and this may explain symptoms of anorectic patients. Failure of adequate salivation may explain the difficulty in swallowing reported by some patients. Inadequate gastric secretions may account for a sense of early fullness. Inadequate pancreatic secretions may lead to sluggish digestion of the first meal of the day and may elucidate the decreased eating of meals later in the day described by many patients.

The tastes of acid or bitter may cause blunting of these physiological reflexes (Kare, 1969). In addition, it is noteworthy that within a population of normal subjects, taste threshold for bitter correlated with food dislikes: subjects with a low taste threshold for bitter reported a higher percentage of dislikes on a standard food list (Fischer and Griffin, 1961; Fischer *et al.*, 1966). By extension, one can suggest that patients with an abnormally low bitter (urea) threshold may have blunting of physiological reflexes to food intake and may experience a dislike for a high percentage of foods.

The mechanism of these taste abnormalities in cancer patients is not understood. It is known that cell mass and cell turnover in normal tissues may be reduced as a systemic effect of malignancy (DeWys and Mansky, 1972; Hutchison *et al.*, 1972). Because taste bud cells have a high rate of cell turnover (Biedler, 1970), they may be vulnerable to depression of cell renewal as a systemic effect of malignancy. This could result in decreased numbers of taste bud cells and might produce an elevated taste threshold, such as was observed for sucrose. Another possible explanation is the development of a deficiency state, such as zinc deficiency, which can cause elevation of taste threshold (Henkin *et al.*, 1971). This is currently under study.

To explain the lowered threshold for bitter taste, one could speculate that circulating substances may provide subthreshold stimulation of taste buds and in this way lower the taste threshold. It is well known that circulating substances can stimulate taste sensation (Greenfield, 1950; Hartridge, 1945). It is also known that stimulation of the taste buds with subthreshold concentrations of one substance can lower the taste threshold for another substance as much as 32-fold (Griffin and Fischer, 1960). Many amino acids elicit a bitter taste (Solms, 1969; Kirimura *et al.*, 1969; Halpern *et al.*, 1962). Perhaps altered blood levels of amino acids or other substances in the presence of malignancy provide subthreshold stimulation of the taste buds and thus lower the threshold for bitter taste.

Another speculation is a conditioned alteration in the bitter threshold in response to toxicosis which may have followed ingestion of bitter substances. It is known that bitter receptors have evolved as a protection against bitter poisons (Garcia and Hankins, 1975). Perhaps in the cancer patient, altered metabolism induced by the patient's tumor leads to the development of toxins or deficiencies on ingestion of bitter foods. Perhaps the bitter threshold can be adjusted to protect against such adverse metabolic effects of certain foods.

Future studies should seek better delineation of the time course of development of these taste abnormalities and the time course of the effect of altered taste sensation on caloric intake. (We have the impression that patients can cope with altered taste sensation for a variable period of time but then coping fails and caloric intake falls.) Investigations should also attempt to delineate other factors such as personality and motivational determinants which may determine why some patients with abnormal taste sensation have reduced caloric intake while others do not. And finally, research to elucidate the pathophysiology of these alterations may permit the development of specific intervention strategies.

ACKNOWLEDGMENT

This investigation was supported by Contract NO1-CP-65779 from the National Cancer Institute.

REFERENCES

Biedler, L. M. (1970). Physiological properties of mammalian taste receptors. *Taste Smell Vertebr. Ciba Found. Symp., 1969* pp. 51–67.

DeWys, W. D. (1970). Working conference on anorexia and cachexia of neoplastic disease. *Cancer Res.* **30,** 2816–2818.

DeWys, W. D. (1974). Abnormalities of taste as a remote effect of a neoplasm. *Ann. N. Y. Acad. Sci.* **230,** 427–434.

DeWys, W. D., and Mansky, J. (1972). Studies of the specificity of the systemic aspect of gompertzian tumor growth slowing. *Proc. Am. Assoc. Cancer Res.* **13,** 104.

DeWys, W. D., and Walters, K. (1975). Abnormalities of taste sensation in cancer patients. *Cancer* **36,** 1888–1896.

Fischer, R., and Griffin, F. (1961). "Taste-blindness" and variation in taste-threshold in relation to thyroid metabolism. *J. Neuropsychol.* **3,** 98–104.

Fischer, R., Griffin, F., and Rockey, M. A. (1966). Gustatory chemoreception in man: Multidisciplinary aspects and perspectives. *Perspect. Biol. Med.* **9,** 549–577.

Garcia, J., and Hankins, W. G. (1975). The evolution of bitter and the acquisition of toxiphobia. *Olfaction Taste, Proc. Int. Symp., 5th, 1974.*

Greenfield, I. (1950). Sodium succinate as a test of circulatory efficiency. *Ann. Intern. Med.* **32,** 524–527.

Griffin, F., and Fischer, R. (1960). Differential reactivity of saliva from "tasters" and "nontasters" of 6-n-propylthiouracil. *Nature (London)* **187,** 417–419.

Halpern, B. P., Bernard, R. A., and Kare, M. R. (1962). Amino acids as gustatory stimuli in the rat. *J. Gen. Physiol.* **45,** 681–701.

Hartridge, H. H. (1945). The importance of taste and smell in nutrition. *J. Physiol. (London)* **103,** 34P–35P.

Henkin, R. I., Gill, J. R., and Barter, F. C. (1963). Studies on taste thresholds in normal man and in patients with adrenal cortisol insufficiency: The role of adrenalcorticalsteroids and of serum sodium concentration. *J. Clin. Invest.* **42,** 727–735.

Henkin, R. I., Schechter, P. J., Hoye, R., and Mattern, C. F. T. (1971). Idiopathic hypoguesia with dysguesia, hyposmia and dysosmia: A new syndrome. *J. Am. Med. Assoc.* **217,** 434–440.

Hill, A. B. (1961). "Principles of Medical Statistics," p. 172. Oxford Univ. Press, London and New York.

Hutchison, D. J., Philips, F. S., Schmid, F. A., and Sodergrenand, J. E. (1972). Inhibition of host proliferative tissue by L 1210 leukemia: The inverted effect of methotrexate (MTX). *Proc. Am. Cancer Res.* **13,** 18.

Kare, M. R. (1969). Digestive functions of taste stimuli. *Olfaction Taste, Proc. Int. Symp., 3rd, 1968* pp. 586–592.

Kirimura, J., Shimizu, A., Kimizuka, A., Ninomiya, T., and Katsuya, N. (1969). The contributions of peptides and amino acids to the taste of foodstuff. *J. Agric. Food Chem.* **17,** 689–695.

Nicolaidis, S. (1969). Early systemic responses to orogastric stimulation in the regulation of food and water balance: Functional and electrophysiological data. *Ann. N.Y. Acad. Sci.* **157,** 1176–1203.

Pavlov, I. P. (1910). "The Work of the Digestive Glands" (transl. by W. H. Thompson.), 2nd ed. Griffin, London.

Solms, J. (1969). The taste of amino acids, peptides and proteins. *J. Agric. Food Chem.* **17,** 686–688.

U.S. Department of Agriculture. (1963). "Composition of Foods." USDA, Washington, D.C.

DISCUSSION

Pangborn: As compared to normals, does the composition of saliva in cancer patients differ?

DeWys: Ongoing studies are looking at the mineral content and also the various proteins in the saliva. It is suggested that there are some differences, primarily in protein content.

Cantor: In one of your earlier slides, I noted several people who had an aversion to meat. I assume that it was cooked. In addition, there was aversion to coffee and also cocoa.

DeWys: It is cooked meat. They just are unable to eat certain kinds of meat. These patients with an aversion to coffee or cocoa were all included in the meat aversion group.

Cantor: It is interesting that the processes which these products undergo are all roasting or cooking, and that many of the significant flavors associated with coffee, cocoa, and meat are artifacts; also, there is a time factor in flavor produced and a fine line between acceptability and bitterness.

DeWys: That is correct. Indeed, if you have a patient who is able to verbalize, he will distinguish such differences. Some patients will simply say coffee tastes bad; others will say coffee tastes bad because it has a bitter taste.

Naim: Since these patients all get continuous medication, is there not an effect of this medication on sensory quality?

DeWys: It was appreciated that this might be a variable, and we attempted to deal with it as follows: (1) the majority of the patients were tested before they started on a drug treatment program; (2) the majority of the drug treatment programs are given on an intermittent basis; that is, they might have an injection of medication and then not have another treatment until 3 weeks later. Our testing was done immediately before their next treatment. We have done some very limited studies looking at serial taste studies hours and days after the administration of a medication. In the couple that we have looked at there has not been a significant trend. I do not think we have the whole answer.

Nicolaidis: Most of your patients were breast cancer patients, correct? Are all of the conclusions related to these people?

DeWys: Yes. There were 15 out of the total series of 50 with breast cancer. That reflects the very prominent distribution of that disease in the population and the fact that they are referred to a medical oncology program. I could have broken down the data as to the total series and the breast cancer series alone, or I could have broken out any disease category, and we would have found roughly similar distributions either by symptom category or by taste-testing result. There is no effect of disease site in these data.

Brightman: Is there any other literature you are aware of on meat aversion? I have the impression that this was a symptom that trauma patients frequently report.

DeWys: I think much of this is at the level of anecdotal description. There are similarly anecdotal reports of patients with various liver diseases having meat aversion. Again, there is to my knowledge no data on taste in these patients.

Jordan: I get the feeling you are saying that the abnormal taste is preceding a loss of weight or the cachetic state.

DeWys: What we have done is a cross-sectional study. I do not really have a good handle on which came first, whether the patient had a taste abnormality before they lost weight. I think that kind of study has to be done. However, it is not a simple weight loss leading to taste abnormality because of several observations. Within our control population we had several normal volunteers who had been in an obesity weight-reducing program. We also had in the control population several people who had lost weight because of other illnesses, and finally in the tumor population, we had patients who had lost weight because of bowel obstruction or pain or various other abnormalities, and these patients fell within the normal taste threshold. Now, how much of a role these taste abnormalities play in weight loss I think will require longitudinal studies. Some of these taste abnormalities have been appreciated in patients before they knew their diagnoses. For example, Henkin at his taste clinic tells me that he now is diagnosing cancer patients on the basis of their taste abnormalities.

Sharma: Have there been some studies on the cellular dynamics or the turnover rates of the taste buds?

DeWys: That is a speculation; we know there is decreased turnover of the mucosa in the intestine. We have studied turnover of stem cells in the bone marrow. That has decreased. I am speculating that that may apply to the taste buds cell pool as well.

Maller: I am very interested in the taste-testing procedure, in particular, the consistency or reliability of the technique. You are testing these patients over a long period of time. Do you have some index of consistency or reliability of the technique?

DeWys: We have not analyzed that to any great extent, systematically. The major analysis was in that group of patients whose disease did not change. If you have patients in whom other variables are changing, you cannot use that data to test for the reproducibility of the technique.

Maller: What about your control patients? The ones who did not have a disease?

DeWys: We had repeat testing in a number of those patients. I have not looked in detail at the data. It is in our files. You apparently have some basis for making this comment.

Maller: Yes. The details of the technique are not clear. I want to find how reliable a technique it is and also get the details from you. Is it a standardized technique? Can anybody administer the testing?

DeWys: We have it done by one person. I have done some of it, but for the most part it has been done by a technician who is highly motivated and does an excellent job.

Maller: Standardized so that anybody could come in and essentially use the same taste-testing procedure?

DeWys: I think so.

Chapter 20

Palatability and Oral Factors and Their Role in Obesity

HENRY A. JORDAN AND THERESA A. SPIEGEL

Although common sense suggests that oral factors must influence the selection of food and initiation and termination of eating, most studies with human subjects have overlooked the important sensations arising from the oral cavity and their effect on the control of food intake. There is, however, a large body of research in animals which clearly underscores the role of palatability and other oral sensations in the control of food intake.

The relative contributions of oropharyngeal, gastric, and postabsorptive factors in the production of satiety have been successfully explored in several animal species, where satiety is defined as the cessation of food intake (Janowitz and Grossman, 1949; Mook, 1963). Interspecies differences have been marked, however, and therefore understanding hunger and satiety in man in light of these studies is not simple extrapolation.

Investigations with animals have suggested that oropharyngeal factors are indeed important, not only in governing appetite, but also in their contribution to satiety (Berkun *et al.*, 1952; Epstein and Teitelbaum, 1962, Janowitz, 1967; Kohn, 1951; Share *et al.*, 1961). For instance, sham feeding (food eaten orally but never reaching the stomach) in both rats and dogs with esophageal fistulae eventually ceases, but it takes longer than if the food is allowed to reach the stomach (Janowitz and Grossman, 1949; Mook, 1963). Also, various amounts and kinds of food placed directly into the stomach affect both sham and normal feeding (Kohn, 1951; Moyer and Bunnell, 1962; Smith *et al.*, 1962). Thus, the stomach itself does contribute to regulation of intake and to satiety, but this contribution provides greatest accuracy when it is closely associated with oropharyngeal factors. Food by mouth in both the rat and dog (the two species most widely studied) produces greater satiety and more accurate regulation than food ingested directly into the stomach. Janowitz and Grossman (1949), working with dogs, likewise concluded that oropharyngeal factors are fully effective only when closely associated with the foods entering the stomach. It seems clear from this work that both oropharyngeal and gastric factors may exert influences when they are separated and operate in isolation, but that maximum regulation and production of satiety are possible only when the two work together. To determine how man's perception of satiety and the size of his meals are controlled by oropharyngeal and gastric events during a meal, and how these controls are related to the processes regulating energy balance over the past several years, we have conducted a number of studies. At the same time we have observed what happens to obese patients when the role of oropharyngeal factors is not considered in a treatment regimen. Before discussing our experiences with the obese, we will turn to our studies of man's eating behavior in the laboratory, where we have manipulated oral stimulation by varying the type of food and the manner and rate of ingestion and have also isolated the various elements of the ingestive pathway.

First, we eliminated oral factors all together by by-passing the oropharyngeal cavity with a nasogastric tube (Jordan, 1969). The remarkable thing is that intragastric feeding in a single meal can be virtually identical to oral feeding with respect to the rate of intake and the total amount consumed. This would suggest that oropharyngeal factors are unimportant. However, several additional manipulations demonstrate the importance of oral factors in the control of feeding. When the rate of delivery of the diet ingested orally or intragastrically was varied day to day, the variability in meal size was greater in intragastric compared to oral feeding, although the amount consumed was directly related to the rate at which the food was ingested in both oral and intragastric feeding. Thus, oral metering adds to the precision of the control of meal size, but rate of ingestion controls the development of satiety in both oral and intragastric feeding.

We also had subjects simultaneously ingest food through the mouth and through the nasogastric tube. In this condition, they ingested amounts of diet in excess of the base line amounts. Although there was an attempt to compensate for the stomach load by reducing oral intake, this reduction was not sufficient to cause regulation (Jordan, 1969). Since the simultaneous feeding situation increases the rate of ingestion, it was not clear whether this increased intake was due to the increased rate of ingestion or to the need to maximize oral satisfaction in spite of food entering the stomach.

An additional experiment by Spiegel and Jordan was conducted to isolate the effects of rate versus route of ingestion on meal size and satiety in simultaneous feedings as well as to explore the persistence of this phenomenon. Four normal weight males (18–21 years old) were subjects in this experiment. The diet used was a nutritionally balanced liquid food (Metrecal, Mead–Johnson). The basic procedure and equipment for oral and intragastric feeding used in these experiments have been described in detail elsewhere (Jordan, 1969; Spiegel, 1973). Briefly, for orally ingested meals subjects sat at a table with a straw which was connected by Tygon tubing through a Sigmamotor peristaltic pump to a hidden reservoir of diet. To eat, the subject depressed a lever at the table, which activated the pump. The rate of pump delivery of the diet was present by the experimenter. The pattern and amount of ingestion were recorded either automatically on a cumulative recorder or by observation by the experimenter.

Intragastric feeding was accomplished by passing an opaque rubber nasogastric tube before each intragastric meal. The nasogastric tubing was connected through a second pump to a second hidden reservoir. The pump rate was preset by the experimenter, who also recorded the volume ingested per minute from the reservoir.

Meals were 20 min long. Subjects were given the daily paper to read or could bring their own reading material. They rated their degree of hunger and fullness and the taste of the diet at the beginning and end of each meal.

Subjects were weighed daily but were not allowed to see their weight or to weigh themselves outside the laboratory. They were also interviewed daily as to general health, mood, and their feelings of fullness and satiety during and following the meal.

In a first experiment designed to test the persistence of increased intake during simultaneous ingestion, oral–intragastric feeding was compared to oral feeding in two meals per day (lunch and dinner) for 5 or more days each. Breakfast and snacks were *ad libitum* diet. The following three conditions were run at lunch and dinner: (a) 5 consecutive days of only oral feeding to determine meal size and daily caloric intake under base line conditions. (See the first column of Table I for the pump rate for each subject.) (b) 2–5 consecutive days of oral–intragastric controls at the same pump rate to assess the effect of the nasogastric tube itself on intake (blank intragastric condition) and to determine whether subjects consciously reduced oral intake when they thought food was being delivered intragastrically (0% intragastric condition). Subjects were intubated at every lunch and dinner. In the blank intragastric condition they were told that no food

TABLE I

Mean Intakes during Oral and Oral–Intragastric (O–IG) Feeding (5 Days of Each) in Experiment 1

	Pump rate[a]		Lunch[b]		Dinner[b]		Daily total[b]	
Subjects	Oral	O–IG[c]	Oral	O–IG	Oral	O–IG	Oral	O–IG
KG	100	200	1366	1460	1331	1571	3980	3956
BS	150	200	885	1265	1009	973	2425	2651
BM	100	250	968	1343	1078	1450	3039	3502
RW	100	250	1543	1838	1234	1291	3277	4029

[a] In cm^3/min.
[b] In cm^3/or kcal.
[c] Combined oral and intragastric rates are given. In all cases the intragastric pump rate was 100 cm^3/min.

would be delivered intragastrically and only oral food was given. In the 0% intragastric condition subjects were told that food might be delivered intragastrically on some occasions although only oral food was given. (c) 5–8 days of simultaneous oral–intragastric feeding in which subjects were intubated at every lunch and dinner and were told that sometimes food would be delivered intragastrically. In the simultaneous condition, however, food was always delivered intragastrically as well as orally. (See second column of Table I for the pump rate for each subject.)

The results of the oral–intragastric control meals showed that the presence of a blank nasogastric tube did not suppress the subjects' intake, and that none of the subjects was consciously attempting to reduce his oral intake when he thought food might be delivered intragastrically.

The individual intakes from lunch and dinner meals during the oral and oral–intragastric feeding conditions averaged over 5 days each are presented in Table I. The difference between oral and oral–intragastric meals was significant. On the average subjects ate 222 cm^3 more in the oral–intragastric condition compared to the oral feeding condition at lunch and again at dinner every day of the experiment. There was no significant difference between the sizes of lunch and dinner, and there was no significant change in meal size over days. In other words, there was no tendency for subjects to increase or decrease intake at either lunch or dinner in either oral or oral–intragastric feeding conditions as the experiment progressed.

The daily lunch and dinner intakes of BM illustrate the major effects of oral–intragastric feeding, and they are presented in Fig. 1. Following 5 days of oral feeding, BM was run for six meals of oral–intragastric control feeding during which his intake remained at oral base line levels. During oral–intragastric feeding his intake remained elevated, although it immediately returned to oral base line when manipulation was dropped for 2 days and increased when oral–intragastric feeding resumed. In all conditions his lunch and dinner meals were about the same size. The oral–intragastric manipulation was equally effective at lunch and dinner meals throughout the experiment.

The subjects' taste ratings indicated that they like the flavors of diet they chose to eat, and there was no consistent change in taste ratings from the beginning to the end of meals nor with the different intubation conditions.

There were no significant differences in hunger and fullness ratings between oral and oral–intragastric conditions for either meal for the 5-day periods. That is, at the beginning of a meal subjects rated themselves with

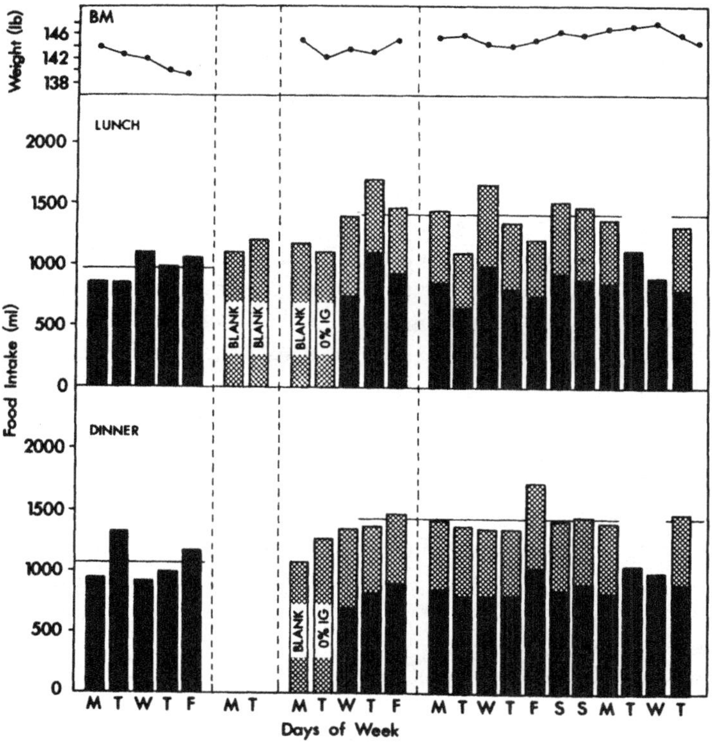

Fig. 1. Weight changes and intakes at lunch and dinner during oral feeding (filled bars), oral–intragastric control feeding (blank and 0% intragastric), and simultaneous oral–intragastric feeding (partially filled and partially hatched bars) of a representative subject.

the same degree of hunger and emptiness whether it was lunch or dinner and whether it was an oral or oral–intragastric meal every day of the experiment. Similarly, when the meal was over, even though subjects had eaten significantly more during an oral–intragastric meal than during an oral meal, they rated themselves no more full and no less hungry regardless of the meal (lunch or dinner) and regardless of the day of the experiment. Subjects, however, sometimes complained of being uncomfortably full after oral–intragastric meals.

Table I also presents the mean daily intakes of each subject during the 5 days of oral feeding and the first 5 days of oral–intragastric feeding. Although every subject ate more on a daily basis during oral–intragastric feeding, this difference was not significant. Since subjects did eat significantly more at lunch and dinner during oral–intragastric feeding, they must have reduced the size and/or frequency of other meals such that a significant difference in daily intake was not obtained. Analysis of the breakfasts and snacks of each of the subjects indicated that three of the subjects (KG, BS, and BM) did reduce both the size and the frequency of breakfasts and snacks. The fourth subject (RW), on the other hand, actually increased the size and frequency of evening snacks. Subjects' weight changes were not significant during the time period involved in this experiment.

This experiment demonstrated that simultaneous oral–intragastric feeding leads to consistent and sustained increases in food intake. Subjects ate significantly more (222 cm^3 on the average) at lunch and again at dinner for 5 days of oral–intragastric feeding compared with 5 days of only oral feeding. Although the hunger and fullness scales were not sensitive to this increase in intake, subjects sometimes complained to the experimenters at the end of oral–intragastric meals that they felt uncomfortably full. The increased intake at lunch and dinner during oral–intragastric feeding did not produce a significant increase in daily intake or a significant weight gain, however. Three of the subjects (KG, BS, BM) ate fewer and smaller breakfasts and snacks than during oral feeding. The fourth subject (RW) increased his intake at breakfasts and snacks in addition to his increased intake at lunch and dinner. These results indicate that the simultaneous oral–intragastric feeding manipulation might lead to increased daily intake and experimental obesity only if used for periods longer than 5 days or if used at every meal and snack.

In the oral–intragastric feeding situation used in this long-term experiment and used by Jordan (1969), not only were subjects fed simultaneously via two routes but they also received the diet 1⅓ to 2½ times faster than when feeding only orally. (See pump rates presented in Table I.) Given that Jordan found that intake during oral or intragastric ingestion alone is directly related to the rate of delivery of the liquid diet, it is possible that subjects ingested increased amounts of diet during simultaneous oral–intragastric feeding (compared to oral only feeding) merely because the diet was delivered more rapidly, not because they were eating simultaneously via two routes. A second, shorter term experiment was conducted to compare rates and routes of ingestion to determine the contribution of each of these two factors to the increased ingestion found in simultaneous feeding. Subjects were fed orally at 100 and 200 cm^3/min, intragastrically at 100 and 200 cm^3/min, and simultaneously orally and intragastrically at 100 cm^3/min each (combined rate of 200 cm^3/min). Thus, it was possible to assess the effect of rate of ingestion on intake (all 100 cm^3/min conditions compared to all 200 cm^3/min conditions), the effect of route of ingestion on intake (oral only compared to intragastric only regardless of rate), and the effect of simultaneous oral–intragastric feeding on intake with rate held constant (oral–intragastric feeding at 200 cm^3/min compared to oral only and intragastric only at 200 cm^3/min).

All four of the subjects were run in the short-term experiment following their experience in the longer study. They came to the laboratory 4 or 5 days each week at the same time every day for lunch. Lunch was always 20 min long and followed the standard procedure outlined earlier. Subjects were run on oral only feeding (pump rate 100 cm^3/min) to determine that their intake was stable (3–4 days). They were then run for four trials in each of the following five conditions in random order: (a) oral only, pump rate 100 cm^3/min; (b) intragastric only, pump rate 100 cm^3/min; (c) oral only, pump rate 200 cm^3/min; (d) intragastric only, pump rate 200 cm^3/min; (e) simultaneous oral–intragastric ingestion, oral and intragastric pump rates each at 100 cm^3/min. RW was the only exception; he was not run in the intragastric 200 cm^3/min condition because he seemed to find the rapid rate of intragastric ingestion distressing. The other subjects tolerated this condition well.

Table II presents the mean intakes obtained in each condition for each of the subjects and the grand mean for each condition. Three independent comparisons were made of the means obtained in the five conditions. The first comparison showed that subjects ate significantly more (mean of 177 cm^3) when the rate was 200 cm^3/min compared to 100 cm^3/min regardless of the route of ingestion. The second comparison showed that there was no significant difference between oral only and intragastric only ingestion, regardless of rate (means of 968 cm^3 for oral ingestion at both rates and 1006 cm^3 for intragastric ingestion at both rates). The third comparison showed that subjects did eat significantly more (135 cm^3) when feeding simultaneously orally and intragastrically at 200 cm^3/min than when feeding orally only or intragastrically only at 200 cm^3/min. Thus, there was a significant effect of rate of ingestion on intake and a significant effect of simultaneous ingestion on intake when the rate of ingestion was held constant. Table II shows, however, that there were individual differences in the response to these two factors. BS was primarily affected by the rate of ingestion, KG by the simultaneous oral–intragastric manipulation, and BM by both the rate and the simultaneous manipulation. RW was not included in this analysis because of his missing data from the intragastric conditions at the higher rate, but he was also affected by both the rate of ingestion and the simultaneous manipulation.

The cumulative curves of intake during meals were also analyzed for each subject. All of the subjects showed the same general patterns of ingestion that they had demonstrated during the longer term study. That is, they ate without pause for some length of time at the beginning of the meal; once they had paused, three of the subjects (BS, BM, RW) returned periodically for more to eat, and KG stopped eating altogether.

Although KG had a slightly different pattern of ingestion than the other subjects, his cumulative intake curves for the five conditions demonstrate the major effects of rate and route of ingestion on patterns of intake within the meal, and his intake patterns are presented in Fig. 2. It is obvious from Fig. 2 that rate of ingestion has a marked effect on intake during the first part of the meal regardless of the route of ingestion. This effect of rate was seen in all of the subjects and is due to the fact that they ate without pause for some minutes at the beginning of the meal. However, all of the subjects paused earlier in the meals at the higher rate. KG paused early enough that his total intake by the end of the meal was not different under the two rate conditions. The second major finding illustrated by Fig. 2 is that the oral–intragastric manipulation affects intake after rate has had its effect. Sub-

TABLE II

Mean Intakes (in cm^3 or kcal Averaged over 4 Days Each) in the Five Conditions in Experiment 2

	Condition				
	100 cm^3/min		200 cm^3/min		
Subjects	Oral	IG	Oral	IG	O–IG
BS	828	780	923	870	939
KG	1158	1143	1160	1123	1369
BM	724	890	1018	1225	1256
Grand mean	903	938	1033	1073	1188

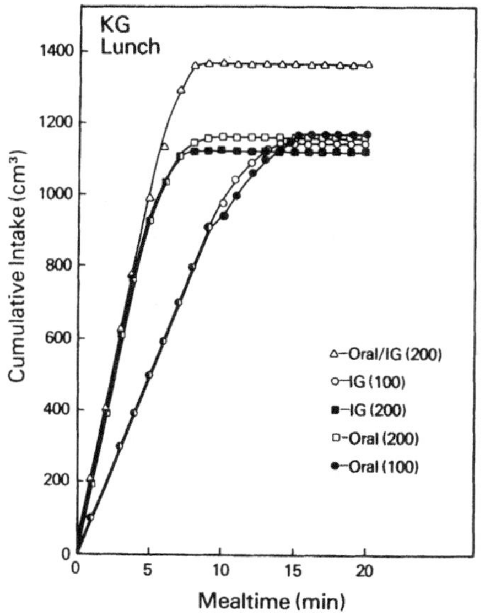

Fig. 2. Mean cumulative intake curves for 4 days each during oral, intragastric, and simultaneous oral–intragastric feeding of one subject. The pump rate for each condition is given in parentheses.

jects continued to eat rapidly or return to eat more in oral–intragastric meals at a time when eating had slowed down or terminated in oral only or intragastric only meals.

This short-term experiment demonstrates that the intake of a liquid diet is directly related to the rate of ingestion whether the route of ingestion is intragastric, oral, or simultaneously oral and intragastric. Simultaneous oral–intragastric ingestion, however, increases intake compared to oral or intragastric ingestion, even when the three conditions are matched for rate of ingestion.

Additional experiments in our laboratory by Thomas Drummond and experiments reported by Wooley and Wooley (1975) support the finding that rate of eating is an important determinant of amount eaten and degree of satiety. Wooley and Wooley (1975) measured the salivary response (a physiological index of satiety) of obese and nonobese to an ice cream dessert 1 hr following a meal (hamburger and coke) eaten slowly or rapidly. They found that both groups of subjects salivated more to the dessert when they had eaten the meal quickly than when they had eaten the meal slowly.

Drummond, using the standard Metrecal diet and 20-min meal situation of Jordan (1969), ran three subjects for one or two meals each at three pump rates of 75, 135, and 185 cm^3/min. He found that intake varied directly with pump rate with an average intake across subjects of 470, 570, and 770 cm^3 for the three pump rates. Another four subjects run on alternate days at pump rates of 90 or 180 cm^3/min ingested 870 and 1150 cm^3 on the average for the two pump rates, respectively. Drummond also found that rate significantly affected intake in the beginning of the meal. Intake after 5 min for the four subjects averaged 600 cm^3 for the fast pump rate compared to 355 cm^3 in the slow pump rate condition.

Drummond then ran five subjects at a constant pump rate (180 cm^3/min) for three to eight meals under the standard conditions and for two to five meals with instructions designed to have subjects pace their ingestion during the meal. Subjects did change their ingestion patterns under pacing instructions by eating for shorter bursts (burst size changed from a mean of 154 to 29 cm^3) and by taking more bursts (an average of 35 per meal, compared to 11 per meal), and their average total intake dropped from 993 to 784 cm^3. Only one of the five subjects did not eat less when pacing his ingestion.

Thus Drummond's results confirm that in subjects of normal weight eating faster leads to eating more, and this effect was seen from the very beginning of the meal. Even more important from the point of view of treating persons with weight problems, subjects can follow instructions which slow their ingestion rate and lead to eating less.

In a final series of experiments in our laboratory Shaw (1973) found that satiety and amount of food ingested by normal male subjects are affected by the type of food and its manner of presentation. Shaw used the liquid Metrecal diet and turkey salad sandwiches. One standard sandwich plus 4 oz of water were equivalent to one 8 oz can of Metrecal in calories (225 cal), volume, and weight (266 g).

In the first experiment Shaw ran seven subjects for four meals each in each of the following conditions: sandwiches served a quarter at a time (plus 4 oz of water after every four quarters); Metrecal served in 2-oz cups; Metrecal by pump at a rate of 160 cm^3/min. Four of the subjects ran in a fourth condition where they received Metrecal in 8-oz cups. The sandwich quarters and 2-oz cups yielded comparable levels of ingestion (683 and 691 cm^3), the 8-oz cup condition yielded significantly higher levels of ingestion (935 cm^3), and the pump yielded the highest level of ingestion (1051 cm^3), significantly higher than the sandwich quarter and 2-oz cup conditions but not significantly higher than the 8-oz cup condition. These results for the individual subjects are presented in Fig. 3. Two of the major differences among the different methods of presentation are the amount of effort and the amount of time it takes to ingest the food, both of which produce markedly different rates of ingestion. Figure 4

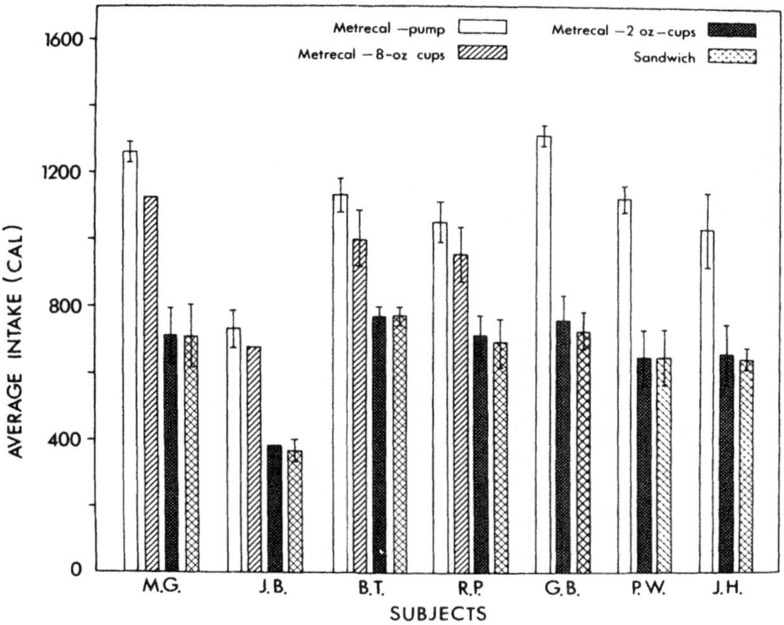

Fig. 3. Meal size (averages) as a function of type of food (Metrecal or sandwiches) and method of delivery (pump, 8-oz cup, or 2-oz cup).

demonstrates the effect of these presentation methods on rate of ingestion in an individual subject. Since the sandwich quarter and 2-oz cups of Metrecal conditions did not differ significantly, it seems that liquid food can provide a degree of satiety comparable to solid food, if it is ingested slowly.

The second experiment of Shaw's showed, however, that combining the sandwiches and Metrecal in a preload followed by a meal leads to greater ingestion than when either type of food is eaten alone. In this experiment preloads amounting to 50% of the subject's base line intake were ingested rapidly 15 min before the laboratory meal. Four subjects were run for four meals each in six different preload-meal conditions. There were two different preloads: sandwich quarters or 2-oz cups of Metrecal and three meal conditions: sandwich quarters, Metrecal by 2-oz cups, or Metrecal by pump at 160 cm^3/min. The results replicated Shaw's first experiment in that the meal factor was significant. (Subjects ate more by pump than by 2-oz cups or sandwich quarters, the latter two conditions yielding equal amounts eaten.) The type of preload did have a significant effect, which one would also expect from the first experiment since they do not yield significantly different meal intakes. However, the preload × meal interac-

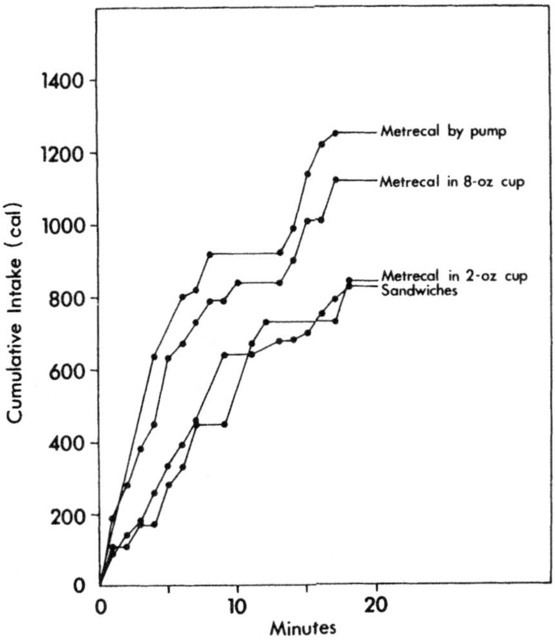

Fig. 4. Cumulative intake curves for single test meals of a typical subject. The final level of intake is predicted by a divergence between the curves as early as the first 5 min.

tion was significant. Subjects ate significantly more when the preload and meal were different types of food. The results for a single representative subject are shown in Fig. 5. The result of this experiment confirms the experience of many of us who find that a sequence of varied foods is more likely to lead to increased intake than is a single food item. This finding is also paralleled in the animal literature (LeMagnen, 1971).

To summarize our laboratory studies, we conclude from the simultaneous oral–intragastric experiments that although subjects control their intake fairly well when eating only intragastrically, when the oral component is added to intragastric ingestion, the increased rate of ingestion and the oral stimulation drive them to increased intakes, an effect which persists for up to 10 days. The studies by Drummond confirm that rate of ingestion is an important determinant of meal size and satiety. Drummond also showed that subjects can slow their rate of ingestion and thus eat less. Shaw's experiments show that when food is presented in a way that requires some time and effort to ingest, intake is reduced, probably because rate of ingestion is slowed. Furthermore, variety of oral stimulation from a sequence of different types of foods leads to increased ingestion.

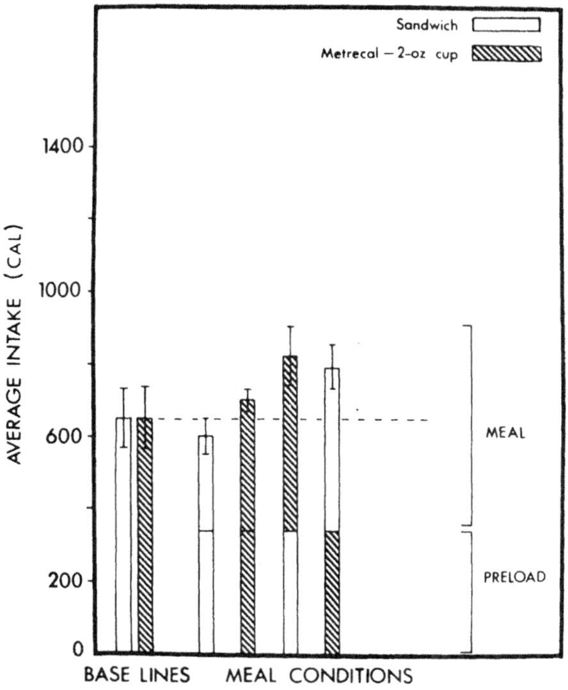

Fig. 5. Effect of type of preload (sandwich or Metrecal by 2-oz cup) on a typical subject's intake during a subsequent test meal (sandwich or Metrecal by 2-oz cup). Greatest intake over base line occurred when there was a change in food from the preload to the meal.

What are the implications of our laboratory experiments for the treatment of obesity? In our clinical population most of our patients have successfully lost weight on numerous dietary programs. Almost universally, however, they are unable to adhere to the restricted diets and regain the lost weight. They have essentially two kinds of problems. First, they are expected to adhere to their diets in the face of a wide variety of palatable foods which are readily available between meals and often within meals. This is summarized by a quote from William Beaumont (1959):

> In the present state of civilized society, with the provocatives of the culinary art, and the incentives of highly seasoned food, brandy and wine, the temptations to excess in the indulgences of the table are rather too strong to be resisted by poor human nature (p. 14).

Second, the manner in which obese persons often ingest food minimizes the amount of enjoyment and oral stimulation obtained and thus leads to a low degree of satisfaction. Let us deal with the second issue first.

Overweight persons on a diet typically find themselves staving off cravings for "forbidden" highly palatable and highly preferred foods by ingesting more acceptable low-calorie foods. Usually, after some time they ingest the preferred food anyway. Now they not only have the calories of the calorically dense food but also the calories from the less preferred foods. Under these circumstances the highly preferred food is often ingested rapidly (and sometimes surreptitiously) with little enjoyment. Even while the obese are eating "allowed foods" their food is ingested rapidly and in conjunction with other activities, another set of circumstances which minimizes the attention paid to the food eaten.

Our experience with these patients and from the laboratory studies varying rate and route of ingestion indicates that food should be eaten under circumstances which maximize oral satisfaction and enjoyment: preferred, palatable foods should be eaten, but slowly with concentration on the taste and texture of the food and with no other activities occurring to distract one's enjoyment. Like Drummond's subjects, many of our patients find that slowing down the rate of eating produces increased satiety with less food. This means that rate of eating has an important effect on meal termination. In addition, the Wooleys have shown that rate of eating affects the duration of satiety following a meal, and this may affect subsequent meal and snack initiation.

The first issue still remains, that is, the easy availability of many palatable foods. Here, the results of Shaw's experiments and our clinical experience indicate that access to these foods must be controlled both during the meal and between meals. Treatment often consists of helping the patient learn to be discriminating and selective in his food choices and to reduce the immediate easy availability of palatable foods, without denying himself these foods altogether.

In other words, treatment must be directed at decreasing the probability that the initiation of eating will occur. If eating does start, the patient must learn to establish conditions where satisfaction is maximized and termination occurs after fewer calories have been ingested.

It is a common belief that obese people have increased need for oral gratification and take great pleasure in consuming large quantities of highly palatable food. Yet our clinical experience has often shown just the opposite. Many patients eat rapidly, restrict themselves to nonpreferred foods in the face of an abundance of highly palatable foods, and feel guilty about eating, all factors leading to decreased satisfaction and increased intake. Many of these same characteristics of eating behavior studies in our laboratory lead to increased intake and decreased satisfaction in normal weight subjects.

REFERENCES

Beaumont, W. (1959). "Experiments and Observations on the Gastric Juice and the Physiology of Digestion." Dover, New York.

Berkun, M. M., Kessen, M. L., and Miller, N. E. (1952). Hunger-reducing effects of food by stomach fistula versus food by mouth measured by a consummatory response. *J. Comp. Physiol. Psychol.* **45,** 550–554.

Epstein, A., and Teitelbaum, P. (1962). Regulation of food intake in the absence of taste. smell, and other oropharyngeal sensations. *J. Comp. Physiol. Psychol.* **55,** 753–759.

Janowitz, H. D. (1967). Role of gastrointestinal tract in regulation of food intake. *Handbook Physiol., Sect. 6: Aliment. Canal* **1**.

Janowitz, H. D., and Grossman, M. (1949). Some factors affecting food intake of normal dogs and dogs with esophagastomy and gastric fistulae. *Am. J. Physiol.* **159,** 143–148.

Jordan, H. A. (1969). Voluntary intragastric feeding: Oral and gastric contributions to food intake and hunger in Man. *J. Comp. Physiol. Psychol.* **68,** 498–506.

Kohn, M. (1951). Satiation of hunger from food injected directly into the stomach versus food ingested by mouth. *J. Comp. Physiol. Psychol.* **44,** 412–422.

LeMagnen. J. (1971). Advances in studies on the physiological control and regulation of food intake. *Prog. Physiol. Psychol.* **4,** 204–261.

Mook, D. (1963). Oral and postingestional determinants of the intake of various solutions in rats with esophagene fistulas. *J. Comp. Physiol. Psychol.* **56,** 645 and 659.

Moyer, K. E., and Bunnell, B. N. (1962). Effects of stomach distention caused by water on food and water consumption in the rat. *J. Comp. Physiol. Psychol.* **55,** 652–655.

Share, L., Martinuik, E., and Grossman, M. I. (1961). Effect of preload IG feeding in oral food intake in dogs. *Am. J. Physiol.* **201,** 593–598.

Shaw, J. (1973). The influence of type of food and method of presentation in human eating behavior. Doctoral Dissertation, University of Pennsylvania, Pennsylvania (unpublished).

Smith, M., Pool, R., and Weinberg, H. (1962). The role of bulk in the control of eating behavior. *J. Comp. Physiol. Psychol.* **55,** 115–120.

Spiegel, T. A. (1973). The caloric regulation of food intake in man. *J. Comp. Physiol. Psychol.* **81,** 24–37.

Wooley, O. W., and Wooley, S. C. (1975). Short-term control of food intake. *In* "Obesity in Perspective" (G. A. Bray, ed.), p. 49. US Govt Printing Office, Washington, D.C.

DISCUSSION

Bosma: The passage of a nasogastric tube substantially changes the feeding experience in some persons. This may be a very noxious experience.

Jordan: During the first couple of times there may be a little bit of gagging and tearing and so on. After three times, I know more about their anatomy and they have gotten used to the idea, so they swallow at the right time and it really is a very, very easy experience for them.

Bosma: Do subjects continue to swallow air while the nasogastric tube is in place?

Jordan: I do not know. I have looked at these people and my feeling is that they are obviously continuing to swallow saliva, but I do not think they would continue swallowing in the same way that they would if the food were coming into their mouth.

Bosma: Further specification of the oral experience in obese persons might be interesting.

Each of the parameters of oral sensory input is probably liable to aberration of experience and to variation in acquired preferences. Touch, pressure, temperature, and taste deserve separate evaluation.

Jordan: This is one of the things that we really are attempting to do. But we find a paradox; people have said that the obese individual may have a heightened need for oral satisfaction; what we are seeing in our clinical population is that because they are told that certain foods are horrible for them—the oral satisfaction is actually reduced in these people. What is needed then is to enable somebody to get as much satisfaction out of 500 cal as they can out of 1000. Then their body weight starts to come down. Almost everything we do is directed at increasing the awareness of eating and actually getting them to take time and slow down the ingestive process itself. You know, even putting candles on the table, surrounding the eating experience by making it a dining experience, will cause you to eat fewer calories and get more pleasure.

Moskowitz: There is something that is very confusing about your data. From learning theory and as a psychologist I was taught that things which are pleasant are reinforcing. Now if you say that the obese people do not find the oral sensory experience to be reinforcing, then why do they continue to eat? That behavior should extinguish, and in fact, should lead to negative feedback.

Jordan: I have struggled with that myself for the last 10 years with obese people. They come in and say, "I really felt lousy after I did that." The problem is that there may be a very short period of time right after the ingestion where they do, or during the ingestion where they are getting positive reinforcement, and the negative psychological experience comes later. I cannot imagine why these people persist in doing this when it makes them unhappy.

Cagan: This may be an appropriate time to alter the theory to fit the data, rather than the other way around. That is an important point, because the theory does not seem to apply. The signal is not necessarily simple but rather there are multiple signals, and what the obese person is doing somehow sending his or her sensor out to one area of the experience, which is a dominating factor.

Garcia: There are some very interesting data on rats which have striking parallels for what you are doing. Peck showed that in a situation something like yours, the rat, when one adulterates the food with bitterness, will opt for a certain body weight which is a function of the palatability of the food. That is not too surprising, but what is surprising is that if he increases the work load, it works just exactly the same way. The only constant in this situation is always the body weight, and it is almost as if the animal were choosing the body weight. The harder you make him work, the less he will opt for a lower body weight.

Cagan: The terminology can be confusing. Is it correct to say that the term "palatability" is, in fact, being used in somewhat different ways? In experiments on animals, it seems to be used in an operational sense, meaning the amount ingested. In human experiments it appears to be derived from a subjective report.

Jordan: You see, hunger presents the same problem.

Cagan: Your results lead one to suggest that it may be misleading to go from the operational term, as used in animals, to a subjective evaluation by humans, because by your criterion they are going to be cutting down on that which would be deemed less palatable.

Jordan: It is always a problem. You infer hunger in an animal by how much he eats. You infer whether he likes or dislikes by how much he eats. In humans, you can ask him a question, and sometimes they do not go together.

Garcia: You can tell palatability in an animal without looking at quantity. When an animal finds something disgusting, a rat will rub his chin on the ground, a coyote will turn around and act to urinate on it. You can see disgust reactions. It is no problem telling when an infant is offended.

Pfaffmann: There is a very difficult terminology problem here. But it does not mean that we are necessarily stuck in an impass forever. It just means that the methodology needs improvement. Steiner developed the procedure of taking pictures with newborn babies to get their facial expressions following taste stimulation. Grill has been developing a similar kind of methodology to apply to the animal to determine if, in fact, there are mimetic responses that can be seen. Now with videotape you can time the sequence and find that indeed it is possible to get characteristic responses, mimetic and oral responses, to stimuli which will also affect food intake. I would point out that this kind of analysis of other behavior patterns rather than simply ingestion and intake may provide one resolution of some of this circularity.

Cagan: Steiner's results are certainly dramatic in showing the facial expressions elicited by various taste stimuli. I originally raised the question about palatability to focus attention on the possibility that somewhat different phenomena may be involved using the different techniques.

Rosenthal: The trickiness of palatability I think is seriously affected by the collateral point made about the higher acceptance and higher ingestion of combination foods. Now my own guess in gastronomy is that that greatly enhances the palatability, whether it is through eliminating the boredom factor or not, I do not know. But that does affect the question of palatability and there is no question that with the combination, in both mammals and humans, you will get much higher ingestion of food.

Jordan: That is what desserts are all about. We sometimes just tell a patient to have the dessert first, which gets that one out of the way, and they can finish up on the less acceptable food later on.

Schrire: The point you had in your summary suggests that in a society where highly palatable foods are abundant and readily available, the probability that man will overeat and become obese is quite high. That kind of statement should surely be hedged in with the idea that it depends on whether it is socially acceptable or not to eat constantly.

Jordan: The most interesting question for me is not why there are so many obese Americans running around the streets who are restrained eaters. They may be avoiding the social interesting new work coming out of Northwestern that there are large populations of Americans running around the streets who are restrained eaters. They may be avoiding the social stigma of being obese. When you look at their behavior, they are just the same as an obese individual who is trying to diet and lose weight. It really is a puzzle. I do not know how some people are able to do it and stay thin and others not.

Schrire: What if it were socially unacceptable to eat, munch, and kind of feed all the time? My point is that if that were so, people would not do it. You know, it is socially unacceptable at the moment that someone should be sitting here, chewing potato chips or anything like that, which is probably why no one is doing it, though there might be people who want to do it. But it is not going to be done.

Jordan: Well, it was done an hour ago at our coffee break. It is socially acceptable to have six ingestions a day in our society at least.

Nicolaidis: You were talking about overeating when you give a preload or gastric pass over association, but when one looks at the valuate curve, there is no overeating because the valuate is constant.

Jordan: It is too short a period.

Nicolaidis: Regulatory eating.

Jordan: What happens is that in the periods where *ad lib* feeding is still acceptable, in breakfast and in snacks, there is an alteration. The subjects drop their calories there. That is the only place they can climb out from under this phenomenon and they do regulate.

Nicolaidis: Yes, but that is a most important point. First of all, indulgence is great in humans and some animals, proportional to the circumstances. When you go to a wedding, in

France particularly, for two days they do nothing but eat. Such indulgence does not compromise body weight, because there is a compensatory undereating and everything is okay. So, if you say the mouth is very important and oral cues are very important in altering food intake, it is not totally true.

Jordan: I did not mean to say this was the single most important thing, because it is not.

Nicolaidis: Would you define "palatability?"

Jordan: To me, when I am talking about palatability in my patients I am really talking about satisfaction.

Nicolaidis: You mean the reaction to something sensory?

Jordan: When they say, "I really enjoyed this ingestion."

Nicolaidis: So, to you, palatability is the response, not the stimulus. For other participants here, palatability is a characteristic of the food that is the stimulus.

OVERVIEW

Chapter 21

The Role of the Chemical Senses in Nutrition

SAMUEL LEPKOVSKY

INTRODUCTION

Chemical compounds that act as sensory stimuli or flavors excite the sensory receptors in the nose, mouth, and pharynx. The nerves linked with the sensory receptors transmit the excitation to the brain as electric impulses, where they are appreciated and expressed as a sensation. The nerves do not transmit the stimuli in their original form. The nerve acts as an indifferent conductor, similar to a telephone wire. As an example, the sensation "red" has nothing to do with light as a wavelength of 80 mμ; the odor of camphor has nothing to do with its chemical composition (von Buddenbrock, 1958).

Survival is considered the first law in nature, and survival is not possible without proper nutrition. "All animals that have been studied, those without alimentary tracts as well as those which have, recognize food, spurn food when it is superabundant, and put forth extra efforts to get it when it is rare. Hence, whatever be the machinery which may fix the priorities in rats, the comparable patterns seem the endowments of all animals, whether or not they possess specialized neuromuscular or alimentary systems" (Adolph, 1947).

Current concepts about nutrition are at variance with nutrition biologically. In nutrition we do not feed the animal. In nutrition we are feeding the cells, the living elements in the animal. Life began in the primordial sea

(Oparin, 1938), and the composition of the sea water set the conditions of life which have not changed substantially since life began (Baldwin, 1949). "The living elements of the body are water inhabitants" (Cannon, 1932); the primordial sea became the internal environment and is referred to as the extracellular fluids. The living elements, the cells, live in ". . . an internal environment . . . with a controlled unchanging atmosphere which made the organism independent of outside conditions" (Smith, 1953). The animal body acts as a vehicle to feed the cells. The role of the chemical senses in nutrition will be explored in (1) hunger, appetite, or urges for food; (2) food-seeking behavior; (3) recognition of food; (4) selection of food; (5) intake of food; (6) digestion; (7) elimination of metabolic wastes and excesses of nutrients ingested; and (8) satiety.

HUNGER, APPETITE, AND URGES FOR FOOD IN ANIMALS

Survival of the animal depends upon the delivery to the cells of the essential nutrients needed for homeostasis, energy, growth, and reproduction. The usage of the terms "hunger" and "appetite" have been so varied that ". . . many have advocated their banishment for the sake of linguistic peace" (Janowitz, 1958). In general, hunger is caused by depletion (Janowitz, 1958; Grossman, 1955) and repletion brings satiety. It is difficult to understand why corpulent animals get hungry even though the amount of depletion is an insignificant fraction of the calories in the fat depots. In contrast, complete starvation for two days or longer is followed by absence of hunger, despite the considerable depletion of bodily reserves.

The needs for nutrients initiate metabolic activities which are transferred to the brain where they are appreciated and are expressed as appetite or urges for food. The nature of these metabolic processes are unknown, as indicated by the following examples: (1) Animals such as the woodchuck or ground squirrels prepare for hibernation by hyperphagia and obesity (Lyman, 1954). (2) Migratory birds, before flight, become hyperphagic and store the correct amount of fat to supply the energy needed for the flight (King and Farner, 1933; Odum, 1960). (3) During estrus, rats are hyperactive and eat less. During diestrus, the rats are less active and eat more (Brobeck *et al.*, 1947).

In general, appetite decreases in animals while they are engaged in excessive physical activity. A compensatory increase in food intake occurs on the following day in man (Edholm *et al.*, 1955) and rat (Stevenson *et al.*, 1966; Thomas and Miller, 1958; Premack and Premack, 1963). This may be due to slow-acting metabolic adjustment of food intake to energy expenditure. The role of sensory stimuli plays little or no part in these processes.

HUNGER, APPETITE, AND URGES FOR FOOD IN HUMAN BEINGS

The sensory stimuli of food appear to play an important role in the regulation of food intake in the human being. Appetite and hunger are sensations and may be beyond definition, since ". . . each animal, including the human being, lives in a world outlined by his own senses" (Kare, 1960). "The sensory world of every being is sealed to him. Nobody can reveal his sensations to others" (von Buddenbrock, 1958). The chemical senses may play a role in the appetite of the human being. As an example, a marine chef was preparing a meal for over a thousand men. He was frying a couple of onions. When asked how he proposed to divide a couple of onions among so many men, he replied that he had no intention of feeding the onions to the men, but he had noticed that when the odor of frying onions pervaded the mess hall the men ate better.

FOOD-SEEKING BEHAVIOR

"Drive" sets the animal to search for food. This requires locomotion, exploration, and examination. Food is often on the hoof and requires chasing. It may require fighting and fleeing so that the searcher does not become food for the animals. This phase of nutrition does not involve the chemical senses.

RECOGNITION OF FOOD: THE CHEMICAL SENSES

The universality of the chemical senses is best explained by the fact that in order to eat the organism must possess some chemical knowledge. ". . . the lowly amoeba . . . must be able to distinguish a grain of sand from an alga or some digestible form of life" (von Buddenbrock, 1958).

More is known about the food-sensing mechanism in the coelenterate, *Hydra littoralis,* which lives on the water flea, *Daphnia. Hydra* will not eat dead water fleas; they will eat wounded fleas after puncturing the flea with their nematocysts. The juice which oozes out contains reduced glutathione which acts as a "marker" or "feeding hormone." This is "recognized" by *Hydra* and the wounded flea becomes acceptable as food (Loomis, 1955). The function of the chemical senses has become more complex in higher animals and most of all in human beings.

SELECTION OF THE NEEDED DIETARY NUTRIENTS

The chemical senses play a key role in the selection of the nutrients needed. Richter (1942) postulated that an organism is sensitive to many of its physiological needs and will develop motivated behavior appropriate to the fulfillment of these needs. Bodily needs have a "biochemical core" (Kubie, 1948), and alimentary behavior deals with the translation of bodily needs to behavior. Selection of food is made possible by the taste of food.

Intrinsically, food tastes neither good nor bad. Taste is determined not only by the sensory properties of food, but also by bodily needs and internal biochemistry. Human subjects were asked to choose between 5% and 30% sucrose for a long drink. They chose 5% sucrose, finding that 30% sucrose was sickeningly sweet. After insulin injection, the subjects' blood sugar levels decreased to 40–50 mg% and they chose 30% sucrose; it no longer tasted sickeningly sweet. The taste of sucrose was altered by the action of insulin (Mayer-Gross and Walker, 1946). This was repeated with rats (Jacobs, 1958).

In World War II, American soldiers in a German prison camp were subsisting on a ration 200 calories short of their basal needs. They were ultimately saved by Red Cross food parcels. Trading of the parcels' contents went on among the men, and a point system took form which pretty well reflected the nutritional value of the food. A can of powdered whole milk topped the list at 150 points. Meat followed at 120. There was no point listing for cheese, jam, sugar, and chocolate bars (Englander, 1945). While the men were prisoners, profound changes had taken place in their internal chemistry, as a result of which powdered milk, previously unacceptable, became highly acceptable. Remembering this low acceptability rating of powdered milk among these men as soldiers before their imprisonment, Englander (1945) explored this phenomenon and learned that the powdered milk became the most acceptable item because ". . . it satisfied—even more than the chocolate—the prisoners' craving for something rich to eat."

Deficiency of specific nutrients alters the taste of these nutrients, especially in animals. "A negative feedback appears operant in alimentary behavior, by which peripheral stores relay information concerning their inventory to hypothalamic regulatory mechanisms" (Magoun, 1963). Well-fed pigs usually laid down or strolled about the yard. Those deficient in calcium spent a large part of their time attempting to root out a fragment of cement or mortar from the brickwork, while those suffering from a deficiency of vitamin A were stimulated to search for blades of grass or weeds which might be growing in the cracks of the floor (Drummond, 1934).

Rats were offered two diets: one deficient in thiamin and one containing the vitamin. The rats ate these diets at random. When the rats became deficient in thiamin, they chose the diet containing thiamin and avoided the deficient diet (Harris *et al.*, 1933). Similar results were obtained with riboflavin and pyridoxine (Scott and Quint, 1946). Bovril, meat extract, was then added to a diet containing thiamin. This diet and a diet deficient in thiamin were offered to rats, and they initially ate these diets at random. When the rats became deficient in thiamin, they ate the diet that contained the bovril and thiamin. Bovril was then transferred to the deficient diet and the rats ate the bovril–thiamin-deficient diet and died. The rats had been mentally fooled; they associated their well-being with bovril instead of the thiamin (Harris *et al.*, 1933).

Taste often plays an unusual role in the regulation of food intake. The larvae of the monarch butterfly feed on milkweed containing a digitalis-like compound which is poisonous to bluejays, and they avoid eating them. Larvae which were raised with difficulty on cabbage plants were offered to bluejays and they refused them. Bluejays were deprived of food until they ate some of the cabbage-fed monarch larvae. Two died of starvation rather than eat them. Finally, some of the birds accepted the cabbage-fed monarch larvae. Once the birds overcame their initial hesitancy, they continued to eat them. The bluejays were then again fed the milkweed-fed larvae, and they became sick; they showed violent retching and vomited the partially digested insects. Thereafter, the bluejays again rejected the larvae of the monarch butterflies (Brower *et al.*, 1967).

The roundworm, microstoma, uses nematocysts (Lashley, 1938, cited by Stellar, 1960) or stinging cells to capture prey, but it must acquire these stinging cells from *Hydra*. After the stinging cells have been depleted from the body of microstoma, a physiological need for these cells arises, and it develops appropriate appetitive behavior and ingests *Hydra* voraciously. After microstoma obtains a full compliment of stinging cells, it ceases to ingest *Hydra* even in the complete absence of other sources of food (Stellar, 1960).

EATING: GASTROINTESTINAL TRACT

Oropharyngeal Factors

The mouth is the gateway through which food enters into the animal. As World War II soldiers put it, "Food is not a food until it is eaten." During chewing, many new flavors arise, and these mix with the food flavors, forming changing flavor profiles (Sharon, 1965). The sensory stimuli increase the flow of saliva, which promotes swallowing. These flavor profiles are a source of pleasure to the human being. Oropharyngeal factors may also function in the rat, but less so in the dog, since it swallows the food without chewing it. Epstein (1967) showed that rats regulated food intake without the use of oropharyngeal factors. Rats were taught to feed themselves intragastrically by bar pressing, thus bypassing the mouth.

Gastrointestinal Factors

The sensory stimuli of food increase the flow of gastric juice (Janowitz *et al.*, 1950; Beaumont, 1830, see Cushing, 1935), pancreatic juice (Behrman and Kare, 1968), bile, and intestinal juice (Harper, 1971).

As an example, a woman without an esophagus was fed intragastrically through a tube. The amount of gastric juice secreted into the stomach was measured by withdrawing the gastric juice from the stomach through the same tube. Oatmeal gruel was presented to the patient, who merely chewed it, then spat it out. The gruel evoked little secretion of gastric juice above the basal level. When the patient was given a meal of her own choice, a large amount of gastric juice was secreted (Janowitz *et al.*, 1950).

The effect of the increase in the flow of digestive juices in the digestion remains to be fully determined. This was tested with human beings (Smith *et al.*, 1920). Food served to human subjects was rendered unpalatable by mixing together meat, biscuits, jelly, and other meal components. The plate on which the food was served was smeared with animal charcoal, and the table was dirty and strewn with dirty dishes. A little indole was sprinkled under the table. The food was so unpalatable that one subject vomited shortly after eating. Yet there was little difference in the digestibility of the palatable or unpalatable foods as measured by nitrogen retention. Smith *et al.* (1920) stated: "This short test indicates that flavor is not the outstanding dietetic asset that some people would have us believe."

In contrast, Wolf and Wolff (1943) report the case of a patient who had to be tube-fed directly into the stomach. The food failed to stimulate his appetite, and he was markedly underweight. But when he took the food into his mouth and chewed it before he introduced it into the stomach, his appetite improved. He felt better and gained weight. Presumably, taking the food into the mouth permitted the sensory stimuli of the food to activate the receptors in his mouth and pharynx and promoted better nutrition.

Sensory stimuli unrelated to food can influence food utilization. Rats were divided into two groups: One group was unhandled and the other was overhandled, the sensory stimuli arising from handling. There was no difference in food intake between the two groups, but the overhandled rats gained more weight even though they were more active (Ruegamer and Bernstein, 1954). The food eaten by the overhandled rats was therefore better utilized than that eaten by the unhandled rats. They had more active thyroids than the overhandled rats, and this may account in part for the greater efficiency of the utilization of the food by the overhandled rats.

While the gastrointestinal tract is important in the regulation of food intake, the chemical senses (flavors) play little or no role in this process. Food as it exists in nature cannot be used by the cells. Before it can be used it is digested and reduced to its unit structures which can be used by the cells.

After food is swallowed, it is temporarily stored in the stomach and digestion is initiated. Entry of gastric chyme into the duodenum is regulated by gastric emptying (Thomas, 1957), and therefore participates in the regulation of food intake. Gastric emptying is controlled by the upper intestine (Lepkovsky *et al.*, 1971) and prevents overloading of the intestine. Overloading of the intestine leads to diarrhea (Lepkovsky *et al.*, 1971) or "dumping syndrome." When deprived of the stomach, as in gastrectomized rats (Tsang, 1938; Lepkovsky *et al.*, 1975) or human beings (Wangensteen and Carlson, 1931), the upper intestine nevertheless regulates entry of food into the duodenum by controlling food intake. Food intake is regulated largely by inhibitory mechanisms in the gastrointestinal tract. This has been illustrated by work with the blowfly (Dethier and Bodenstein, 1958). "It would appear that hunger in the fly can be equated with absence of stimulating fluid in the foregut, in other words, *absence of inhibition.*" When the recurrent nerve which connects the foregut to the brain is cut, the flies become hyperphagic, suck continuously, and the gut distends and may burst.

Similar results were reported by Wiersma for the crab (1961): ". . . after the decapod's brain is removed . . . the brainless animal will continue feeding even when the stomach is completely filled and as a result may eventually burst." It appears that the sensory stimuli are not involved in the passage of the food through the intestinal tract.

SATIETY

Unpleasant sensations accompany hunger in human beings, and it may be assumed that this also occurs in animals. Eating abolishes hunger pains in human beings, and it may be inferred that this happens in animals. This may be looked upon as a negative pleasure. Schopenhauer represented pleasure as an absence of pain (cited by Sherrington, 1947). After the hunger sensations are abolished animals cease to eat, but many human beings continue to eat. It has been generally accepted that human beings derive pleasure from eating after hunger has been abolished and metabolic satiety has been achieved. Brillat-Savarin (1960) aptly stated that it is a "privilege of the human species to eat without hunger and to drink without thirst." Sherrington (1947) pointed out that pleasure is largely experienced in the cerebral cortex but pain is experienced lower in the brainstem. It may be inferred that the more highly developed cerebral cortex as in man is essential to sense pleasures which accompany meals with desirable flavors. This difference in eating behavior between human beings and animals is due, in all probability, to the more highly developed cerebrum of human beings. This is supported by the eating behavior of one human being whose cerebral cortex did not interfere with his choice of food (Katz, 1953). "A young man, 24 years old, had gas poisoning. He was unable to remember anything for longer than two seconds. From this time his feeding was entirely governed by his bodily needs: conscious memories could no longer play any part. He appeared for meals when hunger and thirst made him. He never over-ate but stopped eating when satisfied. Once satisfied, nothing on the table attracted him any longer. He knew that he should eat and drink—of course he did not really know, but so his organism directed."

MONOTONY: FLAGGING APPETITE

Some foods can be eaten continuously without losing appetite for them; bread is a good example. Acceptability for some foods that are consumed too often decreases progressively, and ultimately aversion for them may develop. The army had such an experience with Spam in the Pacific Southwest area of operations. Spam was served too often, and an aversion to Spam was so great that it was eliminated from army rations. After awhile requests for Spam came from the same area of operations.

Monotony may result from neural "fatigue." Animals progressively show decreases in response to stimuli when the same stimulus (including sensory stimuli such as flavors) is repeatedly applied. This phenomenon is variously referred to as "habituation," "sensory accommodation," "effector fatigue," "negative adaptation," and "internal inhibition." This may be illustrated with the following example: If a drop of water falls on the surface of the sea just over the flowerlike sea anemone, the animal contracts vigorously. If a second drop falls within a few minutes, there is less contraction and the response disappears after the third or fourth drop. Sharpless and Jaspar (1965) stated: "This is one of the most pervasive phenomena of the animal kingdom. The ubiquity of the phenomenon is one of the most fundamental properties of the animal. The phenomenon is seen in olfaction. The smell of coffee in the morning is very pronounced before eating. During eating, the smell of coffee becomes much less pronounced.

The effect of repetitive or monotonous diets has been studied with human subjects (Siegel, 1957; Siegel and Pilgrim, 1958). The results emphasized the problem as indicated by the following: (1) Monotony is positively related to the number of times a given food item has been consumed. (2) Highly palatable foods slow the growth of monotony. (3) Foods that are highly palatable for some individuals are not necessarily palatable for others. (4) The course of monotony is affected by personality factors. (5) The environment is a factor in the acceptability of food. (6) Monotony is overtly expressed in lowered food acceptance. (7) Populations and entire ethnic groups subsist on essentially repetitive or monotonous diets.

The role of bread in human nutrition is unexplained. Bread is consumed continuously and in large amounts without people losing their appetite for it. No amount of bread offered to soldiers in World War II was too much. They called bread a "morale builder." This may explain why populations subsist on monotonous diets, because bread comprises a large part of their diet. Indeed, food has been referred to for centuries by the whole Western world as "our daily bread" (The Lord's Prayer).

THE ROLE OF FLAVOR IN HUMAN BEINGS

The sensory properties of food are important for the human being, and they often play a greater role in the acceptability of food than does its nutritional value. This was foreshadowed in ancient times by Hippocrates with his aphorism: "An article of food or drink which is slightly worse but more palatable is to be preferred to such as are better but less palatable" (Lusk, 1922). In the same vein, McCollum (McCollum and Simmonds, 1929) wrote, ". . . The artificially established liking for white flour and white corn meal is an illustration of the failure of the instinct of man to serve as a safe guide in the selection of food. The esthetic sense is appealed to in greatest measure in this case by products of lowest biological values."

The highly developed brain of *Homo sapiens* is at odds with nutritional biology. This is due in part to the importance of sensory stimuli for normal mental processes. Schultz (1965) introduced his book entitled "Sensory Restriction" with: "A changing sensory environment seems essential for human beings. Without it, the brain ceases to function in an adequate way and abnormalities of behaviour develop. In fact as Christopher Burney observed in his remarkable account of his confinement: 'Variety is not the spice of life; it is the very stuff of it.' "

Despite customs, social pressures, prejudices, and so forth, human beings do choose correctly some nutrients according to their needs. "In man we can detect traces of a protective instinct" (Drummond, 1934) and this "protective instinct" has at times, but not always, protected man against deficiencies of certain essential nutrients and may have been at times a factor in man's survival. A child whose adrenal cortex was diseased craved salt, consumed large amounts of salt when he could get it, and survived as long as he had access to salt. He was put in the hospital and fed a hospital diet lacking in salt and soon died (Wilkins and Richter, 1940). Another example was of a 34-year-old patient with Addison's disease who had a marked craving for salt. He often filled his glass of tomato juice half full of salt and covered his steak with an even layer of salt about one inch thick and survived (Richter, 1942).

During the Civil War, Union soldiers in a Confederate prison lived largely on corn mush and were short of salt. "We traded buttons to guards for red peppers, and made our mush or bread or dumplings hot with the fiery pods, in hopes that this would make up for the lack of salt, but it was a failure. One pinch of salt was worth all the pepper pods in the Southern Confederacy. The old adage says that 'hunger is the best sauce for poor food,' but hunger failed to render this detestable stuff palatable" (McElroy, 1957).

Before glycerol was produced biologically, the only source of glycerol for gunpowder was fat. During World War I, fat became scarce and "fat hunger" was common in the population. Fat became delicious in taste. After fat became available it was consumed in excessive amounts. However, after bodily deficits of fat were repleted, it lost its delicious taste and at times became aversive (B. Ben Dor, personal communication).

Culinary practices often proved to be of critical value in the survival of populations. In the near past, pellagra was common on the American side of the Rio Grande River, but in contrast, pellagra was seldom seen on the Mexican side. In using corn to prepare tortillas, Mexicans used slaked lime, which makes the nicotinic acid normally incompletely available in corn more available. When Americans used corn to prepare corn bread, they did not make use of the cullinary practice of using slaked lime and many developed pellagra (Laguna and Carpenter, 1951).

Similarly, in northern Sweden, potatoes, which are the chief source of vitamin C, are eaten boiled and not mashed. Boiling does not destroy vitamin C in the potato, whereas mashing introduces air into the potato and the oxygen present destroys the vitamin C (S. Lepkovsky, unpublished).

DEPRIVATION OF SENSORY STIMULI AND THE TASTE OF FOOD IN MAN

As men moved toward battle areas in World War II, they were separated from many of their sources of sensory stimuli. "Despite the efforts of entertainers and Red Cross workers who . . . risked their lives contributing to the morale of troops, there could be no sports, movies, libraries or other pleasures for fighting men in fox holes . . . To a man in combat or waiting to go into combat, food can assume an importance amazing to people in ordinary circumstances. Rations have proved to have a psychological as well as a physiological effect. They can raise or lower morale" (Isker, 1944).

The battle rations, C and K, were thoroughly tested under civilian conditions and were found acceptable since other sources of sensory stimuli were available to the men in civilian life. In battle areas, however, the chief available source of sensory stimuli was food. The soldiers rejected these ratios as lacking desirable sensory properties. The judgment of the soldiers was expressed thus: "We could undoubtedly survive on these rations a lot longer than we'd care to live" (Lepkovsky, 1959).

Human volunteers subsisting on severely restricted food intake showed the effects of undernutrition. They fatigued easily and could not participate in sports or other activities requiring the expenditure of energy. Social activities were sharply curtailed, since they did not even have the energy for conversation. Interest in activities which are a source of sensory stimuli disappeared; only the interest in food remained, and that was exaggerated. Such undernourished subjects turned to food for sensory stimuli. According to Keys (1948): "Subjects were often caught between conflicting desires to gulp their food down ravenously or to consume it slowly so that the taste and odor of each morsel would be fully appreciated. Toward the end of starvation some men would dawdle for almost two hours over a meal which previously they would have consumed in a matter of minutes."

American soldiers in a German prison camp during World War II were subsisting on a diet 200 cal below that required for maintenance. Food was the passionate preoccupation of most "Kriegies." No one ever tired of talking about food during the day or dreaming of it at night. "Sometimes the dreams took odd twists." A young lieutenant produced a nightmare that he was imprisoned in a can of meat (Englander, 1945).

In tests at Iowa State College, volunteers subsisted on very low protein diets for 16 to 18 days, and developed aversions to anything that tasted sweet, especially sugar. They developed cravings which were more or less nonspecific. Some craved pickles, some lettuce. One subject craved eggs to a point where hydrogen sulfide smelled good to him (P. Swanson, personal communication).

An unusual case of specific hunger was reported by Hoelzel (1944), who subsisted on a low-protein diet for a long period. He craved meat so much that he went on "meat sprees." He rapidly devoured raw steak without tasting it. After a day or two the impulse to devour the raw meat began to lag. It then became apparent that the raw steak had a disagreeable, slimy taste and an objectionable bloody odor.

WHEN IS A SENSORY STIMULUS A FLAVOR? THE ONION

Our senses react to an unlimited number of chemical compounds that have an odor or taste, but few are classified as flavors in the sense that they play a role in nutrition. Coffee and perfume both have acceptable odors, but only coffee is associated with food.

The chemical compounds in food that act as sensory stimuli or flavors can be separated with gas chromatography, column chromatography, paper chromatography, electrophoresis, and other chemical and physical procedures. The chemical structure and physical properties of the pure compounds can be determined. These, however, are not flavors until they can be ultimately evaluated by the brain.

Flavorful foods such as onions, garlic, and leeks have played a key role in human nutrition and have been highly prized by man since the earliest times. The first such record is described in the Bible (Numbers 11 : 1), when the Israelites were in the Sinai Desert subsisting on manna, which apparently lacked desirable flavors. Among the six food items missed most, three were chiefly sources of flavor: onions, garlic, and leeks. The others were fish, melon, and cucumbers.

Dana (1840) wrote of the sailors' craving for onions after a long sea voyage during which they had subsisted on preserved foods. When onions were made available to them, they considered them a ". . . glorious treat. We were perfectly ravenous for them. It was like a scent of blood to the hound." In a German prison camp during World War II American soldiers lived on a monotonous diet and the onion was highly prized. "A gravy, made of onions carefully hoarded from small crop grown in the camp garden during the summer, was served as a special treat on Sunday nights and helped take the curse off the potatoes" (Englander, 1945).

During the Civil War Union soldiers as Confederate prisoners lived on corn mush. While traveling one of the men scanned a vegetable garden. "Andrews espied a handful of onions which had remained ungathered. They tempted him as the apple did Eve . . . he sprang from the car, snatched the onions from their bed, . . . and was on his way back before the guard could make up his mind to fire upon him. . . . The onions . . . were to us more delicious than wine upon lees. They seemed to find their way into every fibre of our bodies and invigorate every organ" (McElroy, 1957).

During World War II General Wingate raided the Japanese behind their lines in Burma, and food for him and his men was dropped from the air. Usually, K rations were dropped to them. Once a week, a "holiday" package was dropped, and among other food items a raw onion was included. This the men appreciated very much, nibbling the onions and making them last 2 or 3 days.

ACKNOWLEDGMENT

This work was supported by the Nutrition Foundation, Grant 424.

REFERENCES

Adolph, E. F. (1947). Urges to eat and drink in rats. *Am. J. Physiol.* **151,** 110–125.

Baldwin, E. (1949). "An Introduction to Comparative Biochemistry." Cambridge Univ. Press, London and New York.

Behrman, H. R., and Kare, M. R. (1968). Canine pancreatic secretion in response to acceptable and aversive taste stimuli. *Proc. Soc. Exp. Biol. Med.* **129,** 343–346.

Brillat-Savarin, J. A. (1960). "The Physiology of Taste." Dover, New York (original publication in 1825).

Brobeck, J. R., Wheatland, M., and Strominger, J. L. (1947). Variations in regulation of energy exchange associated with estrous, diestrous and pseudopregnancy in rats. *Endocrinology* **40,** 65–72.

Brower, L. E., Brower, J. V. Z., and Corvino, J. M. (1967). Plant poisons in a terrestrial food chain. *Proc. Natl. Acad. Sci. U.S.A.* **57,** 893–898.

Cannon, W. B. (1932). "The Wisdom of the Body." Norton, New York.

Cushing, H. (1935). William Beaumont's rendezvous with fame. *Yale J. Biol. Med.* **8,** 114.

Dana, R. H. (1840). "Two Years Before the Mast: A Personal Narrative of Life at Sea," pp. 442–443. Harper, New York.

Dethier, V. G., and Bodenstein, D. (1958). Hunger and the blowfly. *Z. Tierpsychol.* **15,** 129.

Drummond, J. C. (1934). "Lane Medical Lectures," 2nd ed. Stanford Univ. Press., Stanford, California.

Edholm, O. G., Fletcher, J. G., Widdowson, E. M., and McCance, R. A. (1955). The energy expenditure and food intake of individual men. *Br. J. Nutr.* **9,** 286–300.

Englander, D. A. (1945). They dreamed of food. *Liberty Mag.* July 7.

Epstein, A. N. (1967). Feeding without oropharyngeal sensations. *In* "Chemical Senses and Nutrition" (M. R. Kare and O. Maller, eds.), pp. 263–280. Johns Hopkins Press, Baltimore, Maryland.

Grossman, M. I. (1955). Integration of current views on the regulation of hunger and appetite. *Ann. N.Y. Acad. Sci.* **63,** 76–91.

Harper, H. A. (1971). "Review of Physiological Chemistry," 13th ed. Lane Med. Publ., Los Altos, California.

Harris, L. J., Clay, J., Hargreaves, F. J., and Ward, A. (1933). Appetite and choice of diet. The ability of the vitamin B deficient rat to discriminate between diets containing and lacking the vitamin. *Proc. R. Soc. London, Ser. B* **113**, 161–190.

Hoelzel, F. (1944). An explanation of appetite. *Am. J. Dig. Dis.* **11**, 71–76.

Isker, R. A. (1944). "U.S. Armed Forces Operation Studies," No. 1. Office U.S.Q.M. Gen., Washington, D.C.

Jacobs, H. L. (1958). Studies on sugar preference. 1. The preference for glucose and its modifications by injections of insulin. *J. Comp. Physiol. Psychol.* **51**, 304–310.

Jacobs, H. L. (1967). Sensory and metabolic regulation of food intake: Thoughts on a duel system regulated by energy balance. *Proc. Int. Congr. Nutr., 7th, 1966* Vol. II.

Janowitz, H. D. (1958). Editorial. Hunger and appetite. *Am. J. Med.* **25**, 327.

Janowitz, H. D., Hollander, F., Orringer, D., Levy, M. H., Winklestein, A., Kaufman, M. R., and Margolin, S. G. (1950). A quantitative study of the gastric secretory response to sham feeding in a human subject. *Gastroenterology* **16**, 104–116.

Kare, M. R. (1960). Animals' "taste" senses differ from man's. *Nulaid News* October, pp. 22–23.

Katz, D. (1953). "Animals and Men." Penguin Books, Baltimore, Maryland.

Keys, A. (1948). "The Biology of Human Starvation," Vol. 2. Univ. of Minnesota Press, Minneapolis.

King, J. R., and Farner, D. S. (1933). The relationship of fat deposition to Zugunruhe and migration. *Condor* **65**, 200–223.

Kubie, L. S. (1948). Instincts and homeostasis. *Psychosom. Med.* **10**, 15–30.

Laguna, J., and Carpenter, K. J. (1951). Raw versus processed corn in niacin-deficient diets. *J. Nutr.* **45**, 21–28.

Lepkovsky, S. (1959). Potential pathways in nutritional progress. *Food Technol.* **13**, 421–424.

Lepkovsky, S., Bortfeld, P., Dimick, M. K., Feldman, S. E., Furuta, F., Sharon, I. M., and Park, R. (1971). Role of upper intestines in the regulation of food intake in parabiotic rats with their intestines "crossed" surgically. *Isr. J. Med. Sci.* **7**, 639–646.

Lepkovsky, S., Dimick, M. K., Feldman, S. E., and Park, R. (1975). The stomach and upper intestine of the rat in the regulation of food intake. *J. Nutr.* **105**, 1491–1499.

Loomis, W. F. (1955). Glutathione control of the specific feeding reaction of hydra. *Ann. N.Y. Acad. Sci.* **62**, 209–228.

Lusk, G. (1922). A history of metabolism. *In* "Endocrinology and Metabolism" (L. F. Barker, ed.), Vol. 3, pp. 3–78. Appleton, New York.

Lyman, G. P. (1954). Activity, food consumption and hoarding in hibernators. *J. Mammal.* **35**, 545–552.

McCollum, E. V., and Simmonds, N. (1929). "The Newer Knowledge of Nutrition." Macmillan, New York.

McElroy, J. (1957). "This Was Andersonville." McDowell, Obolesky, New York.

Magoun, H. W. (1963). "The Waking Brain," 2nd ed. Thomas, Springfield, Illinois.

Mayer-Gross, W., and Walker, J. W. (1946). Taste and selection of food in hypoglycemia. *Br. J. Exp. Pathol.* **27**, 297–298.

Mickelsen, O. (1975). The nutritional value of bread. *Cereal Foods World* **20**, 308–310.

Odum, E. P. (1960). Premigratory hyperphagia in birds. *Am. J. Clin. Nutr.* **8**, 621–629.

Oparin, A. I. (1938). "The Origin of Life." Macmillan, New York.

Premack, D., and Premack, A. J. (1963). Increased eating in rats deprived of running. *J. Exp. Anal. Behav.* **6**, 209–212.

Richter, C. P. (1942). Physiological psychology. *Annu. Rev. Physiol.* **4**, 561–574.

Ruegamer, W. R., and Bernstein, L. (1954). Growth, food utilization, and thyroid activity in the albino rat as a function of extra-handling. *Science* **120**, 184.

Schultz, D. P. (1965). "Sensory Restriction: Effects on Behavior." Academic Press, New York.
Scott, E. M., and Quint, E. (1946). Self-selection of diet. III. Appetites for B vitamins. *J. Nutr.* **32,** 285–292.
Sharon, I. M. (1965). Sensory properties of food and their function during eating. *Food Technol.* **19,** 35–36.
Sharpless, S., and Jaspar, H. (1965). Habituation of the arousal reaction. *Brain* **79,** 655.
Sherrington, C. (1947). "The Integrative Action of the Nervous System." Yale Univ. Press, New Haven, Connecticut.
Siegel, P. S. (1957). Repetitive element in the diet. *Am. J. Clin. Nutr.* **5,** 162–164.
Siegel, P. S., and Pilgrim, F. J. (1958). The effect of monotony on acceptance of food. *Am. J. Pyschol.* **71,** 756–759.
Smith, C. A., Holden, R. C., and Hawk, P. B. (1920). Is unpalatable food properly digested? *Proc. Soc. Exp. Biol. Med.* **17,** 98.
Smith, H. W. (1953). Comparative physiology of the kidney. *J. Am. Med. Assoc.* **153,** 1512–1514.
Stellar, E. (1960). Drive and motivation. *Handb. Physiol., Sect. 1: Neurophysiol.* **3,** 1501.
Stevenson, J. A. F., Box, B. M., Feleki, V., and Beaton, J. R. (1966). Bouts of exercise and food intake in the rat. *J. Appl. Physiol.* **21,** 118–122.
Thomas, B. M., and Miller, A. T., Jr. (1958). Adaptation to forced exercise in the rat. *Am. J. Physiol.* **193,** 350–354.
Thomas, J. E. (1957). Mechanisms and regulation of gastric emptying. *Physiol. Rev.* **37,** 453–474.
Tsang, Y. C. (1938). Hunger motivation in gastrectomized rats. *J. Comp. Psychol.* **26,** 1–17.
von Buddenbrock, W. (1958). "The Senses." Univ. of Michigan Press, Ann Arbor.
Wangensteen, O. H., and Carlson, H. A. (1931). Hunger sensation in a patient after total gastrectomy. *Proc. Soc. Exp. Biol. Med.* **28,** 545–547.
Wiersma, C. A. G. (1961). Reflexes and the central nervous system. *Physiol. Crustacea* **2,** 241.
Wilkins, L., and Richter, C. P. (1940). A great craving for salt by a child with cortico-adrenal insufficiency. *J. Am. Med. Assoc.* **114,** 866–868.
Wolf, S. G., and Wolff, H. G. (1943). "Capitalized Human Gastric Function: An Experimental Study of a Man and His Stomach." Oxford Univ. Press, London and New York.

DISCUSSION

Moskowitz: I am curious whether anyone has investigated the fantasies that people have about food when they are undergoing starvation and whether they have any desires for combinations of food that they would not ordinarily have were they normally fed. For example, tuna fish and cream cheese and jelly sandwiches?

Lepkovsky: Richter postulated that an organism is sensitive to many of its physiological needs. Taste is determined largely by its bodily needs.

Moskowitz: We have done some work with the staff of the University of Massachusetts. They are very interested in computer menu planning and find that if you ask individuals how much they would like to have of such and such a food item, if they haven't had it for a week, a month or 3 months, you can trace what is called a time preference curve which indicates that items not consumed for about 3 months are really desired, and those consumed the day before are not desired at all. Food such as meat and shellfish, food with a heavy fat content, food that carries the meal such as the entree has a very steep curve, and is really desired if not eaten for a very long period of time. But if you had it yesterday, you don't want it at all. It is like killing somebody by giving them roast duck every day, as opposed to items that do not carry the meal and do not have a high fat or protein content, such as bread, salad, potatoes, and some desserts which have a much flatter function—especially salad—that you can continue to eat every day and never lost preference for them.

Lepkovsky: This question highlights the importance of monotony in food. During World War II, some foods that were offered too often were rejected while others were accepted even though they were offered very often. Bread is a good example. Although large amounts of bread were offered frequently, the acceptability of bread remained high. The work of Olaf Mickelsen has a bearing on this subject. Rats fed a diet high in bread grew poorly because of the deficiency of the essential amino acid lysine in the protein of the bread. In contrast, human beings on a high-bread diet grew well and did not become deficient in lysine (Mickelsen, 1975).

Kinnard: You made a distinction between the extent of feeding between lower animals and man, and I was just wondering if you could draw a similar dichotomy between the sex habits of lower animals and man.

Lepkovsky: That is endocrinology and I am a nutritionist.

Kinard: Did you have anything to do with those rations in World War II?

Lepkovsky: During World War II much research work was done on army rations, especially rations K and C. These rations had low acceptability with combat troops; they soon became monotonous. These rations did not seem to have desirable sensory stimuli (taste).

Moskowitz: If you watch people at a smorgasbord you will find that people tend to eat a lot of the meat first; after they load up they no longer want any more meat. Why is it that with all this satiety, they are still able to have dessert? They are supposed to be satiated, they can't stand the sight or smell of the meat—the fatty things—yet they can still have dessert. There is also a corollary question. Has anybody ever studied people who stuff themselves with starch during a meal to the physical bursting point? Can they still have dessert?

Lepkovsky: Well, again you go back to sensory stimuli. Apparently, the dessert is important to human beings while it is not to animals. It appears that these sensory stimuli are needed even after a meal, due largely to a more highly developed neocortex in the human. The problem is complicated and little understood.

Chapter 22

Socioeconomic Factors in Fat and Sugar Consumption

S. M. CANTOR AND M. B. CANTOR

THE PROBLEM

We are concerned in this paper with examining the influences of social and economic factors on fat and sugar consumption by humans. The juxtaposition of these factors and these food components in a volume on the chemical senses and nutrition implies a problem, and it is well to identify this problem clearly. Stated most simply, it is apparent from statistical data available to us that the changes in consumption patterns of sugar and fat over the past few decades represent the most visible example of change in American and indeed in Western food habits. Fat consumption in Western diets has increased while total carbohydrate has decreased, but within the decreased total carbohydrate, the simple sugars component has increased substantially.

Economic development over the same period is associated with ever-increasing industrialization and urbanization and so far as diet is concerned, with increasing dependence on convenience or ready-to-eat foods. In the United States our economy is already classified as being in a postindustrial stage with concomitant exurbanization progressing rapidly and changes in family structure and life style being particularly apparent. How ever these changes are described, affluence is a commonly used term, and in the conventional wisdom affluence in life style is frequently reflected as richness in diet and also in "eating out." Moreover, richness in diet is frequently associated with fat and sugar in the diet and "eating out" with fast foods and snack foods. The latter also are not only identified with high fat and high sugar but reflect "fast" as part of the life style and, in some respects, reinforce fast living. This matrix of change incorporates an increasing incidence of degenerative and stress-related diseases which, in turn, are associated with overconsumption and the affluent diet, that again means high fat and sugar consumption. These diseases include obesity, cardiovascular disorders, diabetes mellitus, and dental caries.

It is within this dynamic complex that nutritionists are examining nutrients and disease relationships and epidemiologists are seeking statistical data to help identify causal relationships between food components and disease. But as Hegsted (1975) has pointed out "this is an oversimplification" for "we have entered into a new era in nutrition"—from emphasis on essential nutrients in the diet to concern for food-related problems which have little to do with essential nutrients and much to do with factors which affect food intake. These include cultural background, odor and taste, marketing pressures, and continuing industrialization of the food system. Industrialization emphasizes mass markets, creates limited choice, and can accordingly accentuate particular food components. And so this cycle can aggravate the problems.

The increasing recognition of this situation demands efforts to understand more about the food behavior of humans. The goal is to provide more effective ways of equating food consumption and individual need both with respect to energy requirements and nutrient balance. The rewards include an increase in the quality of life and substantial reductions in the cost of public health maintenance.

SOCIAL AND CULTURAL FACTORS

It is well recognized that social and cultural as well as economic factors have been of significance in creating patterns of food behavior. The relationship of particular foods to religious and tribal ceremonies and to social events is well established and recorded not only in historical and literary works but also in the pictorial* arts. Moreover, the ancient technologies and crafts related to the production and processing of foods as well as their preparation bear the stamp of such origins. The critical interdepen-

* This paper was illustrated in oral presentation by a series of reproductions of paintings of food. Such pictures speak eloquently of the intimate relationship of man and food—of feast and famine, of acts of grace, of celebration, of sacrifice, of hunt and harvest, of food as decoration or symbolic of rise and fall, and of the sharing of food as power, pleasure, and pain. They provide humanistic dimension to statistical data.

dence of man and his food dominated the lives of our hunter–gatherer ancestors, and this dominance produced dietary laws as well as intricate and sophisticated homeopathic food and health relationships. Perhaps the best known of the latter are the Hippocratic "heating" and "cooling" food designations which were related to the ebb and flow of body fluids. Modifications of these flourish today in the Ayurvedic medicine of India, and derivations are still influential in Western food systems.

Man migrates and so do his ideas, therefore it is not surprising, although it casts a shadow on nutrition education, to see ancient beliefs emerge in some popular contemporary diets and be as unquestioningly accepted now as in antiquity. Such are some of the deeply rooted and persistent mysteries and practices associated with human food behavior.

In addition to survival and ceremonial and health interrelationships, associations of food and social status and social and economic influence play a significant role in establishing food behavior and inevitably influence food identity as well as quantity consumed. A most revealing study of the influence of food transactions in determining social status has been described by Marriott (1968). In examining the intricacies of caste ranking in an Indian village, a much clearer picture of hierarchical relationships was obtained by understanding village food transactions. Who accepted food of a particular quality and identity from whom described caste ranking and therefore social status most accurately.

Taking a guest to dinner or giving an elaborate dinner is a practice in Western society which has objectives of pleasing and influencing the receiver as well as revealing the donor's affluence as demonstrated by his taste in food and also his taste in its presentation.

It is interesting to speculate on the reflection of richness of setting and presentation surrounding the food on the property of richness in the food itself. This subject and many related ones are elaborated in a study of the progress of civilization in England as revealed in English appetites provocatively titled "Consuming Passions" (Pullar, 1970). While the chief passion is for food, the development of styles of food behavior is effectively related to appetites for extravagant entertainment, multiple pleasuring, varied sexual practices, and indeed to almost anything that can be smelled and tasted.

This rich and eclectic patchwork of historic happenings reveals how inequities in the distribution and consumption of food and other needs and pleasures have been interpreted along the entire spectrum from gluttony to asceticism to define social and religious status, create motivations for consumption or denial, and shape patterns of food habits. Cultural factors weave in and out and show ways in which flavors become popular, are modified, and therefore how particular foods become acceptable and indirectly how nutrition status may be affected. Above all, what is apparent from examinations of food systems such as this one is the almost incredible number of factors which have shaped and reinforced patterns of food behavior. Against such fundamentally established practices, simple interventions such as recommendations for balanced diets and nutrition labeling seem puny weapons indeed.

CONSUMPTION DATA: CHANGE AND SIGNIFICANCE

Changes in the consumption of sugar and fat in the United States (Page and Friend, 1974) over approximately the last 60 years are clearly evident in Fig. 1, which presents trends in consumption of food elements in terms of 5-year moving averages. The comparison base (100) is the period 1909–1913, when these data were first gathered. Total carbohydrate on an average per capita daily basis was about 500 g then, of which 156 g were sugars. Today, total carbohydrate is about 380 g, of which 200 g are sugars. This represents an increase in the 60-year period from 31.5% of sugars to 52.6% of sugars in the total carbohydrate consumed daily.

Dietary fat is seen to have increased by one-fourth in this same period and currently represents about 155 g per capita per day. Protein, for comparison, is nearly 100 g daily, which is no great change from 1909–1913, and total calories reflect only an approximately 5% change in 60 years.

Some qualifications must be established about these data. First, the figures do not represent true consumption but rather average daily disap-

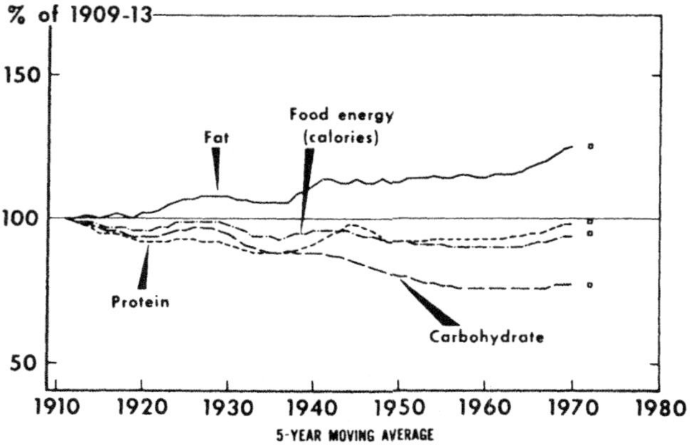

Fig. 1. Per capita civilian consumption of food energy (calories), protein, fat, and carbohydrate, 1909–1913 to 1972. From Sipple and McNutt (1974).

TABLE I

Annual Pounds per Capita Consumption of Sucrose and Corn Sweeteners from Selected Years[a]

Year	Sucrose	Corn	Total
1910–1913	81.3	6.4	87.7
1950	100.4	15.1	115.5
1960	97.6	15.5	113.1
1970	102.5	18.5	121.0
1971	102.4	19.3	121.7
1972	103.0	21.0	124.0
1973	103.2	23.6	126.8
1974	96.5	26.4	122.9
1975	88.7	28.4	117.1
1976 (tentative)	95.1	31.0	126.1

[a] USDA and Corn Refiners Association data.

pearance. They are derived by dividing what is available in the market to consume (what disappears) by a population figure. Actual consumption data are almost nonexistent, although occasional "1-day" surveys have provided some information related to preaggregated population segments (age groups, sex, geographic areas, etc.). Thus, so-called consumption is really the sum of actual consumption and waste. Second, the "sugars" portion of the diet previously cited represents the total of intrinsic sugars, that is, those contained naturally in foods, and those added in preparation, which are largely the sum of sucrose and corn-derived sweeteners.

Usual sugar disappearance data represent sucrose alone, and from these come the familiar "about 100 pounds per year per person" statistics. But as food preparation has moved from the home kitchen to the factory, the use of corn syrups and dextrose has increased. Technologists choose sweeteners on the basis of function and cost. Total sweetener thus becomes the sucrose component plus the corn-derived sweetener component, which gives a sweetener disappearance figure considerably above "100 pounds." Table I shows some selected years, including a projection for sucrose and corn sweeteners for 1976. An upturn is indicated here which reflects both economic (price) and technological [high fructose corn syrup (HFCS) penetration] changes in the sweetener system.

Two events occurred in 1974 which affected sweetener consumption and probably signaled the end of the familiar "100 pounds." The first of these was a fourfold rise in the price of sucrose, which increased the price of sugar-containing products and resulted in an approximate 4 pound per capita drop in annual consumption. The second was the appearance of

HFCS in significant quantities. HFCS is a sweeter corn syrup in which sweetness is increased by isomerizing nearly half of the dextrose contained into fructose. This mixture of fructose, a sweeter hexose produced entirely from corn starch, is about as sweet as invert syrup (liquid sugar) from sucrose, is less costly than invert syrup, and has already achieved a substantial invasion of the liquid sugar market. The affect on sucrose prices generally has been to depress them, since sucrose production internationally was in an expansion phase, and a short term oversupply of sweetener has resulted from the combined events. As the 1976 tentative figure indicates, the elasticity of sweetener consumption has not proved to be very great, and the total usage figure seems to be moving back to a continuation of the rising trend, although this is by no means a certainty. It is generally conceded that the sucrose component in the sweetener system will continue to decline due to replacement by HFCS. However, the rate of decline is heavily dependent on the price of sucrose.

Paralleling the increase in simple sugars in the diet has been the 25% increase in total fat consumed in the same period. However, this change is not regarded as being as significant as the change in the ratio of polyunsaturated fatty acids to saturated fatty acids, which increased 37% between 1909 and 1964 (Antar *et al.,* 1964). The increase in the amount of vegetable fat in the diet is responsible for this. But the data, as this survey concludes, do not fit the hypothesis that low ratios of polyunsaturated fatty acids to saturated fatty acids contribute to the incidence of coronary heart disease. So the carbohydrate changes become more suspect.

That the total fat continues to rise is shown in Table 2, and projections of this trend suggest that per capita consumption of fat may maximize near the present figure (Cantor and Shaffer, 1974). The data in Table II confirm other trends, notably, rising beef consumption, rising cheese consumption, and declining grain consumption. The latter, since 1909–1913, has been responsible for reduced polysaccharide intake and has marked the major difference between Western and developing country food systems, namely, the conversion of grain to animal protein.

While the data referred to are principally from the American diet, changes in the British diet are quite comparable. Grain products in the period 1938–1969 declined from a maximum of nearly 250 pounds per capita (WW II) to less than 170 pounds (basis: market weight of contained products). Sugar increased in the same period from a minimum of 70 pounds per capita (1942) to about 115 pounds and, according to Mount (1975), is currently over 125 pounds. Fat consumption has increased at about the same rate as in the United States.

That sugar and fat relationships appear to be universal is the conclusion

TABLE II

Civilian per Capita Consumption Of Major Food Commodities (Retail Weight) and Civilian Population from Selected Years[a]

Commodity	Consumption (lb) 1960	1967	1973	1974	1975	1976 Forecast as % of 1975
Meats (total)	134.2	125.1	142.3	152.2	145.3	104
Beef	64.3	78.8	81.1	86.4	88.9	106
Pork (excluding lard)	60.4	59.6	57.3	61.9	51.0	103
Poultry products						
Eggs	42.4	40.7	37.3	36.6	35.3	100
Chicken (ready-to-cook)	27.8	36.5	40.7	41.1	40.3	109
Dairy products						
Cheese	8.3	10.1	13.7	14.6	14.5	107
Fats and Oils—total fat content	45.3	49.4	54.3	53.2	53.3	104
Fruits						
Fresh	89.6	79.1	73.8	76.1	80.3	100
Canned	22.6	22.6	21.3	19.8	18.9	106
Grains						
Wheat flour	118	112	109	106	107	100
Coffee	11.6	11.1	10.1	9.4	9.0	97
Legumes						
Peanuts (shelled)	4.9	5.7	6.6	6.4	6.8	101
Dry edible beans	7.3	6.9	6.5	6.7	6.7	98
Sugar (refined)	97.6	97.3	102.1	97.0	88.7	105
Corn sweeteners	13.9	16.4	23.3	26.8	28.4	109
Civilian population (millions)	178.1	195.3	208.1	209.7	211.4	101

[a] From U.S. Department of Agriculture (1976) and Kolodny (1976).

of a study by Wretlind (1974) of world sugar production and European usage. In Fig. 2, a plot of daily sugar consumption against daily fat consumption for a number of European countries shows significant clustering about a line of slope = 1. For purposes of comparison, the point for U.S. consumption has been added. Wretlind concludes that "dietary changes in most countries are characterized not only by an increased sugar content but also by a simultaneous increase in fat content." The interrelationship suggests that higher sugar consumption calls for higher amounts of fat in the diet and also lower amounts of polymeric carbohydrate, notably, starch.

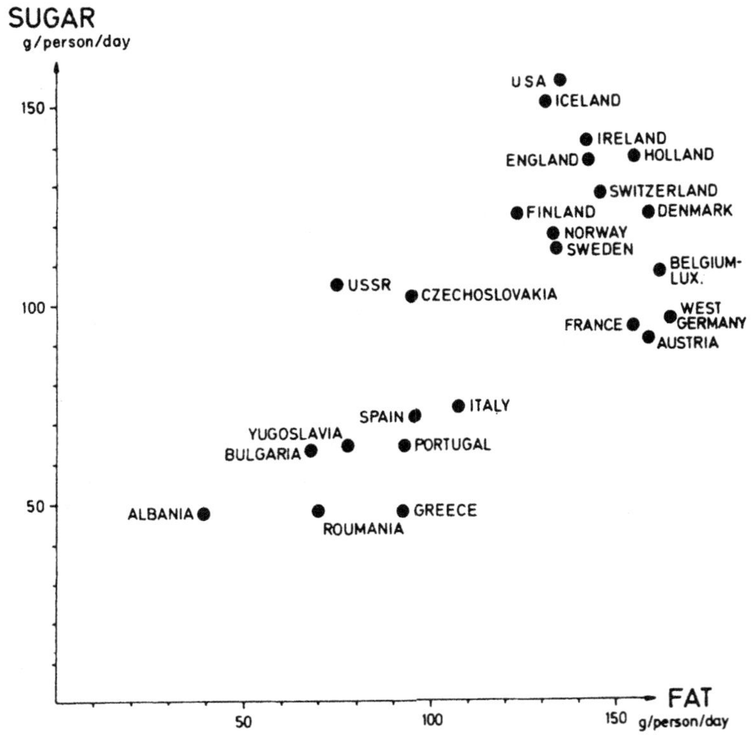

Fig. 2. Interrelation between sugar and fat consumption in some European countries. The values have been obtained from Production Yearbook (1972) and are for the period 1964–1966 for Albania, Bulgaria, Czechoslovakia, Denmark, Greece, Iceland, Roumania, Yugoslavia, and the USSR. All other values are for the period 1969–1970. From Sipple and McNutt (1974).

INDUSTRIALIZATION OF THE FOOD SYSTEM

Simple causal relationships between fat and sugar consumption and degenerative diseases, particularly cardiovascular diseases, elude research efforts, but dietary changes related to total intake seem to be involved at least as indicators. The shift of control of food ingredients from home kitchens to food factories is a faithful reflection of changes in family structure. The need for more family members to contribute to family income in order to enjoy conveniences which rapidly become necessities is characteristic of Western, consumption-oriented societies. But pressure for more consumer goods as proof of claimed improvement in the quality of life is equally evident in socialist societies and surely is a human attribute. Given their availability, convenience items, including foods, will be consumed. Moreover, the demand for these will increase as leisure activities, including television viewing, compete for the liberated time.

The rate at which use of sweeteners in processed foods has increased is shown in Fig. 3. There seems to be no diminution evident in this development, although rising prices of sugar caused some decrease in demand. What is striking is the rapid decline of household use of sucrose as opposed to industrial use. This curve crosses that for increasing use of sugar in processed foods close to the 50 pounds per capita point at about 1950. At the same time, sucrose consumption figures remain constant while total sweetener consumption is rising. These contrasts are explained by the fact that almost all of the corn-derived sweeteners are industrial products and reach the consumer in finished foods. Currently, only about 25% of total sweetener—mostly sucrose—is directly consumed. It would seem therefore that so far as sweeteners are concerned the discretionary choice of consumers has been sharply reduced.

The same would appear to be true of fat consumption judging by figures for total fat as well as polyunsaturated to saturated fatty acid ratios calcu-

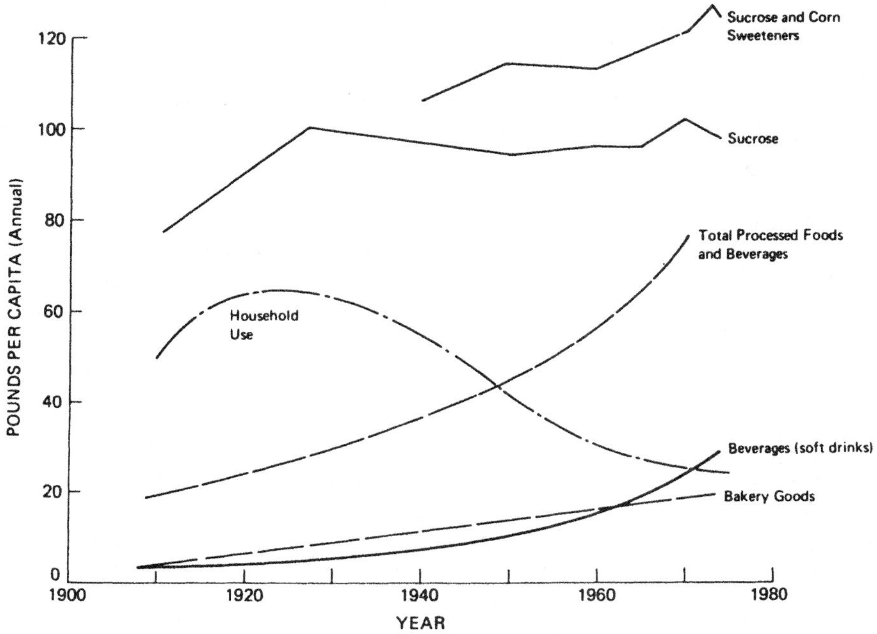

Fig. 3. Use of sugar in selected products, processed foods, and households. From Cantor (1975).

lated from foods available in the marketplace. More and more the consumer is being presented with finished products which carry their own appeal. The amount of fats and sweeteners which might have been under greater discretionary control when more food was prepared in the home is no longer under such alleged control.

FOOD DECISION-MAKING

There is very little evidence to suggest that home cooking in America was conducted with any greater attention paid to nutrition requirements before the advent of our commercialized food system. Food decisions were and are based on ethnic, geographic, status, and other sociocultural factors. Moreover, family and other authority figures, whether right or wrong, dominated decision-making, and these influences continue from one generation to another. The operation of these forces is clearly seen in developing country food studies, where the food system is still in early stages of change, and where emphasis on nutrition has been introduced by outside technical assistance intervention. In a recent study in one Indian southern state, Tamil Nadu (U.S. Agency for International Development, 1973), where dispersion of nutrition information was being emphasized by the state government, a survey revealed the relationship between food knowledge and food practice (Table III). The survey concluded" . . . reported factual knowledge about foods and nutrition is presently derived

TABLE III

Distribution of Responses to Questions on Sources of Knowledge or Information

	Number of responses	
Sources of information	Knowledge about specific nutrients	Information guiding food practices
Culture, traditional authority ("elders")	33	489
Medical practitioners	5	129[a]
Formal schooling	343	0
Practical experience and insights (friends, relatives, self)	88	164
Communications media (books, radio, newspapers, and magazines)	414	4
	—	—
	883	786

[a] Includes 79 responses dealing with herbs and medicines.

almost entirely from formal schooling and public information media, while actual behavior is dictated largely by culture and traditional household authority."

Who makes the decisions about the identity as well as the nutrient and caloric balance of consumer products in a commercialized food system? In a marketing dominated economic system, the consumer is theoretically the dominant figure and the consumer's needs, carefully assessed in the market, are fed back through product development and production channels and emerge to satisfy those needs to the best advantage of the consumer and the "economy." When markets are saturated and highly competitive, organizations vie for market segments and profitability. The consumer's needs become subject to anticipation, enlargement, and to conceptual exercises and promotion. Need may change from real need (a highly flexible term) to "perceived need," and the latter is very much in the eye of the perceiver.

As consumption data, which are really disappearance data, testify, American marketing practices reflect a continuing preoccupation with industrial expansion and increasing production. Expanding the demand structure is essential to absorbing the increased production. In the case of food, the demand for which, from energy and nutrient requirement data, is regarded as virtually inelastic, operation by disappearance statistics can lead to ignoring social and environmental impacts. The results of such a policy can create problems which are both difficult and costly to correct, and which finally attract the attention of protectionist groups and become political issues.

Under such circumstances the government in the form of an enlarged bureaucracy becomes the surrogate decision maker for the consumer. Tight controls are established by legislation, usually after opposing interests have been resolved in hearings and sometimes in courts of law. Frequently, decisions are made in the absence of scientific information because such decisions are forced in the name of public interest. Which is to be preferred is a serious question. A food system in which taste is dominated by perceived need can produce food just as boring as food designed by nutritionists which in so many instances ignores the great breadth of individual food behavior displayed by humans. A system which represents a compromise among pressure groups and has as its sole aim protection of the consumer can be overprotective.

After several years of hearings, committee investigations, special reports, and expert testimony, the Chairman of the Select Committee on Nutrition and Human Needs (1976) of the United States Senate has sponsored legislation which would establish an Office of Food and Nutrition, require an annual report aimed at setting goals for food and nutrition policy, require creation of a national nutrition monitoring or surveillance system, measure the impact of programs introduced through this effort, and conduct a nutrition education program.

It is obvious that in many parts of the developing and developed world, food problems of both famine and feast dominate the public consciousness. Virtually all of the concern, however, is directed at medical and disease considerations, and little attention is paid to learning more about the joys of food consumption and the positive aspects of food behavior which might help us to understand and accept its wide variations.

The history of legislation, which includes education, taxation, and subsidy aimed at limiting or encouraging consumption of particular foods, beverages, or related products, does not reveal any outstanding successes. A timely example is the following. In 1776 Adam Smith classed sugar with rum and tobacco as "commodities nowhere necessaries of life which have become objects of almost universal consumption—and are therefore extremely proper subjects for taxation."

Sugar has not been subjected to direct taxation in modern times but rum and tobacco have. Today one of the National Institutes of Health is devoted to the study of alcoholism, which causes untold misery at untold cost but also provides much pleasure, while tobacco (which many associate pleasurably) and lung cancer seem causally linked to most people. Cigarette consumption increases at increasing cost to the consumer although filters may reduce risk if they indeed control the number of cigarettes consumed. Sugar–better total sweetener consumption appears to be increasing and nutritionists are concerned about sweeteners and degenerative diseases. There is some suggestion in the literature that sweetness (sugars) is addictive (Cantor and Eichler, 1977), while consumer attitudes toward sugar are best described by the use of "sugar" and "sweet" in the language to describe only good or pleasant things, i.e., money, lovers, etc. The "consuming passions" previously referred to lead to both pleasure and pain, and it is in this area of understanding behavior associated with both stimuli that studies in the chemical senses may produce better means of behavior change than legislation.

NUTRITION AND THE CHEMICAL SENSES

The interdisciplinary approach to examining the behavioral aspects of fat and sugar consumption implied by reference to the chemical senses needs to be initiated by understanding the technological and economic pressures which lead to their increasing use. From a technological point of view, sugar and fat provide a food product with enhanced shelf life and reduce distribution costs. Sugars are water retention aids; they slow the drying process. Fat emulsified in the water also serves as a plasticizer, unifies the food components, and acts to retain desirable characteristics, which translate equally into shelf life extension. The result is that a product can stay in the distribution system much longer with freshness preserved, have that much longer to be exposed to the consumer and, therefore, a much better chance of being purchased and consumed. However, fat and sugar are more than functional aids to shelf life; they are equally associated with the richness of food and, therefore, its acceptability. In an affluent society, judging by studies of past affluent societies, the food of the more affluent upper economic classes was associated with the characteristics of richness. Economic wealth as richness dictated richness in other possessions and social activities, among them household furnishings, clothing, entertainment, and, of course, food. Richness of food is a matter of critical proportions and taste. Overrichness is readily detectable as is underrichness. A good cook knows the exact proportions.

Fat and sugar in food products relate also the textural qualities of food, in the vernacular of food technologists and the marketplace, the property of "mouth feel." There is a particular term used in the baking industry which refers to texture and relates to richness. The phrase used is "go away." It describes the formation and dispersion of the bolus as it continues to be chewed, and the avoidance of distribution of the contents of the bolus over the inside of the mouth. The extreme undesirable characteristic is best represented by peanut butter, and standard peanut butter formulas now allow up to 10% sugars. Proper "go away" is perhaps the most important acceptability characteristic of a baked product. Danish pastry, which technologically is produced by adding or "rolling in" fat, has excellent "go away." Low fat products generally tend to have poor "go away." Conversely, very high fat products, because the fat/water emulsion may reverse, can also result in a "go away" which coats the inside of the mouth with fat, obviously an undesirable effect. The right amount of fat is very important, as the ancient practice of larding demonstrated.

Wretlind's (1974) data on the universality of fat and sugar ratios suggest that since fat and sugar most frequently appear together in specific foods, the approximate 1 : 1 ratio (weight basis) may have special significance with respect to texture and therefore acceptability and also quantity consumed. On an energy basis the ratio of fat to sugar is about 2 : 1, and this may also relate to texture and perhaps satiety. An examination of sugar and fat components in foods on a world-wide basis would be most interesting.

The food properties described—sweetness, richness, and texture—all oral stimuli, quite probably have numerous pleasant associations that relate to early life experiences as well as continuing experiences which may reinforce the pleasures. The proper combination of fat and sugar which produces the taste of richness is not only enhanced by associated richnesses sensed differently but revives infantile pleasures of suckling. Textural enjoyment also must have important sexual associations. Recently published information on human sexual practices associates affluence and higher education with fewer taboos against oral sexual practices, and as more data are gathered the prevalence of such practices appears to grow (Rubin, 1976). Thus, all varieties of oral stimulation—food consumption, smoking, speaking, sexual practices—may find a common denominator in the chemical senses and help to explain excesses of consumption. Product development practices in the food industry are perhaps more oriented to such food associations, along with profitability factors, than is generally recognized. Stimulation of a market or invading a market is a common objective, and marketing dominates new product design.

CONSUMPTION MARKETING

More and more of the so-called "creative" people who design products are not in the laboratories and, therefore, least open to technological and scientific constraints. Marketing executives have found that ideas generated by nontechnical people are more realistically associated with markets and are less inhibited by restraints which would concern technical people. As a consequence, new product funds tend to be invested more in services associated with advertising than those of technical groups. The finally chosen product identity, however, is delivered to the technologist who is motivated by the challenge to produce such a product.

The effect of such product development practices on consumption is important. A question which has been presented for consideration is one of flavor recognition. If we define what has been referred to as richness as a concomitant of flavor, then the repeated incorporation of "richness" into a new product would not only provide regular reinforcement for recognizing "richness," but, with all of its omnipresent associations promoted as good, result in increased consumption of fat and sugar.

Is the perception of flavor modified by experience? There is supposedly a safety factor associated with fat consumption and probably also with sweetener consumption. But the statistics, at least on the average, support the conclusion that as preparation of food moves from kitchen to factory, the perception of richness and the continuing emphasis on richness, certainly in snack foods, has contributed not only to reinforcement but also to the resultant increased consumption. Also, it would appear that such increases due to the relative inelasticity of demand for food could seriously unbalance nutrition. This is the quantity displacement of concern to some nutritionists, and the basis for recommending that all commonly consumed food products be nutritionally balanced.

The usual kinds of educational deterrents may not provide an ideal limitation to these consumption increases and, therefore, other methods of behavioral change are needed. What they are is not clear at present. The appearance of greater fat and sugar consumption in affluent Western countries as opposed to a somewhat lesser appearance in socialist countries is interesting to note. There is currently a central planning versus free market debate underway. Central planning would appear to be effective in halting consumption of particular products. On the other hand, the lower consumption in socialist countries may be due to failure of production rather than deliberate planning. Given unrestricted opportunities to acquire rich products, individuals characteristically seem to consume large quantities in either socialist or capitalist countries. In the history of sugar there is some reason to believe that encouraging increased sugar consumption served as a palliative for social unrest.

The degree to which unbridled consumption and cheap or slave labor associations with excesses contributed to the downfall of the Roman Empire is apparently still a historical question and concerns social critics observing Western affluence. It does not appear that our culture will suffer too much from that kind of license—that is, cheap labor. What is perhaps more disturbing is the degree to which the discretionary limits of consumers are being reduced by the system which designs food like any other consumer item. Leadership, and through it instruction, is necessary at the production level if industry is to continue to do the cooking. Certainly, a better understanding of all the factors associated with flavor perception and consumption, including the behavioral factors, is required, and it is in this area particularly that new technologies need development.

What appears to be coming out of work in the chemical senses is complexity rather than simple relations. The challenge is how to apply the knowledge to expand the values and pleasures of the food system and limit quantity rather than to apply prohibitions in the restricted illumination of oversimplified models.

REFERENCES

Antar, M. A., Ohlson, M. A., and Hodges, R. E. (1964). U.S. fat consumption. *Am. J. Clin. Nutr.* **14,** 169–178.

Cantor, M. B., and Eichler, R. J. (1977). "Sweetness: A Super Normal Reinforcer." *Chemtech* **7,** 214.

Cantor, S. M., and Shaffer, G. E. (1974). Pressures for a three commodity sweetener system. *Cereal Sci. Today* **19,** 266.

Cantor, S. M. (1975). Patterns of use. *In* "Sweeteners: Issues and Uncertainties," p. 24. Nat. Acad. Sci., Washington, D.C.

Chairman of the Select Committee on Nutrition and Human Needs (1976). "Nutrition and Health. II. Study of the Impact of Nutritional Health Considerations on Food Policy." U.S. Gov. Printing Office, Washington, D.C.

Deerr, N. (1950). "The History of Sugar," Vol. II, p. 428. Chapman & Hall, London.

Hegsted, D. M. (1975). Health: Summary of evidence. *In* "Sweeteners: Issues and Uncertainties," pp. 86–87. Nat. Acad. Sci., Washington, D.C.

Kolodny, S. (1976). "Sweetener Usage." Abstr. Cent. Meet., Am. Chem. Soc., New York.

Marriott, M. (1968). Caste ranking and food transactions: A matrix analysis. *In* "Structure and Change in Indian Society" (M. Singer and B. S. Cohn, eds.), p. 141. Aldine, Chicago, Illinois.

Mount, J. L. (1975). "The Food and Health of Western Man." Wiley, New York.

Page, L., and Friend, B. (1974). Level of use of sugars in the United States. *In* "Sugars in Nutrition" (H. L. Sipple and K. W. McNutt, eds.), p. 94. Academic Press, New York.

Pullar, P. (1970). "Consuming Passions." Little, Brown, Boston, Massachusetts.

Rubin, L. B. (1976). The marriage bed. *Psychol. Today* **10,** 44.

Sipple, H., and McNutt, K. W., eds. (1974). "Sugars in Nutrition," p. 94. Academic Press, New York.

Smith, A. (1776). "An Inquiry into the Nature and Causes of the Wealth of Nations," Bohn's Standard Library, 1, p. 675. Edinburgh (cited in Deerr, 1950).

U.S. Agency for International Development. (1973). "The Tamil Nadu Nutrition Study," Vol. I, p. 128 (Contract No. AID/nesa-399). Sidney M. Cantor Associates, Inc., 1973.

U.S. Department of Agriculture. (1976). "National Food Situation." USDA, Washington, D.C.

Wretlind, A. (1974). World sugar production and usage in Europe. *In* "Sugars in Nutrition" (H. L. Sipple and K. W. McNutt, eds.), p. 87. Academic Press, New York.

DISCUSSION

Moskowitz: One of your illustrations showed an increasing consumption of fats and sugars. Economically, you can partition this out into a time series where one axis is time and the other consumption. From your experience, is this a combination of two or possibly more factors, one of which is economic income and affluence, but the second one a growing desire—a psychological factor—a kind of national growing sweet tooth? If so, then we can project various kinds of strategies to counteract such a national sweet or fat tooth.

Cantor: One curve which I showed would. No question but that there is an economic factor. That is clearly revealed in the data of the last 3 years with sweetener prices going up. People simply stopped consuming some sweet products. This surprised the sweetener industry somewhat when they found a lot more sweeteners in warehouses than expected. Fat falls into a similar category; there appears to be a price elasticity of demand. Whether the public will become accustomed to giving up a larger share of income for food is uncertain. Our statistics are very deceiving; they contain a large waste factor. Also, the Secretary of Agriculture says, "We spend less than 20% of our income for food." This is true, but it is an average and the limits are very great in our society. Some families spend 80% of their income for food, but not many. Some spend less than 20%. I am not so sure that an increasing "sweet" or "fat tooth" is the case, although it can not be ruled out. There is a literature on a possible addiction mechanism. I think of it more in terms of the richness associated with combinations of fat and sugar, plus the fact that we are moving, more and more, toward mass-produced foods, materials which can be bought and consumed anywhere. It is such products with extended shelf life which constantly intercede peripherally on the consumer. Now the functional value of sugars and fats in prepared foods is extremely important. Sugars are moisture-retaining agents; fats form good emulsions in water. Texture or mouth feel is crucial in consumption—how the product is eaten, how it "goes away" or disperses in the mouth, the sum total of oral satisfaction. Equally important is the fact that in our distribution system an extra day on the shelf at optimum freshness is very valuable. It gives more exposure to the product and the probability of its purchase by the consumer. That is one of the reasons why we are seeing more sugar and fat in some processed foods. They extend shelf life and taste rich and therefore good.

Hegsted: I wanted to ask you about economic analysis of the European countries versus our own, where I understand there are big differences in percent of income spent on food, yet that the fat/sugar point in Western countries is always in about the same place.

Cantor: Europeans, for one thing, are more accustomed to spending a larger portion of their income for food. Their distribution system is still less commercially developed, and perhaps the role of women is somewhat slower in developing. I think we very likely will spend more of our income for food as we become larger purveyors of food to the world and as the energy cost of increased yields is felt. Among the food-plentiful countries there is bound to be some leveling out, but an upward trend in food cost is clear.

Dethier: I would like to make two comments and then ask you to comment on them. I recently saw a graph which plotted consumption of butter against income for various countries in the world, and it fell on a very nice line except for one point, which was the United States, and the interpretation was that falling off of the line at the U.S. point reflected our attitude toward cholesterol. My second point is the removal, a while back, of swordfish from the market because of alleged levels of mercury in it. Can you comment on the effect on consumption of fears that are health related?

Cantor: All I know is that swordfish is beginning to appear in quantity again on the market.

Cagan: That is because museum specimens of swordfish seem to show the same amounts of mercury as contemporary samples.

Cantor: Possibly for that reason, and possibly also because the world supply of fish is down, and fishing is more efficient in many countries. The price of fish has gone up significantly with respect to other foods. To this I would add that people repeatedly warned seem to turn off to the warning. On the matter of butter, we have come through a revolution; we accept margarine now, and not all margarine contains unsaturated fats. That is a fairly recent part of the market, but it is also the smaller part of the market. Margarine has become a major competitor of butter. It is highly acceptable, tastes like butter, and has some superior properties. This is only the beginning of alternative food ingredients; what I call interconverted food products. For example, we have not seen too much penetration of soy into meat, but we will see more of it, also soy as milk. Asian cultures have used soy this way for ages. Food technologists faced with rising costs will make commodity interconversions commonplace. High fructose corn syrup for sugar is a growing reality.

Miller: I have the general sense here that caloric intake in sweeteners is upward. Yet looking at your data, 1970 and onward, it seems to suggest just the opposite—that there has been a reduction. In fact, I was surprised in looking at sweeteners which have been going down somewhat since 1970—which was before the price rise—that the USDA predicted a rise for in 1976. I am curious as to why they predicted that rise, if you happen to know.

Number two, I think you also commented yourself that by 1980 the prediction was a general reduction.

Cantor: Right now, only in the reduction of sucrose. We need to distinguish between sucrose and total sweetener consumption.

Miller: The decrease is just in sucrose.

Cantor: Yes, now that penetration of the sucrose market is being made with starch-derived sweeteners. The total of both has declined since 1973 but now is beginning to rise again. USDA expects sucrose for 1976 to be somewhat above 1975, but still lower than 1970 to 1975. On future projections, there is as much reason to project decrease as increase.

Miller: Your data from 1970 suggest an increase.

Cantor: What was happening there is that the price increase for sucrose was already beginning to be felt, and we were seeing a greater infusion of corn syrup which was used to a large extent to dilute the cost of sucrose in the industrial market. Corn syrup rose rapidly and in 1970 was over 18 pounds per capita, while sucrose was about 102 pounds. By 1973, sucrose was 103 and corn syrup nearly 24. But in 1974 we saw the effects of increased cost which began in 1973–1974, and sugar is still high priced relative to what it was during the period you are talking about. We are still seeing 20¢ per pound sugar. What the USDA is projecting is a drop in sucrose. But it is not the USDA alone; it is the sugar industry itself projecting a drop to 80 pounds per capita annually by the early 1980's. That is because sucrose is becoming the residual sweetener, as corn sweeteners penetrate the sucrose market. The USDA tentative figure for 1976 is over 95 pounds of sucrose.

Mugford: I was very interested in your comments about the waste factor built into any compilations of consumption data. Are there any international data showing waste? One gets the impression that waste varies with changing cultures and we need colloquial definitions of waste.

Cantor: I agree we do not have any significant international data. There are a few U.S. data which suggest that in specific cases waste may be as low as 15% or as high as 35%. Those are the limits one hears. The other day I was talking with someone who had worked on school lunch programs and waste, and in that case it turned out that the products which were distributed to the children on the school lunch program might be wasted as high as 80% and as low as 20%. But the food brought from home was likely to be wasted a lot less.

Chapter 23

The Nimrod* Connection: Myth and Science in the Hunting Model

TIMOTHY PERPER AND CARMEL SCHRIRE

We begin with a few myths about the origin of meat-eating and the origin of man. This might seem inappropriate for a scientific audience, who will surely attribute the appearance of myths here to the presence of the anthropological co-author of this paper. But we must remind you that the other author is a biologist whose training in the formal belief system of Western science should certainly have separated him from the gobbledygook of anthropology. Why then does a sober-sounding discussion of the origins of human dietary behavior kick off with a myth? The reason lies not so much in the different training of the two authors as in the nature of certain belief systems inherent in the problem we are going to analyze. It is our feeling that myths are not at all foreign to science, but that, as in this particular instance, they are an integral part of a model that has been formulated over the years to account for certain aspects of human evolution.

Having mollified our audience to some extent, it might be appropriate for us to expound an exotic myth of some remote peoples, drawn from the vast bank of anthropological fieldwork. Instead we want to relate two Biblical myths: first, that of the origin of man and then, the origin of meat-eating, because it is the structure of these *familiar* tales that underlies the so-called scientific rendition of these questions.

* Genesis X: 8–9.

The first myth describes the cataclysmic effect of an act of eating on the development of man. According to Genesis, God created man on the sixth day, from the dust of the ground (Genesis II: 7). Like all other living animals, man was possessed of a soul, but unlike the rest, man was entrusted with dominion over the other animals, and was placed in the garden of Eden "to dress it and keep it" (Genesis II: 15) and to care for its inhabitants. Finally, when all other forms were operative, God created a "helpmeet" for his primal man and called her "Woman." Perfection was the keynote of Eden, and man lived in harmony with animals, until he became possessed of knowledge and shame and was expelled from Eden into the harsher world beyond. Man lost his primal innocence through a single act of disobedience, an act of eating—not an apple as is widely believed—but simply the fruit of the trees of knowledge of good and evil. Man was transformed, catapulted from innocent childhood to maturity, to bring forth children in sorrow, to eat bread in the sweat of the face, to return to dust, all through the very act of eating forbidden food.

The second myth relates to the origin of meat-eating. After leaving Eden, men multiplied and covered the earth, but since they possessed the fateful knowledge acquired in Eden, life was subject to chaos and turmoil. In order to cleanse the earth and, in a sense, to begin again, the catastrophic flood, known as "Noah's Flood," took place. Examples of each species of animals were saved in an ark so that life could resume when the waters receded. It was only then, perhaps in recognition of the mastery of man over animals, that God allowed Noah to sacrifice to him and thereby gave him the right to eat meat (Genesis IX: Encyclopedia Judaica, 1971).*

Meat-eating was hedged with proscriptions that involved taboos on eating certain animals and preparing certain foods. The most significant of these reflected the original dogma of creation, that *all* living animals are possessed of spirits. The spirit resided in blood so that flesh of living animals was taboo and strict laws of slaughter demanded that the blood be made to flow onto the ground to allow the spirit to return, as dust was rendered unto dust. Meat remained a luxury food, the food of sacrifice to be eaten on joyous occasions like the Sabbath, and avoided on sorrowful ones (Encyclopedia Judaica, 1971).

The structure of the Biblical tale involves the creation of man in

* Interestingly enough, legend has it that Noah did not make the sacrifice himself. He had acquired a "bodily defect" after being mauled by a hungry lion in the ark, and was therefore ineligible to perform the ritual. Instead, his unblemished son, Shem, made the sacrifice. This legend, postdating Genesis by some 1500 years, gains power through its very irony, in that the chief protagonist is prevented from performing the ritual that will transform him and his kind into meat-eaters by the greatest meat-eater of all, the king of beasts (Ginzberg, 1942, Vol. I, pp. 164–167; Vol. VII, pp. 189–190).

Paradise and the fall from grace through an act of eating forbidden food which, in turn, launches man into a harsh world. The scientific version of his origin also sets primal man in Paradise but imposes ecological rather than moral stress upon him, in the form of increasing aridity. His response is to eat a new food, namely meat, and in doing so, he becomes a participant in the new, dry world of competition, bloodshed and warfare. Thus the structure of both versions is similar though the details differ. However, the Biblical version of how man came to eat meat recognizes the potential spiritual conflict involved in the predator–prey relationship of men and animals, and by virtue of this recognition hedges the consumption of meat with prohibitions, proscriptions, and laws.

It is all very well to observe superficial similarities in the Biblical and scientific versions of human evolution, but the question we have to ask is to what extent the scientific version is true to fact, and where, on the other hand, does it ignore the facts and force the scientific evidence to conform to the well-known myth we have described above. In point of fact, we are raising a general problem about scientific models. We feel that they may be derived from two sources, one being objective fact and the other myth. When the factual component is strong enough to shape the model into a form congruent with nature, the logical connectives in the model represent real connections in the real world. However, when there are too few data points to specify unambiguously the real or true connections as they exist in nature, those postulated in the model seem to be isomorphic with links made between similar elements in the myths of the culture of the scientist. Thus, although the data refer to nature, the structure or form of the model is given by myth. Using these observations as guidelines for the analysis of models of human behavioral evolution, we will now examine the proposition that human biological and cultural evolution has been dominated by meat-eating and/or hunting.

The hunting model, which assumes that hunting and meat-eating triggered human evolution and propelled man to the creature he is today, was most explicitly proposed by Washburn and Avis in an essay in 1958. Its basic conclusions have been reiterated frequently since then, but never substantially altered (Campbell, 1974; Laughlin, 1968; Pfeiffer, 1971, 1972; Robinson, 1963; Washburn, 1960; Washburn and Avis, 1958; Washburn and DeVore, 1961; Washburn and Lancaster, 1968; Washburn and Moore, 1974). We will now outline briefly some of the more explicit renditions of the model in order to show its basic structure.

In 1963 a paper appeared by John Robinson, discussing the nature of australopithecine adaptation. Australopithecine, or southern ape-man, is the generic term used to describe a series of hominids, or possible human progenitors which were first recognized as such 50 years ago, from fossil remains in South Africa. Robinson suggested that there were two morphologically distinct but contemporary genera of australopithecines, whose differences reflected dietary variations, in that one was a vegetarian and the other an omnivorous meat-eater. Both were derived from the same vegetarian parent stock which flourished in the Miocene forests. Robinson's version of the origin of man proposed that when the Miocene forests shrank in the late Tertiary desiccation, these animals were forced to find new foods, and, in the same way as modern primates eat meat under stress, some australopithecines became carnivorous.

The response is viewed as a critical point in human evolution, which placed the meat-eating omnivore on the direct line of ancestry to man, in the following fashion (Robinson, 1963):

> . . . the incorporation of meat-eating in the diet, seems to me to have been an evolutionary change of enormous importance which opened up a vast new evolutionary field. The change, in my opinion, ranks in evolutionary importance with the origin of mammals—perhaps more appropriately with the origin of tetrapods. With the relatively great expansion of intelligence and culture it introduced a new dimension and a new evolutionary mechanism into the evolutionary picture, which at best are only palely foreshadowed in other animals. Furthermore it brought into existence the first member of the animal kingdom able to colonize other worlds—or at the very least, able physically to leave this one (pp. 412–413).

A detailed exposition of the idea that hunting transformed the ancestral primate into man is found in the seminal essay by Washburn and Avis (1958). They assume that the rudiments of modern human behavior exist among nonhuman primates, and, furthermore, that human behavior is essentially a modified, developed form of the primate version: "Man's way is built on the ape's way" (p. 435). The authors set about identifying the pressures that birthed human behavior from the primate forms, by constructing a sequence of events that may be summarized as follows. The vegetarian progenitor of men tended to bipedalism which freed the hands for tool-using. Tools, however, were essential in exploiting a food source that became increasingly important in times of stress, namely, hunted meat (Oakley, 1954; Washburn and Avis, 1958). A combination of tool-using and meat-eating were therefore key factors in producing man (Washburn, 1960):

> If one views the first stage in human evolution . . . as the development of bipedal locomotion followed by invasion of savannahs and plains by an ape, all subsequent human change is increasingly dominated by the use of tools. Increase in brain size resulted from the new selection pressures stemming from tool use. Speech, made possible by the larger brain, was correlated with a complicated technological tradition; and the larger and more complicated society was made possible by the larger food supply. Human hunting depended on tools, and hunting brought about greater mobility, the use of clothing, the conquest of the Arctic by a tropical animal, and so the

> peopling of the New World. Increase in brain size was associated with a slowing of the growth rate and a much greater period of dependency. This changed the social life, establishing long-term social relations. Thus the hunting life changed man's psychology and the way of life of the human group (pp. 434–435).

The significance of hunting is said to be threefold, affecting the psychological, social, and territorial behavior of early man. Let us itemize two of the most crucial effects described. The first is a suggestion that hunting contributed largely to the development of human sexual division of labor. Tool-using primates developed larger brains, and with this came a longer period of infantile dependence on the mother. Females were unable to hunt because they were home-bound, feeding and caring for the helpless young. The hunting males therefore assumed responsibility for their families by ranging afar and bringing food back to the home base. The second effect, described in a later paper by Washburn and DeVore (1961), traces the origin of modern patterns of human sexuality to a hunting past. The argument is that ancestral primates must have had estral cycles during which mothers deserted their young. In order for sexual division of labor to exist among hunters, estrus had to yield to constant receptivity. Consequently, hunters could be welcomed home whenever the rigors of the chase allowed, to a family atmosphere reminiscent of what we see around us today. Thus, a commitment to hunting—a behavioral adaptation that may originally have represented nothing more than a way to get a highly prized food—ended up by being the source of the most basic form of human organization, the nuclear family.

Until now, we have dealt with the notion that the development of hunting and meat-eating took place in the earliest stages of human evolution, perhaps as long ago as the Miocene–Pliocene times. Bernard Campbell supports the notion that meat was a food source whose exploitation actually propelled humanity into what they are today, but he sets the scene considerably later in time, around the Middle Pleistocene some ½ million years ago, when it is assumed that cooperative hunting began (Campbell, 1974):

> . . . the final determinant of the direction of human evolution was necessarily environmental and ecological; not obscure, but striking and fundamental. Like all great evolutionary radiations of animals, man's appearance on the earth involved the exploitation of a new food source, and a very rich one, (p. 241).

The new food was meat—not odd scraps of flesh scavenged off the kills of others, nor small prey easily caught and eaten—but meat in large quantities, such as might be landed by bands of hunters preying on herds of large game. Nevertheless, Campbell's hypothetical chain of events that results from constantly eating great chunks of meat is very similar to that that Washburn and others derive from the very earliest human hunting. Campbell assumes that the Middle Pleistocene was a time of stressful drought and that humans responded "adaptively" by preying on large animals. Thereafter the familiar sequence of social institutions is seen to spring from the commitment to meat-eating, including sexual division of labor, food sharing, constant receptivity, and finally, the development of technology and language in order to hunt more effectively.

We leave the pedantic fields of scholarship for a while to view the hunting model in a more poetic setting, in the elegant and evocative prose of Robert Ardrey. In 1960, Ardrey first published an account of the discovery of the australopithecines of Southern Africa in "African Genesis," where in a striking series of metaphors, he showed that the structure of the hunting model is a mirror image of Creation according to Genesis. For Ardrey, the evolution of man began in the lush Pliocene forests of Kenya, a locale which Ardrey calls "Eden." In the forests lived vegetarian primates whose safety was at stake as the forests shrank with the advent of the dry Pliocene. Climatic stress therefore threatened to render these animals into "Eden's outcasts" (Ardrey, 1971), but by an "incredible accident" (Ardrey, 1971) one line mutated to become carnivorous and moved out of the forest to prey on the rich herds that teemed in the Pliocene savanna beyond. The vegetarian primate "Abel" remained in the forest, but his counterpart moved out, and (Ardrey, 1971) ". . . there, faraway under vasty skies, Cain killed for a living" (p. 287).

Hunting, it becomes clear, for Ardrey, is the equivalent of Original Sin. And, in actuality, not only the equivalent, but sin which derives from our origins and which we inherit by reason of our origins. If meat-eating led to hunting and killing, and if we are *therefore* aggressive, we owe our aggression to our origins, and by definition the sin is original, that is, of origin. "African Genesis" begins . . . "Not in innocence, and not in Asia, was mankind born". . . . That, and indeed the very title of his book, should make us aware of why the hunting model seems so familiar, so plausible. The transition from a vegetarian mode of life to that of the hunter is expulsion from Eden. In its train, killing follows hunting, as Cain slays Abel, and carnivorous psychology emerges. No longer are health, happiness, and the fruits of the earth one's lot. Instead one works by the sweat of the brow, as males become economically responsible. In pain, children are brought into this world, and maternal responsibilities with clinging babies appear. Constant receptivity heralds the temptations of knowledge and the sins of Eve. Society struggles with the tendency to prevent successive consortships from destroying the fabric of the family. In Ardrey's words, we are "Cain's Children" (1971, Chapter 11) and violence, warfare, and starvation are our lot.

Over the years, Ardrey reiterated his apocalyptic version of human evolution shaped by hunting, meat-eating, and violence. In "The Social

Contract'' the irrevocable nature of a commitment to hunting emerges (Ardrey, 1974):

> . . . we chose the violent life . . . the ancestral primate went up the ladder to stimulation, excitement, identity . . . we found our daily satisfaction in violence . . . we adapted . . . to the hunt—we became creatures adapted in all ways to the excitements of violent action. Until 5000 years ago there was no other way to survive. And if it was only then that organized warfare became a significant human entertainment, perhaps we may understand it as a substitute for the lost hunting way (pp. 337–338).

In his most recent book, ''The Hunting Hypothesis,'' Ardrey (1976) finally says explicitly what Washburn, Campbell, and others only imply. He lays bare the essence of hunting for the anthropologist, by stating unequivocally that it lies at the center of human behavior today and that its effects were not only powerful but irreversible. In his terms, we hunted because we were human, but more important, we are human because we hunted. The hunting model is the counterpart of a familiar myth, and Ardrey has identified that myth in no uncertain terms: Eden is lost to us forever.

Returning to the more scholarly sources, the hunting model assumes that two closely related processes occurred at the same time. The first, which we call ''historical,'' draws on a body of archaeological evidence to demonstrate that relatively short-term changes prompted major alterations in human behavior. The second and somewhat new series of ''biological'' processes uses a framework of general evolutionary theory that includes studies of changes in gene frequencies over time and space as a consequence of differential survival and reproduction of genetic variants, to promote the study of behavioral evolution. It is not easy to distinguish these processes, but we suggest that they center on transmission and change in traditions learned and relearned throughout human history, and on gradual shifts in material and mental culture. Although debate continues over how to define cultural evolution, there remain the basic data of archaeology and the nature of the evidence it uses to reach its conclusions. Like evolutionary hypotheses, archaeological reconstructions must remain tentative and ultimately, perhaps, unprovable, but in spite of this, the hard facts of human biology and archaeology provide constraints on any reconstruction of the past. We turn now to these sources to see whether they support the hunting model of modern scientists or whether they simply relegate it to being a reworked version of the Biblical creation myth.

The first australopithecine was discovered by Raymond Dart about 50 years ago. He found the specimen in a block of brecciated material that came from a limeworks in Taung, on the edge of the Kalahari desert. In 1925, he published the first announcement of the new ''man-like ape''

(Dart, 1925), which he christened *Australopithecus africanus* in honor of its peculiar and then unique characteristics. After describing these features, Dart (1925) speculated about the behavior of such a creature, living on the inhospitable edge of the Kalahari desert, and thus became one of the first scientists to use archaeological evidence to construct the essence of the hunting model:

> For the production of man a different apprenticeship was needed to sharpen the wits and quicken the higher manifestations of intellect—a more open veldt country where competition was keener between swiftness and stealth, and where adroitness of thinking and movement played a preponderating role in the preservation of the species. . . . Southern Africa, by providing a vast, open country with occasional wooded belts and a relative scarcity of water, together with a fierce and bitter mammalian competition, furnished a laboratory such as was essential to this penultimate phase of human evolution (p. 199).

Dart's idea that environmental stress forced early hominids into a life of violence was not challenged at this time, largely because the controversy that exploded after his announcement centered on the physical features of the specimen and the position of "Dart's baby" (as it came to be called) in the total picture of human evolution. Dart remained free therefore to expound the hunting model, which he did in an increasingly untrammeled form. Some 30 years later, when many more australopithecines had emerged from the limestone breccias of the Northern Transvaal, Dart (1957) was able to confirm their identity as "Man's anthropoid ancestors, whose staple diet was venison and whose whole existence depended on success in rapacious horde-hunting life . . ." (p. 93). He even drew a picture of the life, dominated by activities such as ". . . pulping and crunching a great deal of bone . . . collecting crania of creatures with formidable slashing teeth . . . severing ungulate hind limbs . . . head hunting and bludgeoning . . . decapitation . . ." (p. 87–88).

Wiping aside the gore, we must ask to what extent Dart's vision conformed to the evidence he saw. His earliest rendition of the hunting model appeared in the British journal *Nature* in February, 1925, *a mere few months after he first saw the Taung skull,* "late in 1924" (Dart, 1925, p. 199). Unless we agree that the mere sight of the skull screamed "meat!"—and the childish face lacking fangs or claws appears to have no patent dietary message—we have to conclude that the notion of hunting and meat-eating as powerful deterministic forces in human evolution was *already* there in Dart's mind long before the Taung skull materialized. In fact, the specimen simply seems to have provided the excuse to enunciate the model, a model that we suggest is central to Western beliefs rather than inherent in scientific data.

To what extent, however, do australopithecines actually seem to have hunted and eaten meat? It is true that their broken bones occur together with those of other animals in dense breccias in the South African deposits, but in certain cases they might all represent the prey of carnivores (Brain, 1969), or simply reflect a very slow rate of soil deposition around the bones. The discovery of similar faunal assemblages at Olduvai Gorge prompted Leakey to describe the scene as a "living floor" (Leakey, 1959) as opposed to the carnal chaos envisaged by Dart. The question is largely one of judgment which, of course, reflects the belief system of the scientist. While writers like Isaac (1969) and Brain (1969) regard early man as a well-adapted omnivore, placing no major stress on any particular resource, others such as Clark (1969, 1970), Pfeiffer (1971, 1972), and Speth and Davis (1976) insist that he was first and foremost a hunter. In this way, diametrically opposite views emerge, as for example when Shapiro views *Homo erectus* at Choukoutien as relying on a mixed diet of game, insects, fish, frogs, birds, turtles, and plants (Shapiro, 1974), whereas Grahame Clark (1970) uses the same evidence to conclude baldly that ". . . the men of Choukoutien were above all hunters and meat-eaters" (p. 37) and ". . . Peking man depended largely on venison" (Clark, 1969, p. 37).

The truth of the matter is that time favors the preservation of big bones. Evidence of plant-eating, seed crunching, root gnawing, leaf munching, fish eating, and even egg sucking disappears rapidly over time, while big bones become fossilized. A first impression gained by looking at ancient deposits is a plethora of bones that conjures up thundering herds of meat. But the more balanced viewer recalls the processes of differential decay, and in doing so makes allowance for the absence of the less resilient evidence of a mixed diet.

Let us expand this more balanced view by questioning the plausibility of the ecological and evolutionary assumptions of the hunting model, and asking how likely is it that early hominids were predominately meat-eaters. The few people whose diet is predominately flesh, such as the Arctic Innuit, live in environments with limited land-based vegetation. No one would suggest that man originated at the North Pole, though Robinson, Dart, and Ardrey all stage the emergence of man in harsh, dry regions whose barrenness forced man to hunt in order to survive.

There are several pitfalls in this theory. First, the herds that are supposed to have thronged the arid wastes must have had something to eat. The wide variety of ungulates, carnivores, and primates associated with the extinct South African hominids must have subsisted in rolling, well-vegetated savanna country, far from the marginal zones envisaged by the model builders. This raises a second flaw: is it possible that men deliberately chose to *ignore* plant foods and concentrate on meat, or that they constantly tried to find areas that were well supplied with meat alone?

This would be a silly strategy since survival advantages normally accrue to those individuals who remain in environments with a wide *variety* of foods to provide the flexibility needed to weather shortages of certain resources. Finally, the endless drought envisaged by the hunting modelers seems unlikely. Geological evidence points to gradual fluctuations in rainfall and temperature that would favor movements of predators and prey to more kindly regions but would scarcely force them to remain in barren zones (Isaac, 1969: 7). Indeed, the hominids of Olduvai Gorge Bed I did not live like Bedouins with desert-loving animals, but rather they roamed around streams that flowed into a salty lake, and died in association with the bones of water-loving forms like hippos, crocodiles and fish as well as large birds, carnivores, zebras, and pigs.

The elements in the hunting model discussed above may all be challenged using archaeological data, but what of the more theoretical propositions inherent in the model? First, let us look at the notion that hunting and meat-eating, even infrequently, had a major effect *per se* on human evolution. Modern studies of primates show that they can and do catch prey and eat meat at times of stress. For example, when Thomson's gazelles migrated into an area occupied by baboons, they became primate prey within 3 years (Harding and Strum, 1976). Likewise, Virginia deer have been observed eating fish on occasion (Cahalane, 1961). But whereas no reasonable person would discuss deer evolution in terms of fish-eating, the effect of meat-eating on primates is the keystone of some of the most highly regarded writings on human behavioral evolution! For us, it seems that the ability to catch and eat meat lies well within the behavioral repertoire of nonhuman primates, but that its expression is ecologically regulated.

The second premise of the hunting model is the notion that hunting induced sexual division of labor. If meat were such an important food it is unthinkable that lactating or pregnant females could afford to wait around the home base, grubbing for fruit and vegetables until the hunting male returned with his bounty. Studies of carnivores like wolves, show that both sexes hunt, and *if* humans needed meat, selection would certainly favor the female who hunted, and whose offspring learned to hunt at an early age. These are scarcely the circumstances under which long-term infantile dependence would evolve. Furthermore, if the marginal conditions deemed necessary for the evolution of hunting had pertained, the sexual division of labor would have doomed hominids to speedy extinction through starvation, with women and children chained to the home while man the hunter roamed the arid wastes. We have therefore to conclude that if the object of hunting were to get meat, the development of sexual division of labor would have hampered the entire venture considerably.

A third premise of the model, that constant female receptivity was an outcome of hunting, can now be reassessed. Even if females, as Campbell put it, "learned" to receive males throughout the menstrual cycle (Campbell, 1974 p. 240), this would not ensure reproductive success since the likelihood of impregnation varies during the cycle. Furthermore, if females were home-bound while more males hunted abroad, it is likely that the few nonhunting males who remained at home to guard the females would have enjoyed the most reproductive success, making it, paradoxically enough, reproductively disadvantageous to hunt.

A final premise of the hunting model, that human sociality, cooperation, and intelligence sprang from the need to hunt, is also questionable. The issue is why the ancestral generalized primate who needed to hunt did not evolve the usual carnivorous mammalian adaptations for hunting, such as claws to tear, fangs to slash, molars to crunch, and a keen sense of smell. The answer cannot be that a primate hunter is different from all other hunters in that he is more intelligent than other beasts, because foresight, insight, memory, and powers of abstraction pale to insignificance in a fight to the death, whereas a wolflike loping gait, the jaw reflexes of a hunting cat, and even a wolfish smile loom large. Up against a ravening hyena the ancestral hominid would do better to use claws and jaws than his cerebral cortex! Had an ecologically imposed need for hunting been the primary selection pressure acting on human progenitors one might well have predicted the evolution of more than tool-making, abstract reasoning, and prolonged childhood dependence.

After itemizing all the measured arguments, we are struck anew by the echoes of the hunting model that evoke Genesis rather than Darwin. Human sexual division of labor as seen in the hunting model is explicitly defined throughout the Old Testament. The notion of the hunting *male* as a prime force in human evolution, as opposed to his female counterpart, reflects the very tale of Creation where Eve was created as a helpmeet for Adam. The premise that carnivorous primates had of necessity to develop intelligence rather than the teeth and claws of other meat-eaters stresses the unique nature of primates and matches the Biblical tale, where each animal was created "after its kind."

But finally, and most important of all, is the notion that dominates the theories of Ardrey, Campbell, Dart, Pfeiffer, Robinson, and Washburn, that a commitment to hunting and meat-eating had an *irrevocable* effect on human behavior, and that once man tasted meat he was propelled irreversibly on a trajectory toward humanity as we know it today. It is this very sense of the irrevocable that pervades the Biblical tale of expulsion from Eden, where, as in the hunting model, the act that propelled man to his fate was an act of eating. Implicit in the structure of both events is the notion that eating transforms the eater, that he becomes what he has eaten. In Genesis, Adam and Eve ate forbidden fruit of the tree of knowledge, knew what they were, and in knowing, had to leave Paradise and face the painful realities of the world beyond. In the hunting model, man ate meat in order to survive in the harsh savanna, and by virtue of this strategy became the animal whose subsequent history is etched in a medium of violence, conquest, and bloodshed.

These analyses allow us to see the hunting model for what it is: a mixture of biological facts and evolutionary concepts entangled in the constricting threads of Western myth. It is these mythic notions that distinguish the visions of human evolution we cite from those of modern ecologists and evolutionists for whom human behavior is not dominated by irrevocable actions, but is above all, a matter of constant adaptation, flexibility, and plasticity. The hunting model assumes that hominids adopted a basically dangerous subsistence strategy, and since we survived we must have been shaped by selection pressures generated by the strategy itself. The argument is fraught with tautologies, especially since the most plausible ecological strategy would have been to *avoid* depending on only one foodstuff and to adopt a flexible, mixed diet. In truth, there need have been no single act, no primal trigger, no expulsion from Eden to set off an irreversible series of evolutionary changes. We need not have been shot into existence by one major dietary change. Instead, the gradual accumulation of new genetic material and slow divergence from the ancestral condition would have proved sufficient to make us what we are today (Simpson, 1953).

ACKNOWLEDGMENT

We are most grateful to Mr. S. D. Goitein of Princeton and Dr. T. Schrire of Cape Town for their advice on Biblical matters.

REFERENCES

Ardrey, R. (1971). "African Genesis: A Personal Investigation into the Animal Origins and Nature of Man," Laurel ed., p. 269. Dell Publ., New York.

Ardrey, R. (1974). "The Social Contract: A Personal Inquiry into the Evolutionary Sources of Order and Disorder," Laurel ed. Dell Publ., New York.

Ardrey, R. (1976). "The Hunting Hypothesis." Atheneum, New York.

Brain, C. K. (1969). The probable role of leopards as predators of the Swartkrans australopithecines. *South Afr. Arch. Bull.* **24,** 170–171.

Cahalane, T. (1961). "Mammals of North America," p. 26. Macmillan, New York.

Campbell, B. G. (1974). "Human Evolution: An Introduction to Man's Adaptation," 2nd ed. Aldine, Chicago, Illinois.

Clark, G. (1969). "World Prehistory: A New Outline." Cambridge University Press, London and New York.

Clark, G. (1970). "The Stone Age Hunters." McGraw-Hill, New York.

Dart, R. A. (1925). "*Australopithecus africanus:* The man-ape of South Africa. *Nature* (London) **115,** 195–199.

Dart, R. A. (1957). The osteodontokeratic culture of *Australopithecus prometheus. Transvaal Mus. Mem.* **10.**

Encyclopedia Judaica (1971). Vol. 11, pp. 1162–1163. Macmillan, New York.

Ginzberg, L. (1942). "The Legends of the Jews" (transl. from the German manuscript by H. Szold), Vols. I–VII. Jewish Publication Society of America, Philadelphia, Pennsylvania.

Harding, R. S. O., and Strum, S. C. (1976). The predatory baboons of Kekopey. *Nat. Hist. N.Y.* **85,** No. 3, 46–53.

Isaac, G. L. (1969). Studies of early culture in East Africa. *World Archaeol.* **1,** 1–28.

Laughlin, W. S. (1968). Hunting: An integrating biobehavior system and its evolutionary importance. *In* "Man the Hunter" (R. B. Lee and I. DeVore, eds.), pp. 304–320. Aldine, Chicago, Illinois.

Leakey, L. S. B. (1959). A new fossil skull from Olduvai. *Nature* (London) **184,** 491–493.

Oakley, K. P. (1954). Skill as a human possession. *In* "A History of Technology" (C. Singer, E. J. Holmyard, and A. R. Hall, eds.), Vol. I, pp. 20–22. Oxford Univ. Press, London and New York.

Pfeiffer, J. (1971). Man the hunter. *Horizon* **13,** 28–33.

Pfeiffer, J. (1972). "The Emergence of Man." Harper, New York.

Robinson, J. T. (1963). Adaptive radiation in the Australopithecines and the origin of man. *In* "African Ecology and Human Evolution" (F. C. Howell and F. Bourlière, eds.), pp. 385–416. Aldine, Chicago, Illinois.

Shapiro, H. L. (1974). "Peking Man," Chapter 4. Simon & Schuster, New York.

Simpson, G. G. (1953). "The Major Features of Evolution." Simon & Schuster, New York.

Speth, J. D., and Davis, D. D. (1976). Seasonal variability in early hominid predation. *Science* **192,** 441–445.

Washburn, S. L. (1960). Tools and human evolution. *Sci. Am.* **203,** 62–75.

Washburn, S. L., and Avis, V. (1958). Evolution of human behavior. *In* "Behavior and Evolution" (A. Roe and G. G. Simpson, eds.), pp. 421–436. Yale Univ. Press, New Haven, Connecticut.

Washburn, S. L., and DeVore, I. (1961). Social behavior of baboons and early Man. *In* "Social Life of Early Man" (S. L. Washburn, ed.), pp. 91–105. Aldine, Chicago, Illinois.

Washburn, S. L., and Lancaster, C. S. (1968). The evolution of hunting. *In* "Man the Hunter" (R. B. Lee and I. DeVore, eds.), pp. 293–303. Aldine, Chicago, Illinois.

Washburn, S. L., and Moore, R. (1974). "Ape into Man: A Study of Human Evolution." Little, Brown, Boston, Massachusetts.

SUMMATION

Chapter 24

Concluding Comments

C. PFAFFMANN, V. G. DETHIER, AND D. M. HEGSTED

V. G. Dethier: I should like to make a few comments from the point of view of a biologist who knows very little about nutrition other than that I enjoy ingesting that which results in nutritive well-being. I was struck by the fact that chemical senses and nutrition were placed together as the title of this work. This is a clear assumption that there must be some connection between the two. At the risk of being trite I shall preface my remarks by pointing out that in this universe of ours the second law of thermodynamics still holds: everything is rushing toward great disorder. In order to reverse temporarily the increase of entropy animals must replenish energy by drawing on external sources. This is the process of nutrition. Animals replenish energy by ingesting chemically complex substances, foods, which provide not only energy but matter in the form of a skeleton of carbon atoms, and some essential elements. There is more than one way to meet these needs. Many animals, from slug to man, have illustrated that ecologically, behaviorally, and physiologically each animal occupies its own gastronomic niche. Thus, although feeding has some common denominators—the discrimination of food from nonfood, quality control, and quantity control—details differ.

It is obvious that the specific gastronomic limits of an animal are set by its gene pool. A fine example of specific differences among closely related taxa was given by Burghardt in his account of feeding by reptiles. Similarly, Sato discussed the behavioral and electrophysiological evidence for species differences in taste sensitivity to sweet-tasting substances in mammals. Other examples immediately come to mind. In addition to species variation there is, as several contributors have pointed out, considerable individual variation—not an unusual observation in biological studies. Here as elsewhere one can ask whether this variability is an artifact of the experimental situation or represents a true behavioral polymorphism. This consideration becomes important when we begin to derive answers from comparisons of means of experimental samples and control samples, because then potentially significant individual variations begin to cancel one another. There is much to be said for using each animal as its own control.

We have also been made abundantly aware of the fact that there is a plasticity in all organisms that is imposed upon the genetic background. Indeed the degree of plasticity that is possible is a component of the genetic background. This plasticity, realized by antecedent events and experience, has been touched upon by a number of the contributors. We have seen examples of it at all phylogenetic levels and shall return to it later.

Modifiability and individual variation may be conceived of as a fine tuning which sharpens feeding behavior so that from among the vast array of potential energy sources on the planet each animal can select those that are most appropriate. It is in this context that the chemical senses play a primary role. They provide essential (but not the only) cues to proper foods. The cues themselves may not be of any nutritive value whatsoever. I think it is important to realize that throughout the animal kingdom the initial behavioral component of nutrition is based on cues associated with food and not on the energy or caloric value of the food being ingested. In nature the cues for each particular animal are normally associated with what is nutritionally good. The fact that the individual ends by choosing what is adequate indicates that the *species* makes its choice on a nutritional basis; that is, individual members that are miscued on their food choice do not survive to propagate their error. The importance of chemical stimuli, whether they be nutritional or not, has been repeatedly documented in this book as well as elsewhere. Rogers has presented data, for example, showing that a protein-free diet may be preferred to one with a mixture of amino acids, but an imbalanced one. Gershoff has pointed out that palatability (which involves response to chemical stimuli) can render a deficient diet acceptable.

We are not allowed to forget, however, that chemical cues are not the only ones of value. The importance of motion as a cue for snakes has been described. The extent to which color normally associated with specific beverages of different flavors could mislead subjects when dissociated from the usual flavor has been demonstrated by Beauchamp. Garcia's experiments with learned aversion by the red-tailed hawk showed that the color of a mouse could be used as a cue to its taste. The whole concept of mimicry and prey image is based upon the knowledge that the characteristics of food are frequently cued by its appearance.

Nevertheless, the ultimate determinants for the recognition and/or acceptability of food are its chemostimulatory characteristics. Depending upon the animal concerned these may trigger ingestion or swallowing, influence the quantity of ingested food, or prepare the digestive system to receive food. Two papers dealt with this last, not often appreciated facet of chemosensory stimulation. Nicolaidis' experiments in which olfactory and orodigestive receptors were bypassed showed that utilization of food under these circumstances is impaired. Sensory–vegetative reflexes also lead to anticipatory metabolic changes. Comparable studies by Naim and Kare showed that an appropriate taste stimulus on the tongues of dogs enhanced the volume and protein content of pancreatic secretion.

As is to be expected, chemical stimuli do not exert tyrannical control over feeding response. Since their evolutionary role is to participate in assuring the animal of an energy homeostasis, there are various feedback mechanisms involving internal cues related to the energy pool. One has to be wary of trying to assess feeding solely in terms of the sensory information that is arriving from the outside world. We must constantly remember that the organism is the recipient of an enormous amount of information from diverse sources, that all of it is assembled and integrated, and that final behavioral decisions are made after assessing the totality of information whether it arrives from within or without. This cooperation between the chemical exteroceptors and the different enteroceptors is universal, having been studied in the fly, in the slug as reported here by Senseman, and in vertebrates. The response of an animal to external cues of all sorts is modulated by the state of deprivation, satiation, or dietary deficiency. Although the mechanisms involved in this aspect of feeding were not explored in depth, Rogers and Leung reported on investigations relating to the involvement of amino acids in the control of food intake with special reference to dietary proteins and amino acids. As with other studies of the control of food intake the ultimate solution still eludes us.

There is no doubt that behavioral responses to food are determined by the net result of central neural integration of multiple sources of information from the external environment and the internal milieu. The possibility remains, however, that changing internal states alter directly the responsiveness of the chemosensory systems themselves. Such reports have appeared in the literature from time to time and have not been confirmed. Sharma described experiments suggesting that receptor systems are indeed "tuned" through the agency of centrifugally transmitted information acting at several neural levels including the gustatory papillae themselves.

That centrifugal information does act at the level of second-order neurons was elegantly demonstrated by Paget in the mitral cell responses in the olfactory bulb of hungry and satiated rats. Thus, it is clear that centrifugal control extends far into the periphery (although it must not be forgotten that the bulb is essentially an extension of the brain).

Plasticity has been alluded to earlier. It is clear that the adaptability of feeding systems has been greatly enhanced by their plasticity in the face of changing conditions. Experience exerts a profound influence on the interpretation by animals of the chemical stimuli that they encounter. Experience, dietary imprinting, and learned preferences and aversions were among the topics discussed. (The use of the term imprinting in this connection may be inadvisable because while the phenomenon under discussion resembles imprinting in some respects it differs in others.) As Senseman explained, some of these phenomena (experience, learned aversions) are to be found as early in the evolutionary scale as slugs, and dietary imprinting has been demonstrated in caterpillars. Mugford was unable to observe dietary imprinting in dogs (it had been reported by Kuo) but clearly showed in dogs and cats learned aversions and preferences, the effects of previous dietary history, and social facilitation.

The role of olfaction in learned aversions came under special scrutiny. In cats Mugford showed that odor could provide a discriminative stimulus for normal and toxicosis-induced patterns of food selection. Rogers and Leung showed that in learned aversions with respect to amino acids olfactory cues are of considerable importance, but that in the absence of these cues other senses are utilized. Garcia showed that odor cues are of prime importance in learning to avoid shock, but that elimination of odor cues has no effect upon illness avoidance while elimination of taste does. In general, the role of odor in feeding has received less attention than that of taste. The number of provocative experiments reported in this book should stimulate more research.

One aspect of feeding that has permeated this work is that of palatability. Pfaffmann will discuss this in more detail; however, a few remarks from a biological point of view are relevant. Palatability, whatever it is, pleasantness, unpleasantness, or disgust, is clearly a tremendously important facet of human feeding. Nearly every author has mentioned it in one context or another. One might suppose, biologically speaking, that the olfactory and gustatory stimuli emanating from foods originally evolved to serve as cues, and that in man they have become important for their own sake independent of the relation of energy intake. Sugar, for example, may often be ingested simply for the pleasant sensation of sweetness that it evokes. The experiments of Moskowitz clearly revealed that for some stimuli the sensory intensity is only moderately correlated with hedonic tone. The situation in nonhuman animals at the moment eludes analysis. Watching moving pictures of Garcia's wolves and hawks responding to aversive stimuli, one receives an overpowering impression that the animals thoroughly disliked the taste. Brightman's pictures of the facial expressions of human subjects responding to aversive tastes and vigorously wiping their mouths shout "dislike." Yet, flies, given an aversive taste, will vigorously wipe the proboscis with the feet and regurgitate copiously. Is there a hedonic component to the behavior of the flies? Hawks? Wolves? Where does one draw the phylogenetic line?

C. Pfaffmann: As a biologist, Dr. Dethier has just thrown down the gauntlet on "palatability," and as a person trained in physiology, I probably have similar but, of course, not identical biases about the term. I am also a psychologist, and such terms as "palatability" and others referring to the boundary between behavior and experience must be coped with. I am not so ready to put the term in quite the disrepute that I think Dr. Dethier was suggesting, if I understood him correctly. It seems difficult to keep the term "palatability" out of the discussion no matter how operationally or objectively pure one wishes to be regarding behavior and its determinants. There is no reason that this term cannot be used if it is carefully defined and delimited. In the past era when psychology was dominated by strict scientific operationalism, terms with a subjective connotation were rarely employed—in fact, they were excluded—and certain textbooks were written with almost no mention of what introspection or subjective experience contained.

It is interesting that for quite a time in recent history, in the literature of taste preferences, appetite, and self-selection behavior, there was just one man who used the term "palatability" and the associated concept of hedonic value. Paul Thomas Young had the courage to do so, in spite of the fact that he was something of an outcast in that highly operational behavioristic era. He began developing his hedonic theory of behavior almost 30 years or more ago and referred to this factor as important in determining organisms' actions as well as choice of particular stimuli in a preference test and in feeding generally. Young was criticized because of the basic circularity in the use of this concept. I think this criticism was and is correct and we face a danger in using it. We must try to break out of its inherent logical circularity. This is particularly apparent in studying animals (I dissociate man from other animals in this respect) because until recently the only evidence that a stimulus solution was palatable was whether or not the animal accepted it preferentially. If it did so, one was inclined to say it did so because it liked it. But the only criterion that the organism liked the solution was that it preferred it or took it preferentially. The circularity is obvious.

However, it is not the only case of circularity with which behavioral psychologists have had to deal. The concept of reinforcement suffers from similar difficulties. Yet it has been widely used and in fact is a cornerstone of a good deal of animal learning experimentation and theory. An animal will learn to press a bar when reinforced by food, water, or a variety of other stimuli. Anything that will increase the rate of bar pressing is said to be a reinforcer. One way out of the circularity here is to apply the reinforcer in other situations and to determine their effectiveness therein. At one time when general learning theory assumed that all reinforcers were equally good in all situations, this seemed to be a good way out. Reinforcers were presumably genetically determined, and by and large they could be applied under a variety of circumstances. But this strategy assumed that there was a general learning mechanism in which all stimuli and responses were equal or interchangeable, what might be referred to as "fungibility," that is, with the capability of mutual substitution. Such a generalized view of the learning process, however, has since proved to be inadequate. Learning may be a highly specialized capability so that learning in one situation with one set of stimuli and responses may not transfer or be interchangeable to other circumstances. In fact, it was the conditioned taste aversion which John Garcia discussed that was responsible in large measure for pointing up special cases of learning which were not equally fungible. Taste and visceral distress and disturbance or illness have a special relationship, so that such learning is more easily established than is the connection between illness and exteroceptive stimulation like sounds or flashing lights. Since then a variety of other special relations between certain reinforcers and behavior have been described. For example, hearing the species-specific song will reinforce a bird's tendency to perch, but hearing the song will not reinforce pecking at a key unless a bit of grain is taped to the key. In this case the animal must also be hungry and even then the increase in response rate due to the hearing of the species song is not very great. There are a whole host of other special learning mechanisms which ethologists have emphasized in much of their writings. But it was the conditioned taste aversion paradigm that led to a widespread change in the way psychologists regarded learning mechanisms. Involved in the conditioned taste aversion is not only the connection between gustatory stimulus and visceral response which is selective over the other kinds of exteroceptive stimulation and aversive consequences, there is also a transfer of the hedonic quality to the conditioned taste stimulus. In some of the material presented by Dr. Garcia, a response as if an aversive taste had been placed in the mouth can be observed when the aversively conditioned mouse was presented to the hawk.

As I noted in a previous discussion, videotape camera close-ups are used in our laboratory to observe the oral and mimetic behavior of rats presented with palatable sugar solutions, on the one hand, versus unpalatable quinine solutions on the other. The movements are so distinctive that one can predict with great accuracy which stimulus has been presented to the animal. In addition to the oral responses in the case of the aversive stimulus, quinine, there is the face wiping, chin rubbing, and forepaw shuffling, a species-typical response to strong quinine. Not only is the term "palatability" important in characterizing these responses, the concept of hedonic response, pleasant or unpleasant, seems a necessary construct. Until recently not much attention has been paid to these subsidiary or supportive responses that might, in fact, be direct behavioral indicators of that which we have been calling hedonic. Utilizing and observing them is one way out of the intrinsic circularity in using the term "hedonic" in animal studies. There may be still other responses which can be utilized as behavioral indexes of this most important process.

As far as man is concerned, I for one am satisfied that palatability or hedonic response is a valid concept. The psychophysical experiments described by Dr. Moskowitz with his emphasis on the breakpoint in his psychophysical functions is another important methodological consideration that shows that this is a distinctive judgment over and beyond that of mere sensory magnitude. Quite apart from our anecdotal appreciation that "we all know what we mean when we say something is pleasant," the concept of palatability is becoming methodologically refined and increasingly fitted into our efforts to understand taste and olfactory behavior and food intake and nutrition.

Let me pursue still another aspect of something Dr. Robert Cagan mentioned. He spoke of Ralph Norgren's work in our laboratory, the discovery of a previously overlooked nucleus in the dorsal pons of the mammalian brainstem. To be sure, comparative neurologists, especially C. Judson Herrick, had described the homolog of that nucleus in the fish brain. It is now well established in the mammalian central nervous system interposed between the medullary and thalamic taste nuclei. In the present context this nucleus is of particular interest because it is the source of a bifurcated pathway, one portion going to the classic thalamocortical projections that have been quite well known for some time. Of particular interest is the ventral pathway passing below the thalamus turning anteriorally and laterally to provide some projections to the lateral hypothalamus, and a larger projection to the central nucleus of the amygdala. This is quite distinct from the olfactory amygdala projections, but there may be interactions between these systems beyond this point. Stimulation or ablation of the amygdala can affect food intake or palatability of taste stimuli. In human cases with disease or pathology that requires exposure and stimulation of amygdala and other limbic structures, hedonic aspects of experience are often involved. Pathology may be associated with aberrations of eating behavior, taste, and olfactory experiences. So through the domain of physiological and anatomic studies of gustatory CNS pathways, we see another avenue which raises and validates the concept of palatability.

Dr. Nicolaides asked, "Is palatability a characteristic of the stimulus?" We must be clear that palatability is not in the stimulus alone. There are stimulus properties which activate this response so it is a stimulus response (SR) characteristic. Both S and R must be included and with both we can develop the criteria which permit us to say that a stimulus has a positive or negative hedonic value. I have belabored this point somewhat because, as a psychologist, one of my motivations for continuing to work on the "exotic" chemical senses is that they are "windows" par excellence on the pleasures of sensation. They permit us to control hedonic responses by specific and specified stimulants and for the analysis of motivated behavior. The basic behavioral dimensionality of approach through neurality to aversion is one of the most fundamental responses to stimulation. For the experimentalist, taste provides a unique tool for the analysis of the mechanism of this most fundamental of all biological relations.

Another line of investigation I particularly would like to note is the evidence for the modifiability of receptor function as a result of changes in internal state. We saw good evidence in Dr. Pager's electrophysiological studies of the olfactory bulb. Dr. Engen reviewed in his work further elaborating the modifiability of the hedonic process by internal factors, phenomena which were discovered and called to our attention by Cabanac. Indeed, the increasingly important role of human investigation in this field leads me to raise the question of experimentalist's over-preoccupation with animal models. Perhaps we should regard the human not only as interesting in its own right but as an important human model for animal studies rather than the other way around. Henry Jordan's work is a good example of the application of methods and concepts derived from study of that much maligned white rat in the "unnatural" laboratory for the derivation of the same basic methods that can be applied with profit to the study of man. Man is not a rat, a rat is not a little man, but the methods, with appropriate adjustment, can often be used on one or the other.

I should mention something about species differences as well as individual differences. As a taste electrophysiologist, I was particularly interested in Professor Sato's data because they clarified and answered some of the problems that, in a sense, misled many of us in the early days when we did, in fact, take the rat as our model. The rat's chorda tympani nerve is a relatively restricted preparation. Other small rodents with good sugar sensitivity, like the hamster, are more interesting, but in the last analysis the subhuman primate is really necessary to provide electrophysiological analysis close to human taste. Even the squirrel monkey, which we have studied, is on a different evolutionary line from the hominid primate and is different in certain important respects even though it has good sugar sensitivity. Most nonprimates are in fact rather different in the response of their sweet receptors to nonsugar stimuli. The observations of Professor Sato and his colleagues have filled out the roster of subhuman species at least so far as electrophysiology is concerned, and their correlation with basic approaches at the biochemical and molecular level.

It is encouraging to see the beginnings of serious investigation of the pathology and aberrations of the chemical senses in relation to nutrition, obesity, and other pathological states that have been discussed by Dr. Brightman and Dr. DeWys. It seems to be a sign of approaching maturity of the field when clinical research can be undertaken because we have sufficient background on which to base such approaches. Finally, since I attended the Symposium on Nutrition and the Chemical Senses 10 years ago, I can compare it with the one on which this book is based. I can express a great sense of optimism at seeing how many new facts have brought more order to the field. They are not just miscellaneous new facts that make life more confusing. We are all aware of many problems that still exist, but there is now a better framework for a more reasoned approach to the problem of chemical senses and nutrition. There has indeed been a significant advance in the last 10-year period.

D. M. Hegsted: Nutritionists have always recognized, even though they sometimes appear to forget it, that food is not food unless it is eaten. The importance of the factors that influence consumption will be obvious to most of us. It would seem to an outsider that the whole field of the chemical senses or other factors influencing consumption is somewhat like social anthropology. Social anthropology has an important relationship to nutrition, yet the field is largely descriptive at the moment, and it is uncertain whether it is being described in the most useful ways. The eventual aim, I assume, is to be able to predict consumption and deliberately manipulate consumption levels. The food industry can obviously provide abundant examples of how far we are from those goals.

Nutrition education is generally thought to be a very important activity if we are to modify the food intake patterns of the public for their own good. Nearly all nutrition education is based on the presumption that eating is an intellectual exercise and that knowledge of the nutrient composition of food will determine consumption patterns. It would seem self-evident that if people are going to deliberately modify the kind of food they eat for health reasons, they have to know how to modify it; they have to know enough about the composition of food and what a desirable diet is to make rational choices. Yet, as Dr. Cantor has emphasized, the evidence that traditional nutrition education really works is minimal. This is not to say, of course, that it is useless, but it would appear to be abundantly clear that teaching the facts of nutrition is not a very effective way to modify behavior. More is needed, and I assume it, or some of it, will come from studies of the kind included in this book.

I do want to comment on the changing nature of nutrition since I think this could influence to some degree the kind of problems that investigators should tackle. My generation of nutritionists—from 1920 to the present—saw our mission largely in the following terms: We should identify the essential nutrients, we should determine their biochemical function, and, knowing the biochemical function, we would then be able to determine with reasonable accuracy the human requirement for each nutrient. Having this information, we could evaluate the nutritional adequacy of diets as eaten and improve them or formulate diets which would provide optimal nutrition. This seems reasonably rational, but we are far from accomplishing this mission. Most of the essential nutrients have been identified, certainly, but we are a long way from defining the requirements with any accuracy, and for many nutrients we have no methods or grossly inadequate methods for evaluating nutritional status. This mission still needs more effort.

The major point to be made, however, is that it is now clear that this kind of knowledge will not provide the neat package of information predicted. Admitting that we are still ignorant about the requirements of many nutrients and therefore could be wrong, it still seems abundantly clear that the major nutritional problems of the United States and other developed societies (including the upper class in the poor and developing nations) have little or nothing to do with essential nutrients as usually defined. The primary health problems of our population are coronary heart disease, cancer, hypertension, diabetes, obesity, and dental caries—all of which have an important nutritional component. I do not have the time to provide evidence for this, but these do represent the problem and this is where the primary action must be.

These diseases so far as is known have little or nothing to do with the amounts of essential nutrients consumed. Dietary modification to control these diseases is clearly more difficult than the control of deficiency diseases. With the latter, the principal problem is to demonstrate that a problem exists. Generally speaking, the technology is available to add essential nutrients to diets if we can make an adequate case for the need. In contrast, the nutritional components of the chronic degenerative diseases which kill most Americans and affluent people everywhere relate to the kinds of food habits which appear to be universally desired when economics and the food supply permit—excessive consumption of food in general, excessive meat, fat, cholesterol, sugar, salt, etc. Clearly, it is more difficult to get people to limit their intake of things that they like than to get them to eat more of things they like.

I also wish to comment on animal models, particularly to emphasize that the very factors that have made the young rat such a useful model in nutrition studies make it a poor model for some food consumption studies. The rat has been so useful and popular because it happens to be of a convenient size, and because it grows rapidly and for a relatively long time. A 50-g rat will grow 5–6 g a day and continue to gain rather rapidly until it reaches 300 g or so. The rapid growth rate dilutes the nutrients in the body at the start and the production of deficiency disease is easy. With the rapid growth rate there is also a very large and efficient transfer of materials from the diet into body tissue. The young rat, for example, may transfer up to 90% or so of the total protein eaten into new tissue protein, and large amounts of the energy consumed is similarly transferred. Admitting that I have no idea of what the signals are, it seems perfectly clear that a rat that does not grow has to quit eating. The ability to dispose of excess energy intakes is severely limited (in contrast to most nutrients) so that a reduction of food intake is required when the animal ceases to grow. This does not appear to be true in man. Recall that a preschool child may also grow only 5–6 g/day, the same absolute rate as a 50-g rat, even though the child weighs 10–30 kg. In contrast to the rat the amount of energy and protein required for new tissue formation is trivial relative to the maintenance requirements.

Baby monkeys grow slowly, much more like the human infant. The protein requirement is low, for example, of the order of 6–7% of the total calories as protein, which is about the same as the protein content of human breast milk. These protein requirements contrast to that of a young rat which is of the order of 15–20% of the diet. We find that protein restriction in baby monkeys of a sufficient degree to completely inhibit growth does not appreciably decrease the calorie intake, which contrasts markedly to what one finds with young rats.

The human infant is said to have a very high protein requirement compared to the adult, 2 g/kg/day compared to about 0.5 g/kg/day. Yet the concentration of protein in the diet required for growth, again 6–7% of the total calories in human breast milk, is similar to the concentration in rice, a very low protein material, and much lower than diets consumed anywhere. By the time one corrects for the difference in energy intake, it is not clear that the infant requires more protein in his diet than that of the adult. Although the work is not complete, our current impression is that the adult monkey requires as high a protein concentration in his diet as the growing monkey. Again this contrasts markedly with rats and most animals where it is clear that protein needs of growing animals, however expressed, are elevated compared to adult animals.

I admit, of course, that the very fact that a growing rat *has* to turn off his appetite if he does not grow or vice versa might be a great experimental advantage if the signals are the same as in man. I suspect, however, that they are not. Many nutritionists have developed misconceptions about the importance of growth on nutrient requirements of infants and children because of unwarranted extrapolation from studies with the rat and other rapidly growing species.

Finally, I wish to comment on energy needs and food consumption. I suspect that major underlying assumptions are incorrect. The equation used to estimate energy needs is essentially as follows:

$$\textit{Energy intake} = \textit{basal metabolic rate} + \textit{specific dynamic action} + \textit{physical activity} \pm \textit{body energy content}$$

The basal metabolic rate is assumed to be essentially constant for an individual, being determined by body size. Specific dynamic action, the excess heat lost after a meal, is assumed to be very small under ordinary conditions. Thus, in an individual who maintains a constant body weight and body composition, the energy intake and the energy expended in physical activity must be exquisitely balanced so they are exactly equal over any appreciable time period. Or stated the other way, if the feedback controls between intake and activity are not precise, an appropriate amount of body tissue will be lost or gained. Most of the literature on obesity, appetite, control of food intake, etc., is based upon these assumptions.

Admitting that there is abundant evidence that food intake is controlled, it is highly unlikely that this control can be the only factor of importance. My weight has not changed by more than a few pounds in 40 years. My intake varies from a few hundred to over 3000 calories per day, determined largely by social influences, not by hunger. Being largely sedentary unless forced to be otherwise, it seems preposterous that intake is the only variable controlling my body weight.

My thesis is that the amount of "useful energy" available from food is under metabolic control, "useful energy" being defined for the moment as ATP production. Current biochemical theory tells us that about a maximum of 38% of the energy in glucose, for example, can be converted to ATP or other high-energy bonds. The remainder of the energy is released as heat in the metabolic processes. There are similar energy yields from fat and somewhat less from protein. There is no reason, however, to believe that ATP production is ever maximal or that it is constant. If we assume, as an example, that the degree of coupling in oxidative phosphorylation is under metabolic control, then we have a ready explanation of how weight can remain constant in the face of differing energy intakes. There are, of course, many other possibilities.

I have already mentioned our experience with protein-deficient monkeys which leads us to believe that such animals must somehow be able to waste energy. The experience of Sims, Danforth, and their coworkers at the University of Vermont who have attempted to fatten human subjects similarly leads them to conclude that energy utilization is variable under different conditions and differs among species.

I predict that we will soon be able to show that a calorie is not a calorie but that it depends upon the source of the energy, the metabolic state of the individual receiving the energy, and idiosyncracies of the subject himself. Much of the literature on the control of food intake will eventually require an extensive reevaluation.

Index

T

A 7
B 8
C 9
D 0
E 1
F 2
G 3
H 4
I 5
J 6

Printed in Dunstable, United Kingdom

91342990R00286